CW00806589

Meteorites: Flux with Time and Impact Effects

Geological Society Special Publications

Series Editors

A. J. FLEET

A. C. MORTON

A. M. ROBERTS

It is recommended that reference to all or part of this book should be made in one of the following ways.

GRADY, M. M., HUTCHISON, R., MCCALL, G. J. H. & ROTHERY, D. A. (eds) 1998. *Meteorites: Flux with Time and Impact Effects*. Geological Society, London, Special Publications, **140**.

BEVAN, A. W. R., BLAND, P. A. & JULL, A. J. T. 1998. Meteorite flux on the Nullarbor region, Australia. *In:* GRADY, M. M., HUTCHISON, R., MCCALL, G. J. H. & ROTHERY, D. A. (eds) *Meteorites: Flux with Time and Impact Effects*. Geological Society, London, Special Publications, **140**, 59–73.

GEOLOGICAL SOCIETY SPECIAL PUBLICATION NO. 140

Meteorites: Flux with Time and Impact Effects

EDITED BY

M. M. GRADY
Natural History Museum, London

R. HUTCHISON
Natural History Museum, London

G. J. H. McCALL
Hon. Fellow, Liverpool University
Hon. Associate, Western Australian Museum

AND

D. A. ROTHERY
Open University, Milton Keynes

1998

Published by

The Geological Society

London

THE GEOLOGICAL SOCIETY

The Society was founded in 1807 as The Geological Society of London and is the oldest geological society in the world. It received its Royal Charter in 1825 for the purpose of 'investigating the mineral structure of the Earth'. The Society is Britain's national society for geology with a membership of around 8500. It has countrywide coverage and approximately 1500 members reside overseas. The Society is responsible for all aspects of the geological sciences including professional matters. The Society has its own publishing house, which produces the Society's international journals, books and maps, and which acts as the European distributor for publications of the American Association of Petroleum Geologists, SEPM and the Geological Society of America.

Fellowship is open to those holding a recognized honours degree in geology or cognate subject and who have at least two years' relevant postgraduate experience, or who have not less than six years' relevant experience in geology or a cognate subject. A Fellow who has not less than five years' relevant postgraduate experience in the practice of geology may apply for validation and, subject to approval, may be able to use the designatory letters C Geol (Chartered Geologist).

Further information about the Society is available from the Membership Manager, The Geological Society, Burlington House, Piccadilly, London W1V 0JU, UK. The Society is a Registered Charity, No. 210161.

Published by The Geological Society from:
The Geological Society Publishing House
Unit 7, Brassmill Enterprise Centre
Brassmill Lane
Bath BA1 3JN
UK
(*Orders*: Tel. 01225 445046
Fax 01225 442836)

First published 1998

British Library Cataloguing in Publication Data
A catalogue record for this book is available from the British Library.

ISBN 1-86239-017-7
ISSN 0305-8719

Typeset by Aarontype Ltd, Unit 47, Easton Business Centre, Felix Road, Bristol BS5 0HE, UK

Printed by Cambridge University Press, UK

Distributors

USA
AAPG Bookstore
PO Box 979
Tulsa
OK 74101-0979
USA
(*Orders*: Tel. (918) 584-2555
Fax (918) 560-2652)

Australia
Australian Mineral Foundation
63 Conyngham Street
Glenside
South Australia 5065
Australia
(*Orders*: Tel. (08) 379-0444
Fax (08) 379-4634)

India
Affiliated East-West Press PVT Ltd
G-1/16 Ansari Road
New Delhi 110 002
India
(*Orders*: Tel. (11) 327-9113
Fax (11) 326-0538)

Japan
Kanda Book Trading Co.
Cityhouse Tama 204
Tsurumaki 1-3-10
Tama-Shi
Tokyo 0206-0034
Japan
(*Orders*: Tel. (0423) 57-7650
Fax (0423) 57-7651)

Contents

List of names and addresses of contributors

M. E. Bailey, Armagh Observatory, College Hill, Armagh, BT 61 9DG, UK
F. J. Berry, Department of Chemistry, Open University, Walton Hall, Milton Keynes, MK7 6AA, UK
A. W. R. Bevan, Department of Earth and Planetary Sciences, Western Australian Museum of Natural Science, Francis Street, Perth, Western Australia 6000
P. A. Bland, PSRI, Open University, Walton Hall, Milton Keynes, MK7 6AA, UK
A. Camara-Zi, Facultad de Ingeniería, Universidad Autónoma de Yucatán, Apdo. Postal No. 150, Cordemex 97111, Mérida, Yucatán, México
R. E. Chavez, Instituto de Geofisica, UNAM, Ciudad Universitaria, Codigo 04510, México, D.F., México
E. Cielaszyk, NSF Accelerator Facility for Radioisotope Analysis, University of Arizona, Tucson, Arizona 85721, USA
S. Cloudt, NSF Accelerator Facility for Radioisotope Analysis, University of Arizona, Tucson, Arizona 85721, USA
M. Connors, Athabasca University, 1 University Drive, Athabasca, Alberta, Canada, T9S 3A3
A. Conway, School of Physics, Glasgow University, Glasgow, Scotland
P. Denton, University of Leicester, University Road, Leicester, LE1 7RH, UK
V. V. Emel'yanenko, Chelyabinsk Technical University, Lenina 76, Chelyabinsk, 454080, Russia
I. Gilmour, PSRI, Open University, Walton Hall, Milton Keynes, MK7 6AA, UK
M. M. Grady, Department of Mineralogy, Natural History Museum, Cromwell Road, London, SW7 5BD, UK
E. Graniel-Castro, Facultad de Ingeniería, Universidad Autónoma de Yucatán, Apdo. Postal No. 150, Cordemex 97111, Mérida, Yucatán, México
R. A. F. Grieve, Geological Survey of Canada, Ottawa, Canada, K1A 0Y3
A. Hallam, School of Earth Sciences, University of Birmingham, Edgbaston, Birmingham, B15 2TT, UK
J. F. Halpenny, Geomatics Canada, 615 Booth Street, Ottawa, Ontario, Canada K1A 0E9
A. R. Hildebrand, Geological Survey of Canada, 615 Booth Street, Ottawa, Ontario, Canada, K1A 0E9
D. W. Hughes, Department of Physics, University of Sheffield, Sheffield, S3 7RH, UK
R. Hutchison, Department of Mineralogy, Natural History Museum, Cromwell Road, London, SW7 5BD, UK
A. J. T. Jull, NSF Accelerator Facility for Radioisotope Analysis, University of Arizona, Tucson, Arizona 85721, USA
R. Kind, GeoForschung Zentrum, Potsdam, Germany
C. Koeberl, Institute of Geochemistry, University of Vienna, Althanstrasse 14, A-1090, Vienna, Austria
N. Macleod, Department of Palaeontology, Natural History Museum, Cromwell Road, London, SW7 5BD, UK
G. D. Mackenzie, University of Leicester, University Road, Leicester, LE1 7RH, UK
P. K. H. Maguire, University of Leicester, University Road, Leicester, LE1 7RH, UK
G. J. H. McCall, 44 Robert Franklin Way, South Cerney, Glos, GL7 5UD, UK
A. C. Milner, Department of Palaeontology, Natural History Museum, Cromwell Road, London, SW7 5BD, UK
W. M. Napier, Armagh Observatory, College Hill, Armagh, BT61 9DG, UK
D. Niehaus, PCI Enterprises, 50 West Wilmot Street, Richmond Hill, Ontario, Canada, L4B 1M5
C. Ortiz-Aleman, Instituto de Geofisica, UNAM, Ciudad Universitaria, Codigo 04510, México, D.F., México
M. Pilkington, Geological Survey of Canada, 615 Booth Street, Ottawa, Ontario, Canada K1A 0E9
C. T. Pillinger, PSRI, Open University, Walton Hall, Milton Keynes, MK7 6AA, UK
D. A. Rothery, Department of Earth Sciences, Open University, Walton Hall, Milton Keynes, MK7 6AA, UK
E. M. Shoemaker, Lowell Observatory, Flagstaff, Arizona, USA
T. B. Smith, Department of Physics, Open University, Walton Hall, Milton Keynes, MK7 6AA, UK
J. G. Spray, Department of Geology, University of New Brunswick, Bailey Drive, Fredericton, New Brunswick, E3B 5A3, Canada

S. E. J. Swabey, Department of Earth Sciences, Open University, Walton Hall, Milton Keynes, MK7 6AA, UK
A. Trejo, Universidad Nacional de Mexico, Mexico, DF
J. Urrutia-Fucugauchi, Instituto de Geofisica, UNAM, Ciudad Universitaria, Codigo 04510, México, D.F., México
M. E. Zolensky, Earth Science and Solar System Exploration Division, NASA Johnson Space Center, Houston TX 77058, USA

Preface

Research on the twin themes of this Special Publication is proceeding at a great pace and, though this volume contains much new and interesting material, no doubt much of it will be upstaged by further discoveries and sophisticated scientific research in the next few years. For instance, only part of the Nullarbor Plain has been systematically searched for meteorites; the yield to date is already about 250 separate meteorites (excluding pairings) as compared with about 20 when I was active in Perth and had to rely on finds by rabbit trappers and the searches of W. H. Cleverly. There must also be many further finds to be made in Antarctica and the Sahara. The immense Chicxulub impact structure at the K–T boundary has at present only been investigated in a preliminary fashion and new finds such as the Morokweng Structure in the Kalahari Desert and the Gliksen structure in Australia (named after my former research student to whom Gene Shoemaker also allocated the name of an asteroid) have only recently been recognized. The only reason for wanting to live to a great age is to experience the excitement of the new discoveries in meteoritics and planetology that will surely emerge in the next decades, and change our concepts.

I am particularly pleased that reviews by palaeontologists have been included in this volume. The problem of mass extinctions can have no simple answer and workers in both the meteoritic field and palaeontology must read and consider deeply each other's output and not form two adversary camps. Despite the evidence of a major impact at the K–T boundary, there is overwhelming evidence to suggest many contributory factors, both of extraterrestrial and terrestrial origin, cause mass extinctions and the oversimplified 'media' image of all manner of life forms 'frying in an immense impact produced holocaust' is impossible to accept on the available evidence. The truth must be more complicated. There are strong and real conflicts in the evidence and only exhaustive further studies will take us further towards a real understanding of mass extinction in the geological record.

There has been much discussion lately about the definition of a geologist and the boundaries of geology as a science. Some favour narrow definitions and some wide definitions embracing planetary science. Geologists are specialist planetologists, by definition primarily concerned with one single planet in the solar system, the one on which they live. However, the science of geology gives them unique insights into the surface development and internal workings of planetary bodies so they must also be involved, albeit alongside scientists of other disciplines, in the study of extraterrestrial bodies and processes, including meteoritics. Meteoritics and study of, say, the Moon, also provide a feedback, giving them insight into the history and workings of their own planet, particularly in its very early, obscured development. So this rather unusual topic, relating to astronomy, planetology and meteoritics, is an entirely proper addition to the growing list of Geological Society Special Publications. It is also a model in interdisciplinarity, bringing together as it does geoscientists and astronomers.

Acknowledgements. Late in 1995, I was asked to act as convenor of a Fermor Lecture Meeting in 1997 on a meteorite-related topic. Although I spent a decade in the 1960s supervising and cataloguing the collections of the Western Australian Museum, encouraging new finds there, studying the Wolfe Creek crater and preparing three books on meteorites and possible impact structures, I was obliged in the 1970s to move on to other geological preoccupations because of the need to secure an adequate salary after I left the University in Perth. So, though I had maintained my interest in meteoritics and planetology, it was necessary to enrol as co-convenors active workers in the field of meteoritics, and I sought out Dr Hutchison and Dr Grady at the Natural History Museum who proposed the theme of 'Flux with Time and Impact Effects'. We completed the team by inviting Dr Rothery of the Open University, a geologist with strong planetological interests. These, my co-convenors, have put in an immense amount of hard work both organizing the Fermor Meeting, and preparing this volume for the publishers. Because of the fact that they were equipped with the logistic base necessary, very much of the load in connection with this publication has fallen on Dr Hutchison and Dr Grady, while I myself have carried out a coordinating role including a final coordinating editor's review of the articles, something very necessary in the case of all such special publications, in order to eradicate inconsistencies, missing or incorrect references, etc. My gratitude to my three co-editors is immense and it has been a pleasure working with them. I must also thank the referees who are listed below.

Joe McCall
December 1997

List of referees

The following carried out peer reviews on the articles submitted for publication:

M. J. Benton
A. W. R. Bevan
P. A. Bland
J. C. Bridges
A. Deutsch
R. Fortey
I. A. Franchi
A. S. Gale
M. J. Genge
I. Gilmour
M. M. Grady
R. A. F. Grieve
I. A. Halliday
D. W. Hughes
G. R. Huss
R. Hutchison
E. J. W. Jones
A. J. T. Jull
C. Koeberl
M. E. Lipschutz
B. Mason
G. J. H. McCall
H. Palme
D. A. Rothery
S. S. Russell
J. Smit
D. Tarling
S. R. Taylor
P. B. Wignall
I. P. Williams
L. Wilson

Meteorites: their flux with time and impact effects

MONICA M. GRADY[1], R. HUTCHISON[1], G. J. H. McCALL[2] & D. A. ROTHERY[3]

[1] *Department of Mineralogy, Natural History Museum, Cromwell Road, London SW7 5BD, UK*

[2] *Honorary Fellow, Liverpool University: Honorary Associate, Western Australian Museum, Perth, UK*

[3] *Department of Earth Sciences, Open University, Walton Hall, Milton Keynes MK7 6AA, UK*

The subject of this Special Publication is the flux with time of meteorites and their effects on this planet. Determining the flux of impacting objects from space to the Earth is difficult and estimates have in the last two decades achieved improved validity as a result of extensive researches on optimum fields of preservation such as Antarctica and desert areas such as the Nullarbor Plain, Australia and the Sahara. The impactors to the Earth come from the asteroid belt, the Moon and almost certainly Mars (the last two spalled off by impacts on those bodies); and also it is believed the cometary clouds of the outer Solar System. In recent years it has been recognized that large impactors from space, though representing rare events in the four thousand million years of geological time recorded in the rocks, have played an important part in moulding the surface of the Earth, producing extensive craters and other structures. The statistics of distribution of such large structures in space and time in the record can also be used, alongside astronomical and theoretical methods, to estimate the flux. These structures are also important because it has been suggested that immense-scale impactors have actually caused abrupt changes in the progressive development of life on Earth, causing mass extinctions. An immense impact structure is recognized in Yucatan, Mexico, on and offshore, of the same date as the major Cretaceous–Tertiary (K–T) mass extinction. The geological, geophysical and geochemical evidence for such an impact event is quite clear, but the possible environmental consequences of an impact at this time are still a matter of debate, and the fossil record, of vertebrates or invertebrates, is by no means unambiguous in support of the catastrophic effects hypothesized widely for this event. Also the evidence in the geological record for impacts at the times of other mass extinctions is slender.

The aim of this collection of papers by international experts in their field, whether astronomy or geology, is to extend the knowledge of the subject of meteorite fluxes and impact effects, and in particular to presenting the somewhat conflicting evidence in relation to mass extinction. The Fermor meeting in London in 1997, liberally supported by the Fermor Fund of the Geological Society and one or two other grants, allowed these experts to meet up, give oral presentations and engage in discussions, and all but this introduction are based on such presentations.

The Fermor Lecture Meeting is called by the Geological Society approximately every three years, in memory of Sir Lewis Leigh Fermor (1880–1954); the theme of the meeting is normally chosen to echo his diverse interests. Fermor joined the Geological Survey of India in 1902, eventually becoming Director (1930–1935). During his 33 years at the Survey, 27 meteorites fell or were found across the sub-continent, and Fermor studied many of them, leading to his ideas on the formation of meteorites and the chondrules within them. Thus a meteoritic topic was chosen as appropriate for the 1997 meeting and it was possible for a geological volume on a meteoritic topic to take its impetus from that meeting. Fermor recognized the importance of catastrophic impact events in his hypothesis that 'meteorites are fragments of the Earth torn off at the time of departure of the moon, or fragments loosed from the parent body of the solar system at the time of its tidal disruption by near approach of another stellar body' (1924). It is now known that impact cratering is a significant geological process, in terms of both planetary and biological evolution. The history of the Earth is traced through the geological record, a record of climate and sea-level changes, resulting in marine transgressions and regressions, and periods of intense tectonic activity, causing

GRADY, M. M., HUTCHISON, R., MCCALL, G. J. H. & ROTHERY, D. A. 1998. Introduction. *In*: GRADY, M. M., HUTCHISON, R., MCCALL, G. J. H. & ROTHERY, D. A. (eds) *Meteorites: Flux with Time and Impact Effects*. Geological Society, London, Special Publications, **140**, 1–5.

earthquakes and increased volcanism. All these features might be considered 'internal' effects, and have been well recognized for many years. There is, in addition, an 'external' effect that can have equally profound consequences for the Earth's history, the effect of bombardment from space. It is only during the past few decades that the importance of impacts as an agent of change has been appreciated. Thus the February 1997 Fermor Meeting of the Geological Society was a two-day discussion of meteorites, their flux and possible effects of impact.

Meteorites do not simply land on Earth out of nowhere, make a crater, then lie around awaiting collection. The number and size of meteorites colliding with the Earth, and the rate at which they arrive, are governed by the flux of material travelling within the Solar System. Crater formation depends on the kinetic energy of the impacting meteorite which is related to its size and composition, but recognition of meteorite craters is a function of their preservation on the Earth's surface. The consequences of meteorite impact might be localized and trivial; but at the other end of the spectrum, impact has been proposed as triggering global environmental changes leading to the catastrophic mass-extinction of species. An important facet of the Meeting was the drawing together of independent investigations into cratering rate, the flux of potential impactors and the search for temporal correlations in cratering, extinction and other global phenomena. As Shoemaker (this volume) said, 'Our knowledge of the impact history of the Earth and Moon and of the present flux of impactors has improved sufficiently to attempt identification of some of the expected variations in the cratering rates.' The conclusion, although not yet on a firm statistical basis, is that the Earth's pulse is determined by the Sun's motion in the galaxy.

The papers in this volume are arranged into four sections which broadly reflect the structure of the Fermor meeting. There is a preceding and unfinished paper by the late Gene Shoemaker, the Fermor Lecturer. Subjects covered are the flux of extraterrestrial material to the Earth, the different methods by which it has been determined, and how it has changed with time, leading on to the effects of impacts on the environment and their documentation in the geological record, focusing specifically on the Cretaceous–Tertiary (K–T) boundary. Interpretation of the fossil record across the K–T boundary, as detailed by palaeontologists, is certainly not as straightforward as geochemists and geophysicists might like to believe.

The flux of extraterrestrial material to the Earth: determination by astronomical and statistical techniques

The cratering record on Earth is put into a wider Solar System context by consideration of the bodies that actually cause the craters. The Moon maintains a more complete record of cratering events than the Earth, and recent systematic imaging of the Moon has allowed a more complete estimate of cratering rates in the past. From direct observation we obtain cratering rates on the Moon over the past 4.0 Ga (**Shoemaker**) and on the Earth over the past 2.0 Ga (**Grieve**; **Shoemaker**). The fluxes of Earth-crossing asteroids and different groups of comets also have been estimated from astronomical observations (**Shoemaker**; **Bailey & Emel'yanenko**). The observed cratering rate over the past 0.2 Ga is apparently about twice that from 3.2–0.8 Ga BP on the Moon and significantly greater than that implied by the observed total flux of potential impactors, which leads to the concept of 'dark' comets. Most periodic comets 'exhibit no activity detectable under the usual circumstances of comet search' (**Shoemaker**), so comets must be overlooked and the flux must be considerably greater than observed. As argued by **Shoemaker**, the dominant agents of crater-formation on both the Earth and Moon are comets rather than asteroids. Even so, the traditional division of minor bodies within the Solar System into comets (icy bodies with randomly-inclined elliptical orbits) and asteroids (rocky or metallic bodies with near-circular orbits close to the plane of the ecliptic) is probably too simplistic: it is now apparent that there is a spectrum of bodies between these end-members, a spectrum that must include dark (or inactive) comets as well as Edgeworth–Kuiper objects.

Bailey & Emel'yanenko use the observed flux of long-period comets to derive an estimated flux of Halley-type comets, by capture of the former into less elliptical orbits. They conclude that dark, unobserved Halley-type comets dominate the flux of potential impactors. If this is the case, then one of the major problems that remains with drawing inferences from astronomical data is that the actual size, or mass distribution of the putative impactors is unknown (**Hughes**). And if the main crater-formers are comets (of whatever activity) as opposed to asteroids, then changes in the cometary flux through time will have a significant effect on the cratering rate on Earth. The flux of long-period comets is largely controlled by galactic influences. Increases in the past flux of comets might result from comet showers produced by tidal disruption of individual comets, or by the motion of the Solar System through the galactic mid-plane resulting in perturbation of the Oort cloud. As detailed by **Napier**, the regular motion of the Sun through the mid-plane, and consequential disruption of the Oort cloud, might well lead to a periodicity in the flux of comets and hence the impact cratering record. **Napier** discusses the galactic environment and the Sun's motion therein, then looks for periodicity in a compilation of precisely dated craters supplied by **Grieve**. 'Synthetic' randomly produced datasets are used to test the strength of the statistical basis of inferred periodicities. **Napier** concludes, like **Shoemaker**, that the cratering record over the past 260 Ma comprises a continuum on which is superimposed a number of peaks with a period of 27 Ma, within error of the 31–32 Ma period discussed by **Shoemaker**. **Napier** goes further, however, and suggests that passage of the Sun (and Earth) through the galactic mid-plane, approximately every 27 Ma, correlates with a variety of 'global geological disturbances' including cratering, extinction, episodic sea-floor spreading, flood basaltic volcanism, orogeny, evaporite deposition and anoxia in the oceans. There can seldom be a set of conclusions more fundamental to our view of Earth history.

The flux of extraterrestrial material to the Earth: determination by terrestrial techniques

Complementing flux estimates of extraterrestrial material from astronomical calculations are methods based on actual numbers of meteorites collected. In order for this technique to be accurate, there are several parameters that need to be taken into careful consideration. Firstly, the collection area must be dry, such that chemical weathering of the meteorites is retarded; both hot and cold deserts provide this environment. A stable surface is also a requirement, so that meteorites can accumulate over tens or hundreds of millennia. Meteorites often break up during atmospheric passage; they might also shatter on contact with the ground. Estimates of meteorite 'pairing', where two or more samples of the same fall are matched, must be undertaken, to gain a valid count of the actual number of individual meteorite falls (**Bland**). In addition, weathering does occur, even in desert environments, and meteorites gradually rust and decay. An assessment of the meteorite decay rate for each desert region must also be obtained before flux estimates are made. Recent studies have considered all these variables, and, using Mössbauer spectroscopy to construct a weathering scale, coupled with known terrestrial ages of the meteorites (from ^{14}C-dating), have estimated an accumulation rate of approximately 81 250 falls (mass > 20 g) per year over the last 50 000 years (**Bland**, **Jull**), concluding that there has been little change in the gross meteorite flux over the recent past.

The Nullarbor region (Western & South Australia) has been identified as an area of significant meteorite accumulation, suitable for the calculation of meteorite fluxes. Collection expeditions over the past decade have yielded several thousand fragments from approximately 250 meteorites (**Bevan**). Analysis of the meteorite types present on the Nullarbor (after elimination of 'pairs') has shown that the population includes a higher proportion of rare chondritic meteorites than is present among the modern observed falls, implying that the stable surface of the Nullarbor, combined with low weathering rates, has allowed preservation of a more complete record of the types of material that bombard the Earth. Even so, iron meteorites are surprisingly rare on the Nullarbor Plain, but include the Mundrabilla meteorite of almost unique type, believed to have fallen about a million years ago, much earlier than any of the collected stony meteorites, and from which several large masses and a great number of small fragments have been recovered from a 90 km long dispersion ellipse (McCall in press '*Yearbook of Astronomy for 1999*').

Antarctica, where meteorite concentration mechanisms operate, is also a special case (**Zolensky**), where some 15 000 meteorites have been recovered over the past 20 years, representing around 1500 individuals. The meteorites have terrestrial ages of up to 1 Ma, and so provide another valuable resource for the estimation of flux rates over the recent past. Although the ability to 'pair' samples in the Antarctic collection is less easy than in the Nullarbor (where meteorite transportation processes are minimal), leading to a greater uncertainty in the number and mass distribution of the meteorites present on the ice, flux estimates based on the Antarctic collection also suggest that there has been little change in the rate of accumulation of extraterrestrial material to the Earth over the past million years.

Craters and impactites

One of the most compelling and obvious pieces of evidence for the arrival of extraterrestrial material on the Earth is the formation of a crater, and studies based on crater identification provided the first estimates for the flux of extraterrestrial material. According to **Grieve**, approximately 160 terrestrial impact craters, ranging up to 300 km in diameter and 2 Ga in age, currently are recognized (although these numbers are continually updated). The process of bombardment by extraterrestrial material has played a significant part in the geological evolution of the Earth: perhaps the most notable episode was that in which the Moon is believed to have formed after collision between the proto-Earth and another, almost equally large, body (**Grieve**). The Moon still retains evidence of episodes of ancient cratering events, a record that has been erased by geological activity on the Earth, but it shows that the cratering rate was much higher before 3.8 Ga (**Shoemaker**, **Grieve**). If the Earth experienced a bombardment of similar magnitude to that experienced by the Moon, then its surface would have suffered wide-scale melting of the upper part of the crust, and removal of any primitive atmosphere in the process of formation, rendering the Earth uninhabitable for any life-forms (**Grieve**). Continued collisions, albeit on a smaller scale, have influenced evolution throughout the geological record. After the formation of the Moon, and towards the end of the period of bombardment, the Earth's atmosphere developed. Again, impacts played their role: it has been postulated that the Earth accreted a significant proportion of its volatiles from the impact of ice- and volatile-rich comets. Initially, intense bombardment presumably frustrated the development of life.

The present-day record of craters formed by the impacts of large bodies, although incomplete, is becoming increasingly well-documented. Even so, the precise mass of extraterrestrial material accreted to the Earth as large objects is problematic: the impactors are destroyed by melting of both the impactor and target rocks as a result of shock compression and frictional heating effects (**Spray**). In some cases, a chemical signature may be left in the impact melt rocks (**Koeberl**). Nonetheless, a cratering rate of between two and three craters (diameter > 20 km) per million years is predicted from observations of craters produced in the last 100 million years (**Grieve**).

One of the craters that has been most intensively studied in recent years is that which has been proposed as the impact-site of the bolide that fell at the end of the Cretaceous, leading to the extinction of a range of many different taxa. Ironically, the crater is not visible on the Earth's surface, but has been traced by extensive seismic surveys of the area around Chicxulub, in the Yucatan Peninsula in the Gulf of Mexico (**Hildebrand**). The crater is buried, but the geophysical surveys indicate its diameter to be between 100 and 320 km (**Maguire**, **Hildebrand**). A crater of these dimensions must have been produced by a bolide some 10 km in diameter, leading to a sequence of global environmental effects. Included within the catalogue of possible effects are an initial darkening of the sky, due to ejected rock dust, followed by a rapid, global, drop in temperature. Alternatively, if the impact was into the ocean, then the amount of water ejected into the atmosphere would wash out the dust, and initiate a period of greenhouse warming, closely followed by further evaporation of the ocean, leading to a rise in global temperatures. In the case of the crater at Chicxulub, impact was into a sedimentary rock formation, including evaporite deposits (i.e. sulphate-bearing rocks), resulting in billions of tonnes of sulphur oxides ejected into the atmosphere (**Grieve**, **Hildebrand**). The energy of the impact also fused atmospheric nitrogen and oxygen into nitrogen oxides. As the temperature fluctuated, sulphur and nitrogen oxides washed out of the atmosphere as acid rain, leading to a change in the acidity of streams, lakes and, eventually, the surface of the oceans. The atmosphere, hydrosphere and biosphere were in turmoil, leading to a decrease in biological productivity, and, ultimately, extinction for many taxa. A number of different forms of carbon were produced including fullerenes and diamonds and **Gilmour** presents a discussion of the implications of these.

Of course, there is another feature of impacts in addition to the crater and the impactor that has sparked much debate: tektites. Tektites have a terrestrial composition, and are terrestrial objects that formed by an extraterrestrial agency. They are glassy objects produced within the cloud of vaporized rock that expands rapidly through the atmosphere immediately after an impact. Tektites exist in several extensive strewn fields across the globe, but there is still doubt as to the 'host' crater or impact event associated with some of the most widely-distributed tektites, particularly those of the largest and youngest Australasian strewn-field. A number of outstanding questions concerning tektites were raised (McCall 1997 in '*Yearbook of Astronomy for 1998*'), among the most important of these being the unacceptability on the basis of field evidence of models involving multiple sources and coalescence for the anomalous large, irregular and layered Muong Nong forms of South

East Asia, which, though commonly regarded as proximal to source, are recorded over distances of 2000 km. Much more attention could well be paid to Chinese occurrences as the source might be hidden under loess deposits in Yunnan or Guangxi provinces. Geochemical studies have dominated recent research on tektites and more work is needed on the physical forms of tektites to establish whether all or only some have suffered ablation on atmospheric re-entry. The anomaly that undigested rock relics have never been found in tektites seems remarkable, also the fact that only a very few of Grieve's list of 150 or so impact structures on the Earth are associated with tektites. Yet three out of the four strewn fields are firmly connected to impact structures (Ries, Bosumtwi, Chesapeake).

Environmental consequences

Although there are at least five mass extinctions throughout the geological record, at only one is there clear global evidence that impact of a large bolide occurred concurrently with the extinction. This is at the end of the Cretaceous, 65 Ma BP. As outlined in the previous section, the environmental changes wrought during the impact were tremendous, and more than sufficient to have had severe consequences for living creatures, both plant and animal, marine and non-marine. However, the K–T extinction is not the largest in terms of the number of taxa extinct (~50–60%): at the end of the Permian, where evidence for a catastrophic impact is less than conclusive, some 60–90% of all marine and land species disappeared. Conversely, there is abundant evidence for huge impacts throughout the geological record without associated mass extinctions (for example, the 100 km diameter crater at Manicouagan in Canada is ~210 Ma old, and the 100 km crater at Popigai in Russia is 35 Ma old). The lack of correlation of mass extinctions with impact structures shows that a single factor cannot alone explain extinctions in the geological record. Reviewing features associated with each of the five major extinctions in the Phanerozoic, **Hallam** concludes that there are several factors that may play a part in any extinction, including volcanism, climatic warming or cooling and marine regressions and transgressions, in addition to bolide impact.

Nonetheless, at the K–T boundary there is a clear indication that a major impact event occurred, delineated by a global layer of iridium-bearing clay, the presence of shocked quartz grains, microtektites etc. Whilst interpretation of the stratigraphic record is fairly clear, with protagonists debating the detail of the record (e.g. the actual dimensions of the crater) rather than the fact that an impact occurred, interpretation of the associated fossil record is not so clear-cut. The K–T extinction was obviously complex: as pointed out by **MacLeod**, many invertebrate species were already in decline towards the end of the Cretaceous, as a result of major climate changes and associated variations in sea-level. In contrast, many vertebrate species, particularly fresh-water taxa, appear to have survived into the Tertiary almost unscathed (**Milner**). It is possible that groups such as turtles, lizards and birds owe their continued existence to the presence of relatively undisturbed niches or havens protected from the major global outfall of the impact.

The keynote lecture of the meeting, on long-term variations in the impact cratering rate on Earth, was delivered by Dr Eugene Shoemaker, and his contribution resulting from this was intended to be the first paper in this volume. Gene Shoemaker was still working on the paper at the time of his tragic death in July of 1997. As a tribute to Gene, and with thanks to Carolyn Shoemaker, we include his unfinished manuscript prior to the first section of the volume. In sorrow and in gratitude for his enormous contribution to the understanding of meteorite cratering, both on Earth and on other bodies in the Solar System, this volume is dedicated to the memory of Eugene M. Shoemaker.

Dr. E. M. Shoemaker (1928–1997), planetary geologist, Fermor lecturer in February 1997, to whom this volume is dedicated.

Long-term variations in the impact cratering rate on Earth

EUGENE M. SHOEMAKER

Lowell Observatory, Flagstaff, Arizona

Extended abstract. Lunar and terrestrial impact craters, the terrestrial stratigraphic record of impact events, and astronomical observations of Earth-crossing asteroids and comets provide the primary evidence for evaluating the past and present flux of bodies impacting Earth. Variations in this flux on $\sim 10^5$, $\sim 10^6$, $\sim 10^7$ and $\sim 10^8$ year time scales are expected on theoretical grounds; $\sim 10^6$, $\sim 10^7$, and $\sim 10^8$ year variations appear to be reflected in the crater and stratigraphic records. The available evidence suggests that, late in geological time, comets have produced about half the impact craters on Earth >20 km in diameter and nearly all craters >100 km in diameter.

Systematic imaging of the Moon by the Clementine mission has enabled a thorough reexamination of the stratigraphic classification of Eratosthenian (3.2 to 0.8 Ga) and Copernican (0.8 Ga to the present) craters (McEwan *et al.* 1998). Further, it is now clear that the age of Copernicus, ~0.8 Ga, probably is close to the Eratosthenian–Copernican boundary. The mean rate of production of craters >20 km diameter in the past 0.8 Ga may have increased by ~40% over the mean rate in the previous 2.4 Ga, although the difference is within the statistical uncertainties (Table 1). A larger increase in the cratering rate is suggested by the spatial density of small craters on the ejecta blanket of Copernicus.

The record of Proterozoic craters in Australia, the most complete on Earth, yields an average Proterozoic cratering rate very close to that estimated from Eratosthenian craters on the Moon, when appropriate scaling and other calculations are used to find the cratering rate on Earth corresponding to that on the Moon (Table 1). In contrast, the Phanerozoic record of craters >10 km diameter on the Mississippi lowland of the United States and the record of craters ≥20 km diameter and younger than 120 Ma on the North American and European cratons (Grieve & Shoemaker 1994) suggest that the cratering rate may have increased late in geological time, perhaps as much as ~50% or more (Table 1). Again, the differences, individually, are well within the statistical uncertainties. However, the consistency between the lunar and terrestrial data sets strengthens the suggestion of a late increase in the long-term cratering rates. Moreover, the cratering rates derived from the late crater records are consistent with the present crater production rate estimated from astronomical observations of Earth-crossing asteroids and comets (Table 1).

The frequency of large craters formed in the last 100 Ma appears particularly anomalous when compared with the 3.2 Ga crater record of the Moon. The young age of the 170 km diameter, 65 Ma Chicxulub crater is especially noteworthy. Only one crater equal to or larger than Chicxulub has been formed in the past 3.2 Ga on the Moon. On the basis of the average 3.2 Ga lunar cratering rate, there would be about a 35% chance that a crater the size of Chicxulub was formed on Earth in the last 65 Ma. The odds that such a crater was formed on the continents is ~11%. Similarly, the chance that the 100 km diameter Popigai crater was formed on the continents in the last 36 Ma is only about 20%; the chance that the 85-km diameter crater Tycho was formed on the Moon in the last 100 Ma is about 25%. While the statistics are small, the ages of these large craters collectively indicate a substantial late increase in the rate of production of large craters.

Short-term variations in the cratering rate are suggested by the distribution of terrestrial crater ages and the stratigraphic record of impact events. The distribution of ages of craters >5 km diameter over the last 220 Ma appears to contain a periodic component with a mean period near 31 or 32 Ma. About half of the ages may be distributed periodically and half randomly. The significance of the apparent periodicity has been challenged, but significance tests for the periodic component are model dependent. The last three peaks in the cratering rate, which occur at ~1, 35, and 65 Ma are strongly reflected in independent data on ages and stratigraphic distribution of impact glass, noble metal anomalies, and shocked quartz grains.

A major problem in assessing the present flux of impacting bodies from astronomical observations lies in the difficulty of determining the sizes of comet nuclei. In addition, there are severe selection effects in the discovery of comets, particularly periodic comets, the vast majority of which exhibit no activity detectable under the usual circumstances of comet search. Within present observational uncertainties, it is entirely possible that comet impact has contributed to the production of about half or more of the craters larger than 20 km on Earth.

Order of magnitude and larger variations in the past flux of comets probably have occurred as a result of comet showers and changes in the motion of the sun through the

SHOEMAKER, E. M. 1998. Long-term variations in the impact cratering record on Earth. *In*: GRADY, M. M., HUTCHISON, R., MCCALL, G. J. H. & ROTHERY, D. A. (eds) *Meteorites: Flux with Time and Impact Effects.* Geological Society, London, Special Publications, **140**, 7–10.

galaxy. Comet showers due to tidal disruption of large, short-period comets have expected half-lives of about 0.5×10^5 years; showers caused by encounter of the sun with stars and other massive objects have half-lives of order 10^6 years. Varying galactic tidal perturbation of the Oort comet cloud modulates the delivery of long-period comets to Earth-crossing orbits, which leads to peaks in the comet flux about every 30 to 35×10^6 years (Matese *et al.* 1995).

The apparent periodic component in the distribution of ages of terrestrial impact structures and impact glass now appears understandable in terms of the variation in the tidal perturbation of the orbits of Oort Cloud comets. Both the mean period and timing of peaks are roughly consistent with plausible times of crossing of the sun through the galactic plane. The non periodic component present in the distribution of crater ages is consistent with the estimated production of craters by impact of Earth-crossing asteroids.

The most likely explanation for a long-term increase in the cratering rate is a decrease in the amplitude of the motion of the sun normal to the galactic plane and a consequent increase in the average flux of comets derived from the Oort Cloud. Combined evidence from lunar and terrestrial craters suggests that a major increase in the long-term comet flux occurred at about 200 to 300 Ma. It is unlikely that this surge is due to a change in the flux of asteroids, as a pulse in the flux of Earth-crossing asteroids decays with a half-life of 2 to 3×10^7 years. A previous long-term surge in the cratering rate may have occurred in the Neoproterozoic.

A broad understanding of the impact cratering history of the Earth-Moon system over the last 4 Ga has been in hand for about two decades. As determined from geological mapping of the Moon and precise age determinations of lunar samples returned by the Apollo missions, the rate of collision with the Moon by large crater-forming bodies declined by more than two orders of magnitude between ~4.0 and 3.2 Ga (Wilhelms 1987). The documented episode of higher cratering rate in this time interval has been referred to as *the late heavy bombardment* (Wasserburg *et al.* 1964). While the provenance of the impacting bodies is still a matter of debate (Wetherill 1989), evidence from the crater record of Mars and Mercury suggests that all the terrestrial planets were heavily cratered during the late heavy bombardment episode. Over the remainder of geological time the average cratering rate on the Moon since 3.2 Ga has been similar to the average late Phanerozoic cratering rate on Earth. Within a factor of about two, these long-term average cratering rates are about the same as the impact crater production estimated from the present near-Earth flux of asteroids and comets (Shoemaker 1983; Shoemaker *et al.* 1979, 1990). The uncertainties in some of the estimates of cratering rate are of the order of a factor of two, and it has commonly been assumed, by default, that the cratering rates on Earth and the Moon have actually been constant for the last 3.2 Ga (e.g. Wilhelms 1987).

A number of theoretical lines of argument, on the other hand, lead to the suspicion that the cratering rates since the late heavy bombardment have not been constant. Catastrophic break-up of very large main-belt asteroids should lead to occasional surges in the flux of Earth-crossing bodies and in the production of impact craters, especially craters with diameters on order 100 km or larger (Shoemaker 1984). The expected frequency of these surges is about once every few hundred Ma. About ten or so major surges in the near-Earth asteroid flux are

Table 1. *Estimated cratering rates on Earth*

Basis	Time interval	Production of craters ≥20 km diameter (10^{-15} km^2 a^{-1})
Eratosthenian craters†	3.2–0.8 Ga	3.7 ± 0.4 (asteroidal velocities) 3.0 ± 0.3 (50 : 50 asteroids:comets)
Proterozoic impact structures in Australia	2.6–0.54 Ga	3.8 ± 1.9
Copernican craters†	0.8–0.0 Ga	5.3 ± 1.8 (asteroidal velocities) 4.3 ± 1.4 (50 : 50 asteroids:comets)
U.S. Mississippi lowland	0.5–0.0 Ga	6.3 ± 3.2
N. American and European cratons‡	0.12–0.0 Ga	5.6 ± 2.8
Astronomical surveys	present	5.9 ± 3.5

† McEwen *et al.* (1998).
‡ Grieve & Shoemaker (1994).

likely to have occurred in the last 3.2 Ga and the time distribution of these surges is expected to be random. Hence the number of large craters produced by impact of asteroids probably has varied from each billion year interval to the next.

The arrival of comets in the neighbourhood of Earth is expected to be time variable. Comets from the Oort cloud are perturbed into Earth-crossing orbits by a combination of galactic tidal forces and gravitational impulses from passing stars. Peaks in the near-Earth flux of Oort Cloud comets should occur at the times of passage of the sun through the galactic plane (Matese *et al.* 1995). These passages recur periodically with an estimated frequency of about once every 30 to 35 Ma. Near the times of tidal perturbation peaks, moderate impulses from passing stars may lead to mild comet showers. About once per few hundred Ma, strong comet showers produced by close stellar encounters may occur, where the comet flux may rise by a factor of 30 to 100 times the mean background flux (Hills 1981; Heister *et al.* 1987). Because they are few in number and probably random, the distribution of strong comet showers over billion year time intervals in unlikely to be uniform. Changes in the sun's trajectory in the Galaxy and encounters with one or more giant molecular clouds that resulted in stripping comets from the outer Oort Cloud may also have changed the long-term average flux of comets near the Earth (Shoemaker 1983).

Showers of short-period Jupiter-family comets might be produced by tidal break-up of very large comets during close approaches to Jupiter. This type of shower could occur if large comet nuclei have negligible strength, as was the case for the precursor of Comet Shoemaker-Levy 9 (Scotti & Melosh 1993; Asphaug & Benz 1994). Break-up of periodic comets the size of Chiron (diameter ~200 km) might occur with a frequency about once per few tens of Ma and should be randomly distributed in time.

Our knowledge of the impact history of the Earth and Moon and of the present flux of the impactors has improved sufficiently to attempt identification of some the expected variations in the cratering rates. For a number of years I have suspected that the long-term average cratering rates may have increased late in geological time, perhaps by as much as a factor of two (Grieve & Shoemaker 1994; Shoemaker *et al.* 1990; Shoemaker & Shoemaker 1996). Hints of this increase come from comparison of the terrestrial cratering rate estimated from the present flux of asteroids and comets and from the Phanerozoic crater record of North America and Europe with the 3.2 Ga crater record of North America and Europe with the 3.2 Ga crater record of the Moon. The number of small craters on the rim deposits of the large young lunar craters Copernicus and Tycho also suggest a late increase in the cratering rate (basaltic volcanism). Several authors have analysed the distribution of published terrestrial crater ages and found suggestive but not compelling evidence for a periodic component in the crater ages with a period of order 30 Ma (Alvarez & Müller 1984; Rampino & Stothers 1986; Shoemaker & Wolfe 1986). The ages of and stratigraphic distribution of known horizons of impact glass, shocked quartz, and noble metal anomalies in Cenozoic deposits may also reflect a rough 30 to 35 Ma periodicity for major impact events (Hut *et al.* 1987; Shoemaker & Wolfe 1986). Farley and Patterson (1995) have recently reported apparent peaks in the delivery of cosmic dust to the sea floor, and traced by ^{3}He, that appear to correspond to the last three peaks in the production of large craters. Improved dating and discovery in the last few years of some very large terrestrial craters has strengthened the evidence for a late increase in the production of large craters and for possible peaks in the cratering rate.

Editorial note: This was as far as Gene Shoemaker had got with his paper at the time of his death in a road accident in Australia on 18 July 1997. We have elected to include the unfinished paper as a record of Gene's most recent thoughts on cometary flux. The abstract is the one distributed at the Fermor meeting, 18–19 February 1997, which was thus written several months previously.

References

ALVAREZ, W. & MÜLLER, R. A. 1984. Evidence from crater ages for periodic impacts on Earth. *Nature*, **308**, 718–720.

ASPHAUG, E. & BENZ, W. 1994. Density of comet Shoemaker-Levy 9 deduced by modelling break-up of the parent 'rubble pile'. *Nature*, **370**, 120–124.

FARLEY, K. A. & PATTERSON, D. B. 1995. A 100 K-yr periodicity in the flux of extraterrestrial ^{3}He to the sea floor. *Nature*, **378**, 600–603.

GRIEVE, R. A. F. & SHOEMAKER, E. M. 1994. The record of past impacts on Earth. *In*: GETIVELS, T. (ed.) *Hazards due to Comets and Asteroids*, University of Arizona, Tucson, 417–462.

HEISTER, J., TREMAINES, S. & ALCOCK, C. 1987. The frequency and intensity of comet showers from the Oort cloud. *Icarus*, **70**, 269–288.

HILLS, J. G. 1981. Comet showers and the steady-state infall of comets from the Oort cloud. *Astronomical Journal*, **86**, 1730–1740.

HUT, P., ALVAREZ, W., ELDER, W. P. *et al.* 1987. Comet showers as a cause of mass extinctions. *Nature*, **329**, 118–126.

MCEWAN, A. S., MOORE, J. M. & SHOEMAKER, E. M. 1998. *Journal of Geophysical Research*, submitted.

MATESE, J. J., WHITMAN, P. G., INNANEN, K. A. & VALTONEN, M. J. 1995. Periodic modulation of the Oort cloud comet flux by the adiabatically changing Galactic tide. *Icarus*, **116**, 255–268.

RAMPINO, M. R. & STOTHERS, R. B. 1986. Geological periodicities in the Galaxy. *In*: SMOLUCHOWSKI, R., BÀHCÀLL, J. N. & MATTHEWS, W. S. (eds) *The Galaxy and the Solar System*. University of Arizona Press, Tucson.

SCOTTI, J. V. & MELOSH, H. J. 1993. Estimate of the size of comet Shoemaker-Levy 9 from a tidal break-up model. *Nature*, **365**, 733–735.

SHOEMAKER, E. M. 1983. Asteroid and comet bombardment of the Earth. *Ann. Rev. Earth Planet Science*, **11**, 461–494.

——1984. Large body impacts through geologic time. *In:* HOLLAND, H.-D. & TRENDALL, A. F. (eds) *Patterns of Change in Earth Evolution*. Springer, Berlin, 77–102.

—— & SHOEMAKER, C. S. 1996. The Proterozoic impact record of Australia. *AGSO Journal of Australian Geology & Geophysics*, **16**, 379–398.

—— & WOLFE, R. E. 1986. Mass extinction, crater ages and comet showers. *In*: SMOLUCHOWSKI, R., BÀHCÀLL, J. N. & MATTHEWS, W. S. (eds) *The Galaxy and the Solar System*. University of Arizona Press, Tucson, 338–386.

——, WILLIAMS, J. G., HELIM, E. F. & WOLFE, R. E. 1979. *In:* GEHRELS, T. (ed.) *Asteroids*. University of Arizona Press, Tucson, Arizona, 253–282.

——, WOLFE, R. E. & SHOEMAKER, C. S. 1990. Asteroid and comet flux in the neighbourhood of the Earth. *In*: SHARPTON, V. L. & WARD, P. D. (eds) *Global Catastrophes in Earth History*. Geological Society of America, Special Papers, **247**, 155–170.

WASSERBURG, G. J., ALBEE, A. L. & LANPHERE, M. A. 1964. Migration of radiogenic strontium during metamorphism. *Journal of Geophysical Research*, **69**, 4395–4401.

WETHERILL, G. W. 1989. Cratering on terrestrial planets by Apollo objects. *Meteorites*, **24**, 15–22.

WILHELMS, D. E. 1987. *The Geological History of the Moon*. USGS Professional Paper, 1348.

Cometary capture and the nature of the impactors

M. E. BAILEY[1] & V. V. EMEL'YANENKO[2]

[1] *Armagh Observatory, College Hill, Armagh BT61 9DG, UK*
[2] *Department of Theoretical Mechanics, South Ural University, Chelyabinsk 454080, Russia*

Abstract: The inclination-averaged capture probability from nearly parabolic orbits to Halley-type orbits with periods $P < 200$ years and $q < 1.5$ AU is 0.0128 for an isotropic long-period source flux with perihelion distances $q < 4$ AU. The observed near-parabolic flux brighter than visual absolute magnitude $H_{10} = 7$ is approximately 0.2 comets $AU^{-1}\,a^{-1}$, and the dynamical lifetime of the captured Halley-type comets is on the order of 0.3 Ma. The predicted steady-state number of such comets is thus about 3000, many times more than the number of observed Halley-type comets. Provided that long-period comets do not totally disintegrate during the capture process (e.g. into streams of small bodies such as cometary meteoroids and dust), this indicates that there should be a large population of undiscovered Halley-type 'asteroids' in the inner Solar System. The predicted number of dark Halley-type objects is sufficient to produce a terrestrial cratering rate comparable with that from observed near-Earth asteroids, and hence could explain the reported Galactic signal seen in the geological and biological records. The assessment of the terrestrial impact hazard should consider the possibility that a significant number of dark Halley-type objects may exist.

The formation of comets, whether or not seen as part of the formation of the Solar System, can be viewed generally as a process of increasing concentration of diffuse interstellar material, first in the interstellar medium and the Sun's parent molecular cloud, then during the collapse phase of the protosolar nebula, leading to formation of a protoplanetary disc, and finally in several distinct accretionary phases in the protoplanetary disc. Recent observations, for example of comet C/1995 O1 (Hale–Bopp), have reinforced the notion that a significant component of cometary dust and ices originates in the interstellar medium, and recent reviews of the comet formation process, involving various stages of coagulation, have been given by Bailey (1991, 1994) and Weidenschilling (1997).

The resulting so-called 'small-bodies', which have sizes ranging from *c*. 1 km up to $>10^2$ km, accumulated in the protoplanetary nebula to form the rock-and-ice cores of the outer planets. Subsequently, a significant fraction of the original nebular material was gravitationally ejected by the growing planets into the supposedly safe storage reservoir known as the Oort cloud (e.g. Fernández & Ip 1983; Fernández 1985). Some objects, formed or scattered into nearly circular orbits beyond Neptune, may have remained in quasi-stable low-eccentricity orbits with semi-major axes in the range 30–100 AU, in the circumplanetary region known as the Edgeworth–Kuiper belt (Edgeworth 1943, 1949; Kuiper 1951; Whipple 1964; Fernández 1980; Duncan *et al.* 1988; Quinn *et al.* 1990; Stern 1996). A review of Edgeworth's interesting life-story has been given by McFarland (1996).

The resulting cometary reservoir has a structure that extends far beyond the planetary system, initially taking the form of a flared disc of low-inclination, nearly circular orbits in the Edgeworth–Kuiper belt and then encompassing a wider distribution of orbits of growing inclination and eccentricity in an extended trans-Neptunian cometary reservoir (Fernández 1985), with semi-major axes ranging up to on the order of 10^3 AU. Beyond this, the system merges into a more spherically symmetrical 'dense inner core' of the Oort cloud (Hills 1981; Bailey 1983; Duncan *et al.* 1987), and finally into the observed 'outer' Oort cloud, which contains orbits of all eccentricities and semi-major axes $>2 \times 10^4$ AU. The outer region, which has a shape determined by the tidal perturbations of the Galaxy, resembles a triaxial ellipsoid and extends roughly half-way to the nearest star, a distance on the order of 10^5 AU from the Sun in each direction (Antonov & Latyshev 1972; Bailey 1977; Smoluchowski & Torbett 1985; Byl 1986).

Observed comets are conventionally divided into two broad classes, namely long-period (comets with periods $P > 200$ years) and short-period (those with $P < 200$ years). The majority of long-period comets have, in fact, very long periods, often measured in thousands or even millions of years, and orbital inclinations that are almost randomly distributed over the sky, consistent with their having an immediate provenance in the almost spherically symmetrical Oort cloud. Short-period comets are often divided into minor classes, of which the principal separations are those denoted Halley-type and Jupiter-family. The former have a wide range of

BAILEY, M. E. & EMEL'YANENKO, V. V. 1998. Cometary capture and the nature of the impactors. *In*: GRADY, M. M., HUTCHISON, R., MCCALL, G. J. H. & ROTHERY, D. A. (eds) *Meteorites: Flux with Time and Impact Effects*. Geological Society, London, Special Publications, **140**, 11–17.

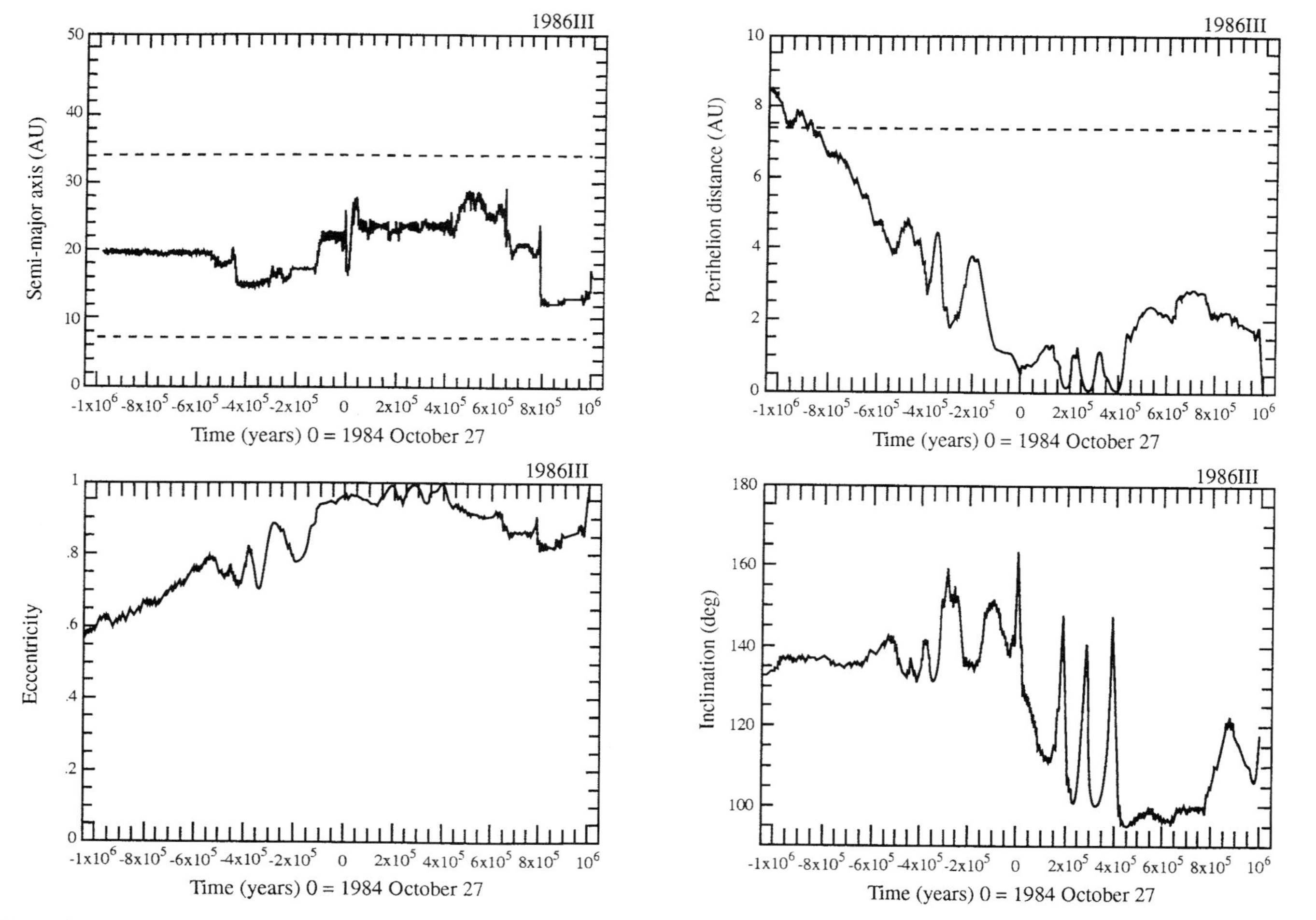

Fig. 1. Dynamical evolution of a test particle with identical initial elements to those of Halley's comet. (Note the periodic and long-term changes in the orbital elements, caused by secular perturbations and mean-motion resonances, and the trend towards the object becoming Sun-grazing in the astronomically near future.)

inclinations and are usually defined to have periods in the range $20 < P < 200$ years and Tisserand parameters $T < 2$; the latter have $P < 20$ years and $T > 2$. The Tisserand parameter, although not a strict constant of the motion, is a roughly conserved quantity in the dynamical evolution of many short-period comets, and the critical value $T = 2$ conveniently separates low-inclination Jupiter-family comets from the somewhat longer period and higher inclination Halley-types (Carusi & Valsecchi 1987; Levison & Duncan 1994).

Table 1. *The relative frequency p_c of capture into Halley-type orbits ($P < 200$ years and $q < 1.5$ AU) from the near-parabolic flux with initial perihelion distances in each of the stated initial ranges (after Emel'yanenko & Bailey 1997)*

q (AU)	p_c
0.0–4.0	0.0128
4.0–6.0	0.0013
6.0–10.5	0.0003
10.5–18.0	0.0002
18.0–31.0	$<10^{-4}$

Evolution

According to this planetesimal picture for comet formation, outlined above, both Jupiter-family and Halley-type short-period comets originate from nearly parabolic orbits in the Oort cloud. The former, however, may also come from regions closer to the planetary system: the dense inner core of the Oort cloud, the trans-Neptunian cometary disc and the Edgeworth–Kuiper belt. Indeed, most researchers nowadays would argue that those inner reservoirs provide the dominant source flux of low-inclination Jupiter-family comets. Recent years in particular have seen much progress on the question of the origin of Jupiter-family comets, especially following the pioneering investigations of Kazimirchak-Polonskaya (1972, 1976) and Everhart (1972, 1973, 1976). Subsequent investigations include those by Duncan *et al.* (1988), Stagg & Bailey (1989), Bailey & Stagg (1990), Quinn *et al.* (1990), Levison & Duncan (1994, 1997), Zheng *et al.* (1996), Duncan & Levison (1997) and Emel'yanenko & Bailey (1997*a*).

During this period, there has been an implicit assumption that Halley-type comets primarily originate from the long-period cometary flux. Recent advances in numerical techniques have now allowed this picture to be placed on a firm footing, taking into account not only the initial random-walk phase of evolution of the cometary energies but also the complex secular perturbations and mean-motion resonances that particularly affect Halley-type cometary evolution (Bailey & Hahn 1992; Hahn & Bailey 1992; Steel & Asher 1992; Asher *et al.* 1994; Bailey & Emel'yanenko, 1996). Figure 1 illustrates some of the relevant dynamical features, in this case involving an initial orbit with elements identical to those of Halley's comet.

Investigations of many such orbits show that Halley-type comets have dynamical lifetimes on the order of 1 Ma, and that they survive as potentially observable Halley-type comets (with orbital periods $P < 200$ years and perihelion distance $q < 1.5$ AU) for *c.* 0.3 Ma. Long-term integrations of the orbital evolution of long-period comets from the Oort cloud, and of their dynamical evolution or 'capture' into Halley-type orbits (Emel'yanenko & Bailey 1997, 1998) allow us to estimate the capture probability, results for which are given in Table 1.

This allows us to calculate the steady-state number of Halley-type comets, assuming a constant near-parabolic flux of 0.2 comets brighter than absolute magnitude $H_{10} = 7$ $AU^{-1}\,a^{-1}$(Bailey & Stagg 1988). Averaging the result over $0 < q < 4$ AU, and ignoring non-gravitational forces (which tend to enhance the probability of capture) and cometary fading or disintegration, gives a steady-state number of Halley-type comets arising from these relatively bright comets alone on the order of $N_{HT} \approx 3000$.

A comet with $H_{10} = 7$ has a diameter in the approximate range 5–15 km (see Hughes 1987; Bailey 1990; Weissman 1990; Bailey *et al.* 1992). The number of Halley-type objects originating from the observed long-period flux of comets larger than this size is therefore on the order of 3000. Outgassing and cometary splitting probably reduce the size of the original comet by a small margin, but unless it totally disintegrates a substantial kilometre-sized remnant seems likely to remain.

Cratering

We adopt the crater scaling law defined by Schmidt & Housen (1988), in which the diameter D_e of the initially excavated crater produced by a projectile of diameter d, density ρ and velocity V impacting a target medium of density ρ_t and surface gravity g_t is given by equation (5) of Bailey (1991), with parameters $A = 0.881$, $B = 0.78$, $C = 0.33$, $D = 0.33$, $E = 0.22$ and $F = 0.43$.

If $D_e > D_t \approx 3$ km, the diameter D_c of the final crater is larger than that of the initially excavated crater owing to slumping. Following

Strom (1987), we adopt the following empirical relation of Croft, namely

$$D_c = (D_e D_t^{-0.015})^{1/0.85} \quad (D_e \geq D_t \approx 3\,\text{km}) \quad (1)$$

Combining these equations yields

$$\left(\frac{d}{1\,\text{km}}\right) = 2.40\left(\frac{D_c}{20\,\text{km}}\right)^{1.090}\left(\frac{\rho}{1\,\text{g cm}^{-3}}\right)^{-0.423} \times \left(\frac{V}{20\,\text{km s}^{-1}}\right)^{-0.551} \quad (2)$$

which shows that a long-period comet or Halley-type object (e.g. $\rho = 0.6\,\text{g cm}^{-3}$, $V = 50\,\text{km s}^{-1}$) has a diameter $d \approx 1.8$ km in order to produce a 20 km diameter crater. Interestingly, the crater scaling law also shows that a similar sized near-Earth asteroid (e.g. $\rho = 2.0\,\text{g cm}^{-3}$, $V = 20\,\text{km s}^{-1}$) would also produce a 20 km crater, the different projectile density almost exactly compensating for the different impact velocity.

The relative number of craters produced by each type of object depends on the respective flux of projectiles with perihelion distances within the Earth's orbit, and the impact probability with the Earth. For near-Earth asteroids, we assume a cumulative diameter distribution of the form

$$N_A(\geq d) \approx 1500\left(\frac{d}{1\,\text{km}}\right)^{-2} \quad (3)$$

which is thought to apply in the approximate range $0.5 < d \leq 10$ km, and a mean terrestrial impact probability $p_A \approx 5 \times 10^{-9}\,\text{a}^{-1}$ per asteroid (see Bailey, 1991). The resulting rate of production of craters with diameter larger than D, assuming $\rho = 2.0\,\text{g cm}^{-3}$ and $V = 20\,\text{km s}^{-1}$, is then

$$\dot{N}_A(\geq D) \approx 2.34 \times 10^{-6}\left(\frac{D}{20\,\text{km}}\right)^{-2.18}\,\text{year}^{-1}. \quad (4)$$

This expression is probably uncertain by about a factor of two. The situation in the case of comets is more difficult to determine, as their numbers and masses are much less well known. Here we follow Bailey *et al.* (1992) in assuming a correlation between visual nuclear magnitude $V(1,0)$ and total magnitude H_{10} of the form $V(1,0) \approx 7.8 + 0.8H_{10}$. This leads to

$$\log\left(\frac{d}{1\,\text{km}}\right) = -0.16H_{10} + 2.333 - \tfrac{1}{2}\log\left(\frac{p_v}{0.04}\right) \quad (5)$$

where p_v is the usual geometric albedo, so that for these assumptions comets with a canonical absolute magnitude $H_{10} = 7$ would have a diameter $d \approx 16.3$ km.

Following Bailey & Stagg (1988), the intrinsic new-comet flux brighter than absolute magnitude H_{10} is approximately

$$\nu_{\text{new}} \approx f\nu_{\text{LP}} \approx f10^{0.27(H_{10}-7)}\,\text{comets AU}^{-1}\,\text{a}^{-1} \quad (6)$$

where $f = 0.2$, which therefore becomes

$$\nu_{\text{new}}(\geq d) \approx 0.2\left(\frac{d}{16.3\,\text{km}}\right)^{-1.69}\,\text{comets AU}^{-1}\,\text{a}^{-1}. \quad (7)$$

These comets evolve dynamically to produce observed long-period comets and unobserved extinct cometary nuclei, the latter also moving in potentially observable long-period cometary orbits. From equation (6), the long-period flux is roughly $f^{-1} \approx 5$ times the new-comet flux, and integrating over the observed $1/a$-distribution to the Halley-type limit $(1/a)_{\text{HT}} = \Delta x = 0.03\,\text{AU}^{-1}$ gives a further factor $\sqrt{2}\,\Delta x/\sigma$, where $\sigma \approx 5 \times 10^{-4}\,\text{AU}^{-1}$ is the mean dispersion in $x = 1/a$ produced by planetary perturbations. The total long-period flux, including both active and extinct or inert members of the cometary $1/a$-distribution, is therefore

$$\nu_{\text{LP}} \approx \nu_{\text{new}}(f^{-1} + \sqrt{2}\,\Delta x/\sigma). \quad (8)$$

These comets cross the Halley-type limit $\Delta x = 0.03\,\text{AU}^{-1}$ and become potentially observable Halley-type objects. Following Emel'yanenko & Bailey (1997, 1998), the new-comet flux $\nu_{\text{new}} = 0.2$ comets $\text{AU}^{-1}\,\text{a}^{-1}$ produces a steady-state population of approximately 3000 Halley-type objects (HTOs) with perihelion distances $q < 1.5$ AU, of which roughly 2000 have $q < 1$ AU. These figures correspond to initial new comets brighter than $H_{10} = 7$; and assuming a mean orbital period $\bar{P} \approx 10^2$ years for HTOs, the corresponding flux of Earth-crossing objects is $\nu_{\text{HTO}} \approx 20\,\text{AU}^{-1}\,\text{a}^{-1} \approx 10^2\nu_{\text{new}}$.

Summing these contributions, the total Earth-crossing flux of long-period and Halley-type cometary objects may be written in the form

$$\nu_c \approx \nu_{\text{new}}\left(f^{-1} + \frac{\sqrt{2}\,\Delta x}{\sigma} + 10^2\right) \approx 200\nu_{\text{new}}\,\text{comets AU}^{-1}\,\text{a}^{-1} \quad (9)$$

where the three terms in parentheses and their values correspond to active long-period comets (5), inactive long-period comets (85), and HTOs (100), respectively.

Using equation (7) to extrapolate the new-comet flux to $d = 1.8$ km (for a crater of final diameter $D_c = 20$ km) and making use of the crater-diameter scaling law together with a mean terrestrial impact probability per revolution of 2.5×10^{-9} (Bailey 1991), thus leads to a 'cometary' cratering rate of the form

$$\dot{N}_C(\geq D) \approx 4.3 \times 10^{-6} \left(\frac{D}{20\,\mathrm{km}}\right)^{-1.84} \mathrm{a}^{-1} \quad (10)$$

comparable with that for the population of near-Earth asteroids. We conclude that even though our adopted cometary mass–magnitude relation may have overestimated the importance of comets, HTOs dominate the observed long-period comet cratering rate by a large margin, and that the total Halley-type comet cratering rate is a significant fraction of that for near-Earth asteroids. For comparison, the *observed* cratering rate is *c.* $3 \times 10^{-6} D_{20}^{-2}\,\mathrm{a}^{-1}$.

Conclusion

We see that within the uncertainties of the argument (which are substantial), a significant fraction of the observed terrestrial craters (roughly one 20-km diameter crater every 200–300 ka) may be produced by 'dark' Halley-type comets, i.e. by extinct or inert cometary nuclei moving on orbits similar to those of observed Halley-type comets. It is interesting that this result provides a possible explanation of the link sometimes suggested between the terrestrial cratering and geological records, and of the roughly 30 Ma solar oscillation cycle perpendicular to the Galactic plane (Clube & Napier 1984, 1996; Matese *et al.* 1995). If the cratering projectiles predominantly originate from the outer Oort cloud, as we suggest, then such a 30 Ma 'periodicity' can be readily understood.

In summary, we predict a still undiscovered population of 'dark' cometary asteroids in orbits similar to those of Halley-type comets. If these objects exist in the predicted numbers (a recently discovered example might be 1997 MD_{10}; Marsden 1997), and if cometary disintegration during their long-term dynamical evolution is not the major cometary end-state, then such objects should make a major contribution to the terrestrial cratering rate. Surveys for Near Earth Objects should consider this possibility, and also the likelihood of finding high-inclination, high-velocity asteroids in regions of the sky not frequented by the population of low-inclination, low-velocity near-Earth asteroids that originates in the main asteroid belt.

This work was supported by DENI and PPARC; we thank the organizers of the Fermor meeting for their patience and help with the manuscript and for support to attend the Fermor meeting.

References

ANTONOV, V. A. & LATYSHEV, I. N. 1972. Determination of the form of the Oort cometary cloud as the Hill surface in the Galactic field. *In*: CHEBOTAREV, G. A., KAZIMIRCHAK-POLONSKAYA, E. I. & MARSDEN, B. G. (eds) *The Motion, Evolution of Orbits and Origin of Comets*. IAU Symp. **45**. Reidel, Dordrecht, 341–345.

ASHER, D. J., BAILEY, M. E., HAHN, G. & STEEL, D. I. 1994. Asteroid 5335 Damocles and its implications for cometary dynamics. *Monthly Notices of the Royal Astronomical Society*, **267**, 26–42.

BAILEY, M. E. 1977. Some comments on the Oort cloud. *Astrophysics and Space Sciences*, **50**, 3–22.

——1983. The structure and evolution of the solar system comet cloud. *Monthly Notices of the Royal Astronomical Society*, **204**, 603–633.

——1990. Cometary masses. *In*: LYNDEN-BELL, D. & GILMORE, G. (eds) *Baryonic Dark Matter*, Kluwer, Dordrecht, 7–35.

——1991. Comets and molecular clouds: the sink and the source. *In*: JAMES, R. A. & MILLAR, T. J. (eds) *Molecular Clouds*, Cambridge University Press, Cambridge, 273–289.

——1994. Formation of outer solar system bodies: comets and planetesimals. *In*: MILANI, A., DI MARTINO, M. & CELLINO, A. (eds) *Asteroids, Comets, Meteors 1993*. IAU Symp. **160**. Kluwer, Dordrecht, 443–459.

—— & EMEL'YANENKO, V. V. 1996. Dynamical evolution of Halley-type comets. *Monthly Notices of the Royal Astronomical Society*, **28**, 1087–1110.

—— & HAHN, G. 1992. Orbital evolution of 1991 DA: implications for near-Earth orbits. *In*: FERNÁNDEZ, J. A. & RICKMAN, H. (eds) *Periodic Comets*. Universidad de la República, Montevideo, 13–24.

—— & STAGG, C. R. 1988. Cratering constraints on the inner Oort cloud. *Monthly Notices of the Royal Astronomical Society*, **235**, 1–32.

—— & ——1990. The origin of short-period comets. *Icarus*, **86**, 2–8.

——, CHAMBERS, J. E. & HAHN, G. 1992. Detection of comet nuclei at large heliocentric distances. *Monthly Notices of the Royal Astronomical Society*, **254**, 581–588.

BYL, J. 1986. The effect of the Galaxy on cometary orbits. *Earth, Moon and Planets*, **36**, 262–273.

CARUSI, A. & VALSECCHI, G. B. 1987. Dynamical evolution on short-period comets. *In*: CEPLECHA, Z. & PECINA, P. (eds) *Interplanetary Matter. Proc. 10th European Regional Meeting in Astronomy*, **2**, 21–28, Prague.

CLUBE, S. V. M. & NAPIER, W. M. 1984. Terrestrial catastrophism – Nemesis or Galaxy? *Nature*, **311**, 635–636.

—— & ——1996. Galactic dark matter and terrestrial periodicities. *Quarterly Journal of the Royal Astronomical Society*, **37**, 617–642.

DUNCAN, M. J. & LEVISON, H. F. 1997. A disk of scattered icy objects and the origin of Jupiter-family comets. *Science*, **276**, 1670–1672.

——, QUINN, T. & TREMAINE, S. 1987. The formation and extent of the solar system comet cloud. *Astronomical Journal*, **94**, 1330–1338.

——, —— & ——1988. The origin of short-period comets. *Astrophysical Journal Letters*, **328**, L69–L73.

EDGEWORTH, K. E. 1943. The evolution of our planetary system. *Journal of the British Astronomical Association*, **53**, 181–188.

——1949. The origin and evolution of the solar system. *Monthly Notices of the Royal Astronomical Society*, **109**, 600–609.

EMEL'YANENKO, V. V. & BAILEY, M. E. 1997. The capture of Halley-type and Jupiter-family comets from the near-parabolic flux. *In*: WYTRZYSZCZAK, I. M., LIESKE, J. H. & FELDMAN, R. A. (eds) *Dynamics and Astrometry of Natural and Artificial Celestial Bodies*. IAU Coll. **165**. Kluwer, Dordrecht, 159–164.

—— & ——1998. Capture of Halley-type comets from the near-parabolic flux. *Monthly Notices of the Royal Astronomical Society* in press.

EVERHART, E. 1972. The origin of short-period comets. *Astrophysical Letters*. **10**, 131–135.

——1973. Examination of several ideas of comet origins. *Astronomical Journal*, **78**, 329–337.

——1976. The evolution of cometary orbits. *In*: DONN, B., MUMMA, M., JACKSON, W., A'HEARN, M. & HARRINGTON, R. (eds) *The Study of Comets: Part I*. IAU Coll. **25**. NASA SP-393, Washington, DC, 445–461.

FERNÁNDEZ, J. A. 1980. On the existence of a comet belt beyond Neptune. *Monthly Notices of the Royal Astronomical Society*, **192**, 481–491.

——1985. The formation and survival of the Oort cloud. *In*: CARUSI, A. & VALSECCHI, G. B. (eds) *Dynamics of Comets: their Origin and Evolution*. IAU Coll. **83**. Reidel, Dordrecht, 45–70.

—— & IP, W.-I. 1983. On the time evolution of the planetary influx in the region of the terrestrial planets. *Icarus*, **54**, 377–387.

HAHN, G. & BAILEY, M. E. 1992. Long-term evolution of 1991 DA: a dynamically evolved extinct Halley-type comet. *In*: HARRIS, A. W. &, BOWELL, E. (eds) *Asteroids, Comets, Meteors 1991*. Lunar and Planetary Institute, Houston, TX, 227–230.

HILLS, J. G. 1981. Comet showers and the steady-state infall of comets from the Oort cloud. *Astronomical Journal*, **86**, 1730–1740.

HUGHES, D. W. 1987. The history of Halley's comet. *Philosophical Transactions of the Royal Society of London, Series A*, **323**, 349–367.

KAZIMIRCHAK-POLONSKAYA, E. I. 1972. The major planets as powerful transformers of cometary orbits. *In*: CHEBOTAREV, G. A., KAZIMIRCHAK-POLONSKAYA, E. I. & MARSDEN, G. B. (eds) *The Motion, Evolution of Orbits and Origin of Comets*. IAU Symp. **45**. Reidel, Dordrecht, 373–397.

——1976. Review of investigations performed in the USSR on close approaches of comets to Jupiter and the evolution of cometary orbits. *In*: DONN, B., MUMMA, M., JACKSON, W., A'HEARN, M. & HARRINGTON, R. (eds) *The Study of Comets: Part I*. IAU Coll. **25**. NASA SP-393, Washington, DC, 490–536.

KUIPER, G. P. 1951. On the origin of the solar system. *In*: HYNEK, J. A. (ed.) *Astrophysics*. McGraw–Hill, New York, 357–424.

LEVISON, H. F. & DUNCAN, M. J. 1994. The long-term dynamical behavior of short-period comets. *Icarus*, **108**, 18–36.

—— & ——1997. From the Kuiper belt to Jupiter-family comets: the space distribution of ecliptic comets. *Icarus*, **127**, 13–32.

MARSDEN, B. G. 1997. 1997MD10. *Minor Planet Electronic Circular* **1997-P07**.

MATESE, J. J., WHITMAN, P. G., INNANEN, K. A. & VALTONEN, M. J. 1995. Periodic modulation of the Oort cloud comet flux by the adiabatically changing Galactic tide. *Icarus*, **116**, 255–268.

MCFARLAND, J. M. 1996. Kenneth Essex Edgeworth—Victorian polymath and founder of the Kuiper belt? *Vistas in Astronomy*, **40**, 343–354.

QUINN, T., TREMAINE, S. & DUNCAN, M. 1990. Planetary perturbations and the origin of short-period comets. *Astrophysical Journal*, **355**, 667–679.

SCHMIDT, R. M. & HOUSEN, K. R. 1988. Crater size estimates for large-body impact. *In*: *Global Catastrophes in Earth History*. Snowbird abstracts, LPI Contrib. **973**. NASA CR-183329, N89-21381, 162–163.

SMOLUCHOWSKI, R. & TORBETT, M. V. 1984. The boundary of the solar system. *Nature*, **311**, 38–39.

STAGG, C. R. & BAILEY, M. E. 1989. Stochastic capture of short-period comets. *Monthly Notices of the Royal Astronomical Society*, **241**, 507–541.

STEEL, D. I. & ASHER, D. J. 1992. The past and future orbit of (extinct comet?) 1991 DA. *In*: FERNÁNDEZ, J. A. & RICKMAN, H. (eds) *Periodic Comets*. Universidad de la Republica, Montevideo, 65–73.

STERN, S. A. 1996. The historical development and status of Kuiper disk studies. *In*: RETTIG, T. W. & HAHN, J. M. (eds) *Completing the Inventory of the Solar System*. ASP Conf. Ser. **107**, 209–222.

STROM, R. G. 1987. The solar system cratering record: Voyager 2 results at Uranus and implications for the origin of impacting bodies. *Icarus*, **70**, 517–535.

WEIDENSCHILLING, S. J. 1997. The origin of comets in the solar nebula: a unified model. *Icarus*, **127**, 290–306.

WEISSMAN, P. R. 1990. The cometary impactor flux at the Earth. *In*: SHARPTON, V. L. & WARD, P. D. (eds) *Global Catastrophes in Earth History*. Geological Society of America, Special Paper, **247**, 171–180.

WHIPPLE, F. L. 1964. The evidence for a comet belt beyond Neptune. *Proceedings of the National Academy of Sciences of the USA*, **52**, 565–594.

ZHENG, J. Q., VALTONEN, M. J., MIKKOLA, S. & RICKMAN, H. 1996. Orbits of short-period comets captured from the Oort cloud. *Earth, Moon and Planets*, **72**, 45–50.

Galactic periodicity and the geological record

W. M. NAPIER

Armagh Observatory, College Hill, Armagh BT61 9DG, UK

Abstract: Precisely dated impact craters, major geological disturbances and mass extinction peaks were used to investigate statistically the hypothesis that global terrestrial phenomena are triggered by the Galactic environment through bombardment episodes. Strong temporal correlations were found between events in these datasets, and in aggregate they were found to have a periodicity of *c.* 27 ± 1 Ma corresponding to the half-period of the Sun's vertical oscillations within the Galactic disc. Global disturbances appear to be forced in quasi-periodic fashion through a combination of impacts and prolonged climatic stress, the latter as a result of stratospheric dusting by debris from very large short-period comets.

It was first proposed by Napier & Clube (1979) that a wide range of terrestrial processes, including mass extinctions of species, ice ages and plate tectonic processes, might ultimately be induced by the Galactic environment. The assumed intermediary between Galaxy and Earth was the Oort comet cloud, which is periodically disturbed when the Solar System encounters spiral arms and molecular clouds as it journeys through the Galaxy. These disturbances dislodge comets from the Oort cloud, some of which are thrown in towards the planetary system to yield bombardment episodes of a few million years' duration. Much of the damage done to the biosphere was considered to be due to the impacts of planetesimals of *c.* 10 km diameter. The rationale for the hypothesis was the then recent finding first, that the Oort comet cloud is unstable in the Galactic environment, and second, that the collision rates are high enough for biotic and geological trauma to be reasonably expected consequences of such bombardment episodes. The Galactic hypothesis has been developed in long series of papers by Clube & Napier (e.g. 1982, 1984, 1986, 1996) and independently by Rampino with various colleagues (e.g. 1984, 1986, 1992, 1994, 1997).

What is widely considered to be 'hard evidence' for a cosmic input at the Cretaceous–Tertiary (K–T) boundary, in the form of the famous iridium anomaly, was discovered by Alvarez *et al.* (1980), who likewise interpreted the mass extinctions occurring at that boundary as the result of the impact of a 10 km asteroid. The search for cosmic signatures at the K–T and other mass extinction boundaries has since become a thriving if controversial industry. If the major reservoir of large impactors is say the asteroid belt, then no temporal structure is expected and the Earth becomes essentially a uniformitarian stage on which the drama of random impacts is occasionally enacted. If, on the other hand, Oort cloud disturbance is involved, then a more continuous control is envisaged, involving not simply the prompt effects of occasional cometary impacts, but also effects deriving from climatic changes consequent on stratospheric dusting. This is because the mass distribution of comets is top heavy, most of the flux into the inner planetary system coming in the form of rare, giant comets whose disintegration leads to multiple impacts and prolonged stratospheric dusting. When such a comet (more than *c.* 100 km, say) disintegrates in a short-period, Earth-crossing orbit, the optical depth of the stratosphere may be significantly reduced for millennia (Bailey *et al.* 1994; Clube *et al.* 1997). On the Galactic hypothesis, the terrestrial record thus becomes a probe reflecting the past history of our journey around the Galaxy. A crucial discriminant between the Alvarez *et al.* (1980) and Napier & Clube (1979) hypotheses is that, on the latter, Galactic periodicities are expected in the terrestrial record.

In fact, claims that the terrestrial record shows a periodicity of *c.* 30 Ma have been made intermittently over the past 70 years or so, going back at least to Holmes (1927). This periodicity has been claimed for such diverse phenomena as climate, sea-level variations, mass extinctions, geomagnetic reversals and global volcanic episodes. From time to time, it was noted that this interval is similar to that between successive passages of the Sun through the plane of the Galaxy, as it oscillates vertically ($\frac{1}{2}P \approx$ 26–33 Ma) while orbiting the centre of the Galaxy in a *c.* 230 Ma period. However, no viable mechanism connecting Galaxy to Earth was apparent. The first suggestion that the *c.* 30 Ma 'Holmes cycle' might be externally forced came in 1979, when Seyfert & Sirkin claimed that terrestrial impact craters were formed during 'impact epochs' recurring at 28 Ma intervals. They also held that these epochs were correlated with a wide range of global disturbances, such as basalt floodings, orogenies and sea-level variations.

NAPIER, W. M. 1998. Galactic periodicity and the geological record. *In*: GRADY, M. M., HUTCHISON, R., MCCALL, G. J. H. & ROTHERY, D. A. (eds) *Meteorites: Flux with Time and Impact Effects*. Geological Society, London, Special Publications, **140**, 19–29.

They ascribed these to the creation of mantle plumes caused by gigantic impacts but had no explanation for the proposed periodicity.

In 1984, Raup & Sepkoski revived the Fischer & Arthur (1977) claim of a periodicity in the mass extinction record as evinced by marine fossils, finding that the marine fossil record shows evidence of a periodicity 26.5 Ma. Further claims of a similar cratering periodicity followed (Table 1). Napier (1989), however, claimed a 16 Ma periodicity, a result supported by Yabushita (1991) for smaller craters, although others have disputed the existence of a cratering periodicity at all (e.g. Grieve *et al.* 1988; Heisler & Tremaine 1989; Grieve & Pesonen 1996), partly on the grounds that the claims made are not justified by the accuracy and precision of the data.

At about the same time, it became clear that the vertical Galactic tide, proportional to the density of the ambient material, is the dominant continuous force acting to throw long-period comets into the inner Solar System (e.g. Byl 1983). The comet influx, thus, peaks as the Sun passes through the plane of the Galaxy during its vertical oscillations (Napier 1987; Bailey *et al.* 1990; Matese *et al.* 1995; Clube & Napier 1996). Local stellar kinematics (Bahcall & Bahcall 1985) reveals that this Galactic clock ticks at a rate somewhere in the range 26–33 Ma, encompassing the Holmes cycle. It therefore appears that there is indeed a potential mechanism connecting Galaxy and Earth, in the form of cyclically recurring bombardment episodes or impact epochs.

A Galactic periodicity present in the impact cratering record would constitute direct evidence that the Galaxy does exert a controlling influence on a range of global terrestrial processes (Napier & Clube 1979; Clube & Napier 1984, 1996; Rampino & Stothers 1984; Rampino *et al.* 1997); and because the solar oscillation period is a function of the ambient density, there is the prospect that the Earth could be used as a test particle to probe *inter alia* the dark matter content of the Galactic disc.

These periodicity claims, however, raise a number of questions. For example:

Table 1. *Some periodicities claimed for terrestrial impact craters*

Author	Period (Ma)
Seyfert & Sirkin (1979)	28
Alvarez & Muller (1984)	28
Rampino & Stothers (1984)	31
Shoemaker & Wolfe (1986)	31
Yabushita (1991)	16, 31

- What is their status, particularly that of the disputed impact cratering periodicity?
- Can the frequently quoted value of 31 ± 1 Ma for the cratering record be reconciled with the 26.5 Ma periodicity claimed for the mass extinction record?
- What is the status of the 15 Ma cycle in the geomagnetic record claimed by Mazaud *et al.* and supported by Napier (1989)?
- Can the best-fit phase derived for these various periodicities (*c.* 10 Ma) be reconciled with the recent passage of the Solar System through the plane of the Galaxy (say 0–4 Ma BP)?

The discussion of these issues (in later sections) can be read independently of the next section, which comprises an outline of the theoretically expected comet flux modulations by the Galactic environment.

Comet flux modulation by the Galaxy

The strength T of the continuous Galactic tide acting on the Oort cloud is given by

$$T = -4\pi G\rho(z)\Delta z$$

where Δz is the instantaneous difference in height above the plane between Sun and comet, and $\rho(z)$ is the effective mass density of perturbers responsible for the tide at altitude z. It is readily shown that, to a first approximation, the flux of comets thrown into the inner planetary system is linearly proportional to T and so to $\rho(z)$. Thus knowing the vertical trajectory of the Solar System and the run of ambient density, the variation of comet flux with time may be found (Matese *et al.* 1995; Clube & Napier 1996). Flux amplitudes in the range 2:1 to 4:1, with periodicities in the range 26–33 Ma, are readily attained. The assumption of continuous action will generally fail at some distance within which the action is dominated by discrete impulses from individual perturbers. However, Monte Carlo trials (see below), in which the Sun drifts through a field of low-mass perturbers (such as black dwarf stars) with a Schwarzschild velocity distribution, reveal that this stochastic component of the Galactic force field also yields a cyclicity with the same range of amplitude.

In addition to the flux cyclicity caused by variations in ambient potential, a second, velocity-dependent peak may in principle occur 90° out of phase with the first. This arises because, at the peak of its orbit, the Sun's speed relative to its surroundings is at a minimum. The decline is only *c.* 1.25 km s^{-1}, and, so long as the principal perturbers have a velocity dispersion greatly in excess of this, the out-of-plane enhancement is negligible. However, in the circumstance where the Galactic disc had a dark matter component comprising many small, magnetically supported molecular clouds, this interpulse might become noticeable and, in combination with the in-plane effect, yield a *c.* 15 Ma period detectable in a very complete record. There is no independent evidence for such such structures in the Galaxy; but such a cycle has been claimed both for the geomagnetic reversal record (Mazaud *et al.* 1983) and the impact cratering one (Napier 1989).

A pure periodicity in comet flux is not expected, however. The Sun's Galactic orbit kinematically resembles that of the B stars, whose velocity dispersion is observed to increase with age, with a relaxation time *c.* 100 Ma (Wielen 1977). This diffusion in velocity may be attributed in large part to giant molecular clouds (Lacey 1984). However, the geological data span a Galactic year (*c.* 230 Ma) or more, and over this interval the assumption of adiabaticity (Matese *et al.* 1995) is not expected to hold.

A middle-of-the-road estimate is that the Solar System penetrates one giant molecular cloud of mass $5 \times 10^5 M_\odot$ every 750 Ma. The mass distribution $n(m)$ of molecular clouds is of the form $n(>M) \propto M^{-0.8}$ down to at least 5000 $M_\odot$. Because of this top-heavy distribution, the bulk of the orbital change caused by clouds will occur in one or two jumps over say a 200 Ma interval. A grazing encounter ($p = 20$ pc) at 15 km s^{-1} with a nebula of mass $5 \times 10^5 M_\odot$ will violently disrupt the Oort cloud, and will also change the the Sun's velocity by *c.* 6 km s^{-1}. Over a 250 Ma interval there is an expectation of a velocity shift $\overline{\delta V} \approx 3.5$ km s^{-1}, while at 50 Ma intervals $\overline{\delta V} \approx 0.5$ km s^{-1}. For $w_o = 7$ km s^{-1} and $P = 27$ Ma, the expected perturbations at (750, 250, 50) Ma intervals translate into maximal phase perturbations $\delta\phi_m \approx (6.1, 4.0, 0.6)$ Ma. The period also changes slightly: a vertical velocity perturbation of 2 km s^{-1} has mean recurrence time of *c.* 185 Ma and may yield a period change of *c.* 1.4 Ma depending on the model adopted for the local Galactic disc.

One such encounter appears to be identifiable (Bailey *et al.* 1990; Clube & Napier 1996). The Sun has recently (10 ± 2 Ma BP) passed through a ring of gas and stars, the Gould's Belt complex, which appears to have a mass of *c.* $2 \times 10^6 M_\odot$. If $V \approx 15$ km s^{-1} and the effective $p \approx 100$–200 pc, a recent phase shift in the range $0 \lesssim \delta\phi_o \lesssim 9$ Ma is implied; in essence the solar orbit may, within approximately the last 10 Ma, have undergone an indeterminate but possibly large phase shift. An impulsive phase shift of more than 2 Ma will generally be accompanied by a strong surge of long-period comets into the inner planetary system.

In addition to these phase and period shifts, relatively frequent encounters with small molecular clouds may cause loss cone flooding and hence bombardment episodes without causing much change in the underlying period P and phase ϕ. The Sun is expected to pass within 20 pc of a 5000 $M_\odot$ nebula apprxoimately every 18 Ma. It may be shown that such a perturbation, acting on long-period comets with perihelia just beyond the planetary system, competes with the Galactic tide in flooding the cometary loss cone, although the effect on the solar orbit is negligible. Thus the periodicity is expected to be noisy, and the bombardment of the Earth may even be described as episodic-cum-periodic. Stochasticity will arise not only from molecular cloud encounters but also from penetrating stars, the asteroid belt (which may induce surges: V. Zappala, pers. comm. 1997) and the recently discovered Edgeworth–Kuiper belt. And, of course, internal processes may create global disturbances without exogenous forcing. No precise predictions for all these noise components are available at present, and the procedure adopted below is to take an empirical approach to the correlation–periodicity questions without too much theoretical preconception.

Periodicity in the terrestrial data

Three datasets are examined here, namely the compendium of Rampino & Caldeira (1992), the list of mass extinction peaks given by Raup & Sepkoski (1984), and the impact crater catalogue of R. A. F. Grieve to epoch January 1997 (pers. comm. 1997).

The spectral estimator used here is the unwindowed periodogram. Independently of frequency tested, its statistic, the unprocessed 'power' I, has a mean value $\bar{I} = 2$ for random, uniform data and is distributed roughly as $\exp(-I/2)$. The power I is often considered to be an unreliable spectral estimator (Thomson 1990; Newman *et al.* 1994), primarily because of its bias and inconsistency. However, numerical experiments provide a robust way round these limitations: synthetic datasets may be successively constructed in such a way as to simulate every significant feature of the real dataset, except the periodicity under test. If these artificial datasets are analysed in identical fashion to the real one, a distribution function $n(I)$ is obtained and the power I for the real dataset compared with it. Any significant difference can, logically, only be due to the periodicity under test. To the extent that non-stationarity, inconsistency, poor convergence and edge effects matter, they affect the real and synthetic datasets equally and so their effects are imbedded in the $n(I)$ distribution. Thus, one applies a differential technique, looking for any signal by examining the offset between real and simulated data. This numerical approach is extremely flexible; for example, a 'recovery time' following a mass extinction event may readily be built into the simulations.

The hypothesis under test, 'there is a Galactic periodicity', is to be tested against the null one, 'there is not'. A careful specification of the latter requires that such features as broad temporal trends, clustering or interdependence of the various data should be incorporated in the synthetic datasets. Such datasets may be constructed by adding, to the age of each recorded event, a random perturbation small enough to preserve such features but large enough to smear out the periodicity under test, should it be present. According to Bahcall & Bahcall (1985), local stellar kinematics yields a range of half-periods for the solar vertical oscillation from 26 to 33 Ma, maximum altitude from 49 to 93 pc, current vertical velocity $w_o = 7 \pm 1\,\mathrm{km\,s^{-1}}$, and the time since last passage through the plane of the Galaxy $\phi_o \approx 2 \pm 2$ Ma BP.

Figure 1a shows the power spectrum of the Rampino–Caldeira geological events but excluding the mass extinction peaks. The high peak has period and phase given by $(P, \phi_o) = (27.2, 6.2)$ Ma. The secondary peak at 12.1 Ma has phase $\phi_o \approx 2.9$ Ma. These peaks have power I given by 9.7 and 9.2, respectively, where $\bar{I} = 2$ for a random (white noise) data distribution.

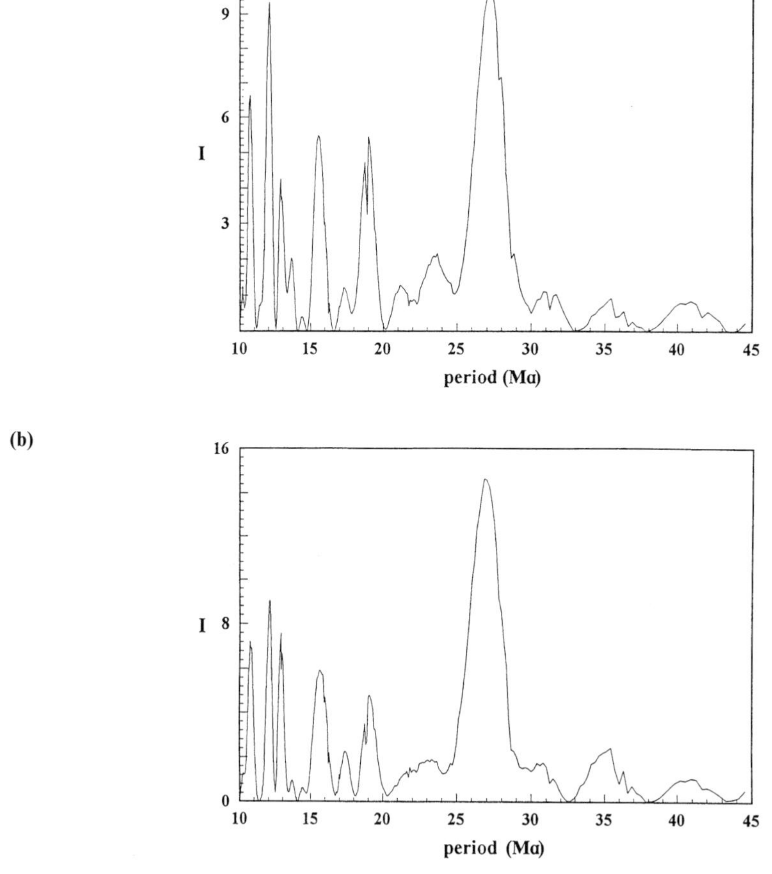

Fig. 1. (**a**) Power spectrum (PS) of the 64 major geological events in the compendium by Rampino & Caldeira (1992). (**b**) PS of the combined geological and biological data, the latter represented by the 11 mass extinction peaks of Raup & Sepkoski (1984).

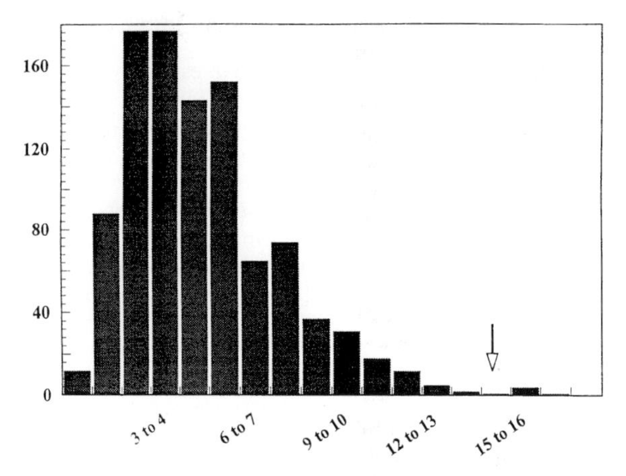

Fig. 2. Peak PS solutions for 1000 random datasets simulating the overall behaviour of the Rampino & Caldeira (1992) dataset with regard to non-stationarity, etc.

The result of adding in the Raup–Sepkoski extinction peaks is shown in Fig. 1b. The chief signal has increased to $I = 14.7$ whereas (P, ϕ_o) stay much the same: (26.9, 7.7) Ma, respectively. The secondary peak has declined to $I = 7.6$, and its $(P, \phi_o) = (12.9, 0.3)$ Ma. In these analyses a rough correction has been made for edge effects as described by Lutz (1985).

Figure 2 shows the power distribution for 1000 sets of randomized data simulating the overall age distribution of the actual Rampino–Caldeira dataset (including the mass extinctions). As the artificial datasets are constructed so as to closely follow the distribution of the real one, systematics such as edge effects and secular trends are automatically taken care of. It can be seen that the 'observed' peak is attained by chance only a few times in 1000, and so in this dataset the hypothesis 'there is a Galactic periodicity' is preferred over the hypothesis 'there is none such' at a confidence level of about 99.5%. This confirms the periodicity claim of Rampino & Caldeira (1992).

Two types of standard error are associated with these results; namely, that associated with the mathematical fit to the given dataset (as in Clube & Napier 1996), and that arising because the given dataset is only one of many possible ones which might have emerged from the same underlying physical process: another planet would have encountered another set of molecular clouds. The standard deviation associated with the latter is much the larger of the two.

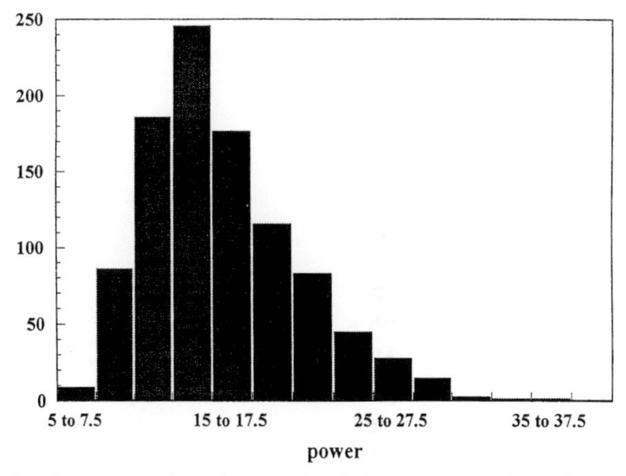

Fig. 3. Power distribution for 1000 random datasets simulating the overall behaviour of the Rampino & Caldeira (1992) dataset with regard to non-stationarity, etc., but with inbuilt $P = 28.0$ Ma, $\phi_o = 4$ Ma and dispersion $\Delta = 10$ Ma.

To find the permissible range of Solar System (P, ϕ_o) which could yield those observed (26.9, 7.7 Ma), trials were conducted in which 'synthetic-periodic' datasets (in batches of 1000) were constructed so as to simulate the overall age distribution of the Rampino–Caldeira set. To construct these synthetic datasets, data were randomly extracted from a rectangular probability distribution of half-width $\Delta = 10$ Ma around each peak of the artificial cycle. Each synthetic dataset was then analysed in identical fashion to the real one, and the peak signal was recorded.

Figure 3 shows the distribution of peak power for one such set of 1000 trials. It appears that $\Delta = 10$ Ma is a middle-of-the-road estimate for the intrisic dispersion (defined as above) of the real periodicity. In these runs (P, ϕ_o) are generally recovered, with no sign of bias induced by edge effects or secular trends. There is, nevertheless, a significant class of harmonic solutions around $P \approx 12$–13 Ma. Figure 4 reveals that the P and ϕ_o obtained are strongly correlated, longer derived periods being associated with smaller phases. The spread in period is nevertheless tight, say 27 ± 1 Ma, whereas the phase is much less well determined, covering the range 4 ± 4 Ma. It seems, however, that there is no obvious conflict between the expected $\phi_o = 2.5 \pm 2$ Ma of the last Galactic plane crossing on the one hand, and the 'best-fit' phase of *c.* 6–8 Ma: the phase is simply an ill-determined quantity, sensitive to shot noise fluctuations. The probable occurrence of a significant recent phase shift because of Gould's Belt penetration adds to the uncertainty in expected phase. However, the uncertainty is such that an out-of-plane peak is, statistically, equally acceptable.

Terrestrial impact craters

A similar exercise was carried out with the 28 terrestrial impact craters listed in Table 2. These were culled from a compilation by R. Grieve, craters associated with iron bolides, and those with quoted age uncertainties $\sigma > 10\,\mathrm{km\,s^{-1}}$, being excluded (for a given periodicity P, little information is obtained from data with age uncertainties $\sigma \lesssim P/2$). Three craters less than 10 Ma old were first excluded from the study, to avoid bias which might arise through the relative completeness of knowledge of very young, large craters. A power spectrum of the 28 data with the ages was constructed in which the data were weighted inversely by formal dating precision (except for the Chicxulub crater whose extremely precise dating was 'softened' to ± 0.2 Ma). A clear, well-defined peak $I \approx 8.2$ was obtained for $P \approx 13.4$ Ma, $\phi_o \approx 9.7$ Ma. This peak was remarkably stable to progressive truncation of the data down to less than *c.* 100 Ma: the 20 craters in the age range 10–12 Ma still have $P \approx 13.4$ Ma with $\phi_o \approx 9.7$. Thus the bulk of the power resides in craters younger than 120 Ma, older craters neither enhancing nor degrading the signal.

If the three youngest craters are included, the spectrum becomes becomes much messier, and a weaker peak appears at $P \approx 12.5$ and $\phi_o \approx 1$ Ma, again holding steady as the data are progressively truncated. If unweighted ages are employed the power drops considerably and two peaks $I \approx 6.5$–7 now appear at *c.* 13.5 and 18 Ma, which are respectively one-half and two-thirds harmonics of a 27 Ma period. Using an earlier compilation of Grieve & Shoemaker (1994), with 28 impact

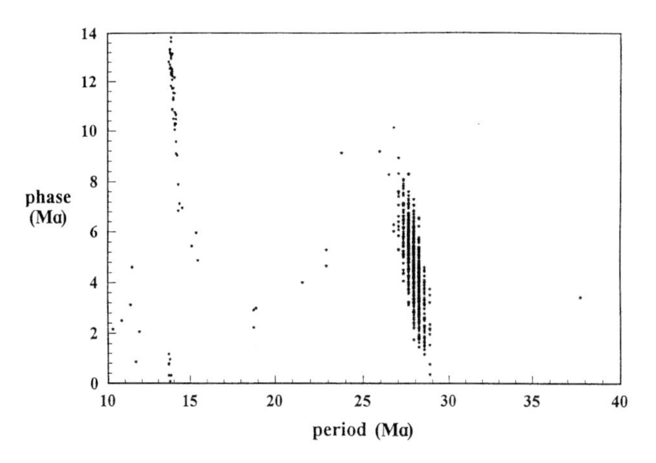

Fig. 4. Period-phase solutions for the 1000 synthetic 'Rampino–Caldeira' datasets.

Table 2. *Craters employed in the analysis*

Crater	Age (Ma)	Diameter (km)
Zhamanshin	0.9 ± 0.1	13.5
El'gygytgyn	3.5 ± 0.5	18
Bigach	6.0 ± 3.0	7
Ries	15 ± 1	24
Haughton	23 ± 1	24
Popigai	35 ± 5	100
Chesapeake Bay	35.5 ± 0.6	85
Wanapetei	37 ± 0.2	7.5
Mistastin	38 ± 4	28
Logoisk	40 ± 5	17
Kamensk	49 ± 0.2	25
Montagnais	50.5 ± 0.76	45
Ragozinka	55 ± 5	9
Marquez	58 ± 2	13
Chicxulub	64.98 ± 0.05	170
Kara	73 ± 3	65
Ust-Kara	73 ± 3	25
Manson	73.8 ± 0.3	35
Lappajärvi	77.3 ± 0.4	23
Boltysh	88 ± 3	24
Dellen	89.0 ± 2.7	19
Steen River	95 ± 7	25
Carswell	115 ± 10	39
Mien	121.0 ± 2.3	9
Tookoonooka	128 ± 5	55
Gosses Bluff	142.5 ± 0.8	22
Puchezh-Katunki	175 ± 3	80
Rochechouart	186 ± 8	23
Manicougan	214 ± 1	100
Araguainha Dome	247.5 ± 5.5	40
Siljan	368.0 ± 1.1	52
Kaluga	380 ± 10	15

craters culled by the above criteria, a weak peak at $P = 30.4$ Ma, $\phi_o = 4.1$ Ma was obtained, along with a secondary peak at 12.6 Ma.

The formal significance level of the 28-crater peak may be obtained by comparison with suitable synthetic datasets. It was found that the signal has only *c.* 90% confidence level and so the null hypothesis of randomness cannot be rejected with confidence; further, these trials cannot discriminate between a noisy periodicity of say 13.5 Ma and one of 27 Ma. The marginal or negative answers which some have obtained for the existence of a cratering periodicity (Grieve *et al.* 1988; Heisler & Tremaine 1989, etc.) are therefore supported by these results. However, one has simply asked of the dataset, Is there a periodicity? In that case, the waveband explored is much wider than when one asks the more specific question, namely: Is there a cratering periodicity in the specific (P, ϕ) range associated with the geological and mass extinction datasets?

To explore this latter, more specific, question, 1000 synthetic datasets were constructed to simulate the observed cratering one, with inbuilt periodicity $P = 27$ Ma and phase $\phi_o = 4$ Ma, and with half-width $\Delta = 10$ Ma. Power spectrum analyses of these revealed that in a significant number of cases the strongest solutions are ($13 \lesssim P \lesssim 15$ Ma, $8 \lesssim \phi_o \lesssim 12$ Ma), around the main harmonic (Fig. 5). The weighted solution for the real cratering data ($P \approx 13.4$ Ma, $\phi_o \approx 9.7$ Ma) lies in this range, and so may be a harmonic of the basic 27 Ma cycle. The data are sufficiently sparse that they could be fitted by a periodicity anywhere in the range 26–33 Ma; there is thus no contradiction between, say, the claimed 26 Ma cycle of Raup & Sepkoski (1984), and the 30 Ma impact cratering periodicity claimed by others (Table 1). However, construction and comparison of figures such as Figs 4 and 5 reveals that overlapping periodic solutions for the geological, biological and impact cratering data occur in only the narrow range $P = 27 \pm 1$ Ma, although the phase remains somewhat indeterminate.

The probability that the cratering solution would by chance lie on the principal harmonic of 27 Ma may be estimated from Fig. 5 (and is of order of a few per cent), but a more instructive way to exploit the geological and biological datasets as templates is to test the craters for correlation with global disturbances. Stratigraphic markers of impact such as shocked quartz, microtektites and large craters have been argued to correlate positively with extinction events at the genus level (Rampino & Haggerty 1994), and Matsumoto & Kubotani (1996) found the mass extinction and impact crater records to be correlated with 93–97% significance. The latter workers' statistic Q, measuring the relative closeness of the points in one dataset to corresponding points in the other, is adjusted so that $Q = 1$ for perfect coincidence, $Q = -1$ for perfect avoidance, and $Q = 0$ for no correlation between the series. The Q statistic is flawed in that it does not register the coincidence of a datum with the cluster as a whole, but only with individual data within the cluster. This shortcoming was handled below by taking the weighted mean of clustered data.

Figure 6 shows the result of comparing the Raup & Sepkoski (1984) extinction peaks with the cratering dataset in Table 1. Very young craters were excluded from this analysis but the recently discovered 70 km Morokweng impact structure dated at 144 ± 4 Ma was included. The Q distribution for analogous but uncorrelated datasets was obtained from Monte Carlo simulations. The figure confirms that there is

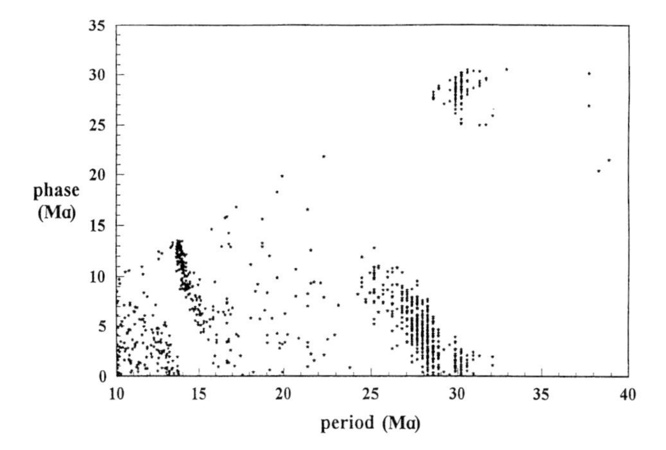

Fig. 5. Period-phase diagram for synthetic impact crater datasets, as discussed in the text. The (P, ϕ) 'fed in' are (27, 4) Ma; those obtained from the real data are (13.4, 9.7) Ma. These solutions are seen to be consistent, the latter possibly being a one-half harmonic.

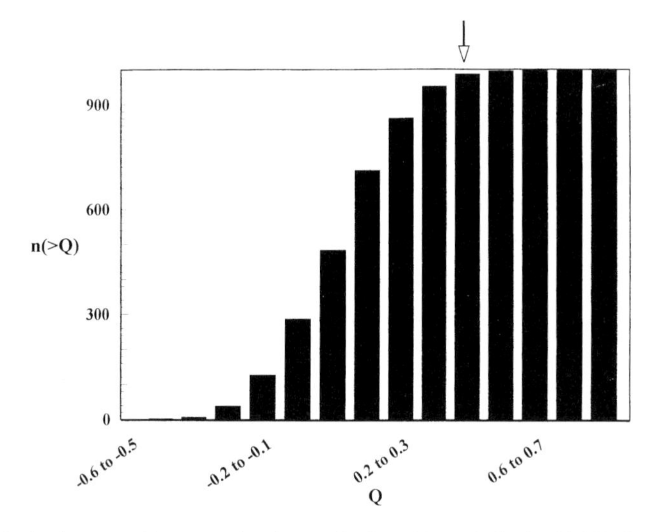

Fig. 6. Correlation between 11 mass extinction peaks from Raup & Sepkoski (1984) and 27 large, well-dated impact craters. The histogram illustrates the cumulative Q distribution for 1000 trials on suitably randomized data, and the arrow represents the Q value for the real data. The observed Q is exceeded by chance in only *c.* 2% of cases.

a significant correlation ($\bar{Q} = 0.462$, confidence level $C \approx 98\%$) between the two time series. The correlated mass extinctions and craters are listed in Table 3. Fifteen of the 27 craters used in this analysis appear to be significantly associated with mass extinctions.

Figure 7 shows the outcome of a similar comparison between the Rampino & Caldeira dataset and the impact cratering. Again, a significant correlation appears ($\bar{Q} = 0.307$, $C \approx 96\%$), but now the comparison dataset, the record of major geological events over the past *c.* 260 Ma, has a Galactic periodicity with a very high confidence level (say *c.* 99%). These events include the deposition of evaporites, sudden sea-floor spreadings, orogenies, flood basalt outpourings and anoxic episodes.

To sum up, these trials indicate that impact craters correlate with global geological disturbances, which in turn show evidence of a

Table 3. *Mass extinctions and correlated impact craters*

Extinction	Craters	*Q*
36.6	35.0 ± 5.0	
	35.5 ± 0.6	
	37.0 ± 0.2	
	38.0 ± 0.4	0.579
65.0	64.98 ± 0.05	0.992
91.0	88.0 ± 3.0	
	89.0 ± 2.7	
	95.0 ± 7.0	0.692
113.0	115.0 ± 10	0.692
144.0	142.5 ± 0.8	
	144.0 ± 4.0	0.974
176.0	175.0 ± 3.0	0.636
193.0	186.0 ± 8.0	0.000
216.0	214.0 ± 1.0	0.758
245.0	247.5 ± 5.5	0.758

periodicity of *c.* 27 Ma. The latter is in the range expected from systematic Oort cloud disturbance by the Galactic environment.

Discussion and conclusions

The main results to emerge from this study are:

- mass extinctions correlate strongly in age with large terrestrial impact craters;
- they likewise correlate well with a variety of global geological disturbances;
- a strong periodicity of *c.* 27 Ma is found in connection with all these global events when taken as a whole.

Correlation does not, of course, prove cause-and-effect, and these empirical results leave one free (say) to devise a hypothesis in which mass extinctions cause impacts. However, the exogenous (impact crater) connection is clearly a severe problem for models of the *coup de grâce* variety, or which ascribe mass extinctions to purely internal processes such as climate or sea-level change.

It is likely that the 13.5 Ma harmonic solutions which emerge for the relatively sparse impact cratering data are the '15 Ma' cycle which Mazaud *et al.* (1983) claimed for the geomagnetic reversal record and Napier (1989) found in an earlier cratering dataset. A 13–15 Ma cycle could in principle exist but it would be indistinguishable from a harmonic of 27 Ma. Likewise, the apparently differing periods derived from various datasets in the past, in the range *c.* 26–33 Ma, appear in fact to be mutually consistent, simply because of the intrinsic variability associated with small, noisy datasets. The observed periodicity 27 ± 1 Ma requires the Galactic disc to have *c.* 40% of its mass in the form of dark matter with a scale height 50–60 pc (Clube & Napier 1996).

There is clear evidence of clustering in the ages of the impact craters (Table 3), consistent with the astronomical picture in which terrestrial disturbances are caused by bombardment episodes of a few million years' duration. The mass distribution of comets is given by $n(\geq m) \propto m^{-0.83}$, with no known upper mass limit. Thus the flux is dominated by Chiron-sized comets which may individually have masses of more than *c.* 10 000

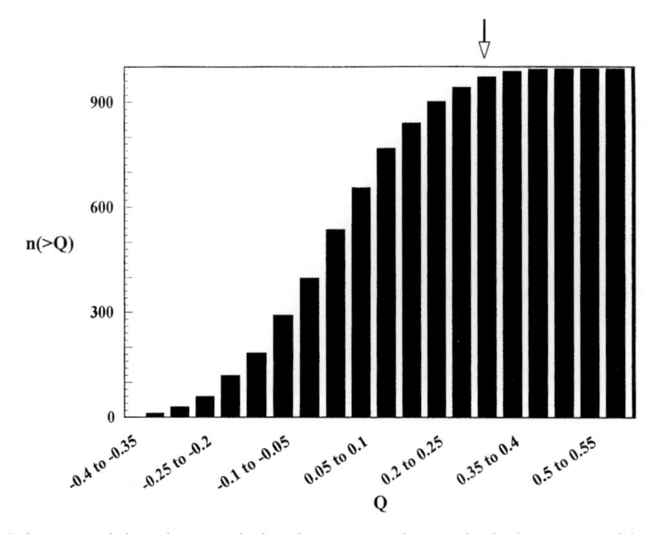

Fig. 7. As Fig. 6, but examining the correlation between major geological events and impact craters.

times that of Halley's Comet. The disintegration of such a monster in the near-Earth environment would lead to multiple impacts and prolonged stratospheric dusting. Mass extinctions and major geological events are thus seen to take place in an environment which is both climatically and impact disturbed (Napier & Clube 1979; Clube & Napier 1984, 1986, 1996; Bailey *et al.* 1994; Clube *et al.* 1997). There is no reason to expect a single, dramatic impact to be correlated one-to-one with mass extinctions, and indeed detailed paleontology and geochemistry do not appear to support this oversimplified view. Statements that 'the dinosaurs were in decline before the Chicxulub impact' or 'mass extinctions correlate with vulcanisms' have little bearing on the above extraterrestrial forcing hypothesis.

Another common misconception (e.g. Hallam, these proceedings) is that the astronomical cyclicity mechanisms were created *ad hoc* in response to the Raup & Sepkoski (1984) claim of a *c.* 26 Ma mass extinction cycle. This is true of the 'Death Star' or Nemesis hypothesis (Davis *et al.* 1984), which is in any case now regarded as untenable (e.g. Bailey *et al.* 1990). However, the basic mechanism of periodic Oort cloud disturbance by the Galaxy was already in place as early as 1979, following the discovery in that decade of the molecular cloud system; specific reference to the possibility of 15 and 30 Ma cycles was made, for example, by Napier (1983). The dominance of the vertical tide in Oort cloud evolution was also established by Byl in 1983, and so the causative chain connecting Galaxy and Earth was effectively in place by that date. The Galactic theory thus has its foundations in the known Galactic environment and is not a *deus ex machina* invoked arbitrarily to account for the Holmes cycle.

In conclusion, the observed correlations between cratering, major geological events and extinctions, the whole united by a *c.* 27 Ma periodicity, appear to support the hypothesis that the Earth is controlled, through bombardment episodes, by the ticking of a Galactic clock.

The author is indebted to M. E. Bailey, S. V. M. Clube and M. R. Rampino for discussions on the above topic, and to R. A. F. Grieve for supplying the impact cratering data.

References

ALVAREZ, L. W., ALVAREZ, W., ASARO, F. & MICHEL, H. V. 1980. Extraterrestrial cause for the Cretaceous–Tertiery extinction: experimental results and theoretical interpretation. *Science*, **208**, 1095–1108.

ALVAREZ, W. & MULLER, R. A. 1984. Evidence from crater ages for periodic impacts on the Earth. *Nature*, **308**, 718–720.

BAHCALL, J. N. & BAHCALL, S. 1985. The Sun's motion perpendicular to the galactic plane. *Nature*, **316**, 706–708.

BAILEY, M. E., CLUBE, S. V. M., HAHN, G., NAPIER, W. M. & VALSECCHI, G. B. 1994. Hazards due to giant comets: climate and short-term catastrophism. *In*: GEHRELS, T. (ed.) *Hazards due to Comets and Asteroids*. University of Arizona Press, Tucson, 479–533.

——, —— & NAPIER, W. M. 1990. *The Origin of Comets*. Pergamon, Oxford.

BYL, J. 1983. Galactic perturbations of nearly-parabolic cometary orbits. *Earth, Moon and Planets*, **29**, 121–137.

CLUBE, S. V. M. & NAPIER, W. M. 1982. Spiral arms, comets and terrestrial catastrophism. *Quarterly Journal of the Royal Astronomical Society*, **23**, 45–66.

—— & ——1984. Comet capture from molecular clouds: a dynamical constraint on star and planet formation. *Monthly Notices of the Royal Astronomical Society*, **208**, 575–588.

—— & ——1986. Comets and the Galaxy: implications of the terrestrial record. *In*: SMOLUCHOWSKI, R., BAHCALL, J. N. & MATTHEWS, W. S. (eds) *The Galaxy and the Solar System*. University of Arizona Press, Tucson, 260–285.

—— & ——1996. Galactic dark matter and terrestrial periodicities. *Quarterly Journal of the Royal Astronomical Society*, **37**, 617–642.

——, HOYLE, F., NAPIER, W. M. & WICKRAMASINGHE, N. C. 1997. Giant comets, evolution and civilization. *Astrophysics and Space Science*, **245**, 43.

DAVIS, M., HUT, P. & MULLER, R. A. 1984. Extinction of species by periodic comet showers. *Nature*, **308**, 715–717.

FISCHER, A. G. & ARTHUR, M. A. 1977. Secular variations in the pelagic realm. *In*: COOK, H. E. & ENOS, P. (eds) *Deep-water Carbonate Environments*. Society of Economic Paleontologists and Mineralogists, Special Publication, **25**, 19–50.

GRIEVE, R. A. F. & PESONEN, L. J. 1996. Temporal distribution and impacting bodies. *Earth, Moon and Planets*, **72**, 357–376.

——, SHARPTON, V. L., RUPERT, D. J. & GOODACRE, A. K. 1988. Detecting a periodic signal in the terrestrial cratering record. *Proceedings of the 18th Lunar and Planetary Science Conference*, 375–382.

—— & SHOEMAKER, E. M. 1994. The record of past impacts on Earth. *In*: GEHRELS, T. (ed.) *Hazards due to Comets and Asteroids*, University of Arizona Press, Tucson, 417–462.

HEISLER, J. & TREMAINE, S. 1989. How dating uncertainties affect the detection of periodicity in extinctions and craters. *Icarus*, **77**, 213–219.

HOLMES, A. 1927. *The Age of the Earth – an Introduction to Geological Ideas*. Benn, London.

LACEY, C. G. 1984. The influence of massive gas clouds on stellar velocity dispersions in galactic discs. *Monthly Notices of the Royal Astronomical Society*, **208**, 687–707.

LUTZ, T. M. 1985. The magnetic reversal record is not periodic. *Nature*, **317**, 404–407.
MATESE, J. J., WHITMAN, P. G., INNANEN, K. A. & VALTONEN, M. J. 1995. Periodic modulation of the Oort cloud comet flux by the adiabatically changing Galactic tide. *Icarus*, **116**, 255.
MATSUMOTO, M. & KUBOTANI, H. 1996. A statistical test for correlation between crater formation rate and mass extinctions. *Monthly Notices of the Royal Astronomical Society*, **282**, 1407–1412.
MAZAUD, A., LAJ, C., DE SÈZE, L. & VEROSUB, K. B. 1983. 15 Myr periodicity in the frequency of geomagnetic reversals since 100 Myr. *Nature*, **304**, 328–330.
NAPIER, W. M. 1983. The orbital evolution of short-period comets. *In*: LAGERKVIST, C.-I. & RICKMAN, H. (eds) *Asteroids, Comets, Meteors*, Uppsala University, 391–395.
——1987. The origin and evolution of the Oort cloud. *Proceedings of the Tenth European Regional Meeting in Astronomy (Prague)*, Vol. 2, 13.
——1989. Terrestrial catastrophism and Galactic cycles. *In*: CLUBE, S. V. M. (ed.) *Catastrophes and Evolution*. Cambridge University Press, Cambridge, 133–167.
—— & CLUBE, S. V. M. 1979. A theory of terrestrial catastrophism. *Nature*, **282**, 455–459.
NEWMAN, W. I., HAYNES, M.P & TERZIAN, Y. 1994. Redshift data and staistical inference. *Astrophysics Journal*, **431**, 147.
RAMPINO, M. R. & CALDEIRA, K. 1992. Episodes of terrestrial geologic activity during the past 260 million years: a quantitative approach. *Celestial Mechanics and Dynamical Astronomy*, **54**, 143–159.
—— & HAGGERTY, B. M. 1994. Extraterrestrial impacts and mass extinctions of life. *In*: GEHRELS, T. (ed.) *Hazards due to Comets and Asteroids*. University of Arizona Press, Tucson, 827.
—— & STOTHERS, R. B. 1984. Terrestrial mass extinctions, cometary impacts and the Sun's motion perpendicular to the Galactic plane. *Nature*, **308**, 709–712.
—— & —— 1986. Geologic periodicities and the Galaxy. *In*: SMOLUCHOWSKI, R., BAHCALL, J. N. & MATTHEWS, W. S. (eds) *The Galaxy and the Solar System*, University of Arizona Press, Tucson, 241.
——, HAGGERTY, B. M. & PAGANO, T. C. 1997. A unified theory of impact crises and mass extinctions. *Annals of the New York Academy of Sciences*, **822**, 403–431.
RAUP, D. M. & SEPKOSKI, J. J. 1984. Periodicity of extinctions in the geologic past. *Proceedings of the National Academy of Sciences of the USA*, **81**, 801–805.
SEYFERT, C. K. & SIRKIN, L. A. 1979. *Earth History and Plate Tectonics*. Harper Row, New York.
SHOEMAKER, E. M. & WOLFE, R. F. 1986. Mass extinctions, crater ages and comet showers. *In*: SMOLUCHOWSKI, R., BAHCALL, J. N. & MATTHEWS, W. S. (eds) *The Galaxy and the Solar System*, University of Arizona Press, Tucson, 338–336.
THOMSON, D. J. 1990. Time series analysis of Holocene climate data. *Philosophical Transactions of the Royal Society of London, Series A*, **330**, 601.
WIELEN, R. 1977. The diffusion of stellar orbits derived from the observed age-dependence of the velocity dispersion. *Astronomy and Astrophysics*, **60**, 263–275.
YABUSHITA, S. 1991. A statistical test for periodicity hypothesis in the crater formation rate. *Monthly Notices of the Royal Astronomical Society*, **250**, 481–485.

The mass distribution of crater-producing bodies

DAVID W. HUGHES

Department of Physics, The University, Sheffield, S3 7RH, UK

Abstract: The physical phenomena governing planetary cratering can be represented by three equations. The first, the cratering rate equation, originates from the observation of areas of the Earth's land-mass that have been exposed to an extraterrestrial flux for a known time, and the detection, counting and measurement of the resultant craters. The second, the energy–diameter equation, is much more uncertain. It relates the kinetic energy of an impacting body to the diameter of the crater that it produces. The third, the mass distribution equation, is also somewhat uncertain. This describes the way in which the flux of incident bodies varies as a function of their masses. This paper investigates the interrelationship between these three equations and comments on the comet/asteroid ratio among the extraterrestrial bodies responsible for producing the $19 < D < 45$ km craters on Earth.

From time to time comets and asteroids hit planets and satellites and produce craters on their surface. In this paper we are going to concentrate on the craters that have diameters greater than *c.* 20 km. This size range is chosen because the impactors that form these craters on Earth punch through the atmosphere unheeded. Also, these large craters are not formed by the secondary bodies that are ejected by other cratering events. Furthermore, in regions unaffected by plate tectonics, they survive the ravages of erosive weathering for time periods well in excess of 100 Ma. In an astronomical sense the Earth's surface is extremely young. It is non-saturated, inasmuch as new craters do not cover old ones. Similar to many previous studies, the present study finds that the number of craters found on Earth having diameters greater than a certain size, D km is proportional to about D^{-2}. Crater counts on stable regions of the Earth's surface can accurately be expanded to provide values for the rate at which these craters are formed.

Problems start when one endeavours to estimate the energy required to produce a large crater. The biggest nuclear explosion crater on Earth is *c.* 0.37 km across. Here we know the crater size and the energy expended in its production. Extrapolating from this relatively small explosion up to the energy required to form craters between say 4 and 40 km in diameter, where the volume of material that has been affected is around 10^3 and 10^6 times greater, is fraught with danger. But extrapolation is all that one can do. The experimental production of craters in the 4–40 km diameter range is clearly out of the question. In this paper it is suggested that the explosion energy E and the crater diameter D are related by $E \propto D^{3.25}$.

The mass distribution equation describes the way in which the flux of incident bodies varies as a function of their masses. Here again we have a problem. At present, there is no observational way of differentiating between the large craters that have been produced by asteroids and those that have been produced by comets. This underlines the strong probability that the kinetic energy of the impacting object is far more important than, for example, the density and composition of the impactor, or the composition and strength of the surface in which the crater is produced, or whether the impactor has low mass and high velocity or high mass and low velocity. Considering the typical impact velocities of asteroids and comets with Earth and the energy–diameter equation mentioned above, the impacting objects have diameters that are between 1/15 and 1/30 the diameter of the craters that they produce, so 4–40 km craters are produced by 0.1–3 km impactors. Most comets in this size range can be seen easily when they are in the inner solar system. But the inactive asteroids in this size range are much more difficult to find. The mass distribution of the impactors is therefore a matter of considerable debate.

The crater-producing impact process is thus governed by three equations, these giving the rate at which craters of differing diameters are formed on Earth, the relationship between the diameter of a crater and the energy required to produce it, and the mass distribution of the impacting objects. In this paper we discuss each of these equations in turn and discuss their interrelationship.

The cratering rate equation

The first basic equation concerns the cratering rate, Φ_c, this being the number of craters produced on each square kilometre of the Earth's surface per year having diameters greater than a certain size, say D km. This is a difficult quantity to measure and the results suffer considerably from the problems of small number statistics.

HUGHES, D. W. 1998. The mass distribution of crater producing bodies. *In*: GRADY, M. M., HUTCHISON, R., MCCALL, G. J. H. & ROTHERY, D. A. (eds) *Meteorites: Flux with Time and Impact Effects*. Geological Society, London, Special Publications, **140**, 31–42.

Table 1. *Large craters on the North American Craton, listed in order of increasing age; these data have been taken from Hodge (1994)*

Name	Latitude (degrees)	Longitude (degrees)	Diameter (km)	Age (Ma)
Sythemenkat Lake	66	151	12	0.01
Haughton	75	90	20.5	21
Mistastin Lake	56	63	28	38
Montagnais	43	64	45	52
Marquez Dome	31	96	15	58
Manson	44	95	35	74
Eagle Butte	50	111	19	<65
Steen River	60	118	25	95
Carswell	58	110	39	115
Sierra Madera	31	103	13	>100
Wells Creek	36	88	14	200
Manicouagan	51	69	100	212
Saint Martin	52	99	40	220
Clearwater Lake West	56	74	32	290
Clearwater Lake East	56	74	20	290
Charlevoix	48	70	54	357
Kentland	41	87	13	>300
Slate Island	49	87	30	>350
Beaverhead	45	113	15	600
Nicholson Lake	63	103	12.5	>400
Presquile	50	75	12	>500
Sudbury	47	81	200	1850

Thankfully, no Earth dweller has experienced the impact of a large (>1 km) asteroid or comet. These events are so rare that around a hundred thousand generations have to pass before one is experienced. The search for such events entails investigating old, stable regions of the Earth's surface and counting the craters that they contain. Let us follow Grieve & Dence (1979) and analyse the Phanerozoic crater record of the structurally stable part of the North American continent. The general area of interest is between the Rockies in the west and the Appalachians in the east, and the Quachita in the south and the Arctic to the north. Data concerning the known craters in this area are given by Hodge (1994), and he carefully lists their positions, diameters and ages.

Crater erosion is a serious problem and small craters disappear relatively quickly. Bearing this in mind, the flux can only be assessed correctly if one concentrates on craters that are both 'recent' and 'large'. Table 1 lists all the craters larger than 12 km in diameter that have been found on the North American Craton, the list being in order of age. There are two problems with a list of this sort. Craters disappear as a result of erosion; and the smaller craters are erased faster than the larger ones. There must,

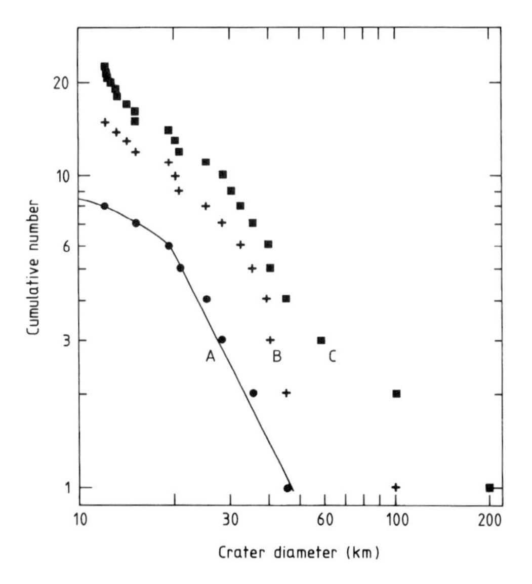

Fig. 1. A logarithmic plot of the cumulative number of Phanerozoic craters on the North American Craton, as a function of diameter. Data A (dots) represent only those craters formed in the last 105 Ma, data B (crosses) only those craters formed in the last 300 Ma, and data C (squares) all the craters listed in Table 1.

however, be a certain exposure age for which it is impossible for craters larger than a certain size to erode below the detection limit. Figure 1 uses the data of Table 1 and shows a logarithmic plot of the number of craters larger than a specific diameter as a function of that diameter, for the area in question. The three curves represent, (A) only those craters formed in the last 105 Ma, (B) only those craters formed in the last 300 Ma, and (C) all the craters listed in Table 1. It is concluded from these data that erosion seriously affects the record of craters smaller than about 19 km, and that, even for craters larger than this size, it is only the record of the last 105 Ma that can be thought of as complete. This unfortunately leaves a rather small dataset, there being only six craters, i.e. Montagnais ($D = 45$ km), Manson ($D = 35$ km), Mistastin Lake ($D = 28$ km), Steen River ($D = 25$ km), Haughton ($D = 20.5$ km) and Eagle Butte ($D = 19$ km).

A linear regression fit to this data (the right-hand linear portion of curve A in Fig. 1) indicates that the number, N, of North American craters larger than diameter D (km) is given by

$$\log N = (3.37 \pm 0.19) - (2.01 \pm 0.13) \log D \quad (19 < D < 45\,\text{km}) \quad (1)$$

The distribution of the eight youngest craters on the North American Craton is shown in Fig. 2. The inner circle is deliberately drawn through the craters Haughton and Marquez Dome. The outer circle is drawn to include some of the well-surveyed Arctic region and the area north of the Quachita. The diameters of the inner and outer circles subtend 44° and 56°, respectively, at the centre of the Earth. Simple trigonometry indicates that the areas of the Earth's surface inside these circles are 7.0×10^6 km^2 and 1.40×10^7 km^2, respectively. Remembering that the Pacific subduction plate and the orogenically new Rocky Mountains region must be excluded, it is estimated that the area over which the crater count has been made is about $(1.05 \pm 0.30) \times 10^7$ km^2. Table 1 indicates that the craters in question have been produced in the last 105 ± 10 Ma. Using a crater of 20 km as our benchmark, equation (1) indicates that 5.7 craters larger than this diameter have been produced during the time in question, on the specified area. The production rate of craters larger than 20 km in diameter is thus $(5.2 \pm 2.4) \times 10^{-15}$ km^{-2} a^{-1}. This figure can be compared favourably with the results of Grieve (1984), who obtained $(5.4 \pm 2.7) \times 10^{-15}$ km^{-2} a^{-1}, and Grieve & Shoemaker (1994), who gave $(5.6 \pm 2.8) \times 10^{-15}$ km^{-2} a^{-1}.

Previous attempts at estimating this production rate had used exposure times that were about three to four times longer than that used above and had thus underestimated the effects of erosion. The value of $(5.2 \pm 2.4) \times 10^{-15}$ km^{-2} a^{-1}

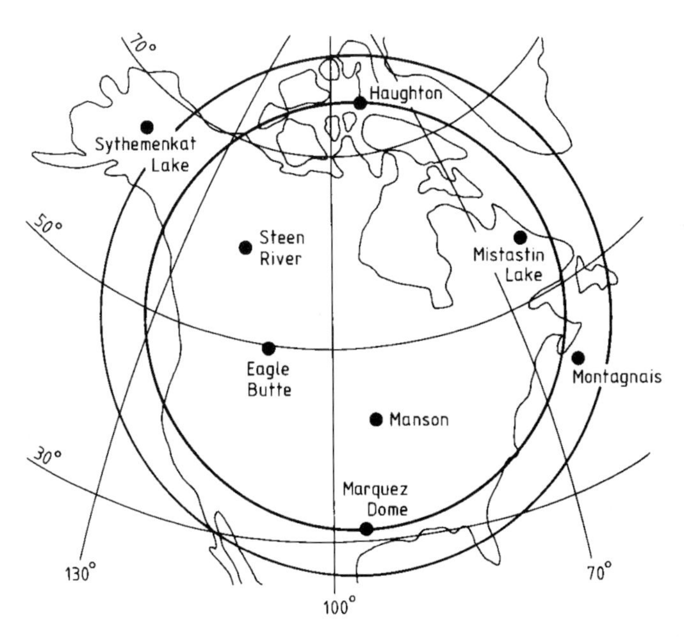

Fig. 2. The distribution of the first eight craters of Table 1 on the North American Craton. The inner and outer circles enclose areas 7.0×10^6 km^2 and 1.40×10^7 km^2, respectively.

($D > 20$ km) given above thus supercedes values such as $(2.6 \pm 0.9) \times 10^{-15}\,\mathrm{km^{-2}\,a^{-1}}$ (Hughes 1981) and $(3.5 \pm 1.3) \times 10^{-15}\,\mathrm{km^{-2}\,a^{-1}}$ (Grieve & Dence 1979).

Equation (1) can easily be replaced by a production rate equation. If Φ_c is the number of craters larger than diameter D km produced on an area of 1 km^2 of the Earth's surface per year then

$$\log \Phi_c = -(11.67 \pm 0.21) - (2.01 \pm 0.13) \log D \quad (2)$$

Equation (2) is strictly valid over a rather restricted range of diameters (typically $19 < D < 45$ km). This range might reasonably be stretched to $1 < D < 500$ km but going any further is fraught with danger. Equation (2) indicates that a crater larger than 1 km in diameter is formed somewhere on the Earth's surface (5.11×10^8 km^2) every 900 years, one larger than 10 km every 94 ka and one larger than 20 km every 380 ka. We must be thankful that two-thirds of the globe are covered with water and, of the remaining land-mass, the frozen wastes, jungles, deserts and mountains account for over 90% of the land area.

After extensive searches for Near Earth Objects, Shoemaker *et al.* (1979) estimated the number of asteroids and comets which have orbits that pass close to Earth. These observations, together with detailed orbital perturbation analysis, were then used to estimate an influx rate, the conclusion being that craters greater than 10 km in diameter are produced every 85 ka, a value that is similar to the one obtained above.

Chyba (1991) noted that the Earth flux represented by equation (2) was about a factor two higher than the expected lunar flux. He also concluded that the lunar cratering rate had been roughly constant over the last 3.5×10^9 years and that this approximate constancy would also apply to the Earth's flux.

In the context of the present paper the quantity -2.01 in equations (1) and (2) is of more importance than the absolute flux. This $D^{-2.0}$ power law between the cumulative crater number and the crater diameter was also obtained by Grieve (1989) when he analysed Earth craters. He also noted that the $D^{-2.0}$ relationship was common to other inner planetary bodies. For large ($D > 3$ km) post-mare lunar craters Hartmann & Wood (1971) found $N \propto D^{-2}$ but Shoemaker (1963) and Baldwin (1971) found $N \propto D^{-1.7}$. Hughes (1993) reanalysed the lunar Earth-facing hemisphere data and found

$$\log N = (5.84 \pm 0.40) - (2.00 \pm 0.19) \log D\,(\mathrm{km})$$

Martian craters have also been found to obey an $N \propto D^{-2}$ law (see Hartmann 1973; Neukum & Wise, 1976), as have the $D > 35$ km Venusian craters (see Schaber *et al.* 1992). Certain of the discrepancies in the value obtained for the power are due to disagreements about the sizes of individual craters and to differences in the diameter range over which the linear approximations are fitted. Some researchers, for example Neukum & Ivanov (1994), have turned their backs on simple linear relationships and have resorted to polynomial fits.

The energy–diameter equation

The second basic equation in the cratering process relates the kinetic energy, E, of the incident body to the diameter, D, of the crater that is formed by the hypervelocity impact explosion. Four old, simple, and much quoted relations are as follows:

$$E = 8.41 \times 10^{23} D^{3.57}\ \mathrm{erg} \quad (D > 1\ \mathrm{km}) \quad (3)$$
(Wood 1979)

$$E = 4 \times 10^{22} D^3\ \mathrm{erg} \quad (4)$$
(Allen 1973)

$$E = 1.45 \times 10^{23} D^3\ \mathrm{erg} \quad (D < 2.4\ \mathrm{km}) \quad (5a)$$
(Dence *et al.* 1977)

$$E = 1.01 \times 10^{23} D^{3.4}\ \mathrm{erg} \quad (D > 2.4\ \mathrm{km}) \quad (5b)$$
(Dence *et al.* 1977)

The scale of the problem is illustrated by the range of parameters quoted in these equations. The relationships are illustrated in Fig. 3. Allen's result (equation (4)) echoes that of Krinov (1963), and was also supported by early nuclear explosion work (see Gladstone 1957). There it was known as the Lampson cube-root law (see Shoemaker 1963).

Much new work has been recently carried out in this field using subscale physical experiments, numerical simulations and scaling laws. Papers by Holsapple (1987) and Schmidt & Housen (1987) should be referred to, as well as the book by Melosh (1989). These works extend the parameterization, and investigate the relationships between the crater diameter and, for example, the impactor density, ρ_i, the surface density of the target body being impacted, ρ_t, and the gravitation field at that surface, g. Melosh (1989) concluded that a reasonable relationship would be

$$D = 1.8 \rho_i^{0.11} \rho_t^{-0.33} g^{-0.22} L^{0.13} E^{0.22} \quad (6)$$

Here L is the diameter of the impacting body and all the quantities in the equation are in MKS units. If one assumes that the impactor is an

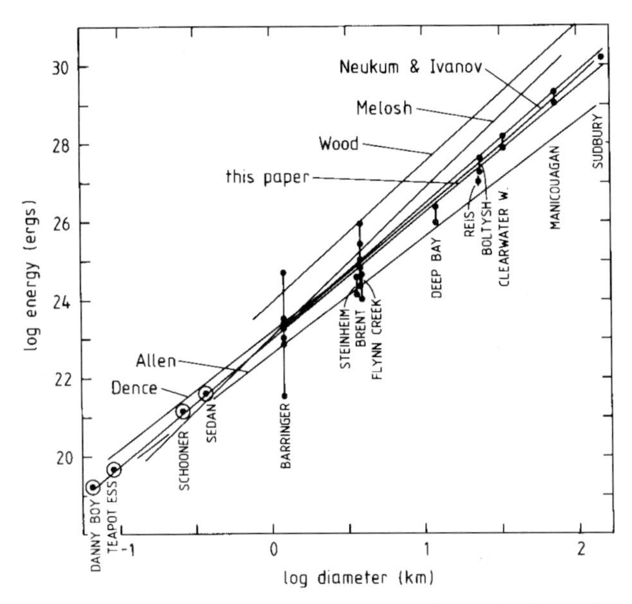

Fig. 3. A logarithmic plot of the energy required to form a crater in the Earth's surface as a function of the diameter of the crater produced. The small craters (represented by circles with central dots) have been produced by nuclear explosions. The large craters (dots) have been produced by the hypervelocity impact of asteroids and comets. It should be noted that, for example, there have been seven estimates of the energy required to produce the Barringer Crater in Arizona. The line marked Wood represents equation (3); Allen, equation (4); Dence, equation (5); Melosh, equation (8); this paper, equation (10). The Neukum & Ivanov line comes from table IV of their 1994 paper.

asteroid with a mean density of $3650\,\mathrm{kg\,m^{-3}}$, equal to the average density of the stony meteorites that fall to Earth (see Sears 1978), and that the impacted surface of Earth has a mean density $3000\,\mathrm{kg\,m^{-3}}$, and $g = 9.81\,\mathrm{m\,s^{-2}}$, and that the mean collision velocity is $20.8\,\mathrm{km\,s^{-1}}$, (see Harris & Hughes 1994) then equation (6), for Earth, becomes

$$D = 0.191 L^{0.13} E^{0.22} \quad (7)$$

Here D is in kilometres, L in metres and E in joules.

This 'Melosh' relationship is also shown in Fig. 3. It should be noted that, for all the complexity of equation (6), the 'Melosh' relationship is equivalent to

$$E = 8.45 \times 10^{22} D^{3.89} \quad (8)$$

over the indicated diameter range, an equation that has much in common with equations (3), (4) and (5) above. Equation (8) can be contrasted with the relationship given in table IV of Neukum & Ivanov (1994). After using the usual scaling relationships they found that $E \propto D^{3.4}$. Their relationship is also shown in Fig. 3.

The reason for the somewhat confusing picture indicated by Fig. 3 and equations (3)–(8) is the fact that large impact craters ($1 < D < 200$ km) have not been experimentally produced on the surface of Earth using known amounts of energy. So equations (3)–(8) are either theoretical or reliant on extrapolations from small-scale laboratory experiments.

The only Earth craters that have been produced, where both the resultant diameter and the causative energy release are well known, are small nuclear test craters. Let us take four well-known examples, these being Danny Boy (diameter 0.066 km), Teapot ESS (0.09 km), Schooner (0.26 km) and Sedan (0.368 km). These were produced (see Nordyke 1977) by just-below-surface and surface nuclear explosions with energies of 0.42, 1.2, 35 and 100 kilotons of TNT equivalent, respectively ($1\,\mathrm{kt} = 4.185 \times 10^{19}$ erg). A least-squares fit to these nuclear test crater data gives

$$\log E\ (\mathrm{erg}) = (23.02 \pm 0.03) + (3.18 \pm 0.03) \log D\ (\mathrm{km}) \quad (0.066 < D\,\mathrm{km} < 0.368) \quad (9)$$

Are we justified in extrapolating this relationship to cover the $D > 20$ km hypervelocity impact craters investigated in this paper and by Grieve & Shoemaker (1994)? One difficulty is that the nuclear craters are bowl shaped and the large impact craters are not. Lunar and terrestrial craters in the size range $10 < D < 300$ km have been modified by the shearing and slumping of surrounding rock. This increases the crater diameter beyond that of the initial bowl-shaped cavity by as much as 35–40% (see Shoemaker 1962; Neukum & Ivanov 1994). Unfortunately, the value of this percentage is dependent on the crater size.

Nuclear explosions also lead to a higher initial pressure than do hypervelocity impacts, and energy-for-energy this could result in slightly different sizes and shapes of craters. A third problem arises from the fact that the potential energy, P, required to remove material from the excavated crater is only a small percentage of the explosive energy, E, that is released when the impactor hits the surface. Crudely, P can be equated to $0.25MgD$, where M is the mass of material excavated from the crater, g is the acceleration of gravity at the planetary surface and D is the crater diameter. The nuclear crater data given above indicate that P/E is around 0.1, so the majority of the available energy is used to generate seismic shock waves, to pulverize the underlying rock, to vaporize and melt some of this rock, and to produce heat, light and sound. From an energetics standpoint the actual crater can be thought of as a minor by-product of the impact event.

The difficulty of estimating the energy required to produce a specific crater is illustrated by the range of values obtained by different researchers. Roddy (1977), for example, estimated that the production of the crater Flynn Creek (diameter 4 km) needed an energy of ‘at least 25 megatons’ (i.e. 1.05×10^{24} ergs). He also quoted estimates of 2.2×10^{21} erg and 4.7×10^{24} erg. The production of the famous Barringer (Meteor) crater (diameter 1.2 km) has been estimated to require (in chronological order) 3×10^{24} erg (Baldwin 1949), 5×10^{24} erg (Öpik 1958), 3.2×10^{23} erg (Bjork 1961), 9.5×10^{22} erg (Innes 1961), $(7 \pm 1) \times 10^{22}$ erg (Shoemaker 1963), 1.8×10^{23} erg (Roddy *et al.* 1975), 2×10^{23} erg (Dence *et al.* 1977), 1.7×10^{23} erg (Greeley 1985) and between 10^{23} and 10^{24} erg (Grieve 1989). Considering that this crater is still in the first flush of youth and has been studied extensively, the required energy range is huge and certainly has not been obtained by just extrapolating from equations similar to equation (9). There is a factor of over a thousand between the largest and the smallest estimate!

Figure 3 is a logarithmic plot of the estimated energy required to produce a crater of diameter D km. These data have been obtained from Dence *et al.* (1977), Nordyke (1977) and Roddy (1977). A linear regression analysis of these data indicates that

$$\log E = (22.94 \pm 0.22) + (3.25 \pm 0.07) \log D \quad (0.09 < D\,\mathrm{km} < 140) \quad (10)$$

Equations (10) and (9) are refreshingly similar, the coefficients being within a standard deviation of each other. As equation (9) has the advantage of relying on real experimental data it will be used in what follows.

The energy released by an impacting body is equal to $0.5mV_C^2$ where m g is the mass of the body and V_C is the collision velocity between body and Earth. Harris & Hughes (1994) found that the mean value of V_C, for the Near-Earth-Asteroid (NEA) population was 20.8 km s^{-1}. Chyba *et al.* (1994) suggested a value of 15 km s^{-1}, Schultz (1988) gave 25 km s^{-1} and Öpik (1976) gave 20 km s^{-1}. Needless to say, the impactor does not have to be an asteroid; it could easily be a comet. For short-period comets, Weissman (1982) found a mean impact speed of 29 km s^{-1}, whereas Schultz (1988) gave 41 km s^{-1}. For long-period comets Schultz (1988) gave 53 km s^{-1}.

If we assume that the majority of the impactors are asteroids and that they have an impact speed of 20.8 km s^{-1} then, on average, $E = 21.6 \times 10^{12} \times m$ erg. Combining this equation with both the cratering rate equation (equation (2)) and the energy–diameter relationship (equation (10)) gives

$$\log \Phi_m = -(5.1 \pm 0.3) - (0.62 \pm 0.04) \log m \quad (6 \times 10^{14} < m < 1 \times 10^{16}\,\mathrm{g}) \quad (11)$$

where Φ_m is the annual flux to each km^2 of the Earth's surface of bodies with masses greater than m g.

If, on the other hand, the impactors are comets, an equal mix of short-period and long-period comets would have a mean impact speed of about 45 km s^{-1} and, on average, $E = 1.01 \times 10^{13} \times m$ erg. Here

$$\log \Phi_m = -(5.5 \pm 0.3) - (0.62 \pm 0.04) \log m \quad (1.2 \times 10^{14} < m < 2.0 \times 10^{15}\,\mathrm{g}) \quad (12)$$

The mass distribution equation

The third basic equation to be considered in this paper governs the mass distribution of the asteroids and comets that hit Earth. Öpik (1976), Kresák (1978*a*, *b*) and Shoemaker (1977) independently started with data from the extensive sky searches that had been undertaken to detect those asteroids and comets which had orbits that brought them closer to the Sun than 1.017 AU (the Earth's aphelion distance). Considering the effects of orbital evolution, these researchers then calculated the expected collision rate as a function of asteroidal and cometary size. (The estimation of the size of an asteroid is not easy because it depends on assumptions as to its albedo and shape. In the case of a comet, one has also to estimate the percentage of the nucleus surface that is active.) Kresák (1978*a*, *b*) concluded that the Earth would collide with a body of diameter, L, greater than 1 km every 1.5–2 Ma. In the $L > 1$ km size range he found that there were 46 times more impacting asteroids than there were impacting comets. Shoemaker *et al.* (1979) concluded that, in a time period of 2 Ma, the Earth would be hit by two Aten asteroids (semi-major axis, $a < 1.0$ AU; aphelion distance >0.9833 AU), four Apollo objects ($a > 1.0$ AU; perihelion distance, $q < 1.0167$ AU) and one Amor ($1.0167 < q < 1.3$ AU), these asteroids being brighter than absolute magnitude 18. This magnitude was taken to be equivalent to their masses being greater than 3.8×10^{11} g. Shoemaker *et al.* (1979) also concluded that, in this mass range, the incident rate of comets was less than 10% the incident rate of asteroids, and that an 18th magnitude impactor would produce a crater of diameter 10 km.

The results from Öpik (1976) are summarized in Table 2. His conclusions differed from those of Shoemaker *et al.* (1979) and Kresák (1978*a*, *b*). Öpik suggested that comets dominated the higher size ranges. In more recent papers we find Schultz (1988) concluding that 7% of the craters on Earth are produced by long-period comets, 26% by short-period comets and 67% by asteroids. Wetherill (1989) derived cratering rates on Earth using Monte Carlo simulations of the collision and perturbation processes that place Main Belt asteroids onto Earth-crossing orbits and concluded that about 30% of Earth's craters were cometary. Shoemaker *et al.* (1994) concluded that 20% of the craters with $D > 20$ km were produced by short-period comets, 20% by long-period comets and 60% by asteroids. The confused state of this topic has been reviewed by Bailey (1991), who concluded that observed comets make a negligible (less than *c.* 10%) contribution to the terrestrial cratering rate at all crater diameters.

The diversity of these percentiles is due to the fact that both NEAs and dormant comets are very hard to discover. Most of the known ones have diameters in the 0.1–10 km range. (Using equation (6), an asteroid of diameter L km produce a crater of diameter D km given by

$$\log D = 1.207 + 0.79 \log L \quad (13)$$

so this range corresponds to Earth craters between 2.6 and 100 km in diameter.)

Most experts in the field agree that all NEAs that have absolute magnitudes smaller than 13.5 have been discovered. This corresponds to the low albedo (0.04) C-class asteroids larger than 12 km and higher albedo (0.15) S-class asteroids larger than 6 km (see Rabinowitz *et al.* 1994). (One of the awkward connotations of the astronomical magnitude scale is that the larger an object is, the more light it scatters, and the smaller becomes its absolute magnitude.) There are about equal numbers of C-class and S-class asteroids in the NEA population (see Luu & Jewitt 1989). If we consider C-class and S-class asteroids of diameters greater than 2 and 1 km, respectively (absolute magnitude 17.7), then over 90% still await detection. This means that most of the asteroids that produce craters in the size range that we are discussing have not been found, and this clearly leads to inexact estimates for both the ratios between the numbers of asteroidal as opposed to cometary craters and for the asteroidal mass distribution index in this size range.

When it comes to cratering potential it is worth noting that an impacting short-period comet has a mean relative velocity of 38.5 km s^{-1} in comparison with a mean velocity of 20.8 km s^{-1}, for an impacting asteroid. Crudely, the diameter of the resulting crater increases as the square-root

Table 2. *The cumulative frequency of asteroidal and cometary impacts suffered by Earth each year as a function of the diameter of the incident body, after Öpik (1976)*

	Diameter (km)					
	0.13	0.52	2.1	8.5	34.0	68.0
Asteroids	2.8×10^{-5}	7.1×10^{-7}	2.1×10^{-8}	9.2×10^{-10}	6.9×10^{-11}	2.1×10^{-11}
Cometary nuclei	1.8×10^{-5}	9.8×10^{-7}	5.3×10^{-8}	2.9×10^{-9}	1.6×10^{-10}	3.7×10^{-11}
Asteroid/comet ratio	1.58	0.72	0.39	0.32	0.43	0.57

of the impact velocity, so a 1 km comet will produce a crater that is about 1.36 times bigger than that produced by a 1 km asteroid.

The mass distribution index, s, of asteroids in a specific size range can be defined (see Hughes 1972) such that the cumulative number of asteroids that are more massive than a specific mass, m, is given by $N = bm^{(1-s)}$, where b is a constant. (A non-cumulative definition is based on the observation that the number of asteroids found to have masses in the range m to $m + dm$, where dm is a constant, is proportional to m^{-s}.) The cumulative flux equations (equations (11) and (12)), which show that a $\log \Phi_m$ v. $\log m$ graph has a gradient of -0.62, indicate also that the mass distribution index of the impacting objects is 1.62. Unfortunately, when this result is compared with the observed asteroidal mass distribution index there is a large discrepancy. Hughes (1982) found that the latter had $s = 2.02 \pm 0.14$, and Hughes & Harris (1994) found that $s = 2.09 + 0.10$ (see Fig. 4). A word of caution, however, needs to be sounded here. The asteroidal mass distribution indices quoted above only really apply to asteroids with diameters in excess of 130 km. These are in the Main Belt and it is highly likely that all the Main Belt asteroids larger than this diameter have been discovered. Many smaller asteroids are, however, waiting to be discovered. It can, for example, be deduced from the data given in Fig. 4 that only about 77% of the 100 km asteroids have been found, this percentage reducing to 16% for the 50 km ones.

The asteroids responsible for producing the $20 < D < 140$ km craters are NEAs and not Main Belt ones. They also have diameters, L, in the $1 < L < 15$ km range (see equation (13)) and their mass distribution index has not been measured accurately.

It is generally proposed (see Hughes 1991) that the present asteroidal population is the product of multiple collisional fragmentation and that a mass distribution index of around 2.0 (this being the index that has been found for the $130 < L < 930$ km range) applies to smaller asteroids too. Extrapolations have been made down to diameters of 0.1 km and even 0.001 km. Morrison and Bowell, however, disagree with this 'constant s assumption' (see Morrison 1992) and have concluded that the mass distribution index varies with diameter. In *The Spaceguard Survey* they have assumed that the mass distribution index of NEAs is 1.87 in the $L < 0.25$ km range, 1 .67 for $0.25 < L < 2.5$ km and 2.43 for $L > 2.5$ km. Unfortunately, data are so incomplete in this region that this suggestion is unverifiable at present.

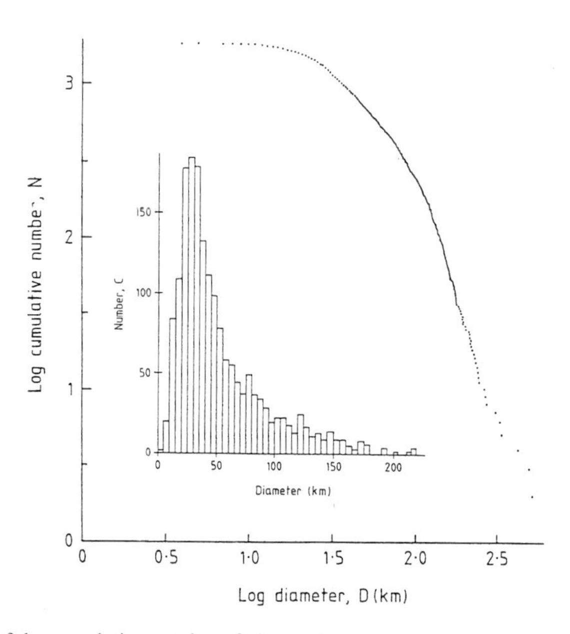

Fig. 4. The logarithm of the cumulative number of observed asteroids larger than a specific diameter is plotted as a function of the logarithm of that diameter. The curve can be seen to be linear for diameters larger than 130 km. The inserted histogram shows the number of asteroids with diameters between L and $L + 5$ km. The 1788 asteroids with known diameters have been used to produce this histogram.

Discussion

We have stressed the fact that the crater-producing impact process is governed by three equations and these can be summarized as follows.

- If we consider the flux of asteroids and comets to the Earth's surface we can define Φ_m as being the cumulative number of impactors with masses greater than m that hit unit area of our planet in unit time. Observations indicate that

$$\log \Phi_m = C_1 - C_2 \log m \quad (14)$$

where C_1 and C_2 are constants over the small mass range responsible for the observed craters. The mass distribution index, s, of the impactors, is equal to $(1 + C_2)$.

- The energy released in the crater production process is the kinetic energy, E, of the impactor and $E = 0.5mV_C^2$, where m is the impactor mass and V_C is its collision velocity with the Earth. In the narrow ($19 < D < 45$ km) size range of equation (2) the comet/asteroid impactor ratio can be taken to be constant and so V_C can also be taken to be constant and independent of impactor mass. The relationship between the diameter, D, of a crater and the energy required to produce it can then be written as

$$\log E \propto \log m = C_3 + C_4 \log D \quad (15)$$

- The rate at which craters are produced on a planetary surface is obtained by combining equations (14) and (15) to give

$$\log \Phi = C_1 - C_2C_3 - C_2C_4 \log D \quad (16)$$

There are now three possible scenarios:

(A) If it is believed that the mass distribution index of the impactors is very close to the asteroidal value of 2.0 (i.e. $C_2 = 1.0$) and that the crater distribution follows a D^{-2} relationship then C_4 must be 2.0. Therefore the equations (4), (5a), (9), (10), (5b), (3) and (8) that give E proportional to D^3, $D^{3.18}$, $D^{3.25}$, $D^{3.4}$, $D^{3.57}$ and $D^{3.89}$ are wrong and E is expected to be proportional to D^2.

(B) If the impactor mass distribution index is close to 2.0 and the relationship between impactor energy and resultant crater diameter is such that E is proportional to $D^{3.18}$ as shown in both Fig. 3 and equation (9), then $C_2 \times C_4$ is also -3.18 and not -2.0, as all crater counting observations on Earth, Moon, Venus and Mars seem to indicate.

(C) If the crater counting is reasonably accurate, i.e. $C_2 \times C_4$ is -2.01 ± 0.13 and the energy–crater diameter relationship is such that C_4 is 3.18 ± 0.03 then C_2 is $2.01/3.18$ i.e. 0.63 ± 0.04, and the mass distribution index of the bodies responsible for producing the craters is 1.63 ± 0.04.

In order of scientific likelihood I would tentatively list these scenarios C, A and B.

Of the three equations discussed in this paper the crater production rate equation (equation (2)) is by far the firmest. This relies on data collected from Earth over the last 105 Ma and does not rely on, for example, the extrapolation of lunar mare data collected over 30 times that time interval.

The second equation concerns the power-law relationship between the kinetic energy, E, of the impacting body and the diameter, D, of the crater that it produces. Unfortunately, as the craters in question are larger than 20 km in diameter, there is absolutely no experimental basis for the required equation. We cannot investigate in the laboratory what happens when a 1 km rock hits our planet's surface at around 21 km s^{-1}. We have either to scale from the effects of the hypervelocity impacts of microparticles, or to extrapolate by factors of 10^5–10^8 from the energies released by nuclear test explosions. Therefore different researchers give E proportional to D^3, $D^{3.18}$, $D^{3.25}$, $D^{3.4}$, $D^{3.57}$ and $D^{3.89}$.

At the heart of the third equation is the mass distribution index of the asteroids and comets that produce the craters. The energy–diameter power laws given in the previous sentence lead to mass distribution indices of 1.67, 1.63, 1.62, 1.59, 1.56 and 1.52, respectively. It was concluded above that a value around 1.62 ± 0.03 was the most likely.

Unfortunately, there are no obvious physical crater characteristics that point to the nature of the producer. When it comes to estimating the asteroid/comet ratio in the flux of say 1 km and 10 km bodies we have to rely on the detection probability of these bodies in the near-Earth environment. Here it is much easier to find the sublimating comets than the merely reflecting asteroids. Not only is the value of the asteroid/comet ratio a matter of considerable debate, there is also no reason to assume that it is absolutely constant over the relevant size range (see, e.g. Öpik (1976) and Table 2). The mass distribution index of the large ($L > 130$ km) asteroids has been estimated by Hughes & Harris (1994) to be 2.09 ± 0.10, but the objects responsible for the crater diameters in Fig. 1 are in the $1 < L < 15$ km range. As far as asteroids go this region is *terra incognita*. This is not, however, the case with comets. Hughes (1990) concluded that the short-period comets with absolute magnitudes less than 10.8 had a mass distribution index of 1.67 ± 0.06. The mass distribution index of the observed long-period comets was found to be the same. Unfortunately, the conversion between the cometary absolute magnitude and the radius and the mass of the cometary nucleus relies on assumptions about the percentage of the nucleus surface that is actively emitting gas and dust, and the mean density of the cometary nucleus. Values for the latter typically vary between 1.3 and $0.2\,g\,cm^{-3}$ dependent on whether the researcher prefers a dirty ice-ball or a dirty snowball paradigm. Bearing these provisos in mind, the 1.67 cometary mass distribution index applies to comets with nuclei masses in the 10^{15}–10^{19} g range, these being capable of producing craters with diameters between about 20 and 200 km, exactly in the range of Fig. 1.

Here we hit a snag. Cometary and asteroidal populations are logarithmic. If one has a cometary population with a constant mass distribution index and adds the comets to an asteroidal population with a constant, but different, mass distribution index, then the combined population has a mass distribution index that varies as a function of mass over certain regions of the mass spectrum. If one imagines that the cometary and asteroidal cumulative fluxes at one specific mass are equal and that the comets have a mass distribution index of 1.67 and the asteroids an index of 2.0, then, at that specific mass, the mass distribution index of the combined population turns out to be 1.835. If one considers objects of higher masses it is found that the comets become more prevalent and the mass distribution index decreases until it eventually becomes indistinguishable from the cometary value. If one decreases the mass below the specific mass value, the asteroids start to dominate and the index increases up to the asteroidal value. This is illustrated in Table 3.

Table 3. *Percentage of comets as a function of the mass distribution of a combined asteroid/comet population*

$\Delta \log m$	s_c	% of comets
−3.5	1.978	6.5
−3.0	1.969	9.3
−2.5	1.956	13.0
−2.0	1.941	18.0
−1.5	1.920	24.2
−1.0	1.895	31.9
−0.5	1.867	40.6
0.0	1.835	50
0.5	1.803	59.4
1.0	1.774	68.1
1.5	1.749	75.9
2.0	1.729	82.0
2.5	1.712	87.0
3.0	1.700	90.7
3.5	1.691	93.5
4.0	1.686	95.4

It is assumed that at a specific mass, where $\Delta \log m = 0$, one has an equal number of comets and asteroids, the comets have a mass distribution index of 1.67 and the asteroids have a mass distribution index of 2.00. These indices ensure that the comets start to dominate the population as the mass increases. The mass distribution index s_c of the combined population, and the percentage of comets in the total comets plus asteroids population is given as a function of the log mass interval from the origin.

It was concluded above that the incident flux of objects (i.e. comets plus asteroids) had a combined mass distribution index, s_{cm} of around 1.62 ± 0.03.

- If the asteroids in the $1 < L < 15$ km diameter range have a mass distribution index of about 2.0 then the combined mass distribution index together with the data in Table 3 indicate that the comets dominate the influx of the objects producing 20–100 km diameter craters, and that asteroids make up no more than some 5% of the incident population.
- If the asteroids in the $1 < L < 15$ km diameter range have a mass distribution index that is very similar to that of the comets in that size range then the combined mass distribution index tells us nothing about the ratio of comets to asteroids in the mass influx. (We remember that Morrison (1992) assumed that the mass distribution index of NEAs was 1.67 for $0.25 < L < 2.5$ km range.)
- If the collision probabilities of the Near Earth Object population are so well known that the comet/asteroid ratio in the $1 < L < 15$ km diameter size range is known with reasonable precision then an analysis similar to that used to produce Table 3 would produce an estimation of the mass distribution index of the incident asteroids.

Conclusions

The rate at which craters are being produced on the Earth's surface is known reasonably accurately. However, many people want to know the rate at which objects of a specific mass strike the surface. Unfortunately, the latter depends on the relationship between the kinetic energy that these objects have just outside the Earth's atmosphere and the diameters of the craters they produce in the Earth's surface. A knowledge of both the velocity distribution of these incident objects and whether they are comets or asteroids is also important. In this paper this flux has been estimated to be

$$\log \Phi_m = -(5.1 \pm 0.3) - (0.62 \pm 0.04) \log m \quad (6 \times 10^{14} < m < 1 \times 10^{16}\,\mathrm{g})$$

assuming that the cratering bodies are asteroids. It is equations of this form that Hughes (1992) extended into the 'meteorite dropping' region. Incident asteroids in the $10^5 < m < 10^{10}$ g range suffer considerable ablation as they are retarded in the Earth's atmosphere and have the tendency to drop meteorites to the surface that are typically 1/20 of their initial mass. An extrapolation of the equation above therefore indicates that a meteorite of mass around 500 tonnes hits Earth every 400 years or so.

It is hoped that the ratio between the number of asteroids and comets of specific sizes that pass close to Earth can be better quantified by extending the excellent work done by such organizations as the joint US Air Force and NASA Project NEAT (Near Earth Asteroid Tracking) and project Spacewatch (see Cox & Chestek 1996; Lewis 1996).

References

ALLEN, C. W. 1973. *Astrophysical Quantities.* Athlone, London, 138.

BAILEY, M. E. 1991. Comet craters versus asteroid craters. *Advances in Space Research*, **11**(6), 43–60.

BALDWIN, R. B. 1949. *The Face of the Moon.* University of Chicago Press, Chicago.

——1971. On the history of lunar impact cratering: the absolute time scale and the origin of planetesimals. *Icarus*, **14**, 36–52.

BJORK, R. L. 1961. Analyses of the formation of Meteor Crater, Arizona: A preliminary report. *Journal of Geophysical Research*, **66**, 3379–3387.

CHYBA, C. F. 1991. Terrestrial mantle siderophiles and the lunar impact record. *Icarus*, **92**, 217–233.

——, OWEN, T. C. & IP, W.-H. 1994. Impact delivery of volatiles and organic molecules to Earth. *In*: GEHRELS, T. (ed.) *Hazards due to Comets and Asteroids.* University of Arizona Press, Tucson, 9–58.

COX, D. W. & CHESTEK, J. H. 1996. *Doomsday Asteroid: Can We Survive?* Prometheus, Amherst NY.

DENCE, M. R., GRIEVE, R. A. F. & ROBERTSON, P. B. 1977. Terrestrial impact structures: principal characteristics and energy considerations. *In*: RODDY, D. J., PEPPIN, R. O. & MERRILL, R. B. (eds) *Impact and Explosion Craters, Planetary and Terrestrial Implications.* Pergamon, New York, 247–276.

GLASSTONE, S. 1957. *The Effects of Nuclear Weapons.* US Atomic Energy Commission, Washington, DC, 198.

GREELEY, R. 1985. *Planetary Landscapes.* Allen & Unwin, London, 39.

GRIEVE, R. A. F. 1984. The impact cratering rate in recent times. *Proceedings of the 14th Planetary Science Conference. Journal of Geophysical Research*, Supplement **89**, B403–B408.

——1989. Hypervelocity impact cratering: a catastrophic terrestrial geologic process. *In*: CLUBE, S. V. M. (ed.) *Catastrophes and Evolution.* Cambridge University Press, Cambridge, 57–79.

—— & DENCE, M. R. 1979. The terrestrial cratering record II. The crater production rate. *Icarus*, **38**, 230–242.

—— & SHOEMAKER, E. M. 1994. The record of past impacts on Earth. *In*: GEHRELS, T. (ed.) *Hazards due to Comets and Asteroids.* University of Arizona Press, Tucson, 417–462.

HARRIS, N. W. & HUGHES, D. W. 1994. Asteroid–Earth collision velocities, *Planetary and Space Science*, **42**, 285–289.

HARTMANN, W. K. 1973. Martian cratering 4. Mariner 9 initial analysis of cratering chronology. *Journal of Geophysics*, **78**, 4096–4116.

—— & WOOD, C. A. 1971. Moon: origin and evolution of multi-ring basins. *The Moon*, **3**, 3–78.

HODGE, P. 1994. *Meteorite Craters and the Impact Structures of the Earth.* Cambridge University Press.

HOLSAPPLE, K. A. 1987. The scaling of impact phenomena. *International Journal of Impact Engineering*, **5**, 343–355.

HUGHES, D. W. 1972. The meteoroid influx and the maintenance of the Solar System dust cloud. *Planetary and Space Science*, **20**, 1949–1952.

——1981. The influx of comets and asteroids to the Earth. *Philosophical Transactions of the Royal Society of London, Series A*, **303**, 353–368.

——1982. Asteroid size distribution. *Monthly Notices of the Royal Astronomical Society*, **199**, 1149–1157.

——1990. Cometary absolute magnitudes – their significance and distribution. *In*: LAGERKVIST, C.-I., RICKMAN, H., LINDBLAD, B. A. & LINGREN, M. (eds) *Asteroids, Comets, Meteors III.* Uppsala University Reprocentralen, 327–342.

——1991. The largest asteroids ever. *Quarterly Journal of the Royal Astronomical Society*, **32**, 133–145.

——1993. Meteorite incidence angles. *Journal of the British Astronomical Association*, **103**, 123–126.

—— & HARRIS, N. W. 1994. The distribution of asteroid sizes and its significance. *Planetar and Space Science*, **42**, 291–295.

INNES, M. J. S. 1961. The use of gravity methods to study the underground structure and impact energy of meteorite craters. *Journal of Geophysical Research*, **66**, 2225–2239.

KRESÁK, L'. 1978*a*. Passages of comets and asteroids near the Earth. *Bulletin of the Astronomical Institute of Czechslovakia*, **29**, 103–114.

——1978*b*. The comet and asteroid population of the Earth's environment, *Bulletin of the Astronomical Institute of Czechslovakia*, **29**, 114–125.

KRINOV, E. L. 1963. Meteorite craters on the Earth's surface. *In*: MIDDLEHURST, B. M. & KUIPER, G. P. (eds) *The Moon, Meteorites and Comets*. University of Chicago Press, Chicago, 183–207.

LEWIS, J. S. 1996. *Rain of Iron and Ice: the Very real Threat of Comet and Asteroid Bombardment*. Helix Books, Addison–Wesley, Reading, MA.

LUU, J. & JEWITT, D. 1989. On the relative numbers of C types and S types among near-Earth asteroids. *Astronomical Journal*, **98**, 1905–1911.

MELOSH, H. J. 1989. *Impact Cratering: a Geologic Process*. Oxford University Press, Oxford.

MORRISON, D. 1992. *The Spaceguard Survey: Report of the NASA Near-Earth-Object Detection Workshop*. Jet Propulsion Laboratories, Pasadena, CA.

NEUKUM, G. & IVANOV, B. A. 1994. Crater size distributions and impact probabilities on Earth from lunar, terrestrial-planet and asteroid cratering data. *In*: GEHRELS, T. (ed.) *Hazards due to Comets and Asteroids*. University of Arizona Press, Tucson, 359–416.

—— & WISE, D. U. 1976. Mars: a standard crater scale and possible new time scale. *Science*, **194**, 1381–1387.

NORDYKE, M. D. 1977. Nuclear cratering experiments: United States and Soviet Union. *In*: RODDY, D. J., PEPPIN, R. O. & MERRILL, R. B. (eds) *Impact and Explosive Cratering, Planetary and Terrestrial Implications*. Pergamon, New York, 103–124.

ÖPIK, E. J. 1958. Meteor impact on solid surface. *Irish Astronomical Journal*, **5**, 14–33.

——1976. *Interplanetary Encounters*. Elsevier, Amsterdam, 49.

RABINOWITZ, D., BOWELL, E., SHOEMAKER, E. & MUINONEN, K. 1994. The population of Earth-crossing asteroids. *In*: GEHRELS, T. (ed.) *Hazards due to Comets and Asteroids*. University of Arizona Press, Tucson, 285–312.

RODDY, D. J. 1977. Tabular comparisons of the Flynn Creek impact crater, United States, Steinheim impact crater, Germany and Snowball explosion crater, Canada. *In*: RODDY, D. J., PEPPIN, R. O. & MERRILL, R. B. (eds) *Impact and Explosion Cratering, Planetary and Terrestrial Implications*. Pergamon, New York, 125–62.

——, BOYCE, J. M., COLTON, G. W. & DIAL, A. L. JR. 1975. Meteor Crater, Arizona, rim drilling with thickness, structural uplift, diameter, depth, volume and mass-balance calculations, *Proceedings of the 6th Lunar Science Conference*, 2621–2644.

SCHABER, G. G., STROM, R. G., MOORE, H. J. *et al.* 1992. Geology and distribution of impact craters on Venus: what are they telling us? *Journal of Geophysical Research*, **97**, 13 257–13 301.

SCHMIDT, R. M. & HOUSEN, K. R. 1987. Some recent advances in the scaling of impact and explosion cratering. *International Journal of Impact Engineering*, **5**, 543–560.

SCHULTZ, P. H. 1988. Cratering on Mercury: a relook. *In*: VILAS, F., CHAPMAN, C. R. & MATTHEWS, M. S. (eds) *Mercury*. University of Arizona Press, Tuscon, 274–335.

SEARS D. W. 1978. *The Nature and Origin of Meteorites*. Adam Hilger, Bristol, 121.

SHOEMAKER, E. M. 1962. Interpretation of lunar craters. *In*: KOPAL, Z. (ed.) *Physics and Astronomy of the Moon*. Academic Press, New York, 283–359.

——1963. Impact mechanics at Meteorite Crater, Arizona. *In*: MIDDLEHURST, B. M. & KUIPER, G. P. (eds) *The Moon, Meteorites and Comets*. University of Chicago Press, Chicago, 301–336.

——1977. Astronomically observable crater-forming projectiles. *In*: RODDY, D. J., PEPPIN, R. O. & MERRILL, R. B. (eds) *Impact and Explosion Cratering, Planetary and Terrestrial Implications*. Pergamon, New York, 617–628.

——, WEISSMAN, P. R. & SHOEMAKER, C. S. 1994. The flux of periodic comets near Earth. *In*: GEHRELS, T. (ed.) *Hazards due to Comets and Asteroids*. University of Arizona Press, Tucson, 313–335.

——, WILLIAMS, J. G., HELIN, E. F. & WOLFE, R. F. 1979. Earth-crossing asteroids: orbital classes, collision rates with Earth and origin. *In*: GEHRELS, T. (ed.) *Asteroids*. University of Arizona Press, Tucson, 253–282.

WEISSMAN, P. R. 1982. Terrestrial impact rates for long- and short-period comets. *In*: SILVER, L. T. & SCHULTZ, P. H. (eds) *Geological Implications of Impacts of Large Asteroids and Comets on the Earth*. Geological Society of America, Special Paper, **190**, 15–24.

WETHERILL, G. W. 1989. Cratering of the terrestrial planets by Apollo objects. *Meteoritics*, **24**, 15–22.

WOOD, J. A. 1979. *The Solar System*. Prentice–Hall, Englewood Cliffs, NJ, 41.

Calculating flux from meteorite decay rates: a discussion of problems encountered in deciphering a 10^5–10^6 year integrated meteorite flux at Allan Hills and a new approach to pairing

P. A. BLAND[1], A. CONWAY[2], T. B. SMITH[3], F. J. BERRY[4], S. E. J. SWABEY[5] & C. T. PILLINGER[6]

[1] *Department of Earth and Planetary Sciences, Western Australian Museum, Francis Street, Perth, WA 6000, Australia*
[2] *Department of Physics and Astronomy, Kelvin Building, University of Glasgow, University Avenue, Glasgow G12 8QQ, UK*
[3] *Department of Physics,* [4] *Department of Chemistry,* [5] *Department of Earth Sciences,* [6] *Planetary Science Research Institute, The Open University, Milton Keynes, MK7 6AA, UK*

Abstract: Meteorite accumulation sites offer the prospect of observing changes in flux over time; however, two main problems must be overcome: calculating a decay constant for samples in an area, and providing accurate pairing data, such that the number of true meteorites (not fragments) can be ascertained. We have used a comparison of meteorite terrestrial age and weathering data to constrain decay constants for hot desert meteorite populations. Meteorites resident in these sites typically have terrestrial ages <50 ka. Our estimates of the flux of meteorites to the Earth are within a factor of 2–3 of independent estimates made by the Meteorite Observation and Recovery Project network, and suggest no change in the flux over the last 50 ka. A similar approach applied to meteorites from Allan Hills, Antarctica, finds much lower levels of weathering, although terrestrial ages for these samples are much longer. We also observe a correlation between terrestrial age and degree of weathering in the Allan Hills meteorites, suggesting a weathering rate 2–3 orders of magnitude slower than values typical of hot desert sites. Given that blue ice regions are subject to some horizontal flow, we propose a model in which the observed oxidation–terrestrial age distribution is largely a result of ice movement, rather than weathering. Comparing oxidation and terrestrial age data in the Antarctic and hot desert accumulations, we estimate a lifetime for the Allan Hills population of 200–300 ka. Given the distance to the nearest ice divide (200 km), this suggests an average horizontal flow rate of 1 m a^{-1}. This approach may allow an estimate of the catchment area for these samples to be made. In addition, we suggest an alternative method for pairing samples which makes use of automated image processing and data analysis of reflected light photomicrographs to acquire a 'textural fingerprint' of a sample, and a genetic algorithm to compare the numerous data variables.

An accurate estimate of the variability in meteorite flux over time would be valuable in throwing light on a number of outstanding questions in our knowledge of the orbital dynamics of asteroid debris, as well as constraining the magnitude of any temporal variation in the flux of the higher-mass micrometeorites, and low-mass ($<10^8$ g) crater-forming events (Fig. 1). One debate in particular that would benefit from an improved understanding of the variability in meteorite flux is whether the Earth receives a random distribution of asteroid debris, or whether some portion of this material formed part of a co-orbital stream (e.g. Wolf & Lipschutz 1995). If part of the flux is derived from objects in co-orbital streams we would expect much more temporal variability (*c*.10^5 years) than if the flux is random (*c*.10^7 years, e.g. Wetherill 1986).

Abyssal sediments have yielded data on variation in the micrometeorite flux over a timescale of a few million years (Murrell *et al.* 1980; Esser and Turekian 1988; Farley & Patterson 1995). Similarly, crater-counting on the Earth and other terrestrial planets (Grieve 1984; Barlow 1990; Shoemaker *et al.* 1990) may provide information on variability of the flux of higher-mass events over a much longer timescale. However, little is known of the variability of the intermediate mass range, meteorite flux, over time.

A number of attempts have been made at determining the present flux of meteorites in the mass range 10 g to 10^6 g (i.e. typical fireballs and meteorite falls). Initial efforts used data on the recovery of eye-witnessed falls in densely populated areas (Brown 1960, 1961; Hawkins 1960; Millard 1963; Buchwald 1975; Hughes 1980).

BLAND, P. A. *et al.* 1998. Calculating flux from meteorite decay rates. In: GRADY, M. M., HUTCHISON, R., MCCALL, G. J. H. & ROTHERY, D. A. (eds) *Meteorites: Flux with Time and Impact Effects*. Geological Society, London, Special Publications, **140**, 43–58.

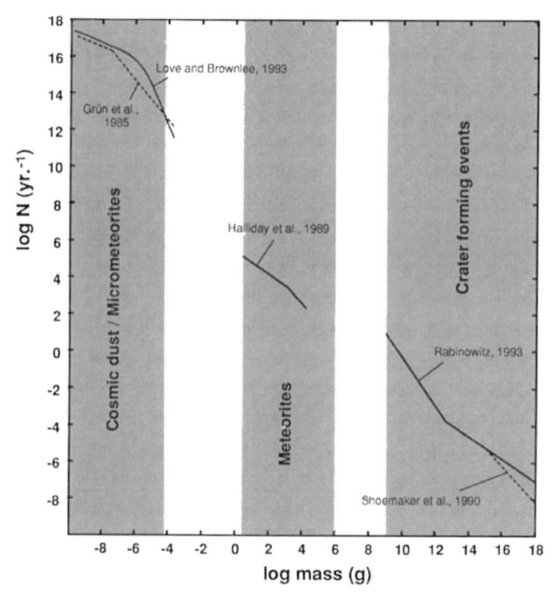

Fig. 1. Number of bodies accreted to the surface of the Earth ($\log N\,a^{-1}$), for a large portion of the mass distribution of extraterrestrial material, as a function of mass (in grams) as determined by previous workers (from Bland *et al.* 1996*b*).

More recently, direct observations of fireball events using camera networks have provided a well-constrained estimate of the present flux of meteorites and its mass distribution. In particular, data from the Meteorite Observation and Recovery Project (MORP) camera network (Halliday *et al.* 1989) suggest a present flux of 8.7 events of mass equal to or greater than 1 kg per 10^6 km^2 per year. These methods have more or less accurately estimated the value for the present meteorite flux rate, however, to resolve temporal variations in flux requires an alternative approach.

There are several locations on Earth where meteorites accumulate: the blue-ice fields of Antarctica, and desert regions such as the Sahara. In principle, such sites may be used to derive information about variations in flux over time. The timescale over which an individual accumulation is active, representing the limit beyond which we are unable to gain information on flux, may be estimated from the terrestrial ages of the meteorites residing there (this information is obtained from the decay of cosmogenic radionuclides). Samples in Antarctica have terrestrial ages up to *c.* 1 Ma (Nishiizumi *et al.* 1989), whereas the upper limit obtained for desert samples is generally around 50 ka (Jull *et al.* 1993) before they are weathered away. Thus, hot desert accumulations should record a *c.* 50 ka time-integrated flux over the lifetime of the site, whereas Antarctic sites may preserve a 100 ka–1 Ma record. Given accurate collection data, and a weathering 'decay rate' for samples, it should be possible to estimate flux. However, early attempts to apply this method produced widely varying estimates. Calculations of the orbital dynamics of asteroid debris indicate that values similar to those of the present day are expected over the relatively short timescale sampled by hot desert accumulations. Initial results from an accumulation site in New Mexico (Zolensky *et al.* 1990), calculated from a terrestrial age frequency distribution, were an order of magnitude higher than MORP estimates (Halliday *et al.* 1989). Two problems limit the wider application of this technique: the first is accurately constraining the decay rate of meteorite samples, and the second is convincingly pairing a population of meteorite finds.

Methods

Deriving a flux estimate from hot desert accumulation sites

Background. To illustrate our method, we will summarize our earlier work on hot desert meteorites before turning to the more complex Antarctic situation. The reader is referred to

Bland *et al.* (1996*a*, *b*) for a complete discussion of the hot desert work.

We make a number of requirements of our data, and several assumptions, to derive a flux calculation from a meteorite population in the manner described above:

- An accurate age for either the accumulation surface and/or terrestrial ages for the meteorite population is required.
- The action of weathering in removing samples must be quantified.
- No other process has acted to remove samples.
- All the meteorite samples within a known catchment area are recovered and a study to pair samples undertaken, so that a 'population density' of meteorites per unit area is known.
- The search area is large compared with the typical strewn field size. Failure to take account of this factor may mean that an area samples an unrepresentative number of falls (see Halliday *et al.* 1991).

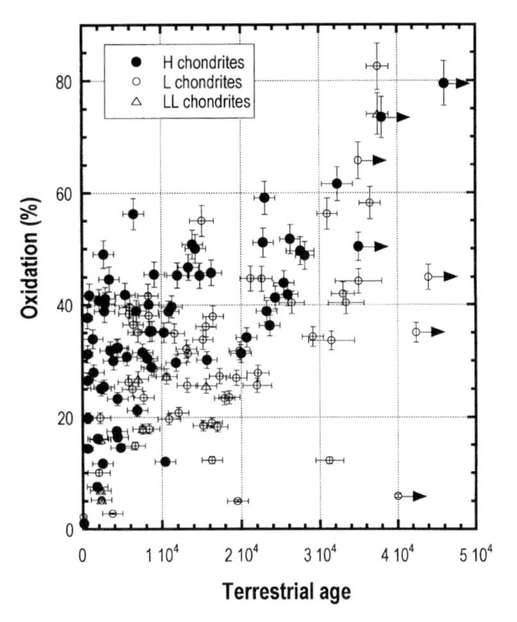

Fig. 2. Percentage of Mössbauer spectra arising from absorptions associated with Fe^{3+}-containing phases (oxidation (%)) against terrestrial age in ordinary chondrites recovered from 'hot' desert regions. The scatter shows the three types of ordinary chondritic meteorite that are recognized (H, high iron group; L, low iron group; LL, low iro–low metal group), and suggests a general increase in oxidation over time and, on average, more weathering in H group chondrites.

Decay constant. The first, and potentially the most problematic of these requirements that we address here is the problem of deriving an overall 'decay rate' for a meteorite population i.e. the rate at which samples are removed from a population by the action of weathering. In recent papers (Bland *et al.* 1996*a*, *b*), we have outlined a technique which combines terrestrial age and weathering data (using Mössbauer spectroscopy) to improve estimates of meteorite decay rate. Mössbauer spectroscopy is effective in understanding the course of oxidation over time, and in quantifying the extent of oxidation, in the most common meteorite type, the ordinary chondrites (OC). In a fresh OC iron is present as Fe^0 or Fe^{2+}, with only a trace of Fe^{3+}. As such, any significant amount of Fe^{3+} may be interpreted as the product of terrestrial weathering. In hot deserts, OCs have been found with 0–85% of iron as Fe^{3+}; in general, oxidation (abbreviated as R) increases over time t (Fig. 2). A plot of the oxidation–frequency distribution (abbreviated hereafter as $R(n)$) shows a peak in the region of 35% for the Saharan population and 45% for the non-Saharan (Fig. 3). We interpret the morphology of this plot as a function of the porosity of OC samples: Fe–Ni and other primary phases in the meteorites are weathered, resulting in volume expansion, until 35–45% oxidation is accommodated by the porosity of the meteorite. After this point, further weathering causes brecciation and the eventual loss of the sample. If $R(n)$ is a function of porosity then we would expect it to remain constant between sites, although $R(t)$ (i.e. oxidation v. time) may vary because of climatic factors. As such, we interpret the difference observed between Saharan and non-Saharan populations as a result of the relative age of these meteorite accumulations: the Acfer site has probably been a stable accumulation site for only the last 20 ka, i.e. insufficient time to show a 'mature' oxidation frequency distribution. Although samples from this region appear to weather at a slower rate than for other hot desert sites (Fig. 4), given a long enough residence time, the $R(n)$ distributions should be comparable.

Two final points: first, we observe that for a meteorite shower, samples separated since their time of fall are weathered to the same degree (Benoit *et al.* 1998), suggesting that variations in the oxidation environment over an accumulation surface are negligible; second, variation in weathering with depth in hot desert meteorites becomes significant only for large (5–10 kg) samples (Bland *et al.* 1995). These factors mean that, given an understanding of the progress of weathering over time for individual sites (from

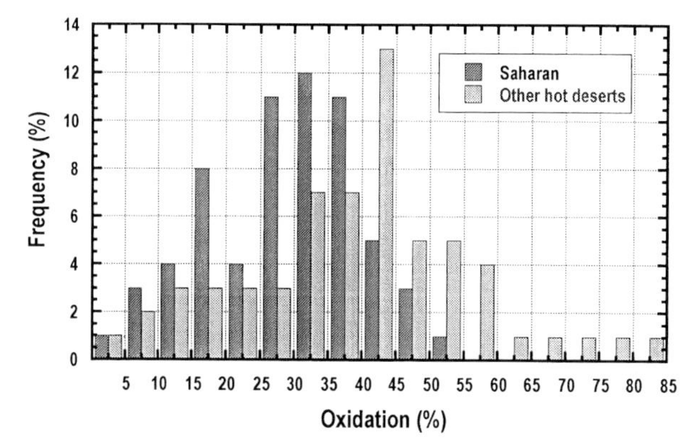

Fig. 3. Oxidation–frequency histogram $R(n)$ comparing Saharan and non-Saharan ordinary chondrite finds; bin widths are 5%.

the oxidation v. terrestrial age plot, $R(t)$; see Fig. 4), and the effect of weathering in disrupting meteorite samples (from the oxidation–frequency plot, $R(n)$; (see Fig. 3)), we are able to derive a decay constant. We have applied this method to three hot desert meteorite accumulations (Roosevelt County, New Mexico; the Nullarbor Region of Western Australia; Reg el Acfer in the Algerian Sahara) that have a large dataset of terrestrial ages, and have obtained well-constrained decay rates for samples at these sites. A full discussion of this method, and details of the modelling of $R(n)$ and $R(t)$, have been given by Bland *et al.* (1996*b*).

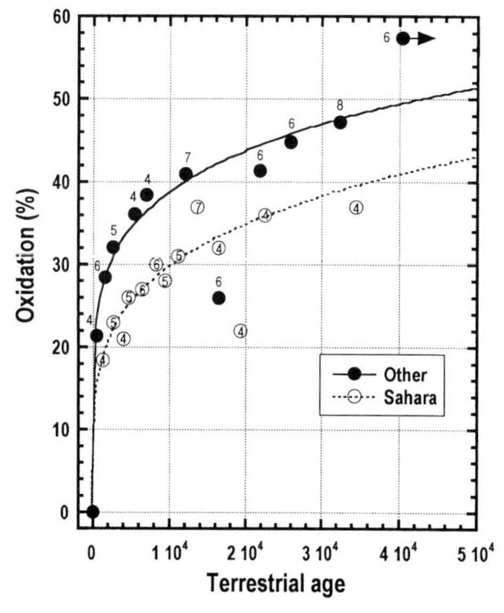

Fig. 4. Oxidation v. terrestrial age $R(t)$ for ordinary chondrites from the Sahara and other hot desert accumulation sites. Data are binned by terrestrial age (the number of points in each bin are shown). The morphology of the distributions suggests that each scatter may be adequately modelled using a power law, and also that the overall weathering rate in the Sahara is lower than elsewhere.

Flux estimate. Accurate decay rates for a population are only half the problem; in addition we need to know the number of individual meteorite samples per unit area. This requires a detailed pairing study of fragments that establishes an estimate of total surviving falls in the region, and careful searching of a defined area so that a large proportion of the surviving samples in a region is recovered. Published pairing work exists for Roosevelt County (Zolensky *et al.* 1990), the Acfer region of the Sahara (Bischoff & Geiger 1995), and the Nullarbor (where samples are routinely paired before classifications are published, e.g. Bevan & Binns 1989). The work by Zolensky *et al.* (1992) indicated that because of the variability in ages of surfaces at Roosevelt County this site was not suitable for a flux estimate. The larger deflation areas, from which all terrestrial age dated samples were recovered, have surface ages of between 53.5 and 95.2 ka (Zolensky *et al.* 1992). This correlates well with the fact that terrestrial ages for Roosevelt County meteorites extend up to, and sometimes beyond, 40 ka (Jull *et al.* 1991). The conformity between surface age and meteorite terrestrial age would suggest that this site is in fact suitable for a flux estimate.

Although only Roosevelt County has been scoured for samples, if we make assumptions about the area that can be effectively searched in a given time, we can arrive at estimates for a population density of samples on the ground in the Nullarbor and the Sahara. These data, combined with decay constants for each site, yield three

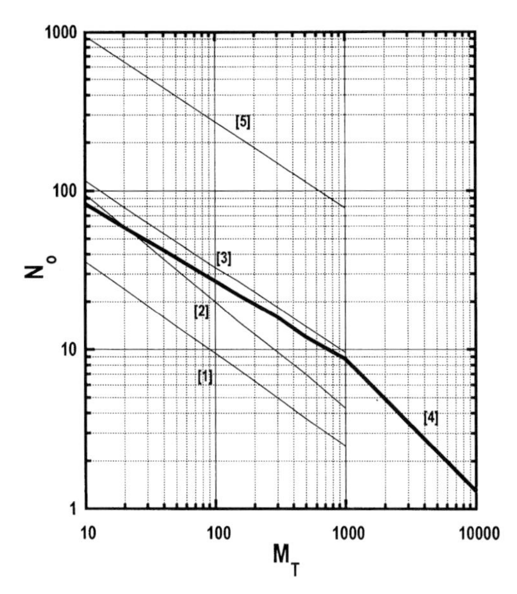

Fig. 5. Our estimates of flux ([1] (Nullabor Region), [2] (Sahara Desert), [3] (Roosevelt County), compared with the best observational estimates of the present flux of meteorites [4] (Halliday *et al.* 1989) and a previous estimate from the RC accumulation site [5] (Zolensky *et al.* 1990). N_0 is the number of events per $10^6\,km^2\,a^{-1}$ for mass larger than M_Tg.

independent estimates of flux (Bland *et al.* 1996*b*) that are within a factor of 2–3 of the estimate of Halliday *et al.* (1989) (Fig. 5). Given the error involved in both our analysis and the camera network data, the results are effectively identical, indicating that the time-integrated flux over the last 50 ka is the same as at the present day.

The approach outlined above appears to be effective in deriving meteorite flux estimates from hot desert populations, which are typically integrated over a *c.* 50 ka timescale. Uncertainties in the model fit to the data, pairing and population density mean that our initial flux estimates made using this method are accurate to a factor of two or three. The magnitude of this uncertainty is thus of a similar order to the MORP fireball study. However, one benefit of our approach is that the uncertainty should be reduced over time as pairing and population density estimates are improved, and more terrestrial age and weathering data allow us to refine the model fit to the data.

The Antarctic situation

Many of the problems that are encountered in deriving a flux estimate from hot desert meteorite accumulations become magnified in the case of the Antarctic populations. Hot desert meteorites generally have fairly precise terrestrial ages (residence times <50 ka allow ^{14}C dating, with a typical error of 1300 years (Jull *et al.* 1989)), and are members of populations that are small enough to allow a fairly accurate attempt at pairing. Once a decay constant is established, it is reasonably straightforward to make an estimate of the integrated flux over the lifetime of a population. None of these factors holds for the Antarctic meteorites: terrestrial ages have large errors (^{36}Cl dating involves errors of *c.*70 ka), populations are far too large to allow pairing of samples with any confidence, and no correlation between weathering and terrestrial age has thus far been established. In addition, a blue ice field is not a static receptacle for incoming samples. As Cassidy *et al.* (1992) and others have discussed, meteorites may well be brought to an accumulation surface from an original fall position far upstream of their collection site. Thus, even estimating the catchment area for Antarctic samples becomes problematic.

The Allan Hills Icefields in Victoria Land and the Yamato Mountains in Queen Maud Land are two of the most productive areas in Antarctica for the recovery of meteorites. The Allan Hills Icefields have been visited continuously by US expeditions since 1976 and in excess of 2000 meteorite samples recovered (Cassidy *et al.* 1992). Regionally, ice flows off the East Antarctic Ice Plateau and diverges through the Mulock and David Glaciers. Locally, however, much of the ice between these two diverging streams follows the MacKay drainage system (Drewry 1982). In sites to the west of the Allan Hills Main Icefield, the general trend of ice is northward (Delisle *et al.* 1989). In this region a small fraction of the ice is trapped behind the Near Western Icefield sub-ice barrier, creating a stranding surface as the ice ablates. Another small fraction on the other side of the channel is diverted to cross the Allan Hills Main Icefield, and the large concentration of meteorites found here suggests that this ice is moving relatively slowly (Cassidy *et al.* 1992). The sub-ice barriers to flow that give rise to these two icefields are plateau-like structures that rise steeply from the average bedrock level and have flat tops (Delisle & Sievers 1991). Although portions of the present surface of the Allan Hills ice may be relatively young (11 ka) (Fireman & Norris 1981), elsewhere much older ice (300 ka) has been identified (Fireman 1987). In addition, Allan Hills meteorites appear to have among the longest terrestrial ages of all Antarctic meteorites (Nishiizumi *et al.* 1989), implying that in the past these concentrations of meteorites must have had catchment areas extending well inland (Cassidy *et al.* 1992). New data for Yamato samples (Michlovich *et al.* 1995) indicate that this blue ice field may also have been an active site of accumulation for a similar timescale to the Allan Hills.

The extremely long survival time of Antarctic samples, in some cases up to 1–2 orders of magnitude greater than for hot desert meteorites (Nishiizumi *et al.* 1989; Welten 1995), suggests that long-term variations in meteorite flux could be observed here. This possibility lies at the heart of a long-running debate, following the observation of a number of features that appear to be unique to the Antarctic population. Some of the differences between Antarctic and non-Antarctic samples are statistical (e.g. an excess of Antarctic H chondrites (Harvey 1989)), and some compositional (a group of old terrestrial age Antarctic H chondrites with anomalous natural thermoluminescence (Benoit *et al.* 1993) and trace element characteristics (Wolf & Lipschutz 1995)). The debate has focused on whether these differences have a genetic origin (caused perhaps through the decay of a stream of asteroidal debris over the lifetime of the blue ice field accumulations) or are caused by the unique weathering environment of Antarctica.The possibility of contributing to this body of work, initially by an enhanced understanding of weathering processes over time, and finally by outlining a method by which an Antarctic flux calculation might be approached, is the focus of the rest of this paper.

Results and discussion

Deriving a flux estimate from Antarctic accumulation sites

Samples. Initially, we apply a similar method to that which has proved effective in the hot desert situation, comparing meteorite terrestrial ages with the amount of weathering observed using Mössbauer spectroscopy in Antarctic OCs. A total of 62 OCs (33 H chondrites and 29 L(LL) chondrites) from the Allan Hills region for which published terrestrial ages exist were analysed using ^{57}Fe Mössbauer spectroscopy. In addition, Burns *et al.* (1995) have published Mössbauer measurements on 33 Antarctic H chondrites from the Allan Hills, Yamato Mountains, Lewis Cliff and Thiel Mountains. Terrestrial ages exist for 23 of these samples. Shinonaga *et al.* (1994) have also analysed 14 H chondrites from Allan Hills and the Yamato Mountains, three of which were terrestrial age dated. This dataset allows some constraint on the rate of Antarctic weathering, and its effect in 'eroding' samples, to be made.

Allan Hills R(t) *plot.* A plot of terrestrial age v. the amount of ferric iron as a percentage of total iron ($R(t)$) for all 84 meteorites for which specific terrestrial ages exist is shown in Fig. 6 (meteorites from all Allan Hills icefields are included here). A plot of grouped data points (grouped by terrestrial age so that 4–5 points fall in each 'bin') is shown in Fig. 7. When compared with the similar plot for the hot desert meteorites (Fig. 2) several features are noteworthy: although the Antarctic population has a much longer terrestrial residence than the hot desert population it shows a significantly lower level of overall oxidation, with only one sample over 35% oxidation (the average hot desert value is 33.5% from analyses of 114 samples); H chondrites appear to weather at a dramatically higher rate than L(LL) chondrites (in the hot desert meteorites a similar feature is observed though the relative difference is less);

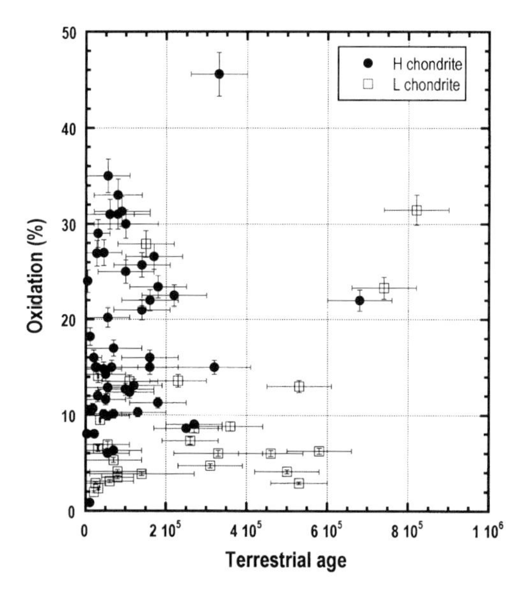

Fig. 6. Percentage of Mössbauer spectra arising from absorptions associated with Fe^{3+}-containing phases (oxidation (%)) against terrestrial age in ordinary chondrites recovered from Antarctica. It should be noted that the much longer terrestrial residence time for these samples, compared with hot desert meteorites (Fig. 2). Although much less well defined than in the hot desert case, the scatter suggests a general increase in oxidation over time, and substantially more weathering in H group chondrites compared with L group.

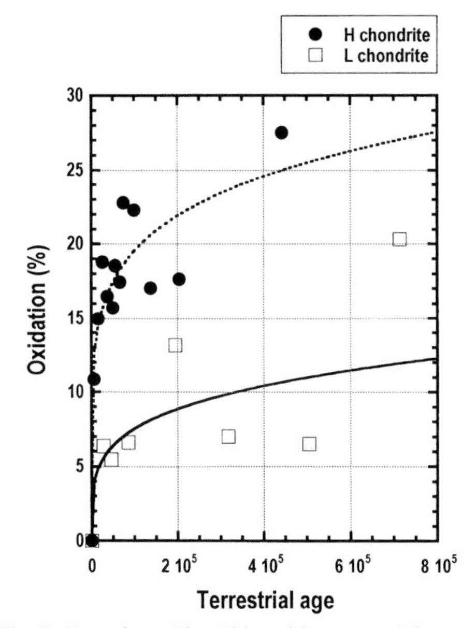

Fig. 7. Data from Fig. 6 binned by terrestrial age so that approximately 4–5 data points fall in each age bin. A broad trend of increasing oxidation over time is observed for both H and L group chondrites.

although we do not observe a strong correlation between terrestrial age and percentage oxidation, there is the suggestion that a weaker correlation does exist: in a similar manner to the hot desert population, there is a relatively 'rapid' period of initial weathering in both H and L(LL) chondrites followed by a more gradual overall increase (see Fig. 7).

Comparing the overall rate of weathering that we derive from Fig. 7 with that for hot deserts (Fig. 2), we find that weathering rates of meteorites in Antarctica are *c.* 2–3 orders of magnitude lower than rates in hot desert regions.

Aside from important unknowns such as the amount of time a sample is exposed on the ice, given the error in determining terrestrial ages for Antarctic meteorites using the ^{36}Cl method it would be surprising if we observed a stronger correlation between oxidation and terrestrial age. What is clear is that the present dataset allows some constraints to be placed on weathering over time.

Allan Hills R(n) *plot.* The above discussion confirms the widely held belief that Antarctic meteorites are less weathered than hot desert finds. In Fig. 8 we highlight this feature by comparing the oxidation–frequency *R*(*n*) plot for hot desert H chondrites with *R*(*n*) for Allan Hills finds. What is apparent is that there is an offset on the order of 20–25% between the peaks of the distributions. When H and L chondrites from Allan Hills are compared in a similar manner (Fig. 9) it is clear that L chondrites in this region are, as discussed above, even less weathered than H chondrites, peaking at 6–9%, compared with *c.* 15% for Allan Hills H chondrites (and 40% for hot desert H chondrites). We have interpreted the hot desert *R*(*n*) distribution in terms of oxidation products filling pore spaces and eventually (when oxidation proceeds to >40%) leading to sample fragmentation. From this, it follows that the *R*(*n*) distribution is a function of the physical properties

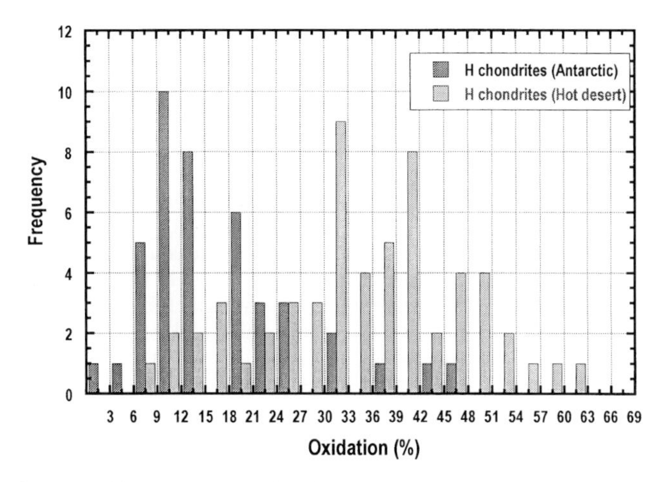

Fig. 8. Oxidation–frequency histogram comparing the H group ordinary chondrite finds from hot desert accumulation sites with H group finds from Antarctica; bin widths are 3%. It should be noted that the peak in oxidation (%) is at a much lower value for Antarctic samples.

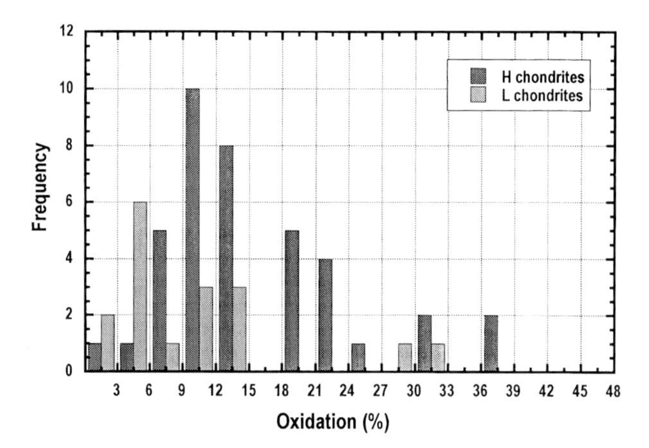

Fig. 9. Oxidation–frequency histogram comparing the H group and L group ordinary chondrite finds from the Allan Hills, Antarctica; bin widths are 3%.

of OCs and the weathering process (relatively independent of climate). Therefore, what process can explain the distribution at Allan Hills?

We are seeking to explain why the Allan Hills $R(n)$ distribution is shifted to lower R values than the $R(n)$ distribution derived from analysis of hot desert meteorites. There are several possibilities. First, the observed distribution may be a result of accelerated physical weathering processes (as opposed to chemical weathering) and/or a dramatic difference in the response of Antarctic meteorites to chemical weathering. Second, the overall distribution may be similar to a hot desert given an ice sheet that is static over the required timescale (the $R(t)$ plot for Allan Hills suggests this would have to be of the order of 1.5–2 Ma for H chondrites, and possibly as much as 10 Ma for L chondrites). However, ice movement means that samples are not allowed to remain in a catchment area for long enough to acquire this 'mature' weathering signature. A third process, wind action, would potentially lead to a similar $R(n)$ distribution to that produced by ice movement.

R(n) *distribution as a result of weathering.* If the observed $R(n)$ distribution is a function of accelerated physical weathering, then freeze–thaw action might be a candidate for the early destruction of samples. Although this process acts in hot deserts, it seems likely that it would be much more effective in a cold desert environment. Expanding and contracting cycles of rocks exposed to diurnal temperature fluctuations and the water phase change are significant factors in the physical weathering of rocks, and may lead to frost shattering (Miotke & von Hodenberg 1980). The process of frost wedging, which is defined as the prying apart by ice growing in rock cracks (Washburn 1980) leads to particle size reduction and eventual disintegration. Although frost wedging varies in intensity depending on the permeability and pore size of rocks (Lautridou & Ozouf 1982), experimental evidence (Lautridou 1975, 1976) indicates that in most cases after 300 freeze–thaw cycles fragmentation virtually stops. After this time it may be considered that ice has effectively penetrated the entire stone.

Schultz (1986) has shown that meteorites exposed on the surface of Antarctic blue ice fields may experience several cycles of freeze–thaw each season (over a 21 day period in the Antarctic summer the interior of a 300 g sample of Allende thawed twice), and Gooding (1985) has suggested that even when buried, at temperatures as low as 263–264 K, thermodynamically 'liquid' water (more than 15 molecular layers thick) should be expected within meteorites. Assuming approximately five freeze–thaw cycles per year, an average chondrite would then be saturated with ice after 60 years. This is a tiny fraction of measured terrestrial ages in Antarctic samples: even if meteorites spent only 1% of their terrestrial lifetimes on the surface of the ice this would still be ample time to saturate the sample. This suggests two things: (1) that freeze–thaw should be successful in introducing water-ice throughout the stone, to the point of saturation, over *c*.100 years; (2) given (1) and much longer terrestrial ages of most samples, freeze–thaw is not a major factor in eroding Antarctic meteorites. In addition, although freeze–thaw is clearly a significant process in shattering

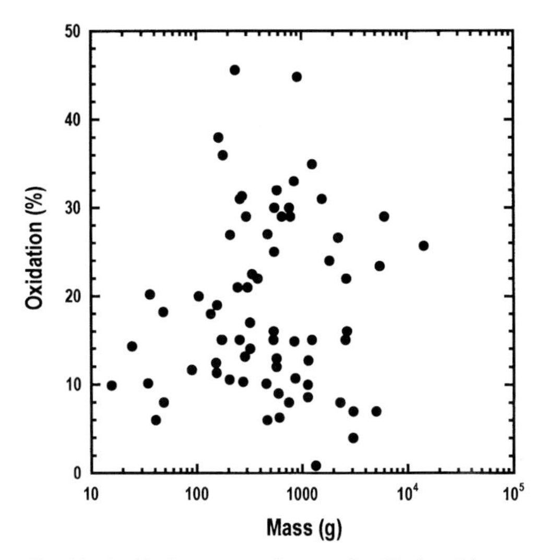

Fig. 10. Oxidation v. sample mass for H chondrite finds from Antarctica.If physical weathering was the primary factor in eroding samples, we may expect some correlation, such that the most weathered samples were also the most 'eroded'. No correlation is observed.

large rock outcrops that possess a jointing pattern, it is not clear that it would be as effective on small samples of meteorite that typically have no preferred fabric.

Another possibility is that the nature of oxidation products produced in weathering an Antarctic meteorite act preferentially to disrupt a sample. This appears intuitively unlikely. Differences in oxide mineralogy are indeed observed between weathered Antarctic and hot desert finds; however, if our interpretation of hot desert weathering is correct, the process by which samples are 'eroded' is largely a function of the specific gravity of oxide phases versus the porosity and mechanical strength of the meteorite. Although the mineralogy is different, the specific gravity of oxide phases are similar and the porosity is the same in hot desert and Antarctic finds. This suggestion also appears unlikely from a comparison of oxidation (per cent) and sample mass (Fig. 10): no correlation is observed. The fact that oxidation and mass are not correlated also causes problems for the idea that physical weathering processes acting at the ice surface are particularly effective in eroding samples: we expect some heightened degree of chemical weathering would presumably be associated with this phenomenon if it took place solely at the ice surface.

R(n) *distribution as a result of ice flow.* The above discussion suggests that weathering processes peculiar to Antarctica may not explain the much lower peak in the $R(n)$ distribution at Allan Hills compared with hot deserts. What is the effect on this weathering profile of a constantly moving ice sheet carrying samples away?

Ice flow at the Allan Hills blue icefields is $<1\,\mathrm{m\,a^{-1}}$ (Annexstad & Schultz 1983); however, the work of Annexstad (1985) suggests that some movement takes place even over apparently stagnant icefields. If average ice flow is only $0.5\,\mathrm{m\,a^{-1}}$ then this equates to 100 km of movement in 200 ka, so even very slow movement might be effective in removing samples from the accumulation site. The effect on the $R(n)$ distribution would be to shift it to low R values: the longer a sample remained at the accumulation site, the more likely it would be for slow ice movement to carry it on, out of the zone of accumulation, and eventually off the continent.

R(n) *distribution as a result of wind action.* Several researchers have noted that small meteorites are easily transported across ice fields by winds, at rates $\gg 1\,\mathrm{m\,a^{-1}}$ (e.g. Zolensky 1998). The degree to which this process acts on larger samples is not clear; indeed, if it were significant we would not expect to find meteorites on open ice at all. Certainly, the type of terrestrial age distribution for samples found at the Allan Hills Main Icefield would not be observed: at this location samples tend to increase in terrestrial age towards the barrier to flow (Nishiizumi *et al.* 1989; Harvey 1990), i.e. meteorites with longer terrestrial ages have been carried further in the ice. This observation is consistent with the Whillans–Cassidy model, and also the model discussed here.

A speculative model. The above discussion suggests that ice flow may be the most significant factor in producing the observed $R(n)$ and terrestrial age distribution at Allan Hills. We now discuss this idea in the context of defining a catchment area for these samples.

Meteorites are carried into an accumulation site from a distance which is less than or equal to the maximum regional drainage length, i.e. the distance from the site to the ice divide separating one drainage system from the next. The time it takes to cover this distance constitutes the first portion of a meteorite's weathering history. As moving ice approaches a barrier it slows down, and a blue ice accumulation surface forms as ice is removed by ablation. The slowly moving blue ice gradually carries samples through the zone of accumulation until the barrier is passed (even a total barrier to flow today is likely to have been overtopped during previous glacial periods). Thus a gradual leakage of samples from an accumulation site is expected, with the meteorites with the oldest terrestrial ages being the most depleted in the population as a whole.

If ice movement is the reason for a cutoff in the $R(n)$ distribution for values greater than *c.* 25–30% oxidation, can a closer look at the $R(n)$ distribution give us any information about the overall rate of movement, and thus the effective lifetime of a population? If it were possible to put some constraints on these parameters then it may be possible to estimate a catchment area for a blue ice accumulation site, a first step in making a calculation of meteorite flux. We use the Allan Hills as an example of how this approach might be applied.

At Allan Hills the distance to the nearest ice divide is 200 km (Drewry 1983). The $R(n)$ plot suggests a cutoff in the region of 25% for H chondrites and perhaps 10–15% for L chondrites. If the $R(t)$ correlation is accurate this would suggest an age of *c.* 200–300 ka for this population (a value similar to the oldest measured ages for Allan Hills blue ice (Firemen 1987)). The fact that this age is younger than some terrestrial ages in the Allan Hills population is not necessarily a cause for concern: old terrestrial age meteorites are found within a very narrow region at Allan Hills (Nishiizumi *et al.* 1989) and thus may indicate a region of trapped and stagnant ice that is not representative of the population as a whole. In addition, Nishiizumi *et al.* (1989) acknowledged that samples for terrestrial age analysis were chosen 'based on low ^{26}Al content and based on special locations in the icefield. We have looked especially for samples with long terrestrial ages.' The age that we calculate for the Allan Hills population, and the length of the drainage area, gives us a maximum horizontal movement rate of the ice integrated over the lifetime of the population, i.e. for ice moving into the Allan Hills on the order of $1\,\mathrm{m\,a^{-1}}$. Although it is necessary to separate ice flow rate in the glacier from ice flow rate in the blue ice accumulation site to obtain an accurate idea of the catchment area for a population, a $1\,\mathrm{m\,a^{-1}}$ average horizontal flow rate appears realistic from the work of Annexstad (1985).

The above discussion is speculative; however, it does illustrate one possible approach to the problem of estimating catchment area, weathering rate, and thus eventually a flux calculation, for Antarctic meteorites. Clearly, more data are required before we could use this method with confidence; however, these preliminary results are promising. Future work will focus on enlarging our dataset for Allan Hills and other blue ice accumulations. If an improved dataset confirms the $R(t)$ correlation, then given the rapidity of the H weathering process an analysis of H chondrite $R(n)$ distributions from individual sites should provide an excellent indicator of population age (over about a 300 ka timescale). There are a number of reasons for caution, however. In this analysis we model a 'static' Antarctic $R(n)$ distribution (i.e. as though meteorites fell on a stagnant ice sheet) by analogy to the hot desert situation; however, we cannot be sure what a 'static' Antarctic $R(n)$ distribution really looks like. Even if weathering is not the chief reason for the differences observed between the hot desert and Antarctic populations, it is likely that some differences in the effect of a given amount of oxidation on an OC sample do exist. In addition, separating the glacial from the blue ice flow components, necessary for an accurate estimate of catchment area, will be difficult. More important though is the problem of pairing: a large unrecognized H chondrite fall would severely skew the $R(n)$ distribution for a site, and any conclusions drawn from it. Performing a reproducible pairing study on any large population of samples is a laborious and time-consuming task. The accuracy of pairing data for Antarctic samples is a major obstacle to deriving well-constrained flux estimates for these populations. We address this problem in the next section.

Automated pairing of meteorite fragments

Background to the pairing problem

In addition to problems for arriving at estimates of flux, the absence of pairing data mean that progress on a variety of other topics (e.g. differences between Antarctic and non-Antarctic meteorites) is slow. At present, pairing studies typically involve consideration of available chemical, isotopic and textural data (i.e. shock level and type) for samples, together with a thin-section examination to determine degree of weathering, distribution and grain size of Fe–Ni, etc. Such studies are laborious, they require an experienced scientist, are limited by the number of variables and/or fragments that an individual can consider at any one time, and they are, in part, qualitative: the relative weight that a worker applies to a data variable is largely a matter of experience. We therefore have adopted a novel approach to this problem, using image processing of meteorite photomicrographs to acquire a digital 'fingerprint' of a sample, combined with a genetic algorithm to automate and standardize the job of comparing data variables for large numbers of samples.

The goal here is twofold. First, given the large numbers of samples involved, we need an easily acquired and non-destructive form of analysis that will provide information of relevance to a pairing study (we choose image processing of polished thin sections in this preliminary study). Second, we seek to evaluate a computational approach to the data analysis (in this case, a genetic algorithm): essentially, the algorithm compares all possible pairs (from any relevant source of data) in the dataset to find the most self-similar groups within the dataset. By choosing a computational approach we hope to make the job of comparing numerous data variables in large numbers of samples less arduous. It will also allow us to test, in some way, the statistical significance of our best paired dataset by comparison with all the other possible pairings that the algorithm assesses (a task beyond the scope of a human operator). In addition, this may give us some feeling for what 'critical mass' of data is required to effectively pair a dataset of meteorite fragments. Finally, and perhaps most importantly, a standardized computational approach to data analysis would allow different populations of samples (e.g. Antarctic v. non-Antarctic) to be compared statistically with confidence.

Method

A textural fingerprint. Initially, digital images of a sample in reflected light are processed to obtain data on the grain size distribution for Fe–Ni and troilite in a scene. We use NIH Image 1.59 for this analysis. Once digital images are acquired the processing is automated using the NIH Image macro language. The data output is as grain size (in pixels), length/width, circularity, etc. for each grain in an image. As we analyse each image for Fe–Ni, troilite and oxides separately, a large volume of data is generated. These data are reduced to give average grain size in an image (and maximum, minimum, standard deviation of grain size, etc.), average ratio of length/width, and overall percentage area, using the macro language in Microsoft Excel. Thus, for each image, three rows of data (one each for Fe–Ni, troilite, and weathering oxides) are generated, each row comprising 20 data variables (see Table 1). This approach means that, once the digital image of a sample is acquired, the image processing and data processing steps are completely automated.

For each meteorite thin section between five and ten images are taken. This gives us better coverage of an area, and an effective way of

Table 1. *'Alpha values' for the multiple scene Acfer data*

Characteristic	Fe–Ni	Troilite	Oxides
Number of grains	2.81679	1.58818	4.69636
Sum of pixel areas	1.89390	1.14346	2.86782
Number >50 pixels	2.80463	1.56082	4.68290
Number >100 pixels	2.57986	1.46870	4.54378
Number >200 pixels	2.03696	1.23948	4.03674
Average area	0.67972	0.49772	0.86351
Maximum area	0.62809	0.37071	0.57616
Median area	0.28047	0.26699	1.14212
SD of area	0.58513	0.35659	0.60265
Ratio (length/width) average	0.82036	0.47878	1.29103
Ratio (length/width) median value	0.71332	0.40927	0.46128
Ratio (length/width) maximum value	0.74062	0.61833	1.02676
Ratio (length/width) minimum value	0.46301	0.16355	0.32066
Ratio SD	0.43040	0.41301	1.33469
Circularity average	1.06326	0.52441	1.80840
Circularity median value	1.06061	0.45032	1.31642
Circularity maximum value	1.05167	0.88302	0.31856
Circularity minimum value	0.67864	0.22813	0.69598
Circularity SD	0.47089	0.48991	0.73738
Grain distribution slope	0.66795	0.32264	1.42770

We seek data variables that show comparatively little variation within a meteorite but a large variation between meteorites. This parameter is summarized using 'alpha values' in which we compare the standard deviation of a particular variable in different images from the same thin section with the total variation observed between all the different thin sections in our dataset. Alpha values >1 are useful for pairing; alpha values <1 are not.

testing our method: we know that two images from the same thin section are 'paired'. Finally, the data are compiled with silicate composition and meteorite type. This dataset involves too many parameters to be easily processed by a human being, so we apply a computational approach relating to knowledge discovery in databases known as a genetic algorithm (see Goldberg 1989).

Automated pairing. The genetic algorithm is a computational approach inspired by Darwin's theory of natural selection: good solutions, in this case good pairings, arise through a process of competition, inheritance of characteristics, and chance. From a pool of fragments, each with measured characteristics from thin sections, we attempt to find significant pairings by searching through the numerous possible pairings with a genetic algorithm. Effectively, the algorithm provides a framework for the data that allows them to be sorted into the most self-similar groupings.

In the simplest case, the genetic algorithm (GA) creates an initial population of trial pairs and, based on similarities in image processing characteristics, Fayalite and Ferrosilite composition, etc., selects the best few (i.e. those most similar to one another). The algorithm then proceeds iteratively, using the best pairs from the previous population as 'parents' in making a new generation, which in turn is treated in the same way, selecting the best few pairs and discarding the rest. To create 'children' from two randomly chosen 'parents', a 'child' has a certain probability of inheriting a fragment from each of its parents, and also a certain probability of accepting a fragment chosen randomly. In this way, inheritance and chance cause the algorithm to proceed through successive generations, maintaining or improving the 'quality' of its parent population.

Problems. A number of problems exist with this basic method, some of which are:

- Deciding the characteristics which should be used;
- Deciding the details of the GA, e.g. probability of inheritance or mutation;
- Deciding what constitutes a good pair;
- Handling possible pairing groups;
- Ending the algorithm.

The first problem may be resolved by examining the statistical properties of the measured characteristics. The usefulness of a given characteristic depends on two things: the measurement accuracy and the variation within the meteorite compared with the variation between different meteorites. If accuracy is poor and/or variation is large then it will be difficult to make a pairing based on that characteristic. We summarize this using 'alpha values' in which we compare the variation of a particular variable in different images from the same thin section, with the total variation observed between all the different thin sections in our dataset. We seek data variables that show comparatively little variation within a meteorite but a large variation between meteorites. The alpha value associated with each fragment is >1 if it might be useful for pairing but <1 if it is not useful. Table 1 gives alpha values for each characteristic, for each of the three grain types derived from our image processing analysis.

The second problem, deciding the probability of inheritance and mutation, may be resolved by testing the algorithm on artificially generated data and assessing how it performs. It turns out, as with many GA applications, that the precise parameter values are not crucial (Conway 1995).

Considering the third and fourth problems together, we seek to go beyond simply finding likely pairs to finding likely groupings of fragments. In principle, it is possible to make the basic member of the GA a variable sized group, but in practice this entails many frustrating complications. So, retaining pairs as the basic element of our population, we associate pairs using 'links'. A link is simply a connection to another pair; in the GA links can be inherited from parents or received at random. So, to fully define the effect of a link we must incorporate it in some form of scoring system. At present, we choose to use a simple sum square scoring system, which does not directly make provision for cross-correlation of characteristics. Given a pair a which consists of fragments $a1$ and $a2$, which is linked to a pair b, the score is given by:

$$S_a = [S(a1, a2) + S_b + S(a1, b2) + S(a2, b1)]/4 \qquad (1)$$

where the scoring function $S(x, y)$ simply calculates the sum square difference in characteristics between fragment x and fragment y:

$$S(x, y) = \text{sum over each characteristic } (i) \; (\{[Ci(x) - Ci(y)]/\sigma_{Ci}\}^2)$$

It should be noted that S_b in equation (1) is calculated in exactly the same way, so that S_a is actually a recursive score containing diluted contributions from other links through b. In this way, linked pairs that rise to the top of the population must also be part of likely groups: any weak link in the chain or bad pair will bring the score down. Many problems arise when using this linking system (for example, avoiding circular links, where $a \rightarrow b \rightarrow c \rightarrow a$) but most can be dealt with relatively easily.

The final problem, ending the algorithm, is more subtle. The essence of the pairing problem is to reduce the fragment population so that there remains only a list of the most likely groupings. To do this, after a group has remained as part of the parent population for a certain number of generations, we remove that group (i.e. remove all its fragments from the fragment pool) and replace the whole group with a representative fragment, whose characteristics are the average of all the group's members. This representative fragment will, it is hoped 'mop up' any remaining fragments that should have been part of the group. The removal of the group now allows for 'less likely' groupings to enter into the parental population. So, in theory, the first groups found should tend to be the most likely pairing groups. Also, as the algorithm proceeds, the pool of fragments will begin to dwindle, until at some point, it is no longer practicable to proceed any further. This then decides the end of the algorithm and any fragments left unpaired or ungrouped must be regarded as lone fragments. There is, of course, no guarantee that the algorithm will have found all the groupings and pairings; in fact, there is no real guarantee that the extracted groups are complete or even correct; only statistical testing of each group can begin to resolve this. To address this latter point we use information derived from multiple-image fragment data (i.e. for one thin section we may have ten different images); in this situation 'pairings' are known because 'different fragments' are just different images from the same thin section.

Results

We have shown that at least some of the parameters that result from the image processing should have significance for a pairing exercise, i.e. alpha >1. However, it is not clear what level of significance, or how many data variables, would be required by a GA to successfully pair a dataset consisting of unknown fragments. To gain some insight into this question we initially test the GA on several artificial datasets, comprising different alphas and running the GA several times with different numbers of pairing characteristics.

Preliminary results on test data. To test the effect of different alphas and different number of pairing characteristics (M), we created four data sets of different alphas (1, 4, 10 and 100), containing ten groups of five fragments (see Table 2). Each set had 16 characteristics, all characteristics being generated to have the desired alpha of that set. On each of these sets the GA was run four times using different numbers of pairing characteristics each time ($M = 1, 4, 8, 16$).

Table 2. *Ability of the GA to pair a test dataset*

M	Alphavalue							
	1		4		10		100	
16	0	1/2	1	3/5	4	5/7	5	7/8
8	0	0/4	1	5/9	2	3/7	1	3/5
4	0	0/5	1	5/8	1	1/5	4	5/7
1	0	0/6	0	0/6	0	0/2	0	0/3

Two numbers are listed for each combination of alpha value and number of pairing characteristics (M). The first is the number of complete groups found by the algorithm; the second gives the total number of correct, but not necessarily complete groups out of the total number of groups found.

The entries in Table 2 list two numbers for each table entry: the first is the number of complete groups found by the algorithm; the second gives the total number of correct, but not necessarily complete groups out of the total number of groups found, e.g. five and 7/8 means that five groups were found in their entirety, and that seven out of the eight groups found were actually real groups or real subgroups.

Tests of Acfer multiple-scene data. To test the GA on real data, we took the 16 characteristics with the highest alphas in Table 1, and used them for pairing. Using the same notation as in Table 2, the result of this run was: zero and 4/27. In this dataset there were actually 13 real groups varying in size from five to ten members; so only four were actually detected and none of them completely.

As the real data lie between the alpha = 1 and alpha = 4 entries on the $M = 16$ row of Table 2 (in Table 1 all 16 characteristics used have alphas between unity and 4.6), it is not surprising that this attempt met with little success. It appears that an overall alpha value between four and ten is required for the GA method to work on 16 pairing characteristics.

Although these results are preliminary, and various substantial improvements could be made to the GA code, it appears that the input data will need to be augmented in some way for the GA to work. There are several possibilities for how this could be achieved:

- Use of other characteristics. In principle, this would mean finding some data variable that is of significantly greater value (i.e. high alpha) in pairing the meteorites. This possibility appears unlikely: if such a variable existed then it would be a simple matter to pair fragments.
- Pre-processing of current input data using principal component analysis or a similar statistical technique, or a neural network. In this situation, each variable is treated as a separate dimension and correlations are sought in this multi-dimensional space, so that all the data are considered as a gestalt. Humans recognize each other's faces in a similar way: a person is not recognizable through a single significant facial feature, but rather through all those features considered together.
- Use of cross-correlation and other statistical properties from multiple scene data in the GA code itself. This would provide for more reliable identification and removal of groups during the search.

Conclusions

Our work indicates a possible approach to the problem of deriving an estimate of flux from Antarctic meteorite populations. Comparing ^{57}Fe Mössbauer spectroscopy data and meteorite terrestrial ages, we find a correlation between the percentage of total iron that exists as Fe^{3+} in the sample and terrestrial age. The overall rate of weathering that we observe is approximately 2–3 orders of magnitude lower than values typical of hot desert accumulation sites. In addition, a comparison of oxidation–frequency distributions between Antarctic and hot desert finds reveals that Antarctic meteorites have a experienced a much lower overall level of weathering. We do not find evidence for action of weathering processes specific to Antarctic to explain this difference, and propose a model that invokes the loss of samples because of ice flow as the principal sink for meteorite samples in Antarctic accumulation sites. Although at a preliminary stage, the observation of a possible correlation between weathering and terrestrial age is intriguing and deserves further study. If an enlarged dataset confirms the correlation between weathering and terrestrial age, the method we discuss may allow an estimate of the lifetime of a population to be made, and of the average flow rates over the area, both of which are necessary first steps to calculating the catchment area for samples.

We also discuss the prospects for automated pairing of meteorite samples using digital image processing of thin sections to acquire a textural 'fingerprint' for a sample, and a genetic algorithm scheme that looks for similarities in the image processing output between samples. Although our initial results do not allow unequivocal pairings to made yet, they are encouraging and suggest that given enough development the GA approach may be effective in automating meteorite pairing.

P.A.B. thanks the Royal Society for support while part of this work was carried out.

References

ANNEXSTAD, J. O. 1985. *Meteorite concentration mechanisms in Antarctic*. LPI Technical Report **86-01**, 23–25.

—— & SCHULTZ, L. 1983. Measurements of the triangulation network at the Allan Hills meteorite icefield. *In*: OLIVER, R. L., JAMES, P. R. & JAGO, J. B. (eds) *Antarctic Earth Science*. Cambridge University Press, Cambridge, 617–619.

BARLOW, N. G. 1990. Application of the inner solar system cratering record to the Earth. *In*: SHARPTON, V. L. & WARD, P. D. (eds) *Global Catastrophes in Earth History*. Geological Society of America, Special Paper, **247**, 181–187.

BENOIT, P. H., CUNNINGHAM, J. M., BLAND, P. A., BERRY, F. J. & PILLINGER, C. T. 1998. Meteorite 'pairing': the recognition of the fragmentation of meteorite finds. *Meteoritics and Planetary Science* (in press).

——, SEARS, H. & SEARS, D. W. G. 1993. The natural thermoluminescence of meteorites 5. Ordinary chondrites at the Allan Hills Ice Fields. *Journal of Geophysical Research*, **98**(B2), 1875–1888.

BEVAN, A. W. R. & BINNS, R. A. 1989. Meteorites from the Nullarbor Region, Western Australia: II. Recovery and classification of 34 new meteorite finds from the Mundrabilla, Forrest, Reid and Deakin areas. *Meteoritics*, **24**, 135–141.

BISCHOFF, A. & GEIGER, T. 1995. Meteorites from the Sahara: find locations, shock classification, degree of weathering and pairing. *Meteoritics*, **30**, 113–122.

BLAND, P. A., BERRY, F. J. & PILLINGER, C. T. 1995. Variation in weathering between the Great Bend meteorite and LEW 85319. *Proceedings of the 26th Lunar and Planetary Science Conference*, 131–132.

——, ——, SMITH, T. B., SKINNER, S. J. & PILLINGER, C. T. 1996*a*. The flux of meteorites to the Earth and weathering in hot desert ordinary chondrite finds. *Geochimica et Cosmochimica Acta*, **60**, 2053–2059.

——, SMITH, T. B., JULL, A. J. T., BERRY, F. J., BEVAN, A. W. R., CLOUDT, S. & PILLINGER, C. T. 1996*b*. The flux of meteorites to the Earth over the last 50,000 years. *Monthly Notices of the Royal Astronomical Society*, **283**, 551–565.

BROWN, H. 1960. The density and mass distribution of meteoritic bodies in the neighbourhood of the Earth's orbit. *Journal of Geophysical Research*, **65**, 1679–1683.

——1961. Addendum: The density and mass distribution of meteoritic bodies in the neighbourhood of the Earth's orbit. *Journal of Geophysical Research*, **66**, 1316–1317.

BUCHWALD, V. F. 1975. *Handbook of Iron Meteorites – Volume 1. Iron Meteorites in General*. University of California Press, Berkeley.

BURNS, R. G., BURBINE, T. H., FISHER, D. S. & BINZEL, R. P. 1995. Weathering in Antarctic H and CR chondrites: quantitative analysis through Mössbauer spectroscopy. *Meteoritics*, **30**, 1–9.

CASSIDY, W. A., HARVEY, R., SCHUTT, J., DELISLE, G. & YANAI, K. 1992. The meteorite collection sites of Antarctica. *Meteoritics*, **27**, 490–525.

CONWAY, A. 1995. *Innovative methods in the prediction and analysis of solar–terrestrial time series*. PhD thesis, University of Glasgow.

DELISLE, G. & SIEVERS, J. 1991. Sub-ice topography and meteorite finds near the Allan Hills and the Near Western Icefield, Victoria Land, Antarctica. *Journal of Geophysical Research*, **96**(E1), 15 577–15 587.

——, —— & SCHULTZ, L. 1989. Radio-echo sounding survey across the Allan Hills Icefield. *Antarctic Journal of the United States*, **24**, 50–52.

DREWRY, D. 1982. Ice flow, bedrock, and geothermal studies from radio echo-sounding inland of McMurdo Sound, Antarctica. *In*: CRADDOCK, C. (ed.) *Antarctic Geoscience*. University of Wisconsin Press, Madison, 977–983.

——1983. The surface of the Antarctic ice sheet. *In*: DREWRY, D. (ed.) *Antarctica: Glaciological and Geophysical Folio, Sheet 2*. Scott Polar Research Institute, Cambridge.

ESSER, B. K. & TUREKIAN, K. K. 1988. Accretion rate of extraterrestrial particles determined from osmium isotope systematics of Pacific pelagic clay and manganese nodules. *Geochimica et Cosmochimica Acta*, **52**, 1383–1388.

FARLEY, K. A. & PATTERSON, D. B. 1995. A 100-kyr periodicity in the flux of extraterrestrial ^{3}He to the sea floor. *Nature*, **378**, 600–603.

FIREMAN, E. L. 1987. Uranium-series dated ice, 100,000 to 350,000 years old, arranged sequentially at Allan Hills. *Antarctic Journal of the United States*, **22**, 75–77.

—— & NORRIS, T. 1981. Carbon-14 ages of Allan Hills meteorites and ice. *Proceedings of the 12th Lunar and Planetary Science Conference*, 1019–1025.

GOLDBERG, D. E. 1989. *Genetic Algorithms in Search, Optimisation and Machine Learning*. Addison Wesley, Reading, MA.

GOODING, J. L. 1985. *Weathering of stony meteorites in Antarctica*. LPI Technical Report **86-01**, 48–54.

GRIEVE, R. A. F. 1984. The impact cratering rate in recent time. *Journal of Geophysical Research*, **89**, B403–B408.

GRÜN, E., ZOOK, H. A., FECHTIG, H. & GIESE, R. H. 1985. Collisional balance of the meteoritic complex. *Icarus*, **62**, 244–272.

HALLIDAY, I., BLACKWELL, A. T. & GRIFFIN, A. A. 1989. The flux of meteorites on the Earth's surface. *Meteoritics*, **24**, 173–178.

——, —— & ——1991. The frequency of meteorite falls: comments on two conflicting solutions to the problem. *Meteoritics*, **26**, 243–249.

HARVEY, R. P. 1989. *Statistical differences between Antarctic finds and modern falls: mass frequency distributions and relative abundance by type*. LPI Technical Report **90-01**, 43–48.

——1990. *Terrestrial age mapping of the Allan Hills Main Icefield and implications for the Whillans–Cassidy model of meteorite concentration*. LPI Technical Report **90-03**, 88–90.

HAWKINS, G. S. 1960. Asteroidal fragments. *Astronomical Journal*, **65**, 318–322.

HUGHES, D. W. 1980. On the mass distribution of meteorites and their influx rate. *In*: MCINTOSH, B. A. & HALLIDAY, I. (eds) *Solid Particles in the Solar System*. Reidel, Dordrecht, 207–210.

JULL, A. J. T., DONAHUE, D. J. & LINICK, T. W. 1989. Carbon-14 activities in recently fallen meteorites and Antarctic meteorites. *Geochimica et Cosmochimica Acta*, **53**, 2095–2100.

——, WLOTZKA, F., BEVAN, A. W. R., BROWN, S. T. & DONAHUE, D. J. 1993. ^{14}C terrestrial ages of meteorites from desert regions: Algeria and Australia (abstract). *Meteoritics*, **28**, 376–377.

——, —— & DONAHUE, D. J. 1991. Terrestrial ages and petrologic description of Roosevelt County meteorites. *Proceedings of the 24th Lunar and Planetary Science Conference*, 667–668.

LAUTRIDOU, J. P. 1975. Les recherches de gelifraction expérimentale du centre de géomorphologie du, C.N.R.S. Les Problèmes Posés par la Gelifraction, Recherches Fondamentales et Appliquées. Le Havre, Foundation Française d'Études Nordiques, 6 Congress International.

——1976. Dix ans de recherches du Centre de Géomorphologie, les expériences de cryoclastic. *Bulletin du Centre de Géomorpholgie*, **21**, 21–28.

——— & OZOUF, J. C. 1982. Experimental frost shattering: 15 years of research at the Centre de Géomorphologie du, C.N.R.S. *Progress in Physical Geography*, **6**, 215–232.

LOVE, S. G. & BROWNLEE, D. E. 1993. A direct measurement of the terrestrial mass accretion rate of cosmic dust. *Science*, **262**, 550–553.

MICHLOVICH, E. S., WOLF, S. F., WANG, M.-S., VOGT, S., ELMORE, D. & LIPSCHUTZ, M. E. 1995. Chemical studies of H chondrites 5. Temporal variations of sources. *Journal of Geophysical Research*, **100**(E2), 3317–3333.

MILLARD, H. T., JR. 1963. The rate of arrival of meteorites at the surface of the Earth. *Journal of Geophysical Research*, **68**, 4297–4303.

MIOTKE, F. D. & VON HODENBERG, R. 1980. Zur Salzsprengung und chemischen Verwitterung in den Darwin Mountains und den Dry Valleys, Victoria Land, Antarktis. *Polarforschung*, **50**, 45–80.

MURRELL, M. T., DAVIS, P. A., JR. & NISHIIZUMI, K. 1980. Deep-sea spherules from Pacific clay: mass distribution and influx rate. *Geochimica et Cosmochimica Acta*, **44**, 2067–2074.

NISHIIZUMI, K., ELMORE, D. & KUBIK, P. W. 1989. Update on terrestrial ages of Antarctic meteorites. *Earth and Planetary Science Letters*, **93**, 299–313.

RABINOWITZ, D. L. 1993. The size distribution of the Earth-approaching asteroids. *Astrophysical Journal*, **407**, 412–427.

SCHULTZ, L. 1986. Allende in Antarctica: temperatures in Antarctic meteorites (abstract). *Meteoritics*, **21**, 505.

SHINONAGA, T., ENDO, K., EBIHARA, M., HEUMANN, K. G. & NAKAMAR, H. 1994. Weathering of Antarctic meteorites investigated from contents of Fe^{3+}, chlorine, and iodine. *Geochimica et Cosmochima Acta*, **58**, 3735–3740.

SHOEMAKER, E. M., WOLFE, R. F. & SHOEMAKER, C. S. 1990. Asteroid and comet flux in the neighborhood of Earth. *In*: SHARPTON, V. L. & WARD, P. D. (eds) *Global Catastrophes in Earth History*. Geological Society of America, Special Paper, **247**, 155–170.

WASHBURN, A. L. 1980. *Geocryology: a Survey of Periglacial Processes and Environments*. Wiley, New York.

WELTEN, K. 1995. *Exposure histories and terrestrial ages of Antarctic meteorites*. PhD thesis, University of Utrecht.

WETHERILL, G. W. 1986. Unexpected Antarctic chemistry. *Nature*, **319**, 357–358.

WHILLANS, I. & CASSIDY, W. 1983. Catch a falling star: meteorites and old ice. *Science*, **222**, 55–57.

WOLF, S. F. & LIPSCHUTZ, M. E. 1995. Chemical studies of H chondrites 4. New data and comparison of Antarctic suites. *Journal of Geophysical Research*, **100**(E2), 3297–3316.

ZOLENSKY, M. 1998. The flux of meteorites to Antarctica. *This volume*.

——, RENDELL, H. M., WILSON, I. & WELLS, G. L. 1992. The age of the meteorite recovery surfaces of Roosevelt County, New Mexico, USA. *Meteoritics*, **27**, 460–462.

——, WELLS, G. L. & RENDELL, H. M. 1990. The accumulation rate of meteorite falls at the Earth's surface: the view from Roosevelt County, New Mexico. *Meteoritics*, **25**, 11–17.

Meteorite flux on the Nullarbor Region, Australia

A. W. R. BEVAN[1], P. A. BLAND[1] & A. J. T. JULL[2]

[1] *Department of Earth and Planetary Sciences, Western Australian Museum of Natural Science, Francis Street, Perth, WA 6000, Australia*
[2] *NSF Accelerator Facility for Radioisotope Analyses, University of Arizona, Tucson, AZ 85721, USA*

Abstract: The Nullarbor Region of Australia is one of the most prolific sites for meteorite recoveries outside of Antarctica. Reported ^{14}C terrestrial ages for chondritic meteorites from the Nullarbor indicate an age range from present day to *c.* 35 ka. There is good evidence to suggest that meteorites are lying on, or close to the surfaces on which they fell and that, physiographically, the region has remained essentially undisturbed for at least the last 30 ka. The Nullarbor can thus provide important data on the flux of meteorites over the period of accumulation. One significant factor influencing flux calculations based on meteorite accumulation sites is the determination of the number of falls represented in the recovered population. In the case of the Nullarbor, a general lack of transportation processes in the region and careful documentation of the distribution of finds allows confident 'pairing' of meteorites. For example, strewn fields of showers that have remained undisturbed for thousands of years are easily recognized and mapped today. Mass distribution statistics confirm that there are few undetected pairs in the population of meteorites so far described from the Nullarbor, and there is no evidence of selection of meteorites of a specific terrestrial age. The Nullarbor is the largest accumulation of meteorites nearest to Antarctica with a terrestrial age range (0–35 ka) that overlaps the Antarctic population (0 to >1 Ma). Analysis of the well-documented population of meteorites from the Nullarbor compared with Antarctica suggests that there are many unrecognized paired meteorites in the Antarctic population and the abundance of small meteorites at that site can be accounted for, at least in part, by the presence of large shower falls of common chondrites. The data suggest that there may also be many unrecognized paired meteorites in collections from the Sahara.

Over the last 30 years there has been a dramatic increase in the number of meteorites recovered world-wide. Although the average number of observed meteorite falls recovered annually remains at around 5–6, the large number of meteorite fragments recovered from the Antarctic ice (currently >16 000) and a realization that some 'hot' deserts of the world also contain an abundance of meteorites accumulated over prolonged periods have yielded a wealth of new material, and opened up many new avenues of research. To date, the most notable arid areas of the world for meteorite recoveries are the Algerian and Libyan Sahara (Bischoff & Geiger (1995) and references therein), Roosevelt County in New Mexico, USA (Zolensky *et al.* (1990) and references therein), and the Nullarbor (Bevan & Binns 1989*a*, *b*; Bevan 1992*a*, 1996; Bevan & Pring 1993). In addition to extending our knowledge of early Solar System materials by providing occasional samples of meteorites hitherto unknown to science, these accumulated collections allow estimates to be made of the flux of meteorites to Earth with time.

The treeless limestone plains of the Nullarbor (Fig. 1) are among the most prolific areas for meteorite recoveries outside Antarctica (Bevan 1992*a*, 1996). The region is coincident with a geological structure, the Eucla Basin, straddling the border between Western Australia and South Australia, that consists of flat-lying limestones of Lower–Middle Miocene age outcropping over a total area of *c.* 240 000 km^2 (Lowry 1970; Lowry & Jennings 1974). Prolonged aridity in the Nullarbor has led to the preservation of meteorites, and barren conditions and pale country rock allow for their easy recognition (Bevan & Binns 1989*a*, *b*). Several thousand fragments from about 260 distinct meteorites, representing *c.* 55% of all meteorites known from Australia, have been described from the Nullarbor to date (Fig. 2), and many hundreds of specimens of potentially new meteorites remain to be classified. Recent systematic searches by joint teams from the Western Australian Museum (WAMET) and EUROMET (a pan-European group of research institutions devoted to meteorite research; Bevan 1992*b*) have recovered more than 600 specimens of meteorites (totalling *c.* 17 kg) and >900 tektites during some 10 weeks of searching on four expeditions between 1992 and 1994 in the Western Australian Nullarbor.

The physiographically stable surface of the Nullarbor combined with low weathering rates

BEVAN, A. W. R., BLAND, P. A. & JULL, A. J. T. 1998. Meteorite flux on the Nullarbor Region, Australia. *In*: GRADY, M. M., HUTCHISON, R., MCCALL, G. J. H. & ROTHERY, D. A. (eds) *Meteorites: Flux with Time and Impact Effects*. Geological Society, London, Special Publications, **140**, 59–73.

Fig. 1. On the Joint WAMET–EUROMET Nullarbor expedition 1992, Professor Christian Koeberl (University of Vienna) finds a small ordinary chondritic meteorite (Loongana 002) on the open Nullarbor Plain in Western Australia. Dr Andrew Morse (Open University) records the event.

over at least the last 30 ka determined from the terrestrial ages of meteorites (Jull *et al.* 1995) provides a model set of conditions for the assessment of the meteorite flux with time. In the Nullarbor Region, the concentration of meteorites appears to be a function only of prolonged aridity and did not involve additional physical concentration processes as in Roosevelt County and Antarctica which complicate and bias estimates of the meteorite flux (e.g. see Zolensky *et al.* 1992). Meteorites in the Nullarbor appear to be lying on, or near, the surfaces on which they fell.

Recently, Bland *et al.* (1996) have used well-documented ordinary chondritic meteorites from the Sahara, Roosevelt County and the Nullarbor to calculate the flux of meteorites to Earth over the last 50 ka. The results from each area (Table 1) are generally in close agreement with each other, and with the present-day flux of meteorites estimated from data acquired from the Canadian regional camera network by the Meteorite Observation and Recovery Project (MORP) (Halliday *et al.* 1989, 1991). From their study of desert meteorite accumulation sites Bland *et al.* (1996) concluded that during the late Quaternary there has been no significant change in the flux to Earth of meteorites in the mass range 10–10^3 g. The flux remains at around 83 falls >10 g per 10^6 km^2 per year (Halliday *et al.* 1989, 1991). However, other features of the population of meteorites from the Nullarbor, notably the constitution and mass distribution of the population, differ markedly from modern falls and require further explanation.

From trace element studies of the most common types of chondrites found in Antarctica, Dennison *et al.* (1986) suggested that Antarctic meteorites and modern falls differ chemically, and may reflect changes in the sampling of meteorite parent bodies with time. Terrestrial ages of Antarctic meteorites are very much greater than most finds from other parts of the world. Nishiizumi (1990) and others have reported terrestrial ages for Antarctic meteorites ranging up to >1 Ma, although most meteorites from Antarctica have terrestrial ages between 11 and 20 ka (Schultz 1986). Harvey & Cassidy (1989) have compared the constitution of meteorite finds from Antarctica with the modern fall frequency. On the basis of non-parametric statistical analysis, they suggested that modern falls and several populations of Antarctic meteorites are probably not good samples of

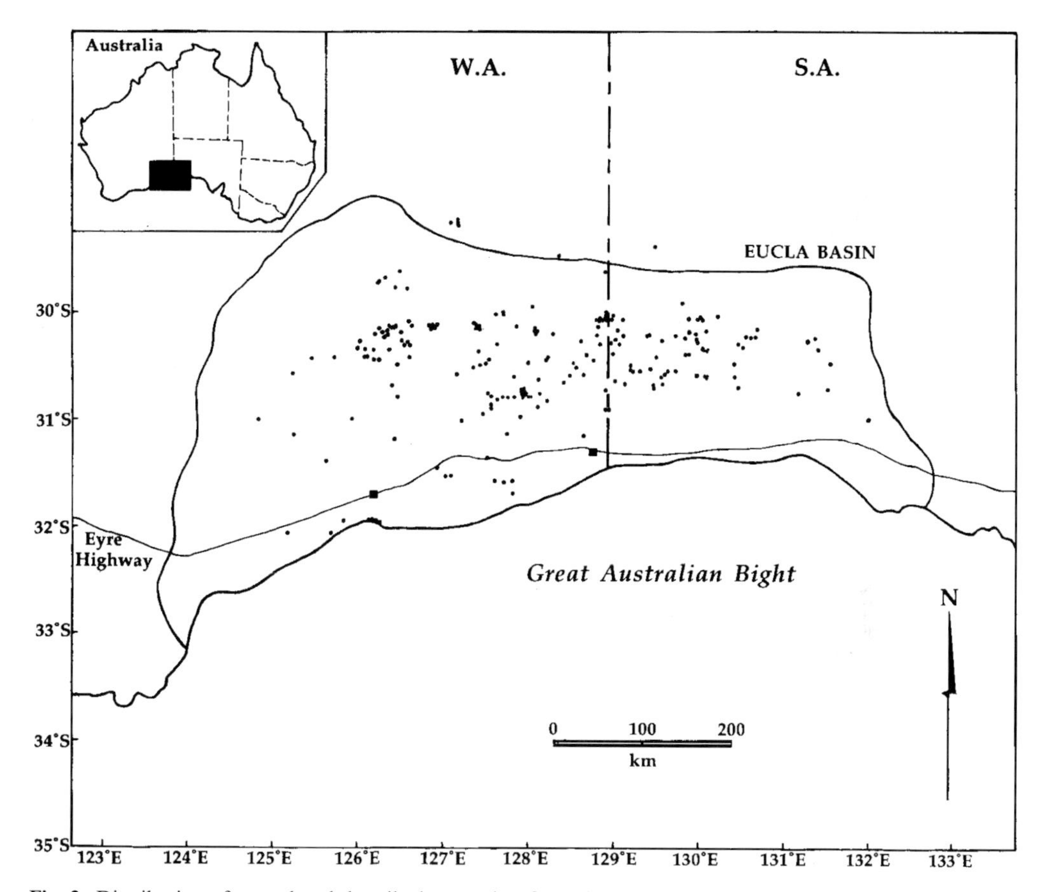

Fig. 2. Distribution of named and described meteorites from the Nullarbor Region.

Table 1. *Estimates of the flux of meteorites to Earth (after Halliday* et al. *(1989) and Bland* et al. *(1996))*

Halliday *et al.* (1989)	83 falls >10 g per 10^6 km^2 per year	MORP
Bland *et al.* (1996)		
RC	116 falls >10 g per 10^6 km^2 per year	50 000 ka
NR	36 falls >10 g per 10^6 km^2 per year	35 000 ka
S	95 falls >10 g per 10^6 km^2 per year	40 000 ka
Average desert*	82 falls >10 g per 10^6 km^2 per year	
Camel Donga (area)	50 falls >10 g per 10^6 km^2 per year	

MORP, Meteorite Observation and Recovery Project; RC, Roosevelt County; NR, Nullarbor Region; S, Sahara.
* Average of the three desert accumulation sites. Camel Donga (area) is an estimate based on meteorite recovery from the strewnfield of the Camel Donga meteorite shower.

a single steady-state meteoritic complex. The conclusions of Harvey & Cassidy (1989) appear to support the view that the flux of meteoritic material to Earth may have been variable over the last 400 ka. However, after including the effects of post-fall processing of meteorites in Antarctica, Cassidy & Harvey (1991) found that there were no significant differences in the mass frequency distributions of major meteorite classes between Antarctic finds and modern falls. Although there is a general consensus that some real differences do exist between Antarctic and non-Antarctic meteorites, the reasons for these differences are not well understood (Koeberl & Cassidy 1991) and may relate either to processes in space (Lipschutz & Samuels

1991), or to terrestrial alteration and sampling bias (Cassidy & Harvey 1991; Huss 1991).

Significantly, the Nullarbor has the largest accumulation of meteorites closest to Antarctica and with a range of terrestrial ages that overlaps those of Antarctic meteorites. The Nullarbor data may serve as a cross-check against the population of meteorites from Antarctica and other accumulation sites, particularly in terms of the mass distribution of fragments and the frequency of meteorite showers. In this paper we review the information on meteorites from the Nullarbor and examine the flux of meteorites on that area by identity of falls, mass distribution and abundance, terrestrial age, and meteorite type.

Recovery and nomenclature

Many of the early meteorite finds from the Nullarbor Region in Western Australia resulted from a recovery programme initiated by staff from the Kalgoorlie School of Mines (e.g. see Cleverly 1993) but also involved Honorary Associates of the Western Australian Museum and private individuals, notably the Carlisle family from Kalgoorlie (Bevan 1992*a*, 1996). No official meteorite collecting programme currently exists in South Australia and much of the material collected from the eastern part of the Nullarbor has been found by itinerant prospectors and rabbit trappers.

The recovery of large numbers of meteorites from an area with few geographical names initially caused problems for meteorite nomenclature. To overcome the problem, Bevan & Binns (1989*a*) devised a system of nomenclature for the Western Australian Nullarbor based on geographically named areas, which was later extended by Bevan & Pring (1993) to the South Australian Nullarbor. In the whole of the Nullarbor Region 74 named areas have now been delineated (47 in Western Australia; 27 in South Australia), and new and distinct meteorite finds take the name of the area in which they are found and a three-digit number (e.g. Reid 010), preferably in chronological order of discovery. A recent minor modification by H. Kruse & F. Wlotzka (pers. comm., 1996) of the geographical boundaries of some named areas (as defined by Bevan & Binns (1989*a*)) in the Western Australian Nullarbor has allowed computerization of Nullarbor meteorite nomenclature. Nullarbor meteorite names are now assigned graphically through GIS format files (J. Grossman, pers. comm., 1997).

In addition to providing a flexible system of meteorite nomenclature, the areas delineated in the Nullarbor, many of which are equidimensional, also provide a framework that can be used for statistical comparison of the density of finds. Several areas of the Nullarbor have already been extensively searched, and this is reflected in the high density of recoveries. However, large tracts of the Nullarbor have yielded only a few recoveries to date, and these areas remain to be searched thoroughly. In these areas it may be many years before the true density of meteorites is known. In the mean time, the well-searched areas provide a baseline for statistical comparison.

The method of searching traditionally employed in the Nullarbor is by foot. Several searchers *c.* 10 m apart walk slowly in one direction scanning the ground. In this way, a team of five cuts a swath *c.* 50 m wide and can carefully search an area of *c.* 1 km^2 for every 20 km walked. Sometimes repeat searches of an area have yielded more fragments of the same or different meteorites, showing that lighting conditions, fatigue, and other human factors can affect collecting efficiency.

Terrestrial ages and weathering 'half-life'

The ^{14}C terrestrial ages of a suite of ordinary chondritic meteorites from the Nullarbor have been measured by Jull *et al.* (1995). The methods employed in age determination have been given by Jull *et al.* (1990 and references therein). In the sample of meteorites from the Nullarbor dated so far, the distribution of ages shows a simple exponential decrease with increasing terrestrial age from the present day to around 35 ka. No chondritic meteorites from the Nullarbor older than *c.* 35 ka have yet been dated. However, Aylmer *et al.* (1988) have measured the ratio $^{26}Al/^{53}Mn$ in the Mundrabilla iron meteorite that suggests a terrestrial age >1 Ma. The absence of very old stony meteorites in the Nullarbor compared with other desert locations may be a statistical problem (Jull *et al.* 1995), although the age of the oldest chondritic meteorite yet dated (33.4 ± 2.3 ka) agrees well with the estimated age (*c.* 30 ka) of the calcareous clay cover of the southern Nullarbor Plain (Benbow & Hayball 1992). If older stony meteorites exist on the Nullarbor they may be buried in the clay cover.

The exponential decrease of the number of meteorites of a particular terrestrial age with time in the Nullarbor population is a good indication that this area of meteorite accumulation has remained substantially undisturbed during at least the last 30 ka. This is supported by the relatively undisturbed nature of the distribution ellipses of shower falls (Cleverly 1972, 1986) such as Mulga (north) (T_{age} 2.7 ± 1.3 ka), Mulga (south) (T_{age} 20 ± 1.3 ka) and Billygoat Donga (T_{age} 7.7 ± 1.3 ka) recognized and mapped in the Nullarbor. The arrested karst topography of the Nullarbor is considered to be eroding slowly (Gillieson & Spate 1992) and the surface has probably been stable for considerably longer than 30 ka. The removal of stony meteorites with increasing terrestrial age in the Nullarbor appears to be simply a function of prolonged weathering, which, in arid regions, depends mainly on the availability of moisture, although it could be accelerated by the presence of chlorine (Buchwald & Clarke 1989). In the Nullarbor, periodic increases in effective precipitation have probably provided the main source of water causing weathering (Bland *et al.* 1998).

Bland *et al.* (1996) used Mössbauer spectroscopy to obtain a quantitative measure of the terrestrial oxidation of ordinary chondrite finds from the Nullarbor. From the terrestrial ages of Nullarbor meteorites and their state of oxidation, Bland *et al.* (1996) have calculated a decay constant (λ) that expresses, mathematically, the effect of meteorite erosion with time. The data for a small population ($n = 18$) meteorites from the Nullarbor give a decay constant for ordinary chondrites of $\lambda = -0.024\ \text{ka}^{-1}$. Combined data from the Nullarbor and Roosevelt County (New Mexico) populations yield a decay rate of $\lambda = -0.047\ \text{ka}^{-1}$ indicating a weathering 'half-life' of stony meteorites in those areas of >15 ka.

Density of falls

For many meteorites recovered from the Nullarbor, collection statistics have been carefully documented. A sample of randomly searched areas of the Nullarbor totalling *c.* 44.9 km^2 yielded samples from 45 apparently distinct meteorites (Bland *et al.* 1996). The minimum density of falls in the Nullarbor is then approximately one meteorite per km^2. Intensive collecting in much smaller areas, such as the strewnfield of the Camel Donga eucrite shower, which covers an area of *c.* 6.5 km^2, has led to the recovery of some 16 non-eucrite meteorite samples representing at least nine additional distinct meteorite falls (Bland *et al.* 1996). Taken together with the Camel Donga meteorite, a minimum density on the ground of 1.5 meteorites per km^2 is indicated in that area.

Mass distribution, mass frequency and pairing of meteorites

As in Antarctica, most stony meteorites recovered from the Nullarbor weigh 100 g or less, with a peak in the mass distribution between 10 and 50 g. However, large masses have also been recovered, such as the largest mass (11.5 t) of the Mundrabilla iron meteorite shower, from which a conservatively estimated total of more than 22 t of material has been collected. Two other large irons (Haig (480 kg) and Watson (93 kg)) have also been recovered from the Nullarbor. Together, these three irons make up more than 99% of the total mass of meteorites recovered from the Nullarbor. Large stones have been recovered only rarely, the largest single mass recorded to date being the Cook 007 H4 chondrite, which is reported to have weighed more than 100 kg (Zbik, 1994).

Huss (1990, 1991) suggested that large shower falls within a population, uncorrected for pairing, bias the cumulative mass distribution of meteorites from some collection areas in Antarctica, notably the Allan Hills Main Ice Field. More recently, Ikeda & Kimura (1992) have also shown that the steeper slope of the mass distribution of Antarctic chondrites compared with modern falls indicates the presence of several unpaired showers in the population. An apparent abundance compared with modern falls of H group (notably H5) ordinary chondrites among meteorites collected from the Allan Hills Main Ice Field yielding a higher ratio of H/L group chondrites (1.6) compared with modern falls (H/L = 0.9) has also been attributed to the presence of shower falls in the population (Huss 1991).

In the population of meteorites from the Nullarbor particular attention has been given to the problem of pairing (Bevan & Binns 1989*b*), and the ratio of H/L chondrites (1.0) is very close to that of modern falls. Unlike Antarctica, where physical concentration processes have mixed the meteorite population and thus made the identification of pairs difficult and uncertain, the stability of the Nullarbor during the accumulation period means that geographical propinquity is a reliable aid to the identification of meteorite fragments from the same fall. At least six shower falls comprising >10 individuals have already been identified in the Nullarbor,

Table 2. *Shower falls identified in the population of meteorites from the Nullarbor Region*

	Class	n	Mass (kg)	T_{age} (ka)
Mulga (north)	H6	>800	>20	2.5 ± 1.3
Mulga (south)	H4	>10	>1	20.0 ± 1.3
Billygoat Donga	L6	>12	>1	7.55 ± 1.3
Forrest 002	L6	>20	>26	
Camel Donga	Aeuc	>600	>30	recent fall
Mundrabilla	Iranom	>1000	>20 000	>1 Ma?

including Mundrabilla (iranom) (De Laeter 1972; De Laeter & Cleverly 1983), Mulga (north) (H6), Mulga (south) (H4), Billygoat Donga (L6), Forrest 002 (L6) and Camel Donga (Aeuc) (Table 2).

Figure 3 shows curves derived from smoothed histograms for cumulative mass distribution of populations of stony meteorites from Antarctica (both Allan Hills and Yamato collections), Roosevelt County and the Sahara compared with modern falls and all incoming meteorites (after Hughes 1981; Huss 1991; Koeberl *et al.* 1992; Bischoff & Geiger 1995). When a large sample of stony meteorites from the WA Nullarbor (including some yet to be described) uncorrected for pairing and including fragments from the recognized shower falls is plotted in the same way, because of the abundance of small meteorites from the showers in the population, the slope of the curve towards higher masses is greater than that for modern falls and mimics the mass distribution curves of the Antarctic meteorite populations. However, when the fragments of Nullarbor shower falls are recalculated to a single mass and their values assigned to higher mass bins, then the curve of the corrected

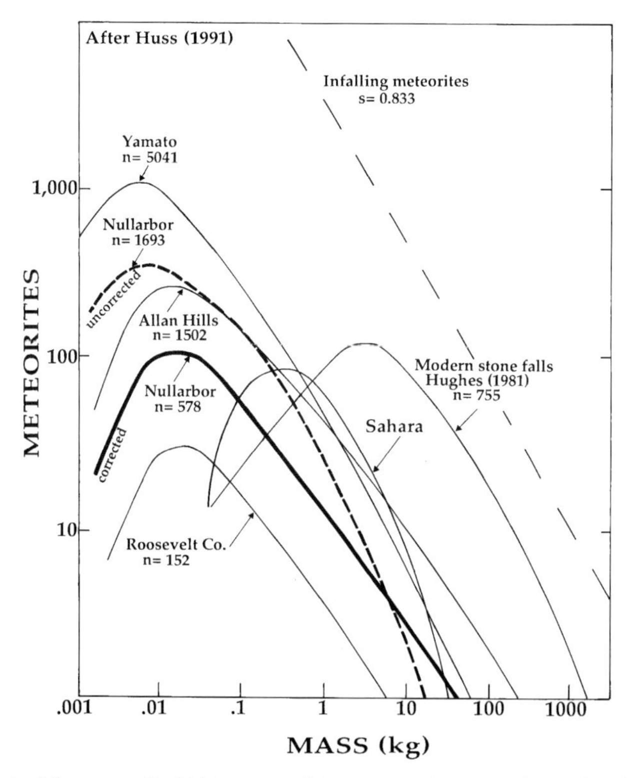

Fig. 3. Curves derived from smoothed histograms of stony meteorite masses in modern falls (after Hughes 1981) the Sahara (after Bischoff & Geiger 1995), Yamato and Allan Hills Antarctic meteorites and Roosevelt County (after Huss 1991). A population of Nullarbor meteorites (dashed line) that contains shower falls and has not been corrected for pairing is shown compared with the same population corrected for pairing.

Table 3. *Mass frequency of stony meteorite groups in Antarctica (A), modern falls (MF), and the Western Australian Nullarbor (N) and Sahara (S) deserts*

Types	Abundance (wt%)			
	A*	MF*	N	S
Ordinary	87.7	85.2	80.2	97.0
Achondrites	11.0	11.2	18.7	0.5
Carbonaceous	1.3	3.6	1.1	2.5
Total	100.0	100.0	100.0	100.0
Mass (kg)	654.3	12587.4	166.0	519.1

* Data from Cassidy & Harvey (1991).

population attains a reduced slope that is much closer to that of modern falls and infalling meteorites (−0.833) (Fig. 3).

The mass frequency for a sample of well-documented stony meteorites from the WA Nullarbor is shown in Table 3 along with the mass frequencies of the Antarctic population, modern falls as calculated by Cassidy & Harvey (1991), and a compilation of data for meteorites recovered from the Sahara. The relative proportions of the masses of chondrites (ordinary and carbonaceous) and achondrites from the Nullarbor are similar to both modern falls and Antarctica, and the data lend strong support to the conclusions of Cassidy & Harvey (1991). An apparent depletion in the masses of achondrites in the Saharan population relative to other desert areas and modern falls is anomalous, although it is probably due to the incomplete collection or description of the population of meteorites from that area.

Mass distribution of shower falls

The mass distributions of fragments from the Mulga (north) (30°11′S, 126°22′E) and Camel Donga (30°19′S, 126°37′E) showers that are in the collections of the WA School of Mines and the WA Museum, respectively, are shown in Fig. 4, and the mapped distribution of fragments from the Camel Donga shower is shown in Fig. 5. The mapped distribution of the Mulga (north) shower (Cleverly 1972) of 781 stones showed a distribution ellipse trending W–E, with larger masses generally found at the eastern end of the strewnfield gradually reducing to small fragments at the western end. More fragments of the shower have been recovered in recent years, bringing the total number of stones to more than 800. Most fragments of Mulga (north) are angular but with smoothed surfaces and many individuals are completely, or almost completely, covered with remnants of primary fusion crust (Cleverly 1972). Other stones have surfaces free of crust or a thin covering of secondary crust, indicating perhaps two episodes of fragmentation of the meteoroid during atmospheric passage.

The distribution of the Camel Donga shower on the ground is more complex than that of the Mulga (north) shower. The shower, as collected, comprised around nine small 'showerettes' each with its own mass distribution of fragments arranged within a larger distribution ellipse trending towards the ENE. A number of large fragments weighing >1 kg were reportedly collected from the ENE end of the strewnfield (A. J. Carlisle, pers. comm., 1987). However, these masses are not held in the collection at the WA Museum although their positions of find were recorded by Mr Carlisle. Overall the whole shower showed a gradual mass distribution, with those showerettes at the ENE end of the shower comprising generally larger fragments than those showerettes towards the WSW end of the shower. The distribution of fragments within the shower is further complicated by a curved distribution such that the ellipses of each individual showerette and the ellipse of the overall shower are distorted towards the S and SE (Fig. 5). At the extreme WSW end of the shower only small fragments, generally <50 g, and tiny glassy spheroids (A. J. Carlisle, pers. comm., 1987) were recovered. The smallest and largest size fractions of the shower are not well represented in the collection of the WA Museum. The majority of the larger fragments examined are completely covered with fresh fusion crust. However, many are highly irregular in shape and show several generations of fusion crust development. A few small fragments show only the incipient 'smoked' development of fusion crust on some freshly broken surfaces (Cleverly *et al.* 1986). In total, an unknown number of fragments were collected from the shower, with the WA Museum holding more than 600 stones.

The mass distributions of accurately weighed fragments of the Mulga (north) and Camel Donga showers (Fig. 4*a*, *b*) show a possible hiatus in fragments with masses around 10–15 g consistent with a complex atmospheric fragmentation history. The mapped distribution of the Camel Donga shower (Fig. 5) clearly indicates a complex, multi-stage history of atmospheric break-up. At least two, and possibly three, episodes of atmospheric fragmentation are indicated by the mass and ground distribution of fragments.

The curious ground distribution of the Camel Donga shower is attributed to high winds during

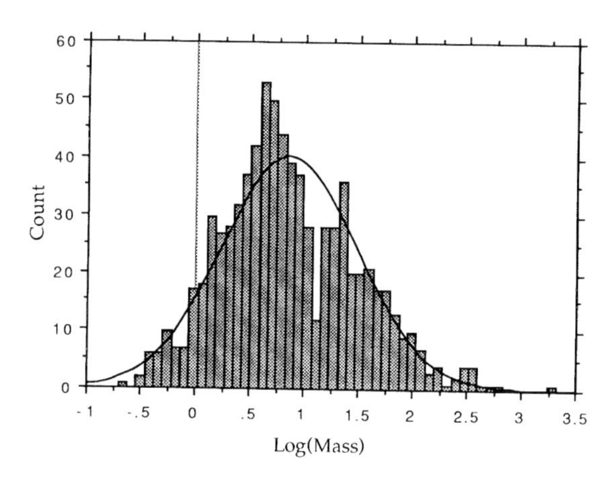

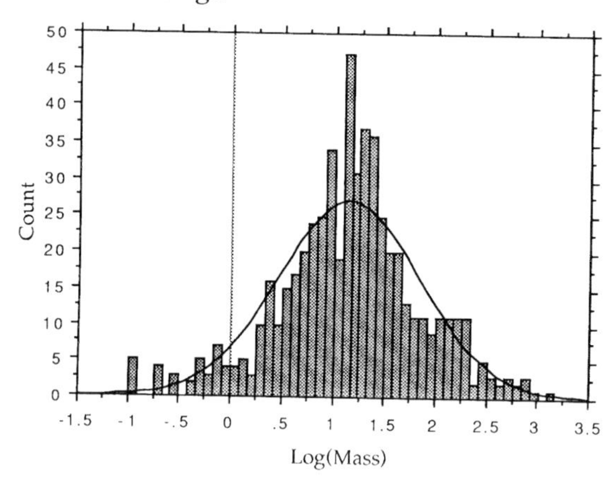

Fig. 4. (**a**) Mass distribution of fragments (population $n = 781$) of the Mulga (north) H6 chondrite shower from the Nullarbor. (**b**) Mass distribution of fragments (population $n = 561$) from the Camel Donga (Aeuc) shower.

the fall of the meteorite. A wind from the W or NW is indicated by the distortion of the lighter fraction from the shower towards the SE. However, the distribution of some of the heavier fragments, notably in the mass range 10^2–10^3 g, across the line of the overall ellipse of the shower does not seem to be explained by wind alone. Several other examples have been recorded of larger masses of meteorite within a multiple fall that have deviated considerable distances from the flight path. McCrosky *et al.* (1971) noted that a 10 kg mass of the Lost City observed meteorite fall was carried some 4 km eastwards relative to the impact point predicted from the ballistic flight data alone. In that case, the deviation was also attributed to wind. A contributory factor in the case of the Camel Donga shower may have involved rotating fragments interacting with a cross wind causing 'swing' similar to that imparted by a bowler to a cricket ball (e.g. see Mehta *et al.* 1983).

Field evidence from the Camel Donga strewnfield indicates that the calcareous clay cover in that area of the Nullarbor was wet at the time of the impact of the fragments. A number of stones were deeply plugged in the clay and some of the larger masses had splashed wet clay from their impact pits. The first stone of the shower was recovered in January 1984 (Cleverly *et al.* 1986) and from its extremely fresh condition the shower probably fell not long before the discovery. Heavy rain in the Nullarbor is sometimes associated with cyclonic or stormy conditions that develop NW of Australia and travel in a

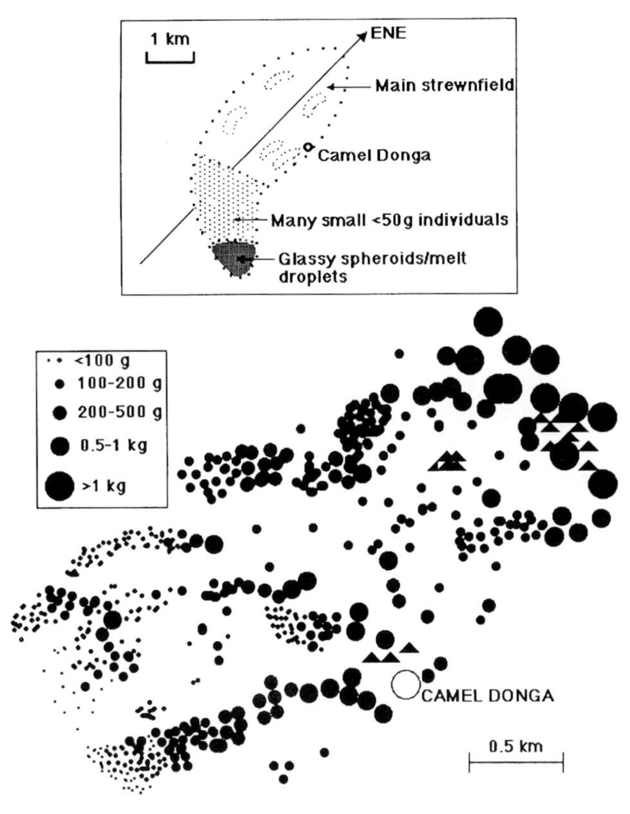

Fig. 5. Ground distribution of larger fragments from the main strewnfield of the Camel Donga (Aeuc) meteorite shower on the Nullarbor Plain in Western Australia. Map includes masses reported by A. J. Carlisle, but which are not in the collection of the WA Museum and are not included in the mass distribution (Figs 4b and 6). Inset: distribution ellipse of the whole shower showing location of the small mass fraction of the shower. Triangles indicate broken fragments.

NW–SE direction across the country towards the Nullarbor. The meteorite shower probably fell during the passage of an atmospheric disturbance perhaps only a few years before the discovery of the first stone. This would also explain why there is apparently no record of what must have been a spectacular visual event over southern Australia.

The mass distributions of the Camel Donga and Mulga (north) showers are shown as smoothed histograms in Fig. 6 and compared with the mass distributions of fragments of a variety of ordinary chondrite types from the Allan Hills Main Ice Field (after Huss 1991). The slope of the curve of the Mulga (north) shower (H6) is very similar to the curve for H5 and H6 chondrites at Allan Hills, and indicates that the latter population does contain at least one large shower fall and supports the conclusions of Huss (1991). The sharp inversion of the curves at lower masses for both the Mulga (north) and Camel Donga showers indicates incomplete collection of the smallest mass fractions of these Nullarbor showers.

The frequency of meteorite types

Using modern falls as a sample, Dohnanyi (1972), Wasson (1974), Harvey & Cassidy (1989), Cassidy & Harvey (1991) and many others have calculated the relative abundances of different compositional types of meteorite in the historic meteorite flux. The 'fall frequency' is the number of each type of meteorite seen to fall expressed as a percentage of a well-documented sample of observed falls. A population (835) of well-documented falls taken from Graham *et al.* (1985) (Table 4), shows that the chondritic meteorites (particularly the ordinary chondrites) are the most abundant types, accounting for 86.2% of observed falls. Irons and stony-irons

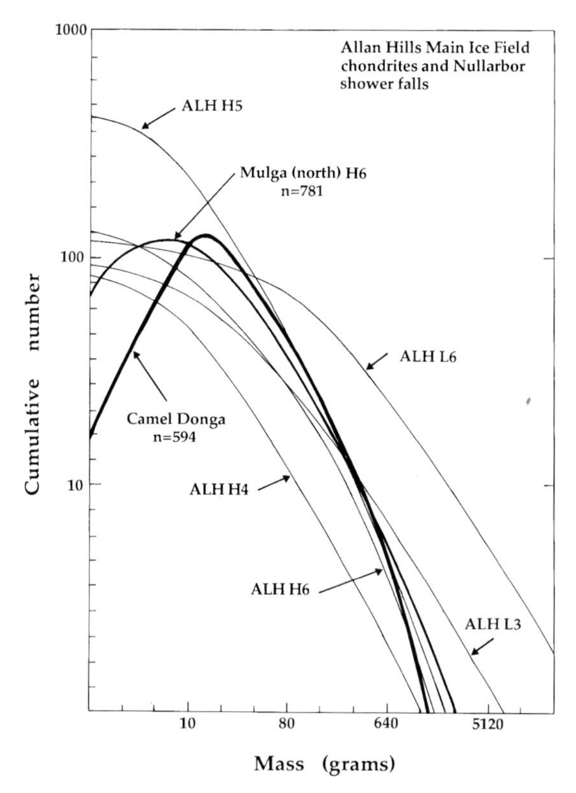

Fig. 6. Mass distribution curves derived from smoothed histograms of stony meteorite masses in populations of ordinary chondrites collected from the Allan Hills Main Ice Field (ALH) in Antarctica (after Huss 1991) compared with mass distribution curves for the Mulga (north) (H6) and Camel Donga (Aeuc) showers found on the Nullarbor Plain in Western Australia. The high mass portions of the curves from the Nullarbor showers that mimic the curves for ALH H5 and H6 collections indicate that unpaired shower falls are represented among these types of ordinary chondrites at this collection site, confirming the conclusions of Huss (1991).

Table 4. *Frequency (%) of meteorite groups in modern falls (MF), world (falls and finds), Australia, Antarctica, Nullarbor and the Sahara*

	MF	World	Australia	Antarctica	Nullarbor	Sahara
Chondrites	86.2	62.3	77.6	91.9	92.7	95.9
Ordinary	80.0	58.5	73.5	87.3	87.3	91.0
Carb. + anom.	4.6	2.9	3.4	3.9	4.6	4.2
Enstatite	1.6	0.9	0.7	0.7	0.8	0.7
Achondrites	7.9	5.4	4.5	6.3	5.4	1.5
Stony-irons	1.1	2.9	2.7	0.5	0.4	0.7
Irons	4.8	29.4	15.1	1.3	1.5	1.9
n	835	2474	468	3930	260	456

Data tabulated from Graham *et al.* (1985), Harvey & Cassidy (1989) and Bevan (1996). Data for the Sahara include the large number of meteorites from the Libyan and Algerian Sahara reported by Bischoff & Geiger (1995) and Grossman (1996), but also include historic recoveries (both falls and finds) from Libya, Algeria, Tunisia, Niger and Mauritania.

are the rarest types seen to fall, accounting for 4.8% and 1.1% of the population, respectively. When the data are recalculated to include a much larger sample of meteorites in collections throughout the world (both observed falls and chance finds, Table 4), whereas the proportions of chondrites (enstatite, carbonaceous and anomalous, and ordinary) and achondrites correspond well to those predicted by the frequency of modern falls, the irons and, to a lesser extent, the stony-irons, are over-represented. The main reasons for the disproportionately high numbers of iron and stony-iron meteorite finds are that the metallic meteorites are more resilient and generally survive longer in the terrestrial environment and, significantly, are more easily recognized as meteorites. The human bias in the collection of metallic meteorites is also seen in the total population of meteorites recovered from Australia, which shows a similar abundance (*c.* 15%) of irons (Table 4). However, taken as a discrete sub-sample, the constitution of the population of meteorites so far recovered from the Nullarbor Region is apparently different from modern falls and shows strong similarities to the population of meteorites from Antarctica (after Harvey & Cassidy 1989) and the Sahara. Statistical analysis by type of the described material from the entire Nullarbor shows that stony meteorites are abundant, with chondrites accounting for 92.7% of the total. Only 1.5% of distinct meteorites so far described from the Nullarbor are irons, which is much less than that (*c.* 5%) represented in the population of modern meteorite falls. As in Antarctica, the lack of irons from the Nullarbor and other desert areas is not easily explained and one possibility is that the deficiency in the Nullarbor may have resulted from human interference (Bevan & Bindon 1996).

In the Nullarbor population four irons (Mundrabilla, Haig, Sleeper Camp 002 and Watson) and one stony-iron (Rawlinna 001) are recorded. Applying the same analysis to the Nullarbor meteorite population as used by Graham & Annexstad (1989) for Antarctica, from the frequency of modern falls, assuming that four irons represent *c.* 5% and one stony-iron represents *c.* 2% of the total number of falls, then the predicted number of distinct meteorites in the Nullarbor population to date works out at 80 from the irons and 50 from the stony-irons. Both of these figures are less than a third of the currently recorded population (260) from the Nullarbor and, assuming that recovered historic falls are a representative sample of the meteorite flux, indicate that the metallic meteorites may be under-represented in the population by a factor of three or more. However, as the data from terrestrial ages and mass distribution of fragments indicate that there are few unpaired meteorites in the population, the alternative that the total population of distinct meteorites described from the Nullarbor has been over-estimated by a factor of three seems very unlikely.

Despite an apparent overall deficiency of irons in the Antarctic population, Wasson (1990) has noted that there is an anomalously high abundance of iron meteorites in the population that do not fit into the 13 well-defined chemical groups. Wasson (1990) attributed this anomaly to the median mass of Antarctic irons, which is about 0.01 of that of non-Antarctic irons. Excluding the numerous small fragments belonging to the Mundrabilla shower, only one distinct iron of small size has been recovered from the Nullarbor to date. An oxidized iron weighing 39.5 g found in 1966 and subsequently named Sleeper Camp 002 has yet to be classified. Interestingly, of the other three irons recovered from the Nullarbor, Mundrabilla is an anomalous ungrouped iron; Watson belongs to the rare group IIE (Olsen *et al.* 1994); and Haig is an unusual low-Ni, high-Ir end-member of the large, common group of irons IIIAB (Bevan *et al.* 1981).

When the metallic meteorites in the Nullarbor meteorite population are disregarded and the frequency of the stony meteorite groups (ordinary chondrites, carbonaceous and anomalous, and achondrites) is recalculated to 100%, then the proportions of each type of meteorite in the population agree remarkably well with similarly recalculated populations of modern falls, combined world falls and finds, and finds from Antarctica and other desert areas (Table 5). This indicates that the population of stones from each of these areas is probably a much better sample of the meteorite complex than populations containing all types of meteorites. Possibly,

Table 5. *Frequency (%) of stony meteorite groups in modern falls, world (falls and finds), Antarctica, Nullarbor and Sahara*

	Falls	World	Antarctica	Nullarbor	Sahara
Ordinary	86.4	87.5	89.4	89.8	94.1
Carb. + anom.	5.1	4.4	4.1	4.7	4.3
Achondrites	8.5	8.1	6.5	5.5	1.6
n	733	1653	3835	253	441
H/L	0.9	1.0	1.6	1.0	1.6
T_{age}	250 years	?	>1 Ma	>35 ka	40 ka

Data sources as for Table 4.

some of the human bias in the recognition of irons, or more correctly the general inability to recognize stones, is affecting the data for modern falls, and the actual percentage of irons in the total meteorite flux is less than 5%, and probably somewhere between the reported frequency in modern falls, and the populations accumulated over longer periods in deserts (1.5%).

As in Antarctica and the Sahara, the population of meteorites from the Nullarbor contains a relatively high proportion of rare, or anomalous types of chondrites. Carlisle Lakes, a previously anomalous and ungrouped chondritic meteorite find from the Nullarbor (Binns & Pooley 1979) belongs to an entirely new group of chondrites including an observed fall, Rumuruti, from Kenya (Schulze *et al.* 1994). The new group of chondrites, known as the 'R' group (after Rumuruti), includes several meteorites from Antarctica and two, Acfer 217 and Dar al Gani 013, from the deserts of North Africa (Rubin & Kallemeyn 1989, 1993, 1994; Bischoff *et al.* 1994; Schulze *et al.* 1994; Grossman 1996). Spettel *et al.* (1992) have suggested that Loongana 001, an unusual chondrite found in 1990 in the Western Australian Nullarbor, is related to the CR group of carbonaceous chondrites. However, Kallemeyn & Rubin (1995) have shown that the meteorite does not belong to any of the established carbonaceous chondrite groups and, along with Coolidge, forms a distinct grouplet of carbonaceous chondrites related to the CV group.

Six other rare carbonaceous chondrites, Mulga (west) (C5/6), Maralinga (CK4), Cook 003 (CK4), Sleeper Camp 006 (CK4), Watson 002 (CK3) and Camel Donga 003 (CK3) have been recovered from the Nullarbor. An additional crystalline CK4 chondrite, Karoonda, has also been recovered from South Australia, east of the Nullarbor and at more southerly latitudes. Karoonda, which fell on 25 November 1930, is the only observed fall of this type of meteorite and provides the CK group name (K for Karoonda). To date, some seven CK or CK-related chondrites have been recovered from the Nullarbor or near environs. A large number of meteorites similar to Karoonda have been recovered from Antarctica (e.g. see Grossman & Score 1996). However, the pairing problem prevents an estimate of the exact number of distinct CK chondrites found in Antarctica. The remaining non-Antarctic examples of CK4–6 chondrites are from southern Australia with only one, Ningqiang (reportedly CK-anom), known from the northern hemisphere. Currently, there are no known CK chondrites from other desert accumulation sites, although one meteorite from the Libyan Sahara, Dar al Gani 055, is reported to have affinities with CV and CK chondrites although its classification is uncertain (Grossman 1996). This apparent concentration of rare meteorite types in the southern hemisphere could be related to the terrestrial time-span of the populations of meteorites sampled in both Antarctica and the Nullarbor, their ease of recognition in those environments compared with the rest of the world, or possibly the product of meteoroid streams that have deposited fragments preferentially in the southern hemisphere as suggested by Kallemeyn *et al.* (1991).

In terms of rare achondrites, no SNC (martian) or lunar meteorites are known from the Nullarbor. Indeed, no martian meteorites are known from Australia as a whole. However, the first non-Antarctic example of a lunar meteorite, Calcalong Creek, was found in the central part of Western Australia (Hill *et al.* 1991). Recently, an additional non-Antarctic lunar meteorite, Dar al Gani 262, was found in the Libyan Sahara, becoming the first discovered in the northern hemisphere (Bischoff & Weber 1997). A new lodranite (an anomalous stony-iron meteorite of rare type) has recently been found in north-western Australia (Wlotzka 1992). The meteorite, named Gibson, is the first meteorite of its kind found in Australia and is one of only two meteorites of its kind recovered outside of Antarctica, the other being the type meteorite, Lodran, which fell in the nineteenth century in Pakistan.

The flux of meteorites with time

Estimates of the flux of meteorites with time derived from the accumulation of meteorites in hot deserts (Table 1) determined by Bland *et al.* (1996) bracket the estimate of the meteorite flux derived from the modern MORP project. Bland *et al.* (1996) have provided the first independent confirmation of the modern meteorite flux of 83 falls >10 g per 10^6 km^2 per year. The flux estimate derived from the Nullarbor population alone is lower at 36 falls >10 g per 10^6 km^2 per year, but the Nullarbor population remains small (260) and has been incompletely collected and described. To date, the Nullarbor data are probably an underestimate of the flux in that region. Collection statistics on the density of meteorites in the Nullarbor from small areas that have been thoroughly searched, such as the Camel Donga strewnfield, suggest that the flux may be $\geq$50 falls >10 g per 10^6 km^2 per year which is much closer to the MORP estimate and other desert areas. Nevertheless, the average meteorite flux derived from desert meteorite populations (combined Nullarbor, Sahara and

Roosevelt County; Table 1) from data provided by Bland *et al.* (1996) is identical to that derived from the MORP by Halliday *et al.* (1989, 1991). The implication is that the flux of meteorites to Earth does not appear to have changed significantly for at least the last 50 ka (Bland *et al.* 1996).

Summary and conclusions

The sample of probably distinct meteorites from the Nullarbor (260) for which classification data are available is similar to that from other desert sites, but an order of magnitude smaller than that from Antarctica. However, the total number of fragments recovered from the Nullarbor compares more closely with the total number of Antarctic fragments. A major problem with the Antarctic meteorite population is determining the actual number of falls that is represented. Whereas the number of different falls represented in Antarctica is uncertain, in the Nullarbor the general lack of transportation processes in the region and documentation of the distribution of finds has allowed 'pairing' of meteorites to at least 90% level of confidence, and most paired meteorites should have been recognized.

Many factors can influence meteorite collection statistics, not least of which is human interference or bias. Moreover, a good understanding of the geomorphological history of the area combined with terrestrial ages of the meteorites and the age of the surface on which the meteorites are found are fundamental to the accurate determination of the flux of meteorites from accumulations in arid areas (Bland *et al.* 1996).

In many respects, the characteristics (fall frequency and mass distribution) of the population of meteorites so far recovered from the Nullarbor are very similar to those of the Antarctic population and appear, superficially, to differ from modern falls. The high abundance of small (<50 g) stony meteorites, low percentage of iron meteorites, and abundance of rare and unique types of chondrite compared with modern falls characterize the Nullarbor and Antarctic populations. However, the Nullarbor meteorite population is only a sub-sample from the much larger catchment area of Australia as a whole. The mass range of stony meteorites found in other parts of Australia spans that found in the Nullarbor and also approaches that of modern falls. In the Nullarbor population, the mass distribution of the recovered material may only be an artefact of the method of searching, which has covered only a small fraction of the entire Nullarbor. In the Sahara, where most meteorites have been found recently by searching from vehicles, the mass distribution (Fig. 3) lies between that of the Nullarbor and modern falls (Bischoff & Geiger 1995).

The question arises: are the apparent differences in mass distribution and frequency of type between the desert meteorite accumulations and modern falls real, or are they simply the result of collection bias? The low frequency of irons in the desert populations is difficult to explain. In Antarctica, meteorites have been collected by experienced personnel, thus eliminating the possibility of any human bias in the collection. Graham & Annexstad (1989) suggested that, in Antarctica, the operation of some terrestrial process may have reduced the number of irons available for collection. However, the data of Nagata (1978) suggesting that irons with masses of 10^3 kg will sink around 100 m into the ice in 1 Ma indicate that their loss in Antarctica by this mechanism may be negligible. Another possibility is that irons are rarer as falls in Antarctica than in other parts of the world. In the Nullarbor, many meteorites have been found by inexperienced personnel and it would be expected that any bias would be towards more, rather than fewer, iron meteorites. A complicating factor is that the Nullarbor has been periodically populated by aboriginal people over at least the last 20 000 years, and by itinerant prospectors and rabbiters within the last 100 years. To date, there is no evidence anywhere in Australia to suggest that early Aborigines collected or utilized meteoritic iron, but the possibility of human interference with the population of meteorites from the Nullarbor cannot be discounted (Bevan & Bindon 1996). The removal of small irons from the population could account for their apparent deficiency in the Nullarbor. Alternatively, and perhaps more likely, the current deficiency of irons in the Nullarbor is a result of incomplete collection in the area. The curves for the mass distribution of meteorites so far recovered from the Nullarbor suggest that there are many more meteorites to be collected from the region.

Normalized mass frequencies and the proportions of stony meteorite groups suggest that there is little, if any, difference between the Nullarbor and Saharan desert meteorite populations, Antarctica and modern falls. The high ratios of H/L chondrites in both the Antarctic and Saharan populations (1.6) compared with modern falls (0.9) and the Nullarbor (1.0) are attributed to the presence of unpaired H group shower falls in the former populations. If differences exist that can be attributed to pre-terrestrial processes then they are subtle and probably related to the occurrence of rare meteorite types in the Nullarbor population and others. This problem may only be resolved when a much larger sample of meteorites has been collected from the Nullarbor, and noble gas and thermoluminescence measurements have also been made.

The authors thank D. West, P. Downes, J. Bevan and K. Brimmell for their assistance in the preparation of the manuscript. We are indebted to the late A. J. (John) Carlisle for data provided by him on the collection and field distribution of meteorites in the Nullarbor, and to the late W. H (Bill) Cleverly for data on the meteorites collected from the Mulga (north) strewnfield. The authors (A.W.R.B. and P.A.B.) acknowledge support from EUROMET for field operations in the Nullarbor in recent years, and particularly thank C. Pillinger and L. Schultz for their organization of the EUROMET contribution.

References

AYLMER, D., BONNANO, V., HERZOG, G. F., WEBER, H., KLEIN, J. & MIDDLETON, R. 1988. Al and Be production in iron meteorites. *Earth and Planetary Science Letters*, **88**, 107.

BENBOW, M. C. & HAYBALL, A. J. 1992. Geological observations of Old Homestead Cave, Nullarbor Plain, Western Australia. *Australian Caver*, **130**, 3–6.

BEVAN, A. W. R. 1992*a*. Australian meteorites. *Records of the Australian Museum, Supplement*, **15**, 1–27.

——1992*b*. 1992 WAMET/EUROMET joint expedition to search for meteorites in the Nullarbor Region, Western Australia (abstract). *Meteoritics*, **27**, 202–203.

——1996. Meteorites recovered from Australia. *Journal of the Royal Society of Western Australia*, **79**, 33–42.

—— & BINDON, P. 1996. Australian Aborigines and meteorites. *Records of the Western Australian Museum*, **18**, 93–101.

—— & BINNS, R. A. 1989*a*. Meteorites from the Nullarbor Region, Western Australia: I. A review of past recoveries and a procedure for naming new finds. *Meteoritics*, **24**, 127–133.

—— & ——1989*b*. Meteorites from the Nullarbor Region, Western Australia: II. Recovery and classification of 34 new meteorite finds from the Mundrabilla, Forrest, Reid and Deakin areas. *Meteoritics*, **24**, 134–141.

—— & PRING, A. 1993. Guidelines for the naming of new meteorite finds from the Nullarbor Region, South Australia. *Meteoritics*, **28**, 600–602.

——, KINDER, J. & AXON, H. J. 1981. Complex shock-induced Fe–Ni–S–Cr–C melts in the Haig (IIIA) iron meteorite. *Meteoritics*, **16**, 261–267.

BINNS, R. A. & POOLEY, G. D. 1979. Carlisle Lakes (a): a unique oxidized chondrite (abstract). *Meteoritics*, **14**, 349–350.

BISCHOFF, A. & GEIGER, T. 1995. Meteorites from the Sahara: find locations, shock classification, degree of weathering and pairing. *Meteoritics*, **30**, 113–122.

—— & WEBER, D. 1997. Dar al Gani 262: the first lunar meteorite from the Sahara. *Meteoritics and Planetary Science*, **32** (supplement), A13–A14.

——, GEIGER, T., PALME, H. *et al.* 1994. Acfer 217 – a new member of the Rumuruti chondrite group (R). *Meteoritics*, **29**, 264–274.

BLAND, P. A., SEXTON, A. S., JULL, A. J. T. *et al.* 1998. Climate and rock weathering: a study of terrestrial age dated ordinary chondritic meteorites from hot desert regions. *Geochimica et Cosmochimica Acta* (in press).

——, SMITH, T. B., JULL, A. J. T., BERRY, F. J., BEVAN, A. W. R., CLOUDT, S. & PILLINGER, C. T. 1996. The flux of meteorites to the Earth over the last 50 000 years. *Monthly Notices of the Royal Astronomical Society*, **283**, 551–565.

BUCHWALD, V. F. & CLARKE, R. S. JR 1989. Corrosion of Fe–Ni alloys by Cl-containing akaganeite (β-FeOOH): the Antarctic meteorite case. *American Mineralogist*, **74**, 656–667.

CASSIDY, W. A. & HARVEY, R. P. 1991. Are there real differences between Antarctic finds and modern falls meteorites? *Geochimica et Cosmochimica Acta*, **55**, 99–104.

CLEVERLY, W. H. 1972. Mulga (north) chondritic meteorite shower, Western Australia. *Journal of the Royal Society of Western Australia*, **55**, 115–128.

——1986. Further small recoveries of the Billygoat Donga and associated meteorites. *Records of the Western Australian Museum*, **12**, 403–406.

——1993. The role of the Western Australian School of Mines in meteorite recovery and research. *Western Australian School of Mines Annual Magazine 90 Years*, 59–65.

——, JAROSEWICH, E. & MASON, B. 1986. Camel Donga meteorite, a new eucrite from the Nullarbor Plain, Western Australia. *Meteoritics*, **21**, 263–269.

DE LAETER, J. R. 1972. The Mundrabilla meteorite shower. *Meteoritics*, **7**, 285–294.

—— & CLEVERLY, W. H. 1983. Further finds from the Mundrabilla meteorite shower. *Meteoritics*, **18**, 29–34.

DENNISON, J. E. LINGNER, D. W. & LIPSCHUTZ, M. E. 1986. Antarctic and non-Antarctic meteorites form different populations. *Nature*, **319**, 390–393.

DOHNANYI, J. S. 1972. Interplanetary objects in review: statistics of their masses and dynamics. *Icarus*, **17**, 1–48.

GILLIESON, D. S. & SPATE, A. P. 1992. The Nullarbor karst. *In*: GILLIESON, D. S. (ed.) *Special Publication 4, Department of Geography and Oceanography University College, Australian Defence Academy*, Canberra, 65–99.

GRAHAM, A. L. & ANNEXSTAD, J. O. 1989. Antarctic meteorites. *Antarctic Science*, **1**(1), 3–14.

——, BEVAN, A. W. R. & HUTCHISON, R. 1985. *Catalogue of Meteorites, with Special Reference to those Represented in the Collection of the British Museum (Natural History)*, 4th edn (revised and enlarged). British Museum (Natural History), London; University of Arizona Press, Tucson.

GROSSMAN, J. (ed.) 1996. Meteoritical Bulletin no. 80. *Meteoritics and Planetary Science*, **31**, A175–A180.

—— & SCORE, R. 1996. Recently classified specimens in the United States Antarctic meteorite collection (1994–1996). Meteoritical Bulletin no. 79. *Meteoritics and Planetary Science*, **31**, A161–A174.

HALLIDAY, I., BLACKWELL, A. T. & GRIFFIN, A. A. 1989. The flux of meteorites on the Earth's surface. *Meteoritics*, **24**, 173–178.

——, —— & ——1991. The frequency of meteorite falls: comments on two conflicting solutions to the problem. *Meteoritics*, **26**, 243–249.

HARVEY, R. P. & CASSIDY, W. A. 1989. A statistical comparison of Antarctic finds and modern falls: mass frequency distributions and relative abundance of type. *Meteoritics*, **24**, 9–14.

HILL, D. H., BOYNTON, W. V. & HAAG, R. A. 1991. A lunar meteorite found outside the Antarctic. *Nature*, **352**, 614–617.

HUGHES, D. W. 1981. Meteorite falls and finds: some statistics. *Meteoritics*, **16**, 269–281.

HUSS, G. R. 1990. Meteorite infall as a function of mass: implications for the accumulation of meteorites on Antarctic ice. *Meteoritics*, **16**, 41–56.

——1991. Meteorite mass distributions and differences between Antarctic and non-Antarctic meteorites. *Geochimica et Cosmochimica Acta*, **55**, 105–111.

IKEDA, Y. & KIMURA, M. 1992. Mass distribution of Antarctic ordinary chondrites and the estimation of the fall-to-specimen ratios. *Meteoritics*, **27**, 435–441.

JULL, A. J. T., BEVAN, A. W. R., CIELASZYK, E. & DONAHUE, D. J. 1995. *^{14}C terrestrial ages and weathering of meteorites from the Nullarbor Region, Western Australia.* Lunar and Planetary Institute Technical Report **95-02**, 37–38.

——, WLOTZKA, F., PALME, H. & DONAHUE, D. J. 1990. Distribution of terrestrial age and petrologic type of meteorites from western Libya. *Geochimica et Cosmochimica Acta*, **54**, 2895–2898.

KALLEMEYN, G. W. & RUBIN, A. E. 1995. Coolidge and Loongana 001: a new carbonaceous chondrite grouplet. *Meteoritics*, **30**, 20–27.

——, —— & WASSON, J. T. 1991. The compositional classification of chondrites: V. The Karoonda (CK) group of carbonaceous chondrites. *Geochimica et Cosmochimica Acta*, **55**, 881–892.

KOEBERL, C. & CASSIDY, W. A. 1991. Differences between Antarctic and non-Antarctic meteorites: an assessment. *Geochimica et Cosmochimica Acta*, **55**, 3–18.

——, DELISLE, G. & BEVAN, A. 1992. Meteorite aus der Wüste. *Geowissenschaften*, **10**, 220–225.

LIPSCHUTZ, M. E. & SAMUELS, S. M. 1991. Ordinary chondrites: multivariate statistical analysis of trace element contents. *Geochimica et Cosmochimica Acta*, **55**, 19–34.

LOWRY, D. C. 1970. *Geology of the Western Australian part of the Eucla Basin.* Geological Survey of Western Australia Bulletin, **122**.

—— & JENNINGS, J. N. 1974. The Nullarbor karst, Australia. *Zeitschrift für Geomorphologie*, **18**, 35–81.

MCCROSKY, R. E., POSEN, A., SCHWARTZ, G. & SHAO, G.-Y. 1971. Lost City meteorite – its recovery and comparison with other fireballs. *Journal of Geophysical Research*, **76**, 4090–4108.

MEHTA, R. D., BENTLY, K., PROUDLOVE, M. & VARTY, P. 1983. Factors affecting cricket ball swing. *Nature*, **303**, 787–788.

NAGATA, T. 1978. A possible mechanism of concentration of meteorites within the meteorite icefield in Antarctica. *Memoirs of the National Institute of Polar Research Special Issue*, **8**, 70–92.

NISHIIZUMI, K. 1990. Update on terrestrial ages of Antarctic meteorites. *In*: CASSIDY, W. A. & WHILLANS, I. M. (eds) *Workshop on Antarctic Meteorite Stranding Surfaces*. Lunar and Planetary Institute Technical Report **90-03**, 49–53.

OLSEN, E., DAVIS, A., CLARKE, R. S. JR *et al.* 1994. Watson: a new link in the IIE iron chain. *Meteoritics*, **29**, 200–213.

RUBIN, A. E. & KALLEMEYN, G. W. 1989. Carlisle Lakes and Allan Hills 85151: members of a new chondrite grouplet. *Geochimica et Cosmochimica Acta*, **53**, 3035–3044.

—— & ——1993. Carlisle Lakes chondrites: relationship to other chondrite groups. *Meteoritics*, **28**, 424–425.

—— & ——1994. Pecora Escarpment 91002: a member of the new Rumuruti (R) chondrite group. *Meteoritics*, **29**, 255–264.

SCHULTZ, L. 1986. Terrestrial ages of Antarctic meteorites: implications for concentration mechanism. *In*: ANNEXSTAD, J., SCHULTZ, L. & WANKE, H. (eds) *Workshop on Antarctic meteorites*. Lunar and Planetary Institute Technical Report **86-01**, 80–82.

SCHULZE, H., BISCHOFF, A., PALME, H., SPETTEL, B., DREIBUS, G. & OTTO, J. 1994. Mineralogy and chemistry of Rumuruti: the first meteorite fall of the new R chondrite group. *Meteoritics*, **29**, 275–286.

SPETTEL, B., PALME, H., WLOTZKA, F. & BISCHOFF, A. 1992. Chemical composition of carbonaceous chondrites from Sahara and Nullarbor Plains. *Meteoritics*, **27**, 290–291.

WASSON, J. T. 1974. *Meteorites, Classification and Properties*. Springer, Berlin.

——1990. Ungrouped iron meteorites in Antarctica: origin of anomalously high abundance. *Science*, **249**, 900–902.

WLOTZKA, F. (ed.) 1992. Meteoritical Bulletin no. 73. *Meteoritics*, **27**, 477–484.

ZBIK, M. 1994. The Cook 007 meteorite: a new H4 chondrite from South Australia. *Transactions of the Royal Society of South Australia*, **118**, 139–142.

ZOLENSKY, M. E., RENDELL, H. M., WILSON, I. & WELLS, G. L. 1992. The age of the meteorite recovery surfaces of Roosevelt County, New Mexico, USA. *Meteoritics*, **27**, 460–462.

——, WELLS, G. L. & RENDELL, H. M. 1990. The accumulation rate of falls at the Earth's surface: the view from Roosevelt County, New Mexico. *Meteoritics*, **25**, 11–17.

^{14}C terrestrial ages of meteorites from Victoria Land, Antarctica, and the infall rates of meteorites

A. J. T. JULL, S. CLOUDT & E. CIELASZYK

NSF Arizona Accelerator Mass Spectrometry Facility, The University of Arizona, 1118 East Fourth Street, Tucson, AZ 85721, USA

Abstract: The results of ^{14}C measurements of 95 meteorites from the Allan Hills region in Antarctica are reported, and terrestrial residence ages calculated. This includes meteorites from the different icefields at Allan Hills and the adjacent Elephant Moraine meteorite stranding area. We determined that terrestrial ages of these Antarctic meteorites can range from recent falls to >40 ka, which is the practical limit for these ^{14}C measurements. The terrestrial age determinations on meteorites from these sites can vary dramatically; the differences between the ages observed from these sites and some of the factors influencing them are discussed. Weathering products found on these meteorites show ^{14}C ages younger than the terrestrial age of the meteorites studied. Calculation of infall rates based on meteorites recovered and their age distributions suggests a minimum infall rate of 40–60 meteorites (>10 g) per 10^6 km^2 per year, in reasonable agreement with the infall rates estimated by Halliday's group based on meteoroid fluxes.

One of the most important collection areas for meteorites is Antarctica. Japanese, European and US search parties have recovered many samples from this continent. The US collection contains over 7800 meteorites (Marvin & Mason 1984; Marvin & MacPherson 1989, 1992; Grossman 1994). The cold and dry conditions allow the storage of meteorites with low rates of weathering and meteorite destruction. The storage time of a meteorite on the Earth's surface, its terrestrial age, can be determined by the decay of cosmic-ray-produced radionuclides. The radionuclide ^{14}C ($t_{1/2} = 5730$ years) is very useful in most meteorite collection areas of the Earth for determining these ages. Some early measurements on large samples (*c.* 100 g) were made by ^{14}C decay counting by Goel and Kohman (1962) and Suess & Wänke (1962). Similarly, Boeckl (1972) used ^{14}C to estimate terrestrial ages of US Prairie-State meteorites, using *c.* 10 g samples, and Fireman (1978) made measurements on Antarctic meteorites using this method. Brown *et al.* (1984) were the first to use accelerator mass spectrometry (AMS) to measure the ^{14}C terrestrial ages of Antarctic meteorites. Subsequent literature on ^{14}C terrestrial age measurements using smaller sample sizes (0.1–0.7 g) and AMS measurements has been summarized by Jull *et al.* (1984, 1989, 1990, 1991, 1993*a*, *b*, 1994*a*) and Cresswell *et al.* (1993). It should be noted that Jull *et al.* (1993*a*) revised some of the earlier measurements of Boeckl (1972). The longer-lived isotopes ^{81}Kr (Freundel *et al.* 1986; Miura *et al.* 1993) and ^{36}Cl (Nishiizumi *et al.* 1979, 1981, 1983, 1989) can determine longer terrestrial ages, beyond the useful range of ^{14}C of *c.* 40 ka. In the case of samples at the limit of ^{14}C age determination, we can sometimes place upper limits on their age by a lower limit determined by the ^{36}Cl age. Nishiizumi *et al.* (1989), Cresswell *et al.* (1993), Jull *et al.* (1993*b*) and Michlovich *et al.* (1995) have shown that the age distributions of meteorites at the Allan Hills and Yamato collection sites in Antarctica can be very different. Nishiizumi *et al.* (1989) reported that many meteorites from the Allan Hills Main Icefield have long terrestrial ages, as determined by ^{36}Cl ($t_{1/2} = 301$ ka). These workers pointed out that the observed ^{36}Cl age distribution may be biased towards older ages, because of selection of some meteorites with low ^{26}Al activity and iron meteorites having long ages. The Yamato site has been known for several years to have much younger meteorite falls (Beukens *et al.* 1988; Nishiizumi *et al.* 1989), although ^{81}Kr ages on eucrites show a wide distribution of ages from recent to 300 ka (Miura *et al.* 1993). In this paper, we investigate the ^{14}C terrestrial ages of a selection of meteorites from the blue-ice regions at Allan Hills and the nearby Elephant Moraine region (see Fig. 1). In this work, we attempted to avoid biasing the sample selection of meteorites towards older or younger terrestrial age. The only possible bias in this selection is two meteorites which had been previously studied by Fireman & Norris (1981), ALHA77256 and ALHA77294.

JULL, A. J. T., CLOUDT, S. & CIELASZYK, E. 1998. ^{14}C terrestrial ages of meteorites from Victoria Land, Antarctica, and the infall rates of meteorites *In*: GRADY, M. M., HUTCHISON, R., MCCALL, G. J. H. & ROTHERY, D. A. (eds) *Meteorites: Flux with Time and Impact Effects*. Geological Society, London, Special Publications, **140**, 75–91.

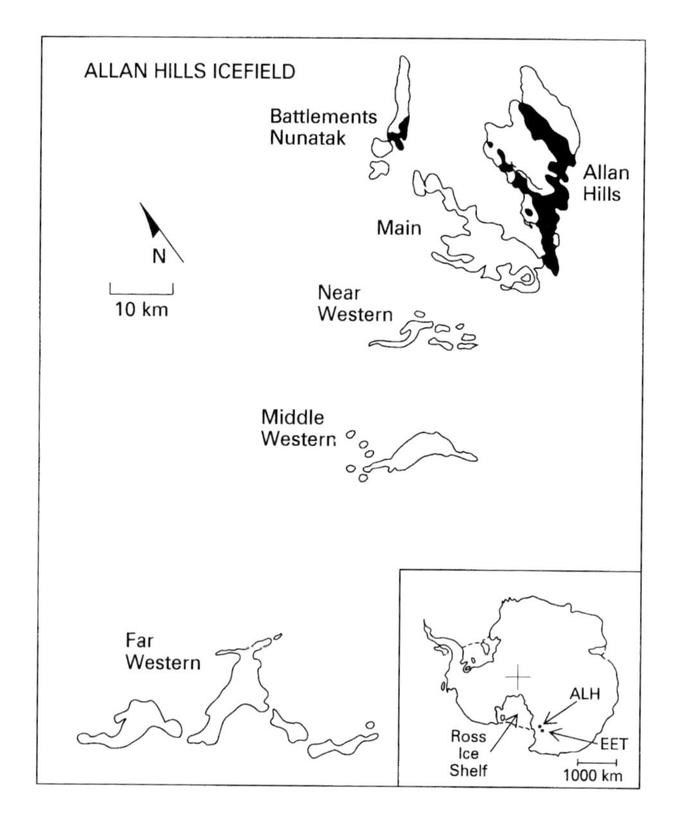

Fig. 1. Relative location of the Allan Hills icefields (from Annexstad & Annexstad 1989). Elephant Moraine lies *c.* 70 km to the north of the Far Western Icefield. The location of the Allan Hills and Elephant Moraine in Antarctica are shown in the inset map.

Production rate of ^{14}C and interpretation of ^{14}C terrestrial ages

The variation in ^{14}C production rate at different depths in meteorites of different sizes has been discussed by Jull *et al.* (1994*b*) and Wieler *et al.* (1996). Recent falls generally show activities of ^{14}C equivalent to a production rate of 38–58 atoms/min/kg (Jull *et al.* 1989, 1993*a*, *b*, 1994*b*). Wieler *et al.* (1996) showed calculations for meteorites of preatmospheric radii from 20 to 45 cm where the saturated activity (or production rate) should vary from about 38 to 52 dpm/kg for an H chondrite. Smaller objects have lower production rates of ^{14}C. In Fig. 2, we show the expected production rates for a sample recovered from a given depth for meteorites of different sizes (Wieler *et al.* 1996). Measurements on the Knyahinya L-chondrite (R (preatmospheric radius) = 45 cm) gave values of 37 at the surface to 58 dpm/kg at the centre of the meteorite. Nearly all ^{14}C is produced from spallation of oxygen, with only about 3% produced from silicon (Sisterson *et al.* 1994). Hence, normalization of the saturated activity observed to the oxygen content works well. We estimated the saturated activity for a given class of meteorite by normalization of the mean value of the ^{14}C content of Bruderheim (51.1 dpm/kg) to the oxygen content of the meteorite (Mason 1979). These values are listed in Table 1. The scatter in measurement on saturated falls suggests that an uncertainty of ±15% should be included in estimates of the terrestrial age to account for uncompensated shielding or depth effects, as well as experimental uncertainty (Jull *et al.* 1993*a*). This variation is confirmed by the study of the depth dependence of ^{14}C in the chondrite Knyahinya (Jull *et al.* 1994*b*), which has an estimated R of 45 cm. Samples from known meteorite falls such as Bruderheim (L6, mean of 51 dpm/kg, Jull *et al.* 1993*a*) Holbrook (H6, 44 ± 1 dpm/kg), Peekskill (L6, 51.1 ± 0.4 dpm/kg, Graf *et al.* 1996), Torino (H6, 42 ± 2 dpm/kg, Wieler *et al.* 1997) and Mbale (L6, 58.1 ± 0.4 dpm/kg) have ^{14}C values comparable with those observed for these two meteorites and with the values in Table 1.

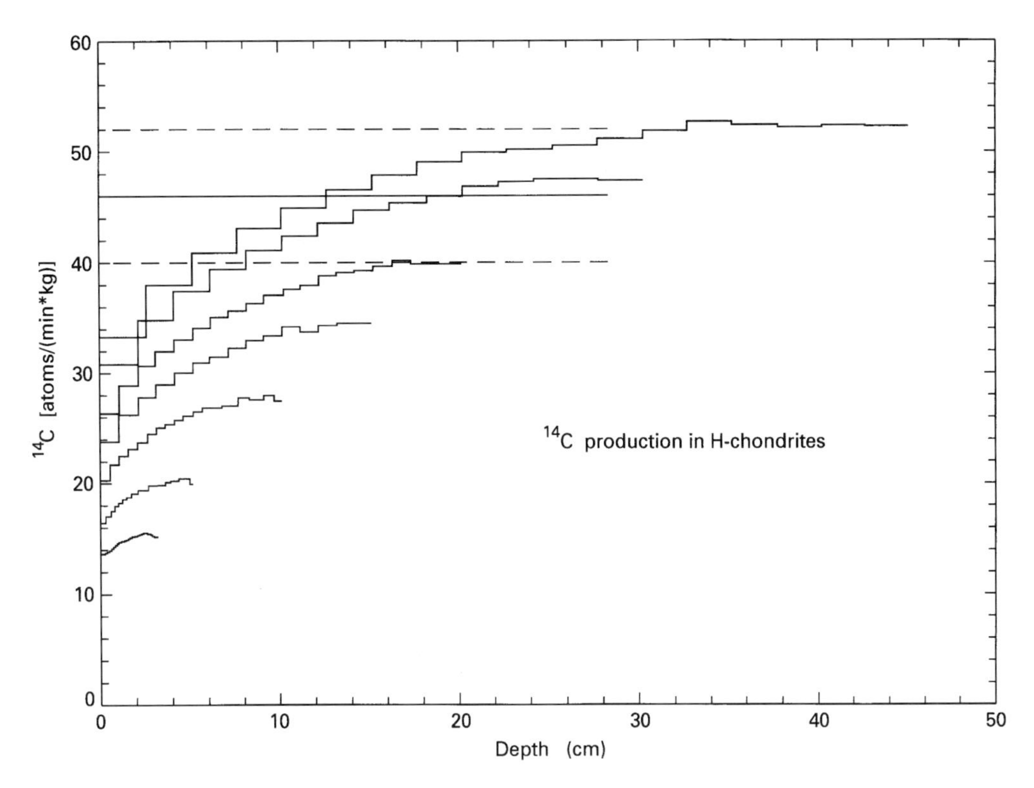

Fig. 2. Production rate of ^{14}C (dpm/kg) as a function of depth for meteorites of different preatmospheric radii, adapted from Wieler *et al.* (1996). The horizontal line indicates the average ^{14}C activity of 46 dpm/kg for H chondrites (Jull *et al.* 1989).

Because many of the meteorites collected from Antarctica are small, <100 g, compared with other finds, we may need additional criteria for determining the production rate of ^{14}C. This information is not available in all cases, but we use rare gas and other radioisotope data (e.g. ^{26}Al and ^{10}Be) to verify that the meteorite appears to have been irradiated as a body with a radius of 20–50 cm. We can also use $^{22}Ne/^{21}Ne$ ratios to estimate the shielding depth of the sample in a meteoroid; Schultz *et al.* (1996) have summarized the available Ne isotopic data. For apparently smaller objects, a lower saturated activity should be used. In many cases where we do not have sufficient information to make this determination, we will quote the result for the standard values listed in Table 1. We calculated the ^{14}C activities in dpm/kg and the terrestrial ages as described by Jull *et al.* (1993*a*).

There are two other possible sources of ^{14}C in meteorites. First, low levels of ^{14}C can be produced by cosmic-ray effects in the meteorite (on the Earth) while the meteorite is near the surface of the ice (Jull *et al.* 1992, 1994*c*; Lal & Jull 1994). We refer to this process as '*in situ* production'. This small, but measurable, effect is *c.* 0.1 dpm/kg or 1×10^6 atoms/g at the latitude and altitude of Allan Hills. Meteorites which apparently contain this low level of ^{14}C are quoted as >40 ka. Second, a different problem arises for those samples that have not been pretreated with acid etching, which give ^{14}C results below about 2 dpm/kg. Samples in this class will include some values which are actually finite ages, and some samples which are really limit ages, but which are contaminated with more recent terrestrial ^{14}C as a result of some weathering products or organic compounds. Karlsson *et al.* (1991) have shown that meteorites often contain weathering products which contain terrestrial carbon younger than the terrestrial-residence age of the meteorite. Levels of about 1.3 dpm/kg could be explained by about 100 ppm modern terrestrial carbon in an old meteorite. Thus, in ordinary chondrites, we can use high levels of carbon as a criterion for assigning apparent 'ages' in the 30–40 ka range to our limit of >40 ka. To reduce the

Table 1. *Calculated saturated activities of ^{14}C in different classes of meteorites of radius 20–50 cm*

Class	Oxygen* (%)	Saturated activity†
CI	42.5	60.2
CM	41.6	59.0 (total sample combustion)
		74.0 (acid-etched)‡
CV	37.0	52.4
CO	35.7	50.6
H	32.8	46.4
L	36.1	51.1
LL	39.0	55.2
E	28.9	41.0
Howardite	43.5	61.6
Diogenite	43.1	61.1
Eucrite	42.6	60.3
Angrite	41.6	59.0
Aubrite	46.3	65.6
Ureilite	39.5	56.0
Shergottite	42.5	61.1
Lunar	45.3	65.2

* Average oxygen content of this class, from Mason (1979).
† This is an average value. As can be seen in Fig. 2, this value has an error of ±15%.
‡ Combustion of acid-etched Murchison (CM2) gives an activity of 74 dpm/kg.

ambiguities for low ^{14}C samples, where possible the meteorite samples are pretreated by an acid-etching step discussed in the experimental section to remove 'weathering' carbonates. Acid-etching of a sample of ALHA78132, shown to contain no ^{14}C in an earlier study (Jull *et al.* 1989), indicated that no additional ^{14}C (0.1 ± 0.3 dpm/kg) was introduced during this procedure.

Age distributions

The distribution of meteorite terrestrial ages can be easily understood in the case of a collection where all meteorites fell directly on the collection area. The meteorites then should eventually disintegrate because of weathering processes and, therefore, the number of meteorites will decrease with increasing age, so there should be more young meteorites than older ones. Meteorites collected from US Prairie States (Jull *et al.* 1993*a*), Libya (Jull *et al.* 1990), Algeria (Wlotzka *et al.* 1994) and Western Australia (Jull *et al.* 1994*a*) appear to obey a first-order exponential decay of number of meteorites with increasing terrestrial age. In this simple case, the number of meteorites per unit time is proportional to

$$\mathrm{d}N_a/\mathrm{d}t = a\,\mathrm{e}^{bt}$$

where N_a is the number of meteorites remaining at time t, a is the infall rate per unit of time, $\mathrm{d}t$, and b is the weathering decay constant for meteorites, assuming a first-order rate law.

However, there are sites such as Roosevelt County, New Mexico (Jull *et al.* 1991) which do not follow this relationship. If there is some transport of the meteorites such as in Antarctic ice, then the age distribution will be distorted. The model of Freundel *et al.* (1986) assumed the age distribution would be shifted to older ages, but transport of meteorites over long distances could have complex results. However, older meteorites could survive longer during transport in an ice flow as proposed by Freundel *et al.* (1986). The total number of meteorites surviving will also be correspondingly lower. Therefore, in most cases it will be difficult to assess the parameters unless a large number of ages are available. The infall rate of meteorites should be a constant, but there is a mass distribution of meteorite falls as has been discussed by Halliday *et al.* (1989). These researchers defined a mass-number scale such that

$$\log N_{\mathrm{i}} = -c \log m_{\mathrm{T}} + d$$

where c is 0.49 and d is 2.41 for falls of <1030 g. For >1030 g, c is 0.82 and d is 3.40. Using this equation, one can calculate that for masses >10 g there should be about 83 meteorites/km^2 per 10^6 years, or about 0.8 meteorites per km^2 per 10^4 years. Huss (1990) has argued for higher infall rates on the basis of observed concentrations at Antarctic collection sites.

Experimental

Some Antarctic meteorite samples are known to contain carbonate weathering products (e.g. Jull *et al.* 1988). For samples where the presence of weathering products was expected, we performed an initial pretreatment to remove weathering carbonates and evaporites. These samples (of about 0.2–0.6 g) were crushed and were initially treated with 85% or 100% H_3PO_4, washed with distilled water and dried. For the second step, the residue was mixed with about 3 g of iron chips (used to enhance combustion) and preheated in air at 500°C for 1 h. For meteorites which were not acid-etched, samples of about 0.2–0.6 g were crushed, mixed with the iron chips and also preheated in air at 500°C for 1 h. Finally, the samples were loaded into an RF induction furnace (Leco Corporation, model HF-10), and heated in a flow of oxygen up to about 1700°C. The rock–iron mixture was completely fused. The gases evolved were passed over MnO_2 and CuO/Pt (Jull *et al.* 1989, 1993*a*) and collected at −196°C, and then excess oxygen was removed. Carbon dioxide was separated from water at −78°C. The CO_2 volume was measured in a known volume using a capacitance manometer, and diluted to 1–2 cm^3 STP with ^{14}C-free CO_2. The gas was then reduced to a graphite, which was subsequently pressed into an accelerator target, and 24 targets were mounted in the accelerator ion source, along with eight standards of graphite prepared from NIST standards. From the isotope ratio of the graphite measured by AMS, the ^{14}C content of the meteorite in atoms/g (or dpm/kg) can be calculated as discussed by Jull *et al.* (1993*a*).

Results and discussion

We present the results of the ^{14}C determinations, and calculated terrestrial ages on a number of meteorites from several Antarctic sites in Table 2. Some published measurements are included for completeness and the samples are grouped depending on the collection site. The ^{14}C content is quoted as dpm/kg and the calculated terrestrial ages using the saturated activities from Table 1 are shown. In specific cases, where there is evidence of irradiation as a very small object in space, the value is in italics. The results show that various sites display different distributions of terrestrial ages, as determined by ^{14}C content. The significance of these results to the sites studied is discussed in the following sections.

Allan Hills Main Icefield

Measurements on seven meteorites collected from the Allan Hills Main Icefield are shown in the first part of Table 2. Four of these samples gave finite ^{14}C ages. Indeed, these four samples are the only ones measured by ^{14}C or ^{36}Cl which have ages clearly <30 ka. Two of these samples were selected for comparison with the earlier work of Fireman & Norris (1981). The result on ALHA77256, a diogenite collected from the centre of the Main Icefield, gives a ^{14}C content of 17.5 ± 0.4 dpm/kg and a terrestrial age of about 10.3 ka, in good agreement with the value measured by Fireman & Norris (1981). The H5 chondrite ALHA77294 is the only meteorite collected from Main Haul Bay to the east of the Main Icefield. This sample, which gives a terrestrial age of about 16.5 ka, has also been measured by Fireman & Norris (1981), though these workers obtained an age of about 29–30 ka. The reason for this discrepancy has so far not been explained. The carbon content we measured was low (86 ppm) so that incomplete removal of weathering products cannot explain the discrepancy. Additionally, shielding cannot explain this large difference in age measurement, as the ^{14}C content differs by a factor of four, which would require a difference in depth of the two samples of about 220 g/cm^2 As this meteorite is only about 400 g, this is clearly not an explanation. Ninagawa *et al.* (1983) estimated a terrestrial age of between 7 and 25 ka from thermoluminescence, which would not be inconsistent with our measurement. However, estimates of terrestrial age by thermoluminescence involve assumption of an average temperature (Benoit *et al.* 1993) and this result must be assumed to have large errors.

Turning to the other samples, the eucritic breccia, ALH81011, gives a limit age as a result of *in situ* production, similar to that observed in other samples. The H5 chondrite, ALHA77262, gives an age of 18 ka, which is in marked contrast to the old ages usually associated with the ordinary chondrites at Allan Hills. This is the only ordinary chondrite from the Allan Hills Main Icefield, except ALHA77294 found east of the Main Icefield, which has a terrestrial age of less than 30 ka.

Because of their limited number, it is relatively easy to understand systematics for achondrites. We studied three ureilites, ALHA77257, ALHA78109 and ALHA81101, that were found in the Main Icefield (Marvin 1992). They have been studied in conjunction with other ureilites, such as Pecora Escarpment 82502, by Takeda *et al.* (1988). The ^{14}C content of ALHA77257 suggests that it contains only ^{14}C produced by cosmic rays *in situ* while the meteorite was on the surface of the ice, a conclusion consistent with the long ^{36}Cl terrestrial age of 360 ± 90 ka measured by Nishiizumi *et al.* (1981). By contrast, the ureilites ALHA78019 and ALHA81101 have ^{14}C ages of about 12 ka and 7.2 ka, respectively. The ALHA81101 meteorite was collected from an area near ALHA77257, but was classified as a different fall by Marvin (1989) on petrological grounds. The ^{14}C data confirm this interpretation. Sample ALHA77257 must then be a distinct fall from ALHA81101 or ALHA78109. The ureilite finds in the Far Western Icefield give interesting results (discussed later): ALH82130 contains 11.7 dpm/kg (i.e. about 13 ka age), that is, similar to ALHA78019 (13.1 dpm/kg) and ALH84136 (14.5 dpm/kg). The Ne-isotope data for ALH82130 indicate it must have been in a near-surface location or a small object (Schultz *et al.* 1996). The ureilite ALH82106 contains 22.4 dpm/kg, similar to the 21.4 dpm/kg found for ALHA81101. It is possible to construct a model with ALH84136 as part of the same meteoroid as ALH82106 and ALHA81101, if ALH82106 and ALH84136 are

Table 2. *^{14}C terrestrial age measurements on Allan Hills meteorites*

Sample	Type	Weight* recovered (g)	Weathering†	Acid etch	C (ppm)	^{14}C (dpm/kg)	Terrestrial age (ka)	^{26}Al‡ (dpm/kg)	^{22}Ne/^{21}Ne§
Allan Hills Main Icefield									
ALHA77256,58	Diogenite	676	A–B	–	83	17.5 ± 0.4	10.3 ± 1.3		1.15
ALHA77257,88	Ureilite	1996	Ae	–	2.4%	< 0.27	>44		1.09–1.14
ALHA77262,42	H4	862	B–Ce	–	95	5.3 ± 0.1	18.0 ± 1.3	47 ± 5	1.11
ALHA77294,38	H5	1351	Ae	–	86	6.8 ± 0.2	16.5 ± 1.3	61 ± 4	1.09
ALHA78019	Ureilite	30	B–C	–	3.7%	13.1 ± 0.9	12.0 ± 1.3		
ALHA81011,60	Eucrite breccia	406	A–B	–	86	0.3 ± 0.1	>41		1.22
ALHA81101	Ureilite	119	A–B	–	0.55%	23.3 ± 0.3	7.2 ± 1.3	35 ± 2	1.04–1.06
ALH88029	H5	29	–	–	506	35.3 ± 0.5	2.3 ± 1.3		
ALH88047	H6	21	–	–	557	44.5 ± 0.6	0.3 ± 1.3		
Near Western Icefield									
ALH77295,27	EH4a	141	B	H_3PO_4	0.10%	35.5 ± 0.4	2.2 ± 1.3	57 ± 4	1.09
ALH77295,47	EH4a	141	B	H_3PO_4	0.23%	31.1 ± 0.3	3.3 ± 1.3		
ALH81015	H5	5489	Be	–	374	2.7 ± 0.1	23.4 ± 1.3		
ALH 81016	L6	3850	Be	H_3PO_4	153	23.7 ± 0.2	6.4 ± 1.3		
ALH81039	H5	206	A–B	–	1015	16.2 ± 0.2	8.7 ± 1.3	56 ± 2	1.14
ALH81041	H4	721	C	H_3PO_4	89	<0.43	>39	52 ± 2	
ALH81099	L6	152	A–B	H_3PO_4	126	0.25 ± 0.14	>40	80 ± 4	
ALH81104	H4	184	C	H_3PO_4	161	0.40 ± 0.11	39 ± 3	57 ± 2	
ALH81119	L4	107	B	H_3PO_4	83	<0.18	>47	36 ± 2	
ALH83001	L4	1569	B	H_3PO_4	29	<0.17	>47	40 ± 1	
ALH84109	H6	246	B–C	H_3PO_4	120	<0.55	>37		
Middle Western Icefield									
ALHA81005	Lunar	31	A	–	153	7.13 ± 0.11	8.3 ± 1.3		
ALH81017	L5	1434	B	–	185	1.27 ± 0.24	30.5 ± 2.0		1.09
ALH81018	L5	2237	B	–	217	1.74 ± 0.09	27.9 ± 1.3		
				H_3PO_4	420	1.56 ± 0.18	28.9 ± 1.6		
				H_3PO_4	138	0.25 ± 0.11	>44		
ALH81022	H4	913	B–C	–	220	0.7 ± 0.2		35 ± 3	
ALH81023	L5	418	B	–	189	1.1 ± 0.2		32 ± 2	
				H_3PO_4	212	1.1 ± 0.2		32 ± 2	
ALH81111	H6	210	B–C	–	174	1.65 ± 0.20	28 ± 2	>610	
				H_3PO_4	139	0.12 ± 0.21	>41		
ALH83004	L6	814	B	–	221	1.24 ± 0.16	31 ± 2	>300	
				H_3PO_4	328	0.73 ± 0.30	>35 ± 4		

ALH83010	L3	395	B	–	1908	2.9 ± 0.1	24.6 ± 1.4		1.12
				H_3PO_4	1812	0.6 ± 0.1	36 ± 2		
ALH83029	H5	96	B–C	–	229	0.88 ± 0.21	33 ± 2		
				H_3PO_4	174	0.57 ± 0.16	>37 ± 3		
ALH83070	LL6	216	A	H_3PO_4	117	41.8 ± 0.3	2.3 ± 1.3		1.11
ALH84007	AUB	706	Ae	–	225	46.4 ± 0.3	2.4 ± 1.3		1.07
				–	249	37.4 ± 0.1	4.6 ± 1.3		
ALH84096	C4	294	A–B	–	0.11%	49.0 ± 0.5	0.55 ± 1.3	<200	
Far Western Icefield									
ALH82101	CO3.4	29	A	–	755	6.0 ± 0.6	16.9 ± 1.5		1.16
ALH82102	H5	48	B–C	–	148	13.9 ± 0.3	11 ± 1.5		1.17
ALH82104	L5	99	A	–	520	19.1 ± 0.1	8.1 ± 1.3	62 ± 2	
ALH82105	L6	363	A–B	–	113	1.89 ± 0.05	27.3 ± 1.3	45 ± 2	
ALH82106	Ure	35	B–C	–	1.86%	22.4 ± 1.5	7.1 ± 1.4	63 ± 6	
ALH82110	H3	39	B–C	–	0.32%	33.9 ± 0.4	2.6 ± 1.3		
ALH82118	H	111	A–B	–	–	7.1 ± 0.5	21‖	59 ± 3	
ALH82123	L6	111	B	–	297	43.1 ± 0.25	1.4 ± 1.3		
ALH82125	L6	178	C	–	77	0.15 ± 0.07	48 ± 4	36 ± 2	
ALH82130	URE	45	B	–	2.25%	11.7 ± 0.15	13.0 ± 1.3	62 ± 5	1.18–1.19
ALH83100	CM2	3019	Be	–	0.36%	10.8 ± 0.8	11.6 ± 1.4		
ALH83101	L6	639	A	–	53	3.5 ± 0.2	21.6 ± 1.4		
ALH83102	CM2	1786	B–Ce	–	0.47%	13.7 ± 0.2	9.5 ± 1.5		
ALH83108	C3O	1519	A	–	456	1.75 ± 0.16	29.6 ± 1.5		
ALH84001	SNC	1931	A–B	–	186	14.7 ± 0.2	11.3 ± 1		1.19–1.23
				–	243	11.8 ± 0.1	3.6 ± 1.3		
				H_3PO_4	216	11.3 ± 0.2	13.9 ± 1.3		
ALH84004	H4	9000		H_3PO_4	182	2.2 ± 0.9	25 ± 4		
ALH84005	L5	12000	A–B	–	73	0.75 ± 0.14	34 ± 2		
ALH84029	C2	120	Ae	total	1.35%	19.3 ± 1.3	15.9 ± 1.4		
				total	1.9%	32.7 ± 0.7	11.5 ± 1.3		
				>500 C	0.28%	5.23 ± 0.44	17.6 ± 1.5		
ALH84058	L6	2003	B	–	94	0.17 ± 0.07	>47 ± 4		
ALH84061	L6	676	B	–	61	0.31 ± 0.05	>42 ± 2	46 ± 2	
ALH84074	H5	758g	A–B	–	68	3.23 ± 0.06	21.6 ± 1.3	33 ± 1	1.27
				H_3PO_4	(632)	4.11 ± 0.09	20.0 ± 1.3		
ALH84076	H5	369	B–C	–	76	0.38 ± 0.05	39.3 ± 1.7	64 ± 3	
ALH84077	H5	276g	B	–	59	0.29 ± 0.06	41.5 ± 2.1	45 ± 2	
ALH84078	H5	283	B–C	–	82	1.32 ± 0.07	29.0 ± 1.4	65 ± 3	
				H_3PO_4	(1078)	3.6 ± 1.2	21 ± 3¶		
ALH84081	LL	612	A	–	83	0.67 ± 0.06	36.5 ± 1.4	46 ± 2	
ALH84082	H5	557	C	–	251	2.7 ± 0.2	23.1 ± 1.5	40 ± 2	
ALH84083	H6			H_3PO_4	117				

Table 2. (*continued*)

Sample	Type	Weight* recovered (g)	Weathering†	Acid etch	C (ppm)	^{14}C (dpm/kg)	Terrestrial age (ka)	^{26}Al‡ (dpm/kg)	$^{22}Ne/^{21}Ne$§
Far Western Icefield									
ALH84084	H4	332g	B	–	170	0.41 ± 0.13	38.6 ± 2.4	59 ± 2	
				–	170	0.70 ± 0.13	34.2 ± 2.0		
ALH84086	LL3	234g	A–B	–	236	12.5 ± 0.16	12.3 ± 1.3	56 ± 2	
ALH84091	H5	215	B–C	–	202	1.33 ± 0.05	29.4 ± 1.3	50 ± 2	
				H_3PO_4	165	0.20 ± 0.22	>39		
ALH84095	L6	277	A–B	–	59	0.56 ± 0.07	37.0 ± 1.6	33 ± 2	
ALH84104	L6	201	B	H_3PO_4	123	0.28 ± 0.43	>35	45 ± 3	
ALH84106	H6	95	B–C	H_3PO_4	238	<0.72	>35	61 ± 4	
ALH84107	LL6	134	A	–	196	39.9 ± 0.3	2.1 ± 1.0		
ALH84109	H6	249	B–C	H_3PO_4	96	<0.94	>35	59 ± 3	
ALH84132	L6	158	B	–	239	2.25 ± 0.07	25.8 ± 1.3	62 ± 4	
				H_3PO_4	494	0.84 ± 0.78	34 ± 7		
ALH84136	Ureilite	84	B	–	1.47%	14.5 ± 0.9	10.7 ± 1.4		
ALH84140	L6	164	C	–	205	1.0 ± 0.7	>28	45 ± 3	
					169	0.80 ± 0.08	34.0 ± 1.5		
				H_3PO_4		0.82 ± 0.21	34 ± 3		
				H_3PO_4	266	<0.62	>36		
ALH84142	L6	78	A–B	–		1.53 ± 0.11	29.0 ± 1.4	54 ± 4	
				H_3PO_4	329	0.82 ± 0.35	34 ± 4		
ALH84153	H6	243	B–C	–	248	13.4 ± 0.2	9.8 ± 1.3		
				–	120	14.4 ± 0.2	9.2 ± 1.3		
ALH84157	H5	89	B–C	–	328	2.5 ± 0.3	24.1 ± 1.7		
				H_3PO_4	232	0.73 ± 0.29	35 ± 4		
ALH85001	EUC	212	A–B	–	912	44.9 ± 0.6	2.5 ± 1.3		
ALH85002	CK4	438	A	–	675	21.5 ± 0.5	8.3 ± 1.3	38.5 ± 1.5	
ALH85007	C2	82	B	–	0.41%	3.20 ± 0.37	24.1 ± 1.6		
				H_3PO_4	0.89%	1.2 ± 2.4 (s)	>25		
ALH85013	C2	130	A	–	0.38%	10.6 ± 0.3	14.2 ± 1.3	34 ± 3	
ALH85019	LL6	633	A	–	90	5.25 ± 0.09	19.5 ± 1.3	69 ± 5	
ALH85022	L6	952	B	–	227	4.08 ± 0.14	20.9 ± 1.3	47 ± 4	
ALH85026	L6	817	A	–	169	15.6 ± 0.6	9.9 ± 1.3		
				–	158	10.2 ± 0.2	13.3 ± 1.3		
ALH85032	H6	424	C	–	157	17.2 ± 0.2	8.2 ± 1.3	56 ± 4	
ALH85033	L4	250	A	–	162	1.43 ± 0.11	29 ± 2		
				H_3PO_4	39	0.77 ± 0.05	34 ± 2		
ALH85036	H6	231	C	–	366	6.07 ± 0.08	16.8 ± 1.3		

ALH85038	H5	125			336	21.0 ± 0.4	6.5 ± 1.3	49 ± 4
ALH85040	L6	96	B	–	149	17.9 ± 0.1	8.7 ± 1.3	49 ± 3
ALH85062	L3	167	B–Ce	–	159	12.8 ± 0.1	11.5 ± 1.3	38 ± 2
ALH85095	L6	33	B	–	234	8.0 ± 0.2	15.3 ± 1.3	
ALH85130	L6	100	B	–	415	31.1 ± 0.2	4.1 ± 1.3	56 ± 5
Elephant Moraine								
EET83213	L3	2727	B	–	297	0.45 ± 0.3	36 ± 3	
EET83225	Ureilite	44	B–C	–	1.48%	19.8 ± 0.4	8.6 ± 1.3	
EET83228	Eucrite	1206	B	–	23	0.36 ± 0.04	42 ± 2	
EET83235	Eucrite	255	B	–	59	0.14 ± 0.06	>47	
EET87517	Ureilite	273	B–C	–	0.78%	4.22 ± 0.16	21.4 ± 1.3	
EET87521	Lunar	31	A	–		0.32 ± 0.09	>42	
EET87536	L6	7526	B–C	–	145	1.12 ± 0.05	31.6 ± 1.4	
EET87549	L6	539	Be	–	276	2.33 ± 0.08	5.5 ± 1.3	
EET87554	L6	1294	B	–	102	10.4 ± 0.1	13.2 ± 1.3	
EET87556	L6	363	A	–		0.70 ± 0.03	35.5 ± 1.3	
				H_3PO_4	453	1.2 ± 0.9¶	>26	
EET87566	L6	388	C	–		0.69 ± 0.04	35.6 ± 1.4	
EET87578	L6	350	A	–		1.24 ± 0.05	30.8 ± 1.4	
EET87720	Urelite	91	Be	–	1.98%	3.13 ± 0.6	23.8 ± 2.0	

* Recovered weight in grams reported by the NSF Antarctic Search for Meteorites (ANSMET).
† Weathering classification from ANSMET. A, light weathering; B, moderate; C, severe weathering; e, evaporite deposits.
‡ ^{26}Al data from Evans *et al.* (1992), Wacker (1993, 1995).
¶ Small sample.
|| E. L. Fireman, unpublished result.
§ Ne-isotope data from compilation of Schultz *et al.* (1995).

near-surface samples, compared with the interior ALH82106 and ALH81101. Rare-gas data for ALH81101 suggest it was in the interior of a relatively large object, with $^{22}Ne/^{21}Ne$ of *c.* 1.05.

Two small meteorites collected by a German–US team in the 1988–1989 field season (Wlotzka 1991) have also been studied. These samples, ALH88029 (H5) and ALH88047 (H6) weighed only 29 and 21 g and gave young terrestrial ages of 2.3 and 0.3 ± 1.3 ka, respectively.

Figure 3 shows a histogram of meteorite terrestrial ages for Main Icefield samples, based on published ^{14}C, ^{36}Cl and ^{81}Kr measurements (Nishiizumi *et al.* 1989) and the data reported here. The diagram of all terrestrial-age data for Allan Hills shows an abundance of old samples, beyond the range of ^{14}C. The four young ages show that there are some more recent falls at the Allan Hills Main Icefield. With two exceptions, all the young meteorites are achondrites. This is in marked contrast to the expected distribution for a field of about 50 km^2, if meteorites all accumulated by direct infall. There must be some error in this estimate of icefield size, as the area of blue ice changes with time (Annexstad 1983), but this is the best estimate available. Because of the age distribution, most meteorites must have been transported a considerable distance under the ice, as suggested by Whillans & Cassidy (1983). The mean age of this transport has been estimated by Freundel *et al.* (1986) to be about 150 ka. The number of directly falling meteorites can be estimated from the meteorites of <20 ka age. We have shown that eight out of 85 meteorites at Allan Hills have low terrestrial ages. About 532 meteorites >10 g (up to 1989) have been collected from the Main Icefield. Thus, we can estimate a minimum infall rate for direct falls only, assuming that not all samples have been collected. This gives us an estimate of *c.* 50 meteorites in the area of the Main Icefield (50 km^2) with ages <20 ka, and an estimated infall rate of *c.* 50×10^{-6} meteorites/km^2 in the last 20 ka based on these meteorites. This is about 60% of the total infall rate of Halliday *et al.* (1989) of 83.2×10^{-6} meteorites/km^2 per year. This difference could be due to other biases in the collection, such as towards larger meteorites of >30 g, but we should also recall that these workers' estimate of infall might be in error by as much as 50% (Halliday *et al.* 1989). However, the number of direct falls actually measured is small. The young ages of the 1988 samples (Wlotzka 1991) suggest that if a truly random sample of Main Icefield meteorites were taken, a larger number of low terrestrial age meteorites might result than expected solely from the summary of Nishiizumi *et al.* (1989), giving a value more in agreement with the infall rate of Halliday *et al.* (1989).

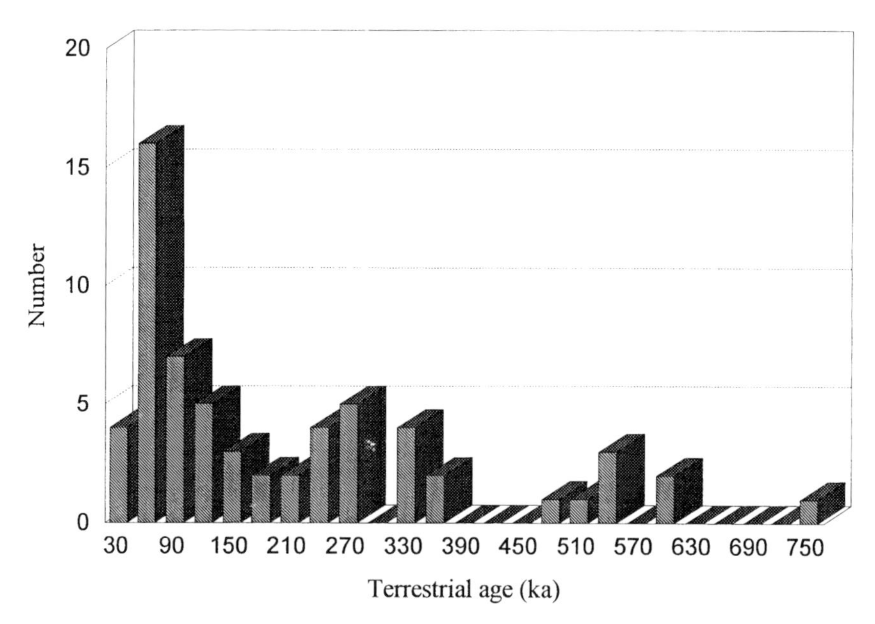

Fig. 3. Terrestrial age distribution of meteorites from the Allan Hills Main Icefield. Data from this work and Nishiizumi *et al.* (1989) are shown. These data include ^{14}C, ^{36}Cl and ^{81}Kr age estimates.

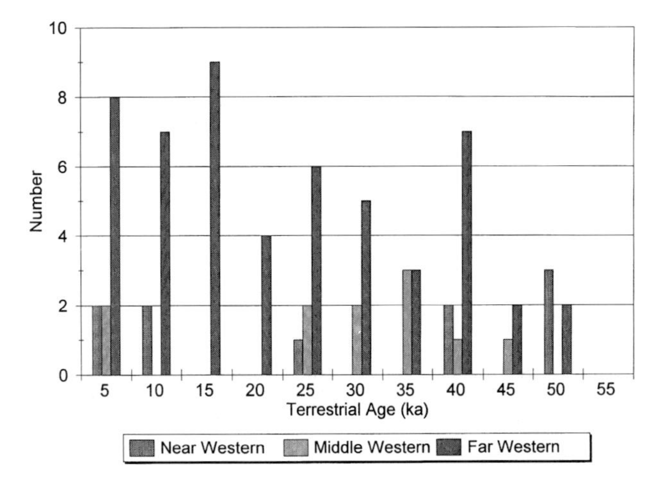

Fig. 4. Terrestrial ^{14}C age distributions of meteorites from the Allan Hills Far, Middle and Near Western Icefields.

Near Western Icefield

We report radiocarbon terrestrial ages for ten ordinary chondrites from the Near Western Icefield in Table 2 and also in Fig. 4. Many Near Western meteorites studied were acid-cleaned before ^{14}C extraction. These results show a significant range of ages, from 2 ka to >47 ka. Five meteorites (ALH81041, ALH81099, ALH81104, ALH81119 and ALH83001) have low levels of ^{14}C at or near the amount expected to be produced by *in situ* production of ^{14}C in rocks exposed at the altitude and latitude of the Allan Hills (Jull *et al.* 1992, 1994*b*; Lal & Jull 1994). Assuming that the selection of samples is random, this suggests that this site has about 50% of the meteorite population with terrestrial ages beyond the limit for ^{14}C. The calculated number of meteorites is too high for all meteorites to have resulted from direct infall in this small area of 6.5 km^2 (Huss 1990), so some transport of meteorites from an area at least twice the size is required to explain the observed terrestrial ages. There is one sample of particular interest in this set of measurements: the EH4 meteorite ALH77295 has a low terrestrial age.

Middle Western Icefield

Results of measurements of 12 samples from the 30-km^2 Middle Western Icefield are given in Table 2 and also in Fig. 4. It should be noted that nine of these meteorites discussed in Table 2 were collected from an area of 16 km^2. These meteorites show a mass distribution with fewer small meteorites than the Near Western Icefield. Two meteorites give ages <5 ka: ALH83070, an LL6 chondrite with little weathering (grade A) and ALH84096, a C4 chondrite with a low terrestrial age of 0.5 ± 1 ka. Two other meteorites of special interest were measured. The aubrite ALH84007 has a young terrestrial age of 2.4 ± 1.3 ka and ALHA81005, a lunar meteorite, has a terrestrial age of 18.3 ± 1.3 ka if we can assume that it was irradiated as a small object in space. If the meteorite had a short Moon–Earth transit time, we would have to assume that the irradiation occurred on the Moon and the terrestrial age could be as low as 7.7 ka. ^{41}Ca data (Nishiizumi *et al.* 1991) indicate that the total time for both transit and terrestrial residence cannot exceed *c.* 80 ka, which allows us to constrain the transit time in space to <62 ka. Other meteorites show ages >20 ka, with several at the limit of ^{14}C measurements.

Far Western Icefield

We have measured ^{14}C terrestrial ages on 56 of the 217 meteorites recovered (before 1990) from the Allan Hills Far Western Icefield, as listed in Table 2. This area of 95 km^2 was well studied. In Fig. 4, we show a histogram of number of meteorites for different terrestrial ages. These results are also compared with the values obtained for the Near and Middle Western Icefields. Only some samples were acid-cleaned, as indicated in Table 2. We attempted to locate paired samples in this collection: there are three potentially paired samples among the meteorites studied. These are ALH83100, ALH83102 and

ALH84029 (CM2), and the ureilites ALH82106, ALH82130 and ALH84136 (McSween 1989). We do not count the two potentially paired CM2 meteorites ALH85007 and ALH85013 (Mason *et al.* 1992) as being paired, because of a difference in ^{14}C of more than a factor of three. Hence, the number of independent samples is 52. Because of the large number of meteorites, we can estimate a weathering decay constant, λ, based on the distribution of ages. We calculate $\lambda = -1.71 \times 10^{-5}$, or the weathering 'mean life' is 58 ka. Bland *et al.* (1996) have shown that meteorites weather in a two-stage process. Initial weathering is rapid, followed by a slower process which we model as a simple decay process. If weathering is the only mechanism for removal of meteorites, we can assume that if infall and decay reach equilibrium, then infall rate equals the weathering decay rate. From this we can calculate that the infall rate at the Far Western Icefield is *c.* 39 per 10^6 km^2 per year. Assuming all samples are recovered, then this value would be low compared with the value of 83 per 10^6 km^2 per year determined by Halliday *et al.* (1989) from meteoroid observations. We can make a second calculation, similar to the one performed for the Main Icefield. Thirty of the 52 independent samples from this icefield have ^{14}C terrestrial ages of <25 ka. Up to 1990, 217 meteorites had been found on the ice at this location (Huss 1990). If we assume that the age distribution is the same for all meteorites, then *c.* 125 meteorites should be <25 ka and the apparent infall rate is $(53 \pm 5) \times 10^{-6}$ km^2 per year. This is much closer to the estimate of Halliday *et al.* (1989). This suggests that the Far Western Icefield collection is likely to consist of direct falls and that there is little transport of meteorites through the ice. The apparently low value may be due to the fact that not all the direct falls have been recovered.

Some determinations of terrestrial ages of meteorites of unusual petrological type from this icefield deserve mention. One is a martian meteorite, ALH84001, that has recently been suggested to contain biogenic material (McKay *et al.* 1996) and that contains carbonates with unusual isotopic compositions of carbon and oxygen (Romanek *et al.* 1994; Jull *et al.* 1995, 1997). The terrestrial age of this meteorite is of more than usual importance, given the interest in this meteorite. Three separate determinations have been made, giving a weighted mean of 13 ka. One different value with a larger error, reported by Jull *et al.* (1994*d*), of 27.9 ± 0.5 dpm/kg ^{14}C, gives an age of 6.5 ± 1.3 ka. However, we have subsequently determined that this value of ^{14}C is too high because of a large amount of carbon released from this particular extraction. A weighted mean of all four determinations gives 12.6 ± 1.6 dpm/kg and excluding the above sample gives 12.2 ± 0.8 dpm/kg, giving a terrestrial age of 13 ka. We also note that ALH84001 has a high $^{22}Ne/^{21}Ne$ ratio of 1.19–1.23, which can be evidence for low shielding. However, Nishiizumi *et al.* (1994) determined a saturated activity of ^{10}Be of *c.* 20.0 dpm/kg, indicating irradiation at sufficient depth that we can use the saturated activity of 61.1 dpm/kg for this meteorite as reported in Table 1. A CK4 meteorite, ALH85002, has an age of 8.3 ± 1.3 ka, significantly different from that reported for ALH84096 from the Middle Western Icefield, classed as a C4 chondrite. (CK meteorites are a group of carbonaceous chondrites named after the Karoonda fall; *Eds.*)

Elephant Moraine

Ten new results using ^{14}C on samples from Elephant Moraine (EET) are reported in Table 1 and Fig. 5. Previously, Jull *et al.* (1989) reported ^{14}C limits for two samples (EETA79004 and EETA79005), which were dated by Freundel *et al.* (1986) to have ^{81}Kr ages of 250 and 180 ka respectively. These measurements showed a small amount of ^{14}C consistent with *in situ* production and we estimated limit ages of >40 ka (Jull *et al.* 1989). Five new samples appear to be at or near the limits for ^{14}C age. The ureilites EET83225, EET87517 and EET87720 were found to have terrestrial ages in the range of ^{14}C dating, 8.7, 21.4 and 23.8 ka. Although EET83225 could be part of the same fall as the ALH ureilites ALH82106, ALH84136 and ALH81101, the other two (EET87517 and EET87720) cannot, based on the ^{14}C data. None can be paired with the ureilite ALHA77257, which has been shown to have a very long terrestrial age (see above). Additionally, the L6 chondrites EETA87536, EETA87549 and EETA87554 have finite ^{14}C ages less than 35 ka. A measurement on the lunar meteorite EET87521 shows a limit age of >42 ka, which is consistent with the exposure history reported by Vogt *et al.* (1993). One other meteorite from this site previously reported is the shergottite EETA79001, which has a terrestrial age of *c.* 12 ka (Jull & Donahue 1988), and Nishiizumi *et al.* (1989) reported a ^{36}Cl age on EET83206, an L6 chondrite, of 45 ± 45 ka, suggesting there are other meteorites within the ^{14}C-dating range. These researchers have also reported similar young values for two small irons found at Elephant Moraine. In any case, the frequency

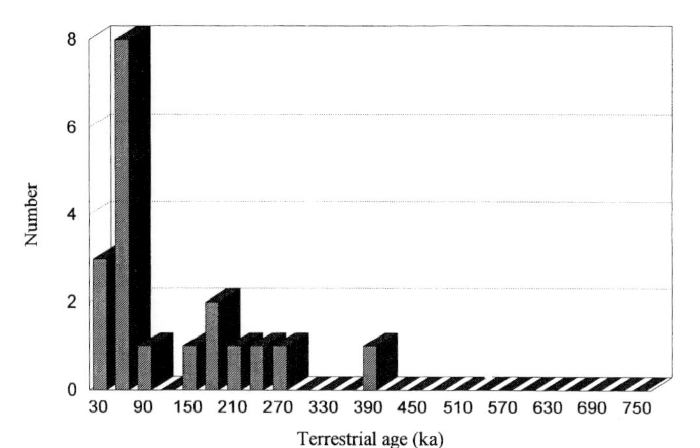

Fig. 5. Terrestrial ^{14}C age distributions of meteorites from the Elephant Moraine Icefield. Both ^{14}C and other results (Nishiizumi *et al.* 1989) are shown for this site.

of finds of <40 ka age at Elephant Moraine is significant. This suggests a meteorite storage regime intermediate between that of the 'old' Allan Hills Main Icefield, and 'younger' sites such as Yamato and the Allan Hills Far Western Icefield, where the majority of meteorites have ages <25 ka (Beukens *et al.* 1988; Nishiizumi *et al.* 1989).

Comparison of the age distributions of meteorites from different icefields

Because of low storage temperatures, we expect Antarctic meteorites to be stored for long periods of time in or on the ice. In Antarctica, we observe samples both within the range of ^{14}C dating, up to 40–50 ka, and beyond. Nishiizumi *et al.* (1989) summarized data on 67 meteorites from the Allan Hills Main Icefield, and most gave ^{36}Cl ages of chondrites of >100 ka, and up to 500 ka in a few cases. Twenty (*c.* 30%) were <70 ka, and using ^{14}C, we can establish that only five meteorites out of the 27 analysed for ^{14}C are <25 ka. The long terrestrial ages of the Main Icefield meteorites are explained by transport of the meteorites in flowing ice over large distances (Nishio & Annexstad 1980; Drewry 1985). By contrast, samples from the Far Western Icefield and the Yamato site show a much younger population of meteorites. At these locations, most of the samples date within the last 40 ka, although Yamato does show a large range of terrestrial ages, with four meteorites of terrestrial age >200 ka (Nishiizumi *et al.* 1989; Michlovich *et al.* 1995). We can summarize the age distributions from different icefields by listing the number with terrestrial ages <25 ka, and categorize sites by the number of such younger falls. With this criterion, the icefields can be ordered in terms of number of falls <25 ka as follows, based on ^{14}C ages only. The results from Yamato are taken from, Cresswell *et al.* (1993), Jull *et al.* (1993*b*), Michlovich *et al.* (1995), and unpublished results from our laboratory: Far Western 65%; Yamato 50%; Middle Western 30%; Elephant Moraine 28%; ALH Main 7%.

In the case of the Far Western Icefield there are few meteorites with less than saturated ^{36}Cl (Nishiizumi *et al.* 1989), which would indicate longer terrestrial ages. Meteorites at sites such as the Far Western Icefield or Yamato cannot have been transported any significant distance in the ice, and probably fell at the location where they were recovered. When pairing is also taken into account, these results allow us to deduce that the distribution of meteorite terrestrial ages can be related to ice flow patterns in the Allan Hills region. In this context, Lindstrom & Score (1994) used statistical arguments to point out that pairing might reduce the number of discrete Antarctic falls by as much as a factor of 2–6. They also suggested that the frequent number of ordinary chondrites in some localities could indicate a pairing factor of five for these mete-orites. We do not believe that our results suggest pairing at such a high level. If the high pairing were the case, our estimates of infall rates would be too high, as our calculations rely on an estimate of the total number of meteorites collected in order to scale our observed age distributions to the whole population. Alternatively, this would imply that only 20–50% of the meteorites which fell at these locations had been

Table 3. *Estimates of weathering and infall rates*

	λ (per ka^{-1})	n (>10 g) (10^{-6} km^2/year)	Reference
Roosevelt County		83	Halliday *et al.* (1989)
Roosevelt County	0.032	116	Bland *et al.* (1996)
Roosevelt County			Jull *et al.* (1991)
Nullarbor Plain	0.024	36	Bland *et al.* (1996)
Sahara	0.011	95–431	Bland *et al.* (1996)
Allan Hills Main		*c.* 50	
Allan Hills Far Western	0.017	53 ± 5	

recovered. We can compare the estimates of infall rate from different areas and the results summarized by Bland *et al.* (1996) in Table 3. Here we can see that the estimates from Allan Hills are not in disagreement with other estimates.

Weathering and stable isotopes. We have also obtained stable-isotope and ^{14}C age information on weathering products from some of the meteorites in this study. These results are shown in Table 4. There is little apparent correlation of $\delta^{13}C$, amount of weathering carbonate and terrestrial age. We also find that the ^{14}C age of the weathering carbonates themselves is invariably much younger than the meteorite's terrestrial age. In Fig. 6, we show $\delta^{13}C$ v. $\delta^{18}O$ for the weathering products. Measurements of $\delta^{13}C$ and $\delta^{18}O$ plot along a trend line as shown in Fig. 5, different from the trend observed for carbonates in meteorites from warmer environments. This is easily understood in terms of different temperatures and sources of water. Weathering products from Antarctic meteorites plot on a trend line parallel to those for desert meteorites, but shifted to lower $\delta^{18}O$. This is the result one would expect for carbonates formed from lighter waters in

Table 4. *Stable–isotope and ^{14}C age data on weathering products*

Sample	$\delta^{13}C$	C* (ppm)	$\delta^{18}O$†	^{14}C age of carbonates (ka)
ALHA77231	−16.188		+7.59	>15.3
ALHA77256	−18.522		+7.63	7.9 ± 1.9
ALHA77262	−8.536		+10.72	
ALHA77295	−5.52	375	+13.36	0.658
ALHA77295	−6.1	299		0.312
ALH81022	−2.289	884	+9.73	5.281
ALH81023	−13.46	57	+12.48	recent
ALH81041	−2.897	49	+11.16	
ALH81041	−2.897	49	+11.16	
ALH81081	−2.897	49	+11.16	4.801
ALH81104	−0.02	232		1.613
ALH81111	−13.2	46	+8.75	
ALH81119	−5.501	134	+13.46	2.875
ALH83001	4.903	194	+11.26	12.1
ALH83010	−0.84	246	+11.3	4.405
ALH83010	−0.841	121	+12.1	
ALH83029	−1.618	150	+15.87	5.05
ALH84004	+1.9	182		
ALH84074	−14.36	236	+10.29	11.5
ALH84083	+21.4	117	+26.39	
ALH84091	−6.943	120	+13.44	5.4
ALH84104	−20.1	151		
ALH84106	−16.4	145		
ALH 84109	−20.1	68		
ALH84142	+14.4	76		0.415
EET87549	−12.70	135	+9.9	

* Amount of carbonate carbon released as CO_2 by phosphoric acid etching of the bulk meteorite.
† Assumes calcite composition of carbonate.

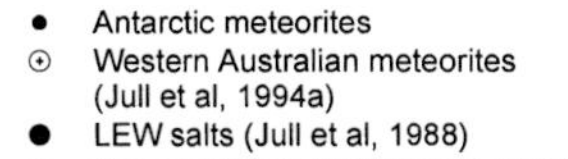

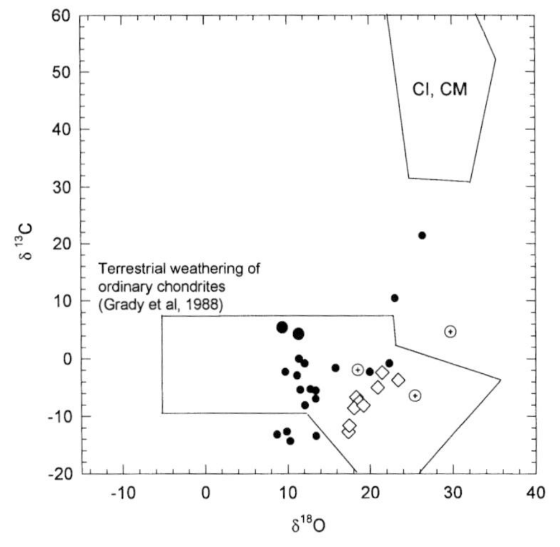

Fig. 6. $\delta^{18}O$ and $\delta^{13}C$ stable-isotopic composition of weathering products from acid-etching of bulk meteorites.

Antarctica. The shift is much lower than would be expected for carbonates in equilibrium with very light (−40‰) ice from Antarctica.

Conclusions

This paper presents new ^{14}C data on meteorites from Allan Hills, Yamato and other sites. The results give further evidence that many meteorites from the sites near the Allan Hills have ages within the range datable by ^{14}C. ^{14}C ages are consistent with those obtained by other methods of terrestrial age determination such as ^{81}Kr, ^{36}Cl and ^{26}Al. We can use the distribution of terrestrial ages to estimate minimum infall rates for meteorites from some of these icefields. These estimates are in the range (40–60) per 10^6 km^2 per year, within a factor of 1.5–2 of the value reported by Halliday *et al.* (1989) based on direct measurements of meteoroid fluxes. The higher estimates of infall rate suggested by Huss (1990), which were based on total numbers of meteorites but without considering their terrestrial age, are not required to explain the observed distributions.

We are grateful to the Meteorite Working Group for provision of the samples. We also thank L. J. Toolin, A. L. Hatheway, L. R. Hewitt, T. Lange and G. S. Burr for technical support of this work. S. Cloudt and E. Cielaszyk were supported in part by the Arizona NASA Space Grant College Consortium. This work was supported in part by NASA grant NAG9-233 and NAGW-3614 and NSF grants EAR 92-03883 and 95-08413.

References

Annexstad, J. O. 1983. *Meteorite concentration and glaciological parameters in the Allan Hills icefield, Victoria Land, Antarctica.* Doctoral dissertation, Johannes-Gutenberg Universität, Mainz.

—— & Annexstad, K. M. 1989. Extension of the Allan Hills triangulation network system. *Smithsonian Contributions to the Earth Sciences*, **28**, 17–22.

Benoit, P. H., Jull, A. J. T., McKeever, S. W. S. & Sears, D. W. G. 1993. The natural thermoluminescence of meteorites VI: carbon-14, thermoluminescence and the terrestrial ages of meteorites. *Meteoritics*, **28** 196–203.

Beukens, R. P., Rucklidge, J. C. & Miura, Y. 1988. ^{14}C ages of Yamato and Allan Hills meteorites. *Proceedings of the NIPR Symposium on Antarctic Meteorites*, **1**, 224–230.

Bland, P. A., Smith, T. B., Jull, A. J. T., Berry, F. J., Bevan, A. W. R., Cloudt, S. & Pillinger, C. T. 1996. The flux of meteorites to the Earth over the last 50,000 years. *Monthly Notices of the Royal Astronomical Society*, **283**, 551–565.

Boeckl, R. S. 1972. Terrestrial age of nineteen stony meteorites derived from their radiocarbon content. *Nature*, **236**, 25–26.

Brown, R. M., Andrews, H. R., Ball, G. C., Burn, N., Imahori, Y., Milton, J. C. D. & Fireman, E. L. 1984. ^{14}C content of ten meteorites measured by Tandem Accelerator Mass Spectrometry. *Earth and Planetary Science Letters*, **67**, 1–8.

Cresswell, R. G., Miura, Y., Beukens, R. P. & Rucklidge, J. C. 1993. ^{14}C terrestrial ages of nine Antarctic meteorites using CO and CO_2 temperature extractions. *Proceedings of the NIPR Symposium on Antarctic Meteorites*, **6**, 381–390.

Drewry, D. J. 1985. *Entrainment, transport and concentration of meteorites in polar ice sheets.* Lunar and Planetary Institute Technical Report, **86-01**, 37–47.

Evans, J.,Wacker, J. & Reeves, J. 1992. Terrestrial ages of Victoria Land meteorites derived from cosmogenic radionuclides. *Smithsonian Contributions to the Earth Sciences*, **30**, 45–56.

Fireman, E. L. 1978. Carbon-14 in lunar soil and in meteorites. *Proceedings of the 9th Lunar and Planetary Science Conference*, 1647–1654.

—— & Norris, T. L. 1981. Carbon-14 ages of Allan Hills meteorites and ice. *Proceedings of the 10th Lunar and Planetary Science Conference*, 1019–1025.

Freundel, M., Schultz, L. & Reedy, R. C. 1986. Terrestrial ^{81}Kr–Kr ages of Antarctic meteorites. *Geochimica et Cosmochimica Acta*, **50**, 2663–2673.

GOEL, P. S. & KOHMAN, T. 1962. Cosmogenic carbon-14 in meteorites and terrestrial ages of 'finds' and craters. *Science*, **136**, 875–876.
GROSSMAN, J. 1994. The Meteoritical Bulletin, No. 76 1994 January: the U.S. Antarctic Meteorite Collection. *Meteoritics*, **29**, 100–143.
GRADY, M. M., WRIGHT, I. P., SWART, P. K. & PILLINGER, C. T. 1988. The carbon and oxygen isotopic composition of meteoritic carbonates. *Geochimica et Cosmochimica Acta*, **52**, 2855–2866.
GRAF, TH, MARTI, K. & XUE, S. 1997. Exposure history of the Peekskill (H6) meteorite. *Meteoritics and Planetary Science*, **32**, 25–30.
HALLIDAY, I., BLACKWELL, A. T. & GRIFFIN, A. A. 1989. The flux of meteorites on the Earth's surface, *Meteoritics*, **24**, 173–178.
HUSS, G. R. 1990. Meteorite infall as a function of mass: implications for the accumulation of meteorites on Antarctic ice. *Meteoritics*, **25**, 41–56.
JULL, A. J. T. & DONAHUE, D. J. 1988. Terrestrial age of the Antarctic shergottite, EETA79001. *Geochimica et Cosmochimica Acta*, **52**, 1295–1297.
——, BEVAN, A. W. R., CIELASZYK, E. & DONAHUE, D. J. 1994*a*. *Carbon-14 terrestrial ages and weathering of meteorites from the Nullarbor Plain, Western Australia.* Lunar and Planetary Institute Technical Report **95-02**, 37–38.
——, CIELASZYK, E., BROWN, S. T. & DONAHUE, D. J. 1994*d*. ^{14}C terrestrial ages of achondrites from Victoria Land, Antarctica. *Lunar and Planetary Science*, **25**, 647–648.
——, DONAHUE, D. J., CIELASZYK, E. & WLOTZKA, F. 1993*a*. Carbon-14 terrestrial ages and weathering of 27 meteorites from the southern high plains and adjacent areas (USA). *Meteoritics*, **28**, 188–195.
——, —— & LINICK, T. W. 1989. Carbon-14 activities in recently fallen meteorites and Antarctic meteorites. *Geochimica et Cosmochimica Acta*, **53**, 2095–2100.
——, ——, REEDY, R. C. & MASARIK, J. 1994*b*. A carbon-14 depth profile in the L5 chondrite Knyahinya. *Meteoritics*, **29**, 649–651.
——, ——, ZABEL, T. H. & FIREMAN, E. L. 1984. Carbon-14 ages of Antarctic meteorites with accelerator and small-volume counting techniques. *Proceedings of the 15th Lunar and Planetary Science Conference, Journal of Geophysical Research*, **89**, C329–C335.
——, EASTOE, C. J. & CLOUDT, S. 1997. Isotopic composition of carbonates in the SNC meteorites Allan Hills 84001 and Zagami. *Journal of Geophysical Research*, **102**, 1663–1669.
——, ——, XUE, S. & HERZOG, G. F. 1995. Isotopic composition of carbonates in the SNC meteorites Allan Hills 84001 and Nakhla. *Meteoritics*, **30**, 311–318.
——, LIFTON, N. A., PHILLIPS, W. M. & QUADE, J. 1994*c*. Studies of the production rate of cosmic-ray produced ^{14}C in rock surfaces. *Nuclear Instruments and Methods in Physics Research*, **B92**, 308–310.
——, MIURA, Y., CIELASZYK, E., DONAHUE, D. J. & YANAI, K. 1993*b*. AMS ^{14}C ages of Yamato achondritic meteorites. *Proceedings of the NIPR Symposium on Antarctic Meteorites*, **6**, 374–380.
——, WILSON, A. E., BURR, G. S., TOOLIN, L. J. & DONAHUE, D. J. 1992. Measurements of cosmogenic ^{14}C produced by spallation in high-altitude rocks. *Radiocarbon*, **34**, 737–744.
——, WLOTZKA, F. & DONAHUE, D. J. 1991. Terrestrial ages and petrologic type of Roosevelt County meteorites. *Lunar and Planetary Science*, **22**, 665–666.
——, ——, PALME, H. & DONAHUE, D. J. 1990*a*. Distribution of terrestrial age and petrologic type of meteorites from the Libyan Desert, *Geochimica et Cosmochimica Acta*, **54**, 2985–2899.
KARLSSON, H. R., JULL, A. J. T., SOCKI, R. A. & GIBSON, E. K. 1991. Carbonates in Antarctic ordinary chondrites: evidence for terrestrial origin. *Lunar and Planetary Science*, **22**, 689–690.
LAL, D. & JULL, A. J. T. 1994. Studies of cosmogenic *in situ* ^{14}CO and $^{14}CO_2$ produced in terrestrial and extraterrestrial samples: experimental procedures and applications. *Nuclear Instruments and Methods in Physics Research*, **B92**, 291–296.
LINDSTROM, M. M. & SCORE, R. 1994. *Populations, pairing, and rare meteorites in the U.S. Antarctic meteorite collection.* Lunar and Planetary Institute Technical Report **95-02**, 43–45.
MARVIN, U. B. 1992. Meteorite distributions at the Allan Hills Main icefield and the pairing problem. *Smithsonian Contributions to the Earth Sciences*, **30**, 113–119.
—— & MACPHERSON, G. J. 1989. Field and laboratory investigations of meteorites from Victoria Land and the Thiel Mountains, Antarctica 1982–1983 and 1983–1984. *Smithsonian Contributions to the Earth Sciences*, **28**, 1–3.
—— & ——1992. Field and laboratory investigations of Antarctic meteorites collected by United States expeditions. *Smithsonian Contributions to the Earth Sciences*, **30**, 1–3.
—— & MASON, B. 1984. Field and laboratory investigations of meteorites from Victoria Land, Antarctica. *Smithsonian Contributions to the Earth Sciences*, **26**, 1–4.
MASON, B. 1979. Cosmochemistry, part 1. Meteorites. *In*: FLEISCHER, M. (ed.) *Data of Geochemistry*, 6th edn. US Geological Survey Professional Paper, **440-B-1.**
——, MACPHERSON, G. J., SCORE, R., SCHWARZ, C. & DELANEY, J. S. 1992. Descriptions of stony meteorites. *Smithsonian Contributions to the Earth Sciences*, **30**, 29–59.
MCKAY, D. S., GIBSON, E. K. JR, THOMAS-KEPRTA, K. L. *et al.* 1996. Search for past life on Mars: possible relic biogenic activity in Martian meteorite ALH84001. *Science*, **273**, 924–930.
MCSWEEN, H. Y. 1989. Antarctic carbonaceous chondrites: new opportunities for research. *Smithsonian Contributions to the Earth Sciences*, **28**, 81–86.
MICHLOVICH, E. S., WOLF, S. F., WANG, M. S., VOGT, S., ELMORE, D. & LIPSCHUTZ, M. E. 1995. Chemical studies of H chondrites 5. Temporal variations of sources. *Journal of Geophysical Research*, **100**, 3317–3333.

MIURA, Y., NAGAO, K. & FUJITANI, T. 1993. ^{81}Kr terrestrial ages and grouping of Yamato eucrites based on noble-gas and chemical compositions. *Geochimica et Cosmochimica Acta*, **57**, 1857–1866.

NINAGAWA, K., MIONO, S., YOSHIDA, M. & TAKAOKA, N. 1983. Measurement of terrestrial age of Antarctic meteorites by thermoluminescence technique. *Memoirs of the National Institute of Polar Research, Tokyo*, Special Issue **30**, 251–258.

NISHIIZUMI, K. ARNOLD, J. R., ELMORE, D. *et al.* 1979. Measurements of ^{36}Cl in Antarctic meteorites and Antarctic ice using a Van de Graaff accelerator. *Earth and Planetary Science Letters*, **45**, 285–292.

——, ——, ——, MA, X., NEWMAN, D. & GOVE, H. E. 1983. ^{36}Cl and ^{53}Mn in Antarctic meteorites and ^{10}Be–^{36}Cl dating of Antarctic ice. *Earth and Planetary Science Letters*, **62**, 407–417.

——, ——, KLEIN, J. *et al.* 1991. Exposure histories of lunar meteorites: ALHA81005, MAC88104 and MAC88105. *Geochimica et Cosmochimica Acta*, **55**, 3149–3155.

——, CAFFEE, M. W. & FINKEL, R. 1994. Exposure histories of ALH84001 and ALHA77005 (abstract) *Meteoritics*, **29**, 511.

——, ELMORE, D. & KUBIK, P. W. 1989. Update on terrestrial ages of Antarctic meteorites. *Earth and Planetary Science Letters*, **93**, 299–313.

——, KLEIN, J., FINK, D. *et al.* 1990. Exposure histories of lunar meteorites MAC88104 and MAC88105. *Lunar and Planetary Science*, **21**, 897–898.

——, MURRELL, M. T., ARNOLD, J. R., ELMORE, D., FERRARO, R. D., GOVE, H. E. & FINKEL, R. C. 1981. Cosmic ray produced ^{36}Cl and ^{53}Mn in Allan Hills – 77 meteorites. *Earth and Planetary Science Letters*, **52**, 31–38.

NISHIO, F. & ANNEXSTAD, J. O. 1980. Studies on the ice flow in the bare ice area near the Allan Hills in Victoria Land, Antarctica. *Memoirs of the National Institute of Polar Research, Tokyo*, Special Issue **17**, 1–13.

ROMANEK, C. S., GRADY, M. M., WRIGHT, I. P., MITTLEFEHLDT, D. W., SOCKI, R. A., PILLINGER, C. T. & GIBSON, E. K. JR 1994. Record of fluid–rock interactions on Mars from the meteorite ALH84001, *Nature*, **372**, 655–657.

SCHULTZ, L., FRANKE, L. & KRUSE, H. 1996. *Helium, neon and argon in meteorites: a data compilation.* Max-Planck-Institut für Chemie, Mainz.

SISTERSON, J. M., JULL, A. J. T., BEVEDING, A. *et al.* 1994. *Nuclear Instruments and Methods in Physics Research*, **B92**, 510–512.

SUESS, H. & WÄNKE, H. 1962. Radiocarbon content and terrestrial age of 12 stony meteorites and one iron meteorite. *Geochimica et Cosmochimica Acta*, **26**, 475–480.

TAKEDA, H., MORI, H. & OGATA, H. 1988. On the pairing of Antarctic ureilites with reference to their parent body. *Proceedings of the NIPR Symposium on Antarctic Meteorites*, **1**, 145–172.

VOGT, S., HERZOG, G. F., EUGSTER, O. *et al.* 1993. Exposure history of the lunar meteorite, Elephant Moraine 87521. *Geochimica et Cosmochimica Acta*, **57**, 3793–3799.

WACKER, J. F. 1993. *Antarctic Meteorite Newsletter*, **16**(2), 23.

——1995. *Antarctic Meteorite Newsletter*, **18**(1), 18.

WHILLANS, L. M. & CASSIDY, W. A. 1983. Catch a falling star: meteorites and old ice. *Science*, **222**, 55–57.

WIELER, R, GRAF, TH., SIGNER, P. *et al.* 1996. Exposure history of the Torino meteorite. *Meteoritics and Planetary Science*, **31**, 265–272.

WLOTKZA, F. 1991. The Meteoritical Bulletin, No. 70. *Meteoritics*, **26**, 68–69.

——, JULL, A. J. T. & DONAHUE, D. J. 1994. *Carbon-14 terrestrial ages of meteorites from Acfer, Algeria.* Lunar and Planetary Institute Technical Report, **95-02**, 72–73.

YANAI, K. 1981. *Photographic Catalog of the Selected Antarctic Meteorites.* National Institute of Polar Research, Tokyo, 11–12.

The flux of meteorites to Antarctica

MICHAEL ZOLENSKY

Earth Science and Solar System Exploration Division, NASA Johnson Space Center, Houston, TX 77058, USA

Abstract: It is premature to report firm conclusions regarding the past meteorite flux rate from the Antarctic record, as variables related to ice flow, meteorite catchment area, weathering and removal rate are not well constrained. The relative population of meteorite types appears to be broadly compatible with the modern fall population, although there are minor differences. In particular, there are populations of small unusual irons, carbonaceous chondrites and lunar samples not found among the modern falls. In addition, slight compositional differences between modern and old (Antarctic) H chondrites may signal a time-dependent change in the flux to Earth of more common meteorites.

Where are meteorites coming from? How fast do these meteorite source regions evolve, as reflected in a changing complexion of meteorite types? What is the danger to life and structures from falling meteorites, and is this threat changing (Beijing Observatory 1988)? Finally, some geochronological techniques rely on a constant flux of extraterrestrial material to Earth; how reasonable is this assumption? To discover the answers to these critical questions we are attempting to determine the meteorite flux to Earth not just for the present day, but back into the past.

Meteorites fall with equal frequency across all corners, nooks and crannies of the Earth. However, the chances of finding them vary considerably according to climatic, geographical, geological and ophthalmic factors. In the ideal case, meteorites falling would be immediately buried, completely protected against any chemical of physical weathering processes, and immediately excavated just before recovery. In the real world the situation is very different. However, across the Earth, preservation and recovery factors are best attained in Antarctica.

Meteorite recovery expeditions over the past 20 years by the National Institute of Polar Research, National Science Foundation, and EUROMET have returned >15 000 meteorite specimens from Antarctica, probably representing >1500 individual falls (Annexstad *et al.* 1995). The population of recovered Antarctic meteorites thus outnumbers the entire harvest from the rest of the Earth. Terrestrial age determinations have revealed that meteorites have been preserved for up to 3 Ma in Antarctica, although most have a terrestrial residence time <100 ka (Nishiizumi *et al.* 1989; Nishiizumi 1995; Welten 1995; Scherer *et al.* 1997). Elsewhere on Earth meteorites are generally completely weathered away in a few tens of thousand years to 100 ka. The population of Antarctic meteorites contains numerous types not found (or rarely so) elsewhere on Earth, such as lunar meteorites, certain irons, and unusual carbonaceous chondrites. Some workers report compositional variations in the H chondrites with time. These reports suggest changes in the source asteroids feeding material to the Earth over the past several hundred thousand years (Michlovich *et al.* 1995). However, many of these unusual meteorites are small in mass, and being so are most easily located on a white, rock-free environment, leading one to speculate that a collection bias is at work (Huss 1991). Therefore, it is desirable to re-examine how the flux and population of meteorites to Antarctica through the ages compares with the modern record. As this goal has not been achieved, the point of this paper is to explain the barriers to this goal, and progress that is being made.

Meteorite flux determinations from fireballs

The premier determination of the number and size distribution of meteorites landing on Earth from fireball data is that by Halliday and coworkers (Halliday *et al.* 1989), and is based upon 11 years of observations by the Canadian Camera Network. They found that approximately nine events each year provide at least 1 kg of meteorites to an area of 1×10^6 km^2. They further found that this same area would be expected to receive about 54 kg of meteorites with individual masses between 0.01 and 100 kg. Those workers speculated that these values should not be in error by more than a factor of two (Halliday *et al.* 1989), with the major uncertainty being the determination of meteorite mass from fireball luminosity. It is unfortunate that this camera network resulted in the recovery

ZOLENSKY, M. E. 1998. The flux of meteorites to Antarctica. *In*: GRADY, M. M., HUTCHISON, R., MCCALL, G. J. H. & ROTHERY, D. A. (eds) *Meteorites: Flux with Time and Impact Effects*. Geological Society, London, Special Publications, **140**, 93–104.

of only one meteorite (Innisfree). Fortunately, similar camera networks in the USA and central Europe also resulted in one recovery each (Halliday *et al.* 1981), which facilitated the calibration of fireball luminosity with resultant meteorite mass as it lands on the Earth. The situation for extraterrestrial material at all masses has been presented by Hughes (1994).

Meteorite population determinations from falls

Without impressing a permanent force of persons into meteorite watch gangs, it is impossible to derive accurate meteorite fluxes from fall observations. But fall information probably does provide the best information on the relative types of meteorites that today pepper Earth (Wasson 1985). Some workers suggest that the modern fall record favours large showers over small, individual falls, but it is not clear what effect this potential bias would have. In any case, we wish to learn how fast the meteorite population changes, i.e. are we sampling different meteorite parent bodies through time?

Flux determinations from finds

To use meteorite finds to calculate flux there are certain criteria that must be satisfied Zolensky *et al.* (1995*b*). The fallen meteorites must have been quickly buried to minimize weathering and destruction. The climate plays a major role, with dry regions being the best; both hot and cold deserts will work. Recent excavation is optimal, again minimizing subaerial weathering. The meteorites must not be removed from the stranding surface by water, wind or any creatures. The meteorite recovery programme has to have been well co-ordinated, guaranteeing essentially complete recovery of all specimens. Acute meteorite recognition skills are necessary for all field party members, as is a certain degree of luck. There must be good age control of the stranding surface or adequate data on the terrestrial residence time of the meteorites themselves. Of course, the weathering criterion is never fully satisfied, as weathering occurs in all geological environments; therefore one must form a good understanding of the local removal rate of meteorites via weathering and other geological processes.

In the past two decades the number of available meteorites has been more than doubled by the recovery of specimens from cold and hot deserts. There have been several attempts to derive past meteorite fluxes to Earth based upon the find record in hot deserts. Roosevelt County, New Mexico, had produced finds representing more than 150 separate falls, with terrestrial ages up to *c.* 100 ka. Bland *et al.* (1996*a*) found that the flux over this period was essentially comparable with that derived from the modern fireball record described above. Bland found similar results for somewhat younger meteorite concentrations in the Sahara and Nullarbor Plain, Australia (Bland *et al.*1996*b*).

Other papers in this volume cover meteorite recovery from hot deserts; here we will be concerned with the coldest, driest, windiest desert, Antarctica.

Antarctica and its special meteorite recovery conditions

Meteorite concentration mechanism in Antarctica

Figure 1 shows the location of meteorite-producing areas in Antarctica. According to the general model (Whillans & Cassidy 1983; Delisle & Sievers 1991; Cassidy *et al.* 1992) (see Fig. 2), as meteorites fall from space onto the Antarctic plateau, they are incorporated into a growing pile of snow and ice within a few years. From this time until they emerge somewhere downslope they are locked into cold ice; subject only to very slow weathering (by the action of 'unfrozen water'; see Zolensky & Paces (1986)) and some dynamic stress as the ice flows towards the periphery of the continent, where it drops into the ocean. However, in some areas the ice has to pass over or around major mountain ranges (often buried but still felt by the ice), and in these areas the ice has to slow down, increasing exposure of the ice to the dry katabatic winds and causing increased sublimation. Over thousands of years meteorites can be highly concentrated at the surface on temporarily stranded, dense, blue ice. In some situations (e.g. the Pecora Escarpment) meteorite stranding zones are not directly downflow of glaciers, but rather appear to have moved laterally for poorly understood reasons, in a sort of 'end run' around the front of an obstruction (J. Schutt, pers. comm. 1991). These stranding surfaces appear to be stable until increased snowfall upslope increases the flow of ice downslope, flushing the meteorites away. It is possible for one stranding surface to form and

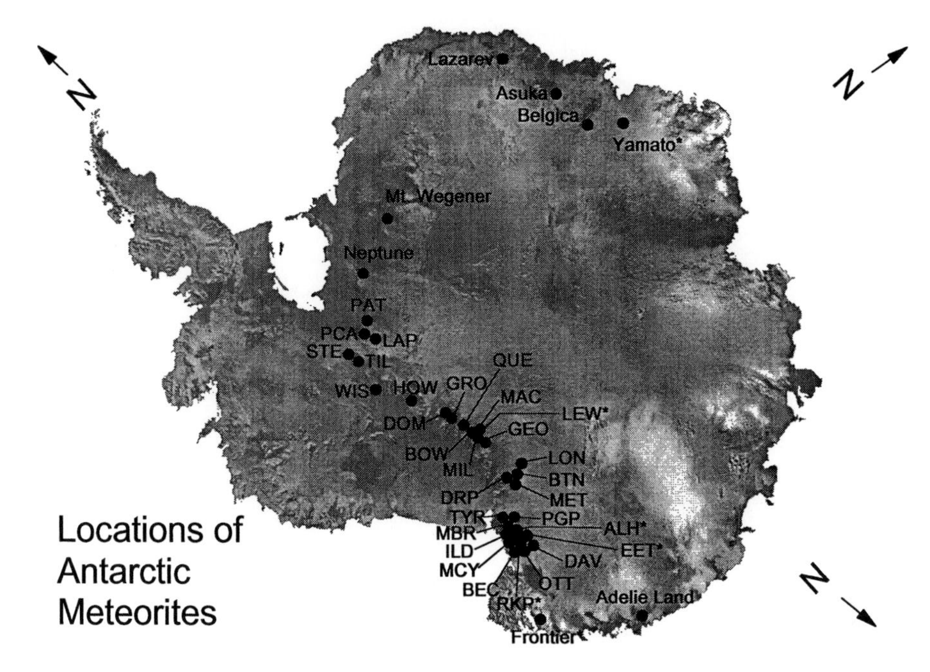

Fig. 1. Galileo spacecraft image of Antarctica, showing locations of meteorite stranding surfaces. Sites with an asterisk (*) are those for which terrestrial age information is shown in Fig. 3. These are: ALH, Allan Hills; Yamato, Yamato Mountains; LEW, Lewis Ice Cliff; Frontier, Frontier Mountain; EET, Elephant Moraine.

be flushed downstream only to re-emerge later. Each stranding surface contains at least two populations of meteorites. The first consists of those which travelled in the ice to the stranding site, and therefore represents a snapshot of the meteorite flux from the past (older meteorites have been flushed away, younger ones are upstream still entrained in the ice). The second population is provided by recent direct falls onto the stranding surface. The presence of at least two chronologically distinct meteorite populations considerably complicates any calculation of meteorite flux based upon statistical considerations. In fact, Huss (1990*b*) has argued that nearly all meteorites in stranding surfaces fell directly there, that the stranding surface is not in fact moving significantly, and that the accumulation times calculated from the terrestrial ages of the meteorites can therefore be used to date the period over which the ice has been stalled. This model is based upon a careful consideration of the meteorite mass–number populations. The conflicts between Huss' model and that by Cassidy, Whillans and Delisle (see Cassidy *et al.* 1992) have not been resolved. This is obviously a major impediment to use of meteorite populations to calculate meteorite flux.

Calculation of Antarctic meteorite flux

In addition, there are numerous other uncertainties which currently impede calculation of meteorite flux from Antarctic collections, most of which cannot be easily circumvented (Harvey & Cassidy 1991). I review these factors below.

Terrestrial age dating of meteorites. In terrestrial age determinations of Antarctic meteorites (principally chondrites) the isotopes ^{36}Cl, ^{81}Kr, ^{10}Be, ^{14}C and ^{26}Al are most effectively employed (Nishiizumi *et al.* 1989; Welten 1995; Wieler *et al.* 1995). There is a measurement gap between ^{14}C ages (<40 ka) and ^{36}Cl ages (>70 ka); Nishiizumi (1995) has suggested that this might be closed in the future by investigations of cosmogenic ^{41}Ca (unfortunately, it is not easy to measure ^{41}Ca in chondrites).

Principally because of the sheer bulk of recovered meteorites, available funding, and finite lifetimes of meteorite investigators, only about 1% of the Antarctic meteorites have had their terrestrial ages determined (Nishiizumi *et al.* 1989; Nishiizumi 1995; Welten 1995; Wieler *et al.* 1995), principally by ^{36}Cl. Indirect means must be found to date the remainder.

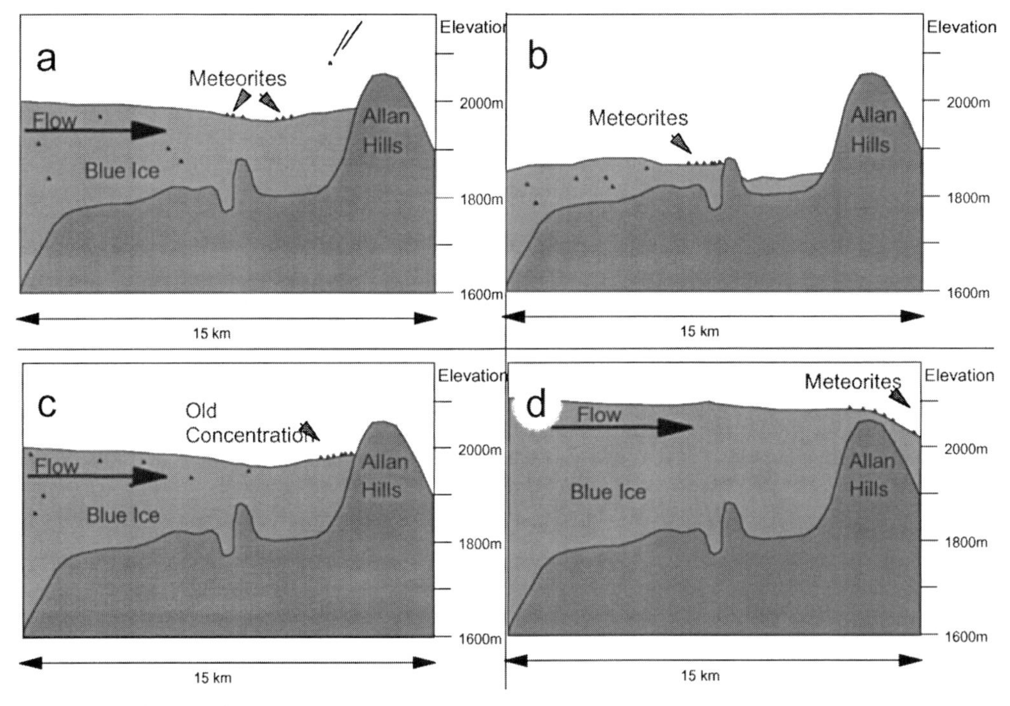

Fig. 2. Meteorite stranding situation at the Allan Hills, adapted after Delisle (1993). (**a**) Present situation at Allan Hills. Views (**b**)–(**d**) show a hypothetical series of schematic illustrations demonstrating how a palaeo-meteorite stranding surface can develop and be flushed away. (**b**) Low ice, interglacial period, with a stranding surface developing behind a cliff that is buried at present. (**c**) As ice volume increases during a glacial period the palaeo-stranding surface can be moved to a downstream position. (**d**) As ice volume increases further the meteorites are liberated, to flow into the ocean further downstream.

In a recent series of papers, Benoit, Sears and coworkers have suggested that natural thermoluminescence (TL) can be used to help address this problem (Benoit *et al.* 1992, 1993, 1994). In this technique, one assumes a mean annual temperature for the storage of Antarctic meteorites, and a 'surface exposure age' as part of the terrestrial age is deduced. According to these TL measurements most of the meteorites spend <50% of their terrestrial history on the surface of the ice in Antarctica, with this conclusion being based upon their low relative TL sensitivity, which in turn is related to their degree of post-exposure weathering. Although this technique cannot be used to obtain absolute terrestrial ages, the calculated exposure ages could be of value in comparing the histories of the different Antarctic stranding surfaces. However, we note that Welten *et al.* (1995) have sounded a dissenting note on the applicability of TL to this problem.

We show results for approximately 280 dated chondrites in Fig. 3 (taken from Nishiizumi *et al.* (1989), Michlovich *et al.* (1995); Welten (1995) and Wieler *et al.* (1995)); and the complex nature of meteorite preservation is well displayed. The Antarctic meteorites display terrestrial ages ranging up to 3 Ma, far in excess of those recovered from elsewhere on Earth, although chondritic terrestrial ages from different Antarctic ice fields have different distributions. The oldest meteorites are generally found at the Main Ice Field of the Allan Hills, and the Lewis Cliff. In particular, the Lewis Cliff Ice Field appears to have the best record of meteorites, as the decay of its meteorite population with terrestrial age is least pronounced (Fig. 3). The Lewis Cliff ice tongue appears to be the oldest of the investigated ice field stranding areas (Welten 1995; Welten *et al.* 1995). At the other end of the scale, terrestrial ages of chondrites recovered from Frontier Mountain are <140 ka, indicating that this is a relatively recent stranding surface.

Various schemes have been proposed for the estimation of terrestrial ages for entire stranding surface meteorite populations. For a given stranding surface one could assume that all

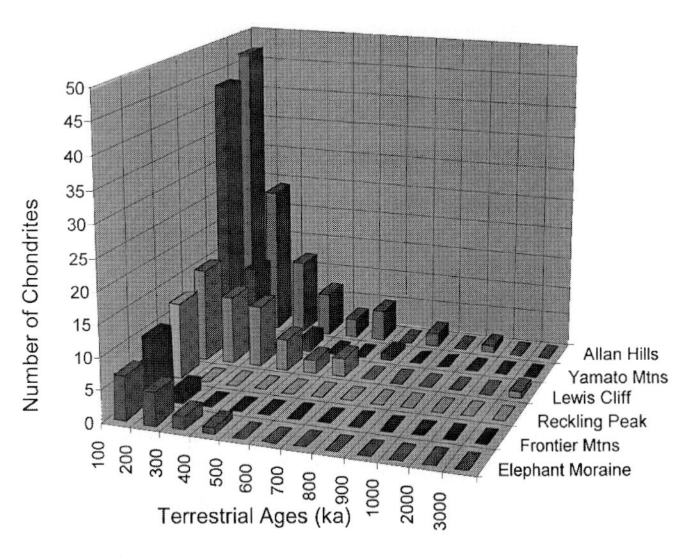

Fig. 3. Terrestrial ages of *c.* 280 Antarctic meteorites, sorted by stranding site. Data are from Nishiizumi *et al.* (1989), Michlovich *et al.* (1995), Welten (1995) and Wieler *et al.* (1995).

other meteorites are younger than the oldest measured meteorite, and that meteorite recovery is 100%, but this is certainly not the case, as explained below. Workers have also attempted to characterize the accumulation age of stranding surfaces by examining the mass distributions and weathering lifetimes of meteorites (Huss 1990*b*). Finally, it would appear that merely measuring the age of a particular blue ice stranding surface and counting up the meteorites presented there would permit one to measure the flux rate of meteorites to that surface, but we have shown above that at least two distinct meteorite populations are always present. Nevertheless, of all factors to be considered in the derivation of past meteorite flux for Antarctica, terrestrial age is one of the best constrained.

Uncertain meteorite catchment area. In other regions of the Earth it is easy to measure the search area, allowing a flux per unit area to be calculated. This is not directly possible in Antarctica because the meteorites exposed today at a stranding surface have fallen over a far larger glacial tributary system. This situation is further complicated by the fact that the ice movement is irregular, and velocity is not constant. Tributary ice streams can provide additional meteorites with an irregular periodicity.

Delisle (Delisle & Sievers 1991; Delisle 1993, 1995) has shown that glacial stages result in thicker coastal ice, whereas interglacials produce thinner coastal ice. The consequences of this are seen in a change of velocity and mass transport of ice with the entrained meteorites. The meteorites at the Allan Hills and possibly Frontier Mountain are suspected to have been buried and re-emerged, as the interglacial sublimation rates are much higher than the glacial sublimation rates. In this picture the meteorite traps are not stagnant regions, but are continually readjusting to the pattern of glacial and interglacial periods.

The uncertain nature of the meteorite catchment area appears to be the greatest impediment to the successful calculation of meteorite flux from Antarctic data, in this author's opinion.

Palaeo-stranding surfaces. During periods of increased snowfall upstream, ice thickness will increase, and a downstream stranding surface (with its population of direct falls) can be flushed further downstream, only to be stranded again and acquire a fresh brood of falls (Delisle & Sievers 1991; Cassidy *et al.* 1992)

Direct infall. Fresh meteorites fall directly onto recovery stranding surfaces, which can cause confusion if they are being added to a recently exposed palaeo-stranding surface.

Recovery rate. One might assume that all meteorites from a given stranding surface can be easily recovered. However, considerable experience indicates that meteorites are often missed because of the wind getting in a searcher's eyes, or burial by a transient snow patch. Meteorite recognition skills also vary considerably from person to person, and luck plays a role. Harvey (1995) has discussed the problem of searching efficiency in some detail, and attempted to quantify a correction factor.

Weathering and differential survival of meteorites. Several factors have been identified which have major effects on the survival of meteorites on the ice. The chemical weathering of meteorites, though minimal in Antarctica because of the cold and aridity, is still significant. Even while stored in ice, the action of thin monolayers of 'unfrozen water' coating meteorites will inevitably cause chemical weathering (Zolensky & Paces 1986). Meanwhile, ice movement can cause shearing and consequent breakage of the entrained meteorites. Finally, once the meteorites have emerged on the stranding surface, they are subject to accelerated chemical weathering on hot days when ice melts round them (the meteorites soak up solar radiation), and physical abrasion from the katabatic winds.

In 'hot' deserts such as the Libyan Desert, Algeria, and Roosevelt County (New Mexico, USA) the mean survival time of chondrites is >10 ka. (Boeckl 1972; Jull *et al.* 1995; Wlotzka *et al.* 1995), although far older meteorites (*c.* 100 ka) are found in some fortuitous regions (e.g. Roosevelt County; see Zolensky *et al.* (1992)).

Certain easily weathered meteorite types will be preferentially lost from the stranding record. Attempts to correct the Antarctic record for weathering effects (Huss 1990*b*; Harvey 1995) have had unknown success. Bland *et al.* (1996*a*) suggested that H chondrites weather faster than L chondrites (as might be expected from the former's greater metal content), although Zolensky *et al.* (1995*a*) found just the opposite for meteorites recovered from the Atacama Desert. Bland *et al.* pointed out that carbonates and hydrous silicates could, in some situations, armour mafic silicates and thereby provide a certain degree of protection against further alteration. Also, the changing temperature and aridity conditions attending regional climatic change would have further complicated this situation.

The main results of weathering will be threefold: (1) to decrease the population of all meteorites and reduce the mass of the survivors; (2) to increase the number of samples, through physical break-up, and subtly change the composition of the survivors (see below), increasing the importance and difficulty of pairing studies; (3) removing from the record especially friable meteorites (e.g. CI1, of which no Antarctic specimens have been located). Various workers have proposed correction factors for weathering, but these appear to be site specific (Boeckl 1972; Zolensky *et al.* 1992; Harvey 1995).

Removal of meteorites by wind. The unrelenting katabatic winds in Antarctica promote rapid ablation (on the order of 5 cm/a at the Allan Hills, Schutt *et al.* 1986; Cassidy *et al.* 1992), and once meteorites emerge from the sublimating ice they are subject to transport by these same winds (Schutt *et al.* 1986; Delisle & Sievers 1991; Harvey 1995). This was elegantly demonstrated by an experiment at the Allan Hills main ice field. During the 1984–1985 field season, the NSF Antarctic Search for Meteorites (ANSMET) field party placed surrogate meteorites (rounded basalt samples of varying masses) onto a 'starting line', one rock 'team' sitting on the ice surface and the second team embedded just below the surface (Schutt *et al.* 1986). This rock race was visited the following year, and the results are given in Table 1. In 1 year all but the three largest buried meteorites had emerged. A number of stones were found to have moved downwind from the starting line, and eight had blown far enough away as to be unfindable.

Aeolian effects are more pronounced in Antarctica than in other meteorite recovery regions, and the experiment described above has shown that specimen size inversely affects specimen movement, and that the mass threshold is a function of wind velocity, which relates to a specific area and may not be the same over different search regions. Small meteorites will be removed from the stranding surface by the action of wind alone, which undoubtedly skews the constitution of the recovered meteorite population.

Meteorite pairing. Many (most?) meteorites fall as showers of individuals, which are subject to later breakage. In addition, physical weathering breaks meteorites into smaller, more numerous specimens. Flux calculations concern the number of individual falls, not total specimens, so that pairing needs to be determined.

Pairing studies for unusual meteorites are well established; however, few of the ordinary chondrites (i.e. 90% of the collection) have received adequate study (because they are generally unexciting to most meteoriticists). Pairing for ordinary chondrites is most often done using comparative petrological, ^{26}Al, TL and noble gas analyses (the last revealing cosmic-ray exposure

Table 1. *Results of wind-driven rock race after 1 year, from Schutt* et al. *(1986)*

Team 1, encased in ice			Team 2, set on surface		
Specimen number	Weight (g)	Distance travelled (m)	Specimen number	Weight (g)	Distance travelled (m)
1	1.01	nf	1	1.17	nf
2	2.82	43.7	2	3.25	nf
3	4.22	nf	3	4.40	168.5
4	4.61	81.0	4	7.62	111.5
5	8.50	nf	5	10.35	108.3
6	11.97	nf	6	12.06	0
7	14.54	nf	7	14.96	nf
8	17.88	2.8	8	20.87	0
			9	26.93	58.6

In Team 1, numbers 9 (25.71 g) to 21 (803 g) had not moved. In Team 2, numbers 10 (35.79 g) to 20 (907 g) had not moved. nf, not found.

histories) (Schultz *et al.* 1991; Benoit *et al.* 1993; Scherer *et al.* 1995).

Unfortunately, as for the determination of terrestrial ages, no workers can analyse all meteorites on a given stranding surface, much less all 15 000 Antarctic specimens. In addition, weathering has been shown to deleteriously affect the noble gas content of chondrites, seriously complicating pairing studies (Scherer *et al.* 1995). Therefore one must derive and apply a correction factor for the pairing of the ubiquitous ordinary chondrites; these have ranged from two to six (Ikeda & Kimura 1992; Lindstrom & Score 1995). Lindstrom & Score (1995) derived a chondrite pairing value of five, based on the known pairing incidences of Antarctic meteorites (see Fig. 4). By applying this simple pairing correction value to the full population of Antarctic chondrites, they were able to address the question of possible differences in the meteorite population through time (see below).

Calculation of Antarctic meteorite flux. Despite uncertainties, workers have attempted to calculate the past meteorite flux for Earth based upon the Antarctic meteorite record. Huss (1990*b*) concluded that the ages of the Antarctic ice fields, as calculated from meteorite mass distributions, indicate an infall rate only slightly higher than the modern rate; he concluded that the flux rate had not changed by more than a factor of two over the past million years.

We conclude that it is premature to attempt direct measurement of the past meteorite flux rate from the Antarctic record, despite the potential value in doing so. Additionally, considerable uncertainties exist in attempts to estimate the most critical variables, which continue to prevent accurate estimation of the flux. To advance this effort, the following studies should be undertaken:

- The pairing of ordinary chondrites needs to be better constrained with real data.
- The nagging issue of the usefulness of TL in estimating meteorite terrestrial ages needs to be settled.
- Investigation of the potential effects of weathering in the differential destruction of meteorite types is needed.
- Most critically, a better understanding of ice flow dynamics at each of the oldest meteorite stranding sites (Allan Hill and Lewis Cliff) is needed, to permit real meteorite fluxes to be calculated for the past 3 Ma.

Have meteorite populations changed through time?

A very contentious subject has been that of potential compositional and mineralogical differences between younger finds and falls v. older finds. An entire workshop was devoted to this subject (Koeberl & Cassidy 1990). In particular, it appears that the important issue of possible observed changes in the H chondrite populations remains unresolved.

In numerous papers, Wolf, Lipschutz and co-workers have argued that there is chemical evidence that Antarctic H4–6 and L4–6 chondrites with terrestrial ages >50 ka (from Victoria Land) are different from those that are younger (Queen Maud Land meteorites and modern falls) (Dennison *et al.* 1986; Michlovich *et al.*

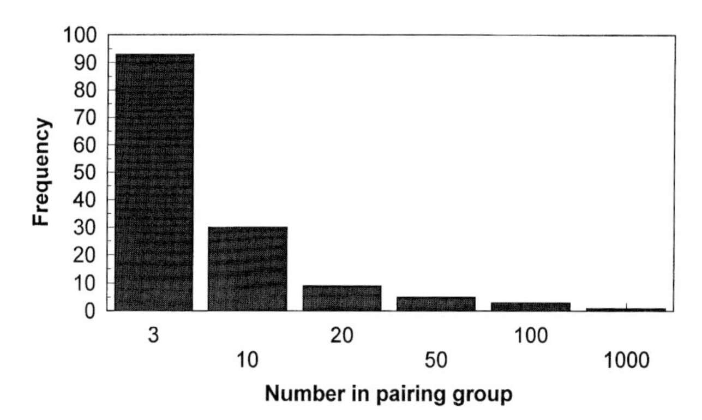

Fig. 4. Histogram of known pairing for ANSMET Antarctic meteorites, from Lindstrom & Score (1995). In the majority of investigated situations individual meteorite finds were shown to pair into groups of between three and ten members, with an average of five.

1995; Wolf & Lipschutz 1995*a*, *b*). A difference between these meteorites would imply a change in the sources providing these meteorites over a timescale deemed too rapid by many dynamicists (Wetherill 1986). Critics suggest that weathering may have redistributed the volatile trace elements being used for the difference arguments, but this view is disputed by some recent studies (Burns *et al.* 1995; Wolf & Lipschutz 1995*c*). Other critics point out that if it takes a sophisticated statistical examination of fully ten trace elements to see a difference between the putative meteorite groups, then they are actually very similar. In a recent test of the Lipschutz–Wolf model, Loeken & Schultz (1995) performed analyses of noble gases in H chondrites from Antarctica (from the Allan Hills and the Yamato Mountains) and modern falls. The researchers were careful to select only Antarctic meteorites analysed by Wolf & Lipschutz. They found no correlation between terrestrial age and exposure age, radiogenic ^{4}He or radiogenic ^{40}Ar over the period of the last 200 ka. One might expect to see such correlations if the population source for Antarctic H chondrite finds and modern H chondrite falls indeed differed. Lipschutz & Wolf steadfastly maintain their position that the apparent population differences are real, and readers will have to decide the issue for themselves.

Lindstrom & Score (1995) have described the population statistics for the Antarctic meteorites recovered by the ANSMET expeditions. By applying a simple pairing correction value to the full population of Antarctic ordinary chondrites, they found that the relative numbers of Antarctic finds match modern fall statistics, so that there is no difference in these populations when they are viewed in this manner. The results of their analysis are given in Table 2. Application of this simple pairing statistic suggests that there are approximately *c.* 1300 separate known Antarctic falls represented in the ANSMET collection. Lindstrom further found that most

Table 2. *Meteorite populations for modern falls and Antarctic finds*

Meteorite type	Modern falls* (%)	Antarctic Finds† (%)
Ordinary chondrites	79.5%	79.5%
Carbonaceous chondrites	4.2	5.2
Enstatite chondrites	1.6	1.7
Achondrites	8.3	8.5
Stony-irons	1.2	0.9
Irons	5.1	4.3
Total meteorites	830	1294

* From Graham *et al.* (1985).
† Numbers are given only for meteorites recovered by the ANSMET programme; data from Lindstrom & Score (1995); corrected for pairing: a pairing value of five was assumed for ordinary chondrites.

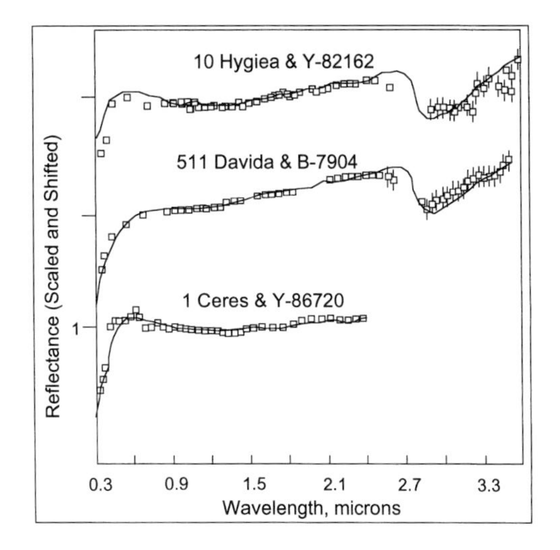

Fig. 5. Comparison of the reflectance spectra of three C-type asteroids (ten Hygiea, 511 Davida and one Ceres) and three unusual Antarctic carbonaceous chondrites (Yamato 82162, Belgica 7904 and Yamato 86720). Asteroid spectra (squares) have been scaled to a best fit to the meteorite spectra (lines). Mineralogical and compositional data indicate that the carbonaceous chondrites have been heated to up to 1000°C on the parent asteroid. These types of carbonaceous chondrites are only known from Antarctica, suggesting that these asteroids provided samples to Earth only in the past. Figure adapted from Hiroi *et al.* (1993, 1996).

rare types of meteorites (except lunar and SNC meteorites) are small in mass.

However, it is undeniable that a significant number of unique meteorites are found principally in Antarctica. These include lunar meteorites (all but two of which have been found in Antarctica), certain irons, CM1 chondrites, and various naturally heated carbonaceous chondrites including Belgica 7904, Yamato 82162 and Yamato 86720. The majority of these unusual meteorites are small, under 1 kg, and it is undeniable that small meteorites are more likely to be recovered in Antarctica than elsewhere on Earth. This is despite the fact that small meteorites can be removed from stranding areas by the katabatic winds; those that remain are generally easily found, particularly in stranding areas devoid of terrestrial rocks. Nevertheless, it is intriguing that none of these types of meteorites have been found outside Antarctica, even in other desert stranding surfaces where small masses are similarly recovered.

Comparison of the reflectance spectra of typical C-type asteroids (see Fig. 5) and the unusual, pre-terrestrially heated Antarctic carbonaceous chondrites shows an improved match over any other meteorites (Hiroi *et al.* 1993, 1996). The carbonaceous chondrites must have been heated to up to 1000°C on the parent asteroid. As these types of carbonaceous chondrites are only known from Antarctica, the suggestion is that these asteroids only provided samples to Earth in the past.

Meteorite masses. To a large degree the arguments for differences between Antarctic (old terrestrial age) and non-Antarctic (generally younger) meteorites comes down to the point that these unusual meteorites are generally small in size. It is certainly a fact that all Antarctic finds have smaller mean masses than comparable meteorites from the modern fall record (Harvey 1990, 1995; Huss 1990*a*, *b*; Harvey & Cassidy 1991). This is shown in Fig. 6. Four explanations for this latter observation are immediately apparent, others are undoubtedly also possible:

(1) Small, unusual meteorites are falling today, but being small are not generally recovered from most locales. Smaller meteorites are relatively easy to locate in Antarctica,

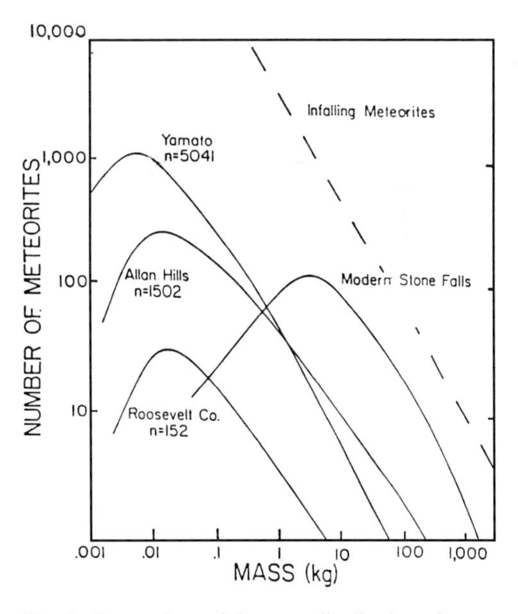

Fig. 6. Comparison of the mass distribution of meteorites in modern falls (Modern Stone Falls), fireballs (Infalling Meteorites), two Antarctic stranding surfaces (Yamato Mountains and Allan Hills) and Roosevelt County, New Mexico. It is observed that the masses of Antarctic meteorites are lower by at least a factor of ten from modern falls. Figure adapted from Huss (1990*a*, 1991).

because of the generally white to blue background and, locally, low number of admixed terrestrial rocks.
(2) Antarctic meteorites were identical to modern falls in mass, possibly different in type, and have been weathered into a steady-state population of smaller mass bits.
(3) Antarctic meteorites are identical to modern falls, but the largest individuals have been preferentially removed from the Antarctic population, leaving only smaller specimens. It should be noted that this would this make the Antarctic find record complementary to the modern fall record (where generally only the largest specimens are recovered). This explanation is, however, totally unsubstantiated by any reasonable mechanism.
(4) The ancient meteorite flux was truly different from the modern one, in type and mass.

The data appear to best support explanation (1).

We are therefore left with subtle indications that the old falls represented by the Antarctic collections have sampled a different meteorite parent body population. These differences are present at the 1–2% level of the total populations, however. Only further work will settle the issue of whether this difference is only apparent, caused by collection efficiency in the Antarctic, or real.

References

ANNEXSTAD, J., SCHULTZ, L. & ZOLENSKY, M. E. (eds) 1995. *Workshop on Meteorites From Cold and Hot Deserts*. Lunar and Planetary Institute Technical Report **95-02**.

BEIJING OBSERVATORY 1988. *Zhongguo gudai tianxiang jilu zongji* (*A Union Table of Ancient Chinese Records of Celestial Phenomena*). Kexue jishu chubanshe, Kiangxu, 63–105.

BENOIT, P. H., ROTH, J., SEARS, H. & SEARS, D. W. G. 1994. The natural thermoluminescence of meteorites. 7. Ordinary chondrites from the Elephant Moraine region, Antarctica. *Journal of Geophysical Research*, **99**, 2073–2085.

——, SEARS, H. & SEARS, D. W. G. 1992. The natural thermoluminescence of meteorites. 4. Ordinary chondrites at the Lewis Cliff Ice Field. *Journal of Geophysical Research*, **97**, 4629–4647.

——, —— & ——1993. The natural thermoluminescence of meteorites. 5. Ordinary chondrites at the Allan Hills Ice Fields. *Journal of Geophysical Research*, **98**, 1875–1888.

BLAND, P., BERRY, F. J., SMITH, T. B., SKINNER, S. J. & PILLINGER, C. T. 1996*a*. The flux of meteorites to the Earth and weathering in hot desert ordinary chondrite finds. *Geochimica et Cosmochimica Acta*, **60**, 2053–2059.

——, SMITH, T. B., JULL, A. J. T., BERRY, F. J., BEVAN, A. W. R, CLOUD, S. & PILLINGER, C. T. 1996*b*. The flux of meteorites to the Earth over the last 50,000 years. *Monthly Notices of the Royal Astronomical Society*, **283**, 551–565.

BOECKL, R. 1972. Terrestrial age of nineteen stony meteorites derived from their radiocarbon content. *Nature*, **236**, 25–26.

BURNS, R. G., BURBINE, T. H., FISHER, D. S. & BINZEL, R. P. 1995. Weathering in Antarctic H and CR chondrites: quantitative analysis through Mössbauer spectroscopy. *Meteoritics*, **30**, 625–633.

CASSIDY, W., HARVEY, R., SCHUTT, J., DELISLE, G. & YANAI, K. 1992. The meteorite collection sites of Antarctica. *Meteoritics*, **27**, 490–525.

DELISLE, G. 1993. Global change, Antarctic meteorite traps and the East Antarctic ice sheet. *Journal of Glaciology*, **39**, 397–408.

——1995. Storage of meteorites in Antarctic ice during glacial and interglacial stages. *In*: SCHULTZ, L., ANNEXSTAD, J. & ZOLENSKY, M. (eds) *Workshop on Meteorites from Cold and Hot Deserts*. Lunar and Planetary Institute Technical Report **95-01**, 26–27.

—— & SIEVERS, J. 1991. Sub-ice topography and meteorite finds near the Allan Hills and the Near Western Ice Field, Victoria Land, Antarctica. *Journal of Geophysical Research*, **96**, 15 577–15 587.

DENNISON, J. E., LIGNER, D. W. & LIPSCHUTZ, M. E. 1986. Antarctic and non-Antarctic meteorites form different populations. *Nature*, **319**, 390–393.

GRAHAM, A. L., BEVAN, A. W. R. & HUTCHISON, R. 1985. *Catalogue of Meteorites*. University of Arizona Press, Tucson.

HALLIDAY, I., BLACKWELL, A. T. & GRIFFIN, A. A. 1989. The flux of meteorites on the Earth's surface. *Meteoritics*, **24**, 173–178.

——, GRIFFIN, A. A. & BLACKWELL, A. T. 1981. The Innisfree meteorite fall: a photographic analysis of fragmentation, dynamics and luminosity. *Meteoritics*, **16**, 153–170.

HARVEY, R. P. 1990. Statistical differences between Antarctic finds and modern falls: mass frequency distributions and relative abundance by type. *In*: KOEBERL, C. & CASSIDY, W. A. (eds), *Workshop on Differences Between Antarctic and Non-Antarctic Meteorites*. Lunar and Planetary Institute Technical Report **90-01**, 43–45.

——1995. Moving targets: the effect of supply, wind movement, and search losses on Antarctic meteorite size distributions. *In*: SCHULTZ, L., ANNEXSTAD, J. & ZOLENSKY, M. (eds) *Workshop on Meteorites from Cold and Hot Deserts*. Lunar and Planetary Institute Report **95-01**, 34–37.

—— & CASSIDY, W. A. 1991. A statistical comparison of Antarctic finds and modern falls: mass frequency distributions and relative abundance by type. *Meteoritics*, **24**, 9–14.

HIROI, T., PIETERS, C. M., ZOLENSKY, M. E. & LIPSCHUTZ, M. E. 1993. Evidence of thermal metamorphism on C asteroids. *Science*, **261**, 1016–1018.

——, ——, ——, & ——1996. Thermal metamorphism of the C, G, B, and F asteroids seen from the 0.7 μm, 3 μm, and UV absorption strengths in comparison with carbonaceous meteorites. *Meteoritics and Planetary Science*, **31**, 321–327.

HUGHES, D. W. 1994. Comets and asteroids. *Contemporary Physics*, **35**, 75–93.

HUSS, G. 1990*a*. Meteorite mass distributions and differences between Antarctic and non-Antarctic meteorites. *In*: KOEBERL, C. & CASSIDY, W. A. (eds) *Workshop on Differences Between Antarctic and Non-Antarctic Meteorites*. Lunar and Planetary Institute Technical Report **90-01**, 49–53.

——1990*b*. Meteorite infall as a function of mass: implications for the accumulation of meteorites on Antarctic ice. *Meteoritics*, **25**, 41–56.

——1991. Meteorite mass distribution and differences between Antarctic and non-Antarctic meteorites. *Geochimica et Cosmochimica Acta*, **55**, 105–111.

IKEDA, Y. & KIMURA, M. 1992. Mass distribution of Antarctic ordinary chondrites and the estimation of the fall-to-specimen ratios. *Meteoritics*, **27**, 435–441.

JULL, T., BEVAN, A. W. R., CIELASZYK, E. & DONAHUE, D. J. 1995. Carbon-14 terrestrial ages and weathering of meteorites from the Nullarbor region, Western Australia. *In*: SCHULTZ, L., ANNEXSTAD, J. & ZOLENSKY, M. (eds) *Workshop on Meteorites from Cold and Hot Deserts*. Lunar and Planetary Institute Technical Report. **95-01**, 37–38.

KOEBERL, C. & CASSIDY, W. A. (eds) 1990. *Workshop on Differences Between Antarctic and Non-Antarctic Meteorites*. Lunar and Planetary Institute Technical Report **90-01**.

LINDSTROM, M. & SCORE, R. 1995. Populations, pairing and rare meteorites in the, US Antarctic meteorite collection. *In*: SCHULTZ, L., ANNEXSTAD, J. & ZOLENSKY, M. (eds) *Workshop on Meteorites from Cold and Hot Deserts*. Lunar and Planetary Institute Technical Report **95-01**, 43–45.

LOEKEM, M. & SCHULTZ, L. 1995. The noble gas record of H chondrites and terrestrial age: no correlation. *In*: SCHULTZ, L., ANNEXSTAD, J. & ZOLENSKY, M. (eds) *Workshop on Meteorites from Cold and Hot Deserts*. Lunar and Planetary Institute Technical Report **90-01**, 45–47.

MICHLOVICH, E. S., WOLF, S. F., WANG, M.-S., VOGT, S., ELMORE, D. & LIPSCHUTZ, M. 1995. Chemical studies of H chondrites. 5. Temporal variations of sources. *Journal of Geophysical Research*, **100**, 3317–3333.

NISHIIZUMI, K. 1995. Terrestrial ages of meteorites from cold and cold regions. *In*: SCHULTZ, L., ANNEXSTAD, J. & ZOLENSKY, M. (eds) *Workshop on Meteorites from Cold and Hot Deserts*. Lunar and Planetary Institute Techhnical Report **95-01**, 53–55.

——, ELMORE, D. & KUBIK, P. W. 1989. Update on terrestrial ages of Antarctic meteorites. *Earth and Planetary Science Letters*, **93**, 299–313.

SCHERER, P., SCHULTZ, L. & LOEKEN, TH. 1995. Weathering and atmospheric noble gases in chondrites from hot deserts. *In*: SCHULTZ, L., ANNEXSTAD, J. & ZOLENSKY, M. (eds) *Workshop on Meteorites from Cold and Hot Deserts*. Lunar and Planetary Institute Technical Report **95-01**, 58–60.

——, ——, NEUPERT, U., *et al.* 1997. Allan Hills 88019: an Antarctic H-chondrite with a very long terrestrial age. *Meteoritics and Planetary Science*, **32**, 769–773.

SCHULTZ, L., WEBER, W. & BEGEMANN, F. 1991. Noble gases in H chondrites and potential differences between Antarctic and non-Antarctic meteorites. *Geochimica et Cosmochimica Acta*, **55**, 59–66.

SCHUTT, J., SCHULTZ, L., ZINNER, E. & ZOLENSKY, M. 1986. Search for meteorites in the Allan Hills region 1985–1986. *Antarctic Journal*, **21**, 82–83.

WASSON, J. T. 1985. *Meteorites. Their Record of Early Solar-System History*. W. H. Freeman, New York, 226–227.

WIELER, R., CAFFEE, M. W., & NISHIIZUMI, K. 1995. Exposure and terrestrial ages of H chondrites from Frontier Mountain. *In*: SCHULTZ, L., ANNEXSTAD, J. & ZOLENSKY, M. (eds) *Workshop on Meteorites from Cold and Hot Deserts*. Lunar and Planetary Institute Technical Report **95-01**, 70–72.

WELTEN, K. C. 1995. *Exposure histories and terrestrial ages of Antarctic meteorites*. PhD thesis, University of Utrecht.

——, ALDERLIESTEN, C., VAN DER BORG, K. & LINDNER, L. 1995. Cosmogenic beryllium-10 and aluminum-26 in Lewis Cliff meteorites. *In*: SCHULTZ, L., ANNEXSTAD, J. & ZOLENSKY, M. (eds) *Workshop on Meteorites from Cold and Hot Deserts*. Lunar and Planetary Institute Technical Report **95-01**, 65–70.

WETHERILL, G. W. 1986. Unexpected Antarctic chemistry. *Nature*, **319**, 357–358.

WHILLANS, I. M. & CASSIDY, W. A. 1983. Catch a falling star: meteorites and old ice. *Science*, **222**, 55–57.

WLOTZKA, F., JULL, A. J. T & DONAHUE, D. J. 1995. Carbon-14 terrestrial ages of meteorites from Acfer, Algeria. *In*: SCHULTZ, L., ANNEXSTAD, J. & ZOLENSKY, M. (eds) *Workshop on Meteorites from Cold and Hot Deserts*. Lunar and Planetary Institute Technical Report **95-01**, 72–73.

WOLF, S. F. & LIPSCHUTZ, M. E. 1995*a*. Applying the bootstrap to Antarctic and non-Antarctic H-chondrite volatile-trace-element data. *In*: SCHULTZ, L., ANNEXSTAD, J. & ZOLENSKY, M. (eds) *Workshop on Meteorites from Cold and Hot Deserts*. Lunar and Planetary Institute Technical Report **95-01**, pp. 73–75.

—— & ——1995*b*. Yes, meteorite populations reaching the Earth change with time. *In*: SCHULTZ, L., ANNEXSTAD, J. & ZOLENSKY, M. (eds) *Workshop on Meteorites from Cold and Hot Deserts*. Lunar and Planetary Institute Technical Report **95-01**, 75–76.

—— & ——1995*c*. Chemical studies of H chondrites – 7. Contents of Fe^{3+} and labile trace elements in Antarctic samples. *Meteoritics*, **30**, 621–624.

ZOLENSKY, M. E. & PACES, J. 1986. Alteration of tephra in glacial ice by 'unfrozen water'. *Geological Society of America, Abstracts with Programs*, **18**, 801.

——, MARTINEZ, R. & MARTINEZ DE LOS RIOS, E. 1995*a*. New L chondrites from the Atacama Desert, Chile. *Meteoritics*, **30**, 785–787.

——, RENDELL, H., WILSON, I. & WELLS, G. 1992. The age of the meteorite recovery surfaces of Roosevelt County, New Mexico, USA. *Meteoritics*, **27**, 460–462.

——, SCHUTT, J. W., REID, A. M., JAKES, P., MARTINEZ DE LOS RIOS E. & MILLER, R. M. 1995*b*. Locating new meteorite recovery areas. *In*: ANNEXSTAD, J., SCHUTZ, L. & ZOLENSKY, M. E. (eds) *Workshop on Meteorites From Cold and Hot Deserts*. Lunar and Planetary Institute Technical Report **95-02**, 78–80.

Extraterrestrial impacts on earth: the evidence and the consequences

RICHARD A. F. GRIEVE

Geological Survey of Canada, Ottawa, Ontario, Canada K1A 0Y3

Abstract: The terrestrial record of impact events is incomplete and evolving. There are inherent biases in ages, distribution and sizes of known impact events that result from the high levels of endogenic activity on the Earth. Nevertheless, an estimated terrestrial cratering rate of $(5.6 \pm 2.8) \times 10^{-15}\,km^{-2}\,a^{-1}$ for impact structures with diameters >20 km and younger than 120 Ma can be calculated and is compatible with astronomical observations. The most obvious evidence of impact is the occurrence of 156 impact structures known as of the end of 1996. Few impact structures, however, are sufficiently pristine to provide great detail concerning their original morphology. Some basic morphometric parameters, however, can be estimated. Confirmation of an impact origin for particularly terrestrial structures comes generally from the recognition of diagnostic shock metamorphic effects in the target rocks. Some 16 impact events are currently recognized in the stratigraphic column but it would appear that many others await discovery. The best documented and only global example of such events is at the Cretaceous–Tertiary (K–T) boundary. Although a number of killing mechanisms have been proposed for the attendant mass extinction, the nature of the target, in this case containing sulphates, may be the reason for the devastating effect of this event on the biosphere. In early Earth history, the high impact flux was probably a significant factor in the modification of the atmosphere, biosphere and hydrosphere and the collision of a Mars-sized object with the proto-Earth may have been responsible for the formation of the Earth's Moon. Prompted by the association of the K–T event with a global mass extinction, it has been proposed that other mass extinctions and geological phenomena in the Phanerozoic are impact related, possibly through periodic cometary showers. At this time, there is little or no evidence for this association, although model calculations and the terrestrial cratering rate suggest that impact cannot be ignored as a forcing function for transient changes in the Earth's atmosphere and climate. Although unequivocal evidence linking impact to climatic changes remains to be discovered, it is a statistical certainty that if human civilization exists for time-scales of hundreds of thousands of years it will be severely affected or possibly destroyed by an impact event. Given the stochastic nature of impact events, however, this could happen sooner rather than later.

The systematic study of the terrestrial impact record is a relatively recent endeavour in the Earth sciences. The first terrestrial impact crater was recognized in the early 1900s (Barringer 1906). By the late 1930s, the number of recognized terrestrial structures of impact origin had risen to about 20 (Boon & Albritton 1938) but there was considerable controversy over the origin of several (e.g. Bucher 1936). The recognition, however, of diagnostic indicators of shock metamorphism, such as shatter cones (e.g. Dietz 1947, 1968), microscopic planar deformation features in tectosilicates and other minerals, high-pressure polymorphs, etc. (e.g. see papers in French & Short (1968)), at certain terrestrial structures, established reliable criteria for the occurrence of extreme dynamic pressures and, hence, an impact event. These criteria have been developed further by observation and experiment over the intervening years and, despite occasional challenges, have stood the test of time and new observations (French 1990).

Considerable impetus for the study of terrestrial impact structures arose as the result of the planetary exploration programmes of the USA and former USSR. Impact craters were found to be ubiquitous landforms on the terrestrial planets and defined entire geological terranes on those bodies that had retained portions of their earliest crusts (Head & Solomon 1981). It was recognized that terrestrial impact structures were essentially the sole source of ground truth data on large-scale impact events, with respect to the nature and spatial distribution of impact-related lithologies and structure, particularly in the third dimension. Although details of the terrestrial impact record served as analogues for the nature of impact cratering on the other terrestrial planets (e.g. see papers in Roddy *et al.* (1977)), impact was not generally regarded as a process of much importance to the evolution of the Earth.

This state of affairs changed dramatically with the discovery of evidence, in the form of enhanced platinum group elemental abundances (Alvarez *et al.* 1980; Ganapathy 1980; Smit & Hertogen 1980), for the occurrence of a major impact at the Cretaceous–Tertiary (K–T) boundary 65 Ma ago. The potential association of major impact events with mass extinctions had

GRIEVE, R. A. F. 1998. Extraterrestrial impacts on Earth: the evidence and the consequences. *In*: GRADY, M. M., HUTCHISON, R., MCCALL, G. J. H. & ROTHERY, D. A. (eds) *Meteorites: Flux with Time and Impact Effects*. Geological Society, London, Special Publications, **140**, 105–131.

been suggested previously (e.g. de Laubenfels 1956; Urey 1973). Similarly, the geochemically anomalous nature of K–T boundary sediments had also been noted earlier (Christensen *et al.* 1973). What differed in 1980 was that the state of knowledge regarding impact processes on Earth had reached such a level that the evidence offered for an impact event at the K–T boundary was readily accepted by a section of the geoscience community.

Although originally contentious to many in the larger community, the involvement of a major impact at the K–T boundary and the related mass-extinction event in the biosphere is now generally accepted (e.g. for the tenor of the development of the K–T debate, see papers in Silver & Schultz (1982), Sharpton & Ward (1990) and Ryder *et al.* (1996)). As a result of the discovery of the K–T impact as an agent for global change, there has been a substantial increase in interest in the terrestrial impact record and major terrestrial impacts have been called upon as causative agent for a variety of geological and geophysical phenomena, including other extinctions (e.g. Raup, 1990; Rampino & Haggerty 1994), geomagnetic reversals (e.g. Pal & Creer 1986), and flood basalt volcanism (e.g. Stothers & Rampino 1990). Some of these proposed associations, however, are controversial and speculative.

The terrestrial impact record is incomplete and evolving. It is characterized by inherent biases, largely related to the high level of geological activity on the Earth. These biases must be accounted for if the record is to be used to make general statements and inferences regarding the record itself and/or its consequences for Earth evolution. This contribution provides an overview of some of the salient points of the known terrestrial impact record and comments upon some of the outstanding issues. It is not intended to be a comprehensive review but provides instead an introduction to the more detailed literature.

General character of the record

At the time of writing, the known impact record consists of 156 impact structures or crater fields, in the case of some small impacting bodies, which broke up in the atmosphere (Melosh 1981), and some ten events, which are recorded in the stratigraphic record, some of which are related to known impact structures. Because of the effects of erosion, the terrestrial impact record contains a mixture of topographic forms and it is more appropriate to use the more generic term impact structures than impact craters, which by definition imply a negative topographic form. The most recent published list of known terrestrial impact structures is in Grieve *et al.* (1995). The discovery rate of new impact structures is about five per year, with the rate of discovery having increased substantially over the last two decades (Fig. 1). A current listing of known terrestrial impact structures can be found on the world wide web at http://gdcinfo.agg.nrcan.cg.can.

The spatial distribution of known terrestrial impact structures is biased towards the cratonic areas of North America, Australia and Scandinavia through to western Russia and the Ukraine (Fig. 2a). This bias is a result of two factors. These are relatively stable cratonic areas with low rates of erosional and tectonic activity over extended periods of geological time. They are thus the best available surfaces for the preservation of impact structures in the terrestrial geological environment. They are also areas where there have been active programmes to search for and study impact structures. Twenty-five years ago the known distribution was almost entirely dominated by discoveries in North America, Europe and Australia (Fig. 2b). Although these areas have continued to add to their inventory of

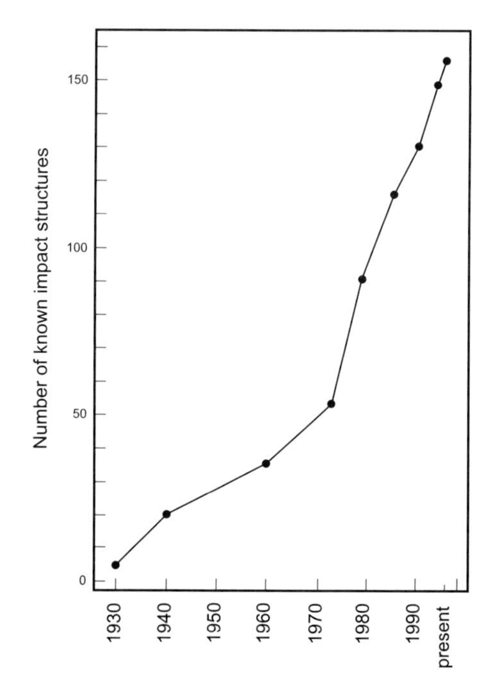

Fig. 1. Variation in cumulative number of known terrestrial impact structures with time. (Note more rapid discovery rate in the last two decades.)

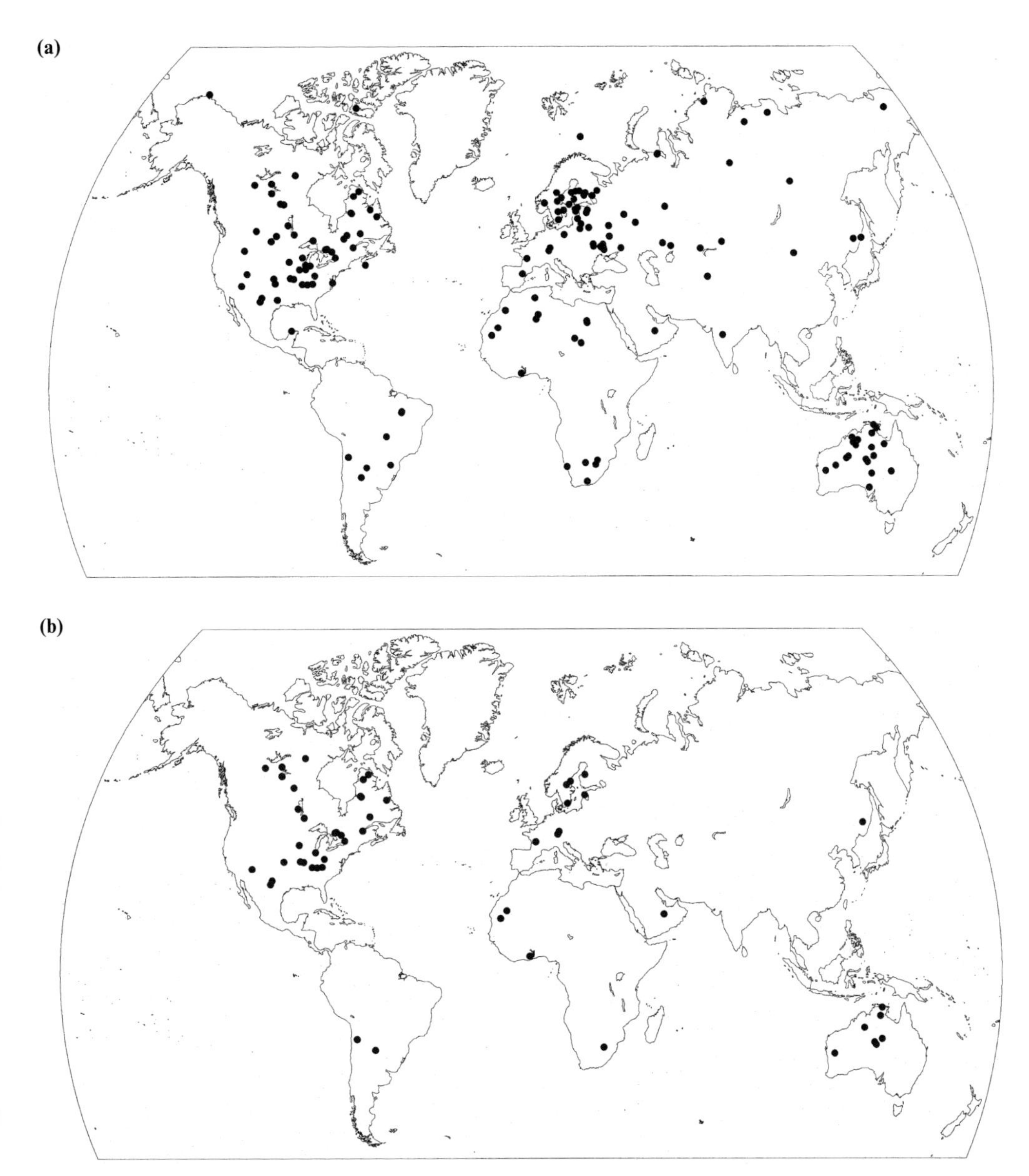

Fig. 2. (**a**) Spatial distribution of currently known terrestrial impact structures. (**b**) Spatial distribution of known impact structures in 1972.

known structures, there have been numerous additions to the known record throughout the world (see Fig. 2a and b), particularly in the former USSR and, to a lesser extent, Africa.

There is also a more subtle change in the manner in which additional terrestrial impact structures are being identified. Twenty-five years ago, the identification of terrestrial impact structures was largely the province of a relatively small number of specialists, who had impact structures within their sphere of primary interest. Currently, it is just as likely that the recognition of a new addition to the list of impact structures will be made by a member of the larger Earth science community. Although it cannot be proven, the higher recognition factor for impact-related phenomena achieved in recent years (Fig. 1) has been substantially compounded by the high scientific and public profile accorded to the K–T boundary event

Table 1. *Basic characteristics of impact structures mentioned in the text*

Crater name	Location	Latitude	Longitude	Age (Ma)	Diameter (km)
Acraman	Australia	S32°1′	E135°27′	>450	90
Barringer	USA	N35°2′	W111°1′	0.049 ± 0.003	1.186
Beaverhead	USA	N44°36′	W113°0′	*c.* 600	75
Boltysh	Ukraine	N48°45′	E32°10′	88 ± 3	24
Bosumtwi	Ghana	N6°30′	W1°25′	1.03 ± 0.02	10.5
Brent	Canada	N46°5′	W78°29′	450 ± 30	3.8
Chesapeake Bay	USA	N37°17′	W76°1′	35.5 ± 0.6	85
Chicxulub	Mexico	N21°20′	W89°30′	64.98 ± 0.05	170
Dellen	Sweden	N61°48′	E16°48′	89.0 ± 2.7	19
Eagle Butte	Canada	N49°42′	W110°30′	<65	10
East Clearwater	Canada	N56°5′	W74°7′	290 ± 20	26
Gardnos	Norway	N60°39′	E9°0′	500 ± 10	5
Gosses Bluff	Australia	S23°49′	E132°19′	142.5 ± 0.5	22
Goyder	Australia	S13°29′	E135°2′	>65	7
Haughton	Canada	N75°22′	W89°41′	23 ± 1	24
Jänisjärvi	Russia	N61°58′	E30°55′	698 ± 22	14
Kelly West	Australia	S19°56′	E133°57′	>550	10
Manicouagan	Canada	N51°23′	W68°42′	214 ± 1	100
Mistastin	Canada	N55°53′	W63°18′	38 ± 4	28
Mjølnir	Norway	N73°48′	E29°40′	143 ± 20	40
Montagnais	Canada	N42°53′	W64°13′	50.5 ± 1.5	45
Morokweng	South Africa	S26°28′	E23°32′	145 ± 0.8	70
Popigai	Russia	N71°40′	E111°40′	35.7 ± 0.2	100
Ries	Germany	N48°53′	E10°37′	15 ± 1	24
Slate Islands	Canada	N48°40′	W87°0′	<350	30
Söderfjärden	Finland	N62°54′	E21°42′	*c.* 600	5.5
Sudbury	Canada	N46°36′	W81°11′	1850 ± 3	250
Tookoonooka	Australia	S27°7′	E142°50′	128 ± 5	55
Vredefort	South Africa	S27°0′	E27°30′	2023 ± 4	300
West Clearwater	Canada	N56°13′	W74°30′	290 ± 20	36
Zhamanshin	Kazakhstan	N48°20′	E60°58′	0.9 ± 0.1	13.5

and its attendant impact structure, Chicxulub (Table 1; Hildebrand *et al.* 1991).

Active terrestrial geological processes, such as erosion and sedimentation, result in characteristics of the terrestrial impact record that are not as dominant on other planetary bodies. For example, *c.* 35% of known impact structures are buried by post-impact sediments. Such structures are particularly prevalent in the former USSR, where there are large areas of platform sediments, which have been subjected to systematic coverage by relatively closely spaced geophysical, particularly gravity, data. Most buried impact structures were detected initially as geophysical anomalies and later drilled, for scientific or economic purposes, which confirmed their impact origin. Recent examples include Morokweng in South Africa (Table 1; Corner *et al.* 1997) and Tookoonooka in Australia (Table 1; Gostin & Therriault 1997). Two impact structures, Mjølnir in the Barents Sea (Table 1; Gudlaugsson 1993) and Montagnais off Nova Scotia, Canada (Table 1; Jansa & Pe-Piper 1987), are completely submerged beneath the sea (Fig. 2a). They occur on continental shelves. No impact structures are known from the true ocean floors. This reflects the relatively young age and the generally poor resolution of geological knowledge of the ocean floors. Meteoritic debris, however, is known over a distance of at least 600 km in the south-east Pacific, where a *c.* 0.5 km body impacted in the Late Pliocene but apparently failed to make a crater in the ocean floor (Kyte *et al.* 1988).

Individual impact structures can have complex post-formation histories involving burial, exhumation and erosion. If unprotected from erosion, the geological signature of a 20 km diameter impact structure may be rendered unrecognizable and removed from the record in as little as 120 Ma (Grieve 1984). This is a general statement, and recognizable lifetimes for individual structures are variable and dependent on local geological history. For example, post-impact burial can result in the preservation of relatively small structures of considerable age, e.g. Söderfjärden (Table 1; Lehtovaara 1992), KellyWest (Table 1; Shoemaker & Shoemaker 1996) and others. There is, however, a general bias in the ages of known terrestrial impact structures. More than 55% are <200 Ma old (Grieve & Pesonen 1996). This reflects the problems of preservation and, to a lesser extent, recognition in the highly active geological environment of the Earth.

The sizes of known terrestrial impact structures range up to *c.* 300 km (Vredefort, Table 1). Although previously listed as 140 km, recent analyses of various attributes of this highly eroded structure suggest that its original diameter was closer to *c.* 300 km (Therriault *et al.* 1997). This illustrates another characteristic of the terrestrial impact record. In most cases, original rim diameters (Table 1) are reconstructed estimates. For highly eroded structures with minimal information, there can be considerable uncertainty in these estimates. For example, the original diameter of Goyder (Table 1) is estimated at between 7 and 25 km (Haines 1996). A similar problem can occur with buried structures, where rim diameter estimates are based on the interpretation of geophysical data; for example, estimates for the diameter of Chicxulub (Table 1) range from *c.* 170 km (Hildebrand *et al.* 1991) to *c.* 300 km (Sharpton *et al.* 1993). The more conservative diameters are adopted here (Table 1).

There is also a bias in the sizes of known terrestrial impact structures. At larger diameters, the cumulative size–frequency distribution can be approximated by a power law (Fig. 3), similar to the production distribution observed on the other terrestrial planets (Basaltic Volcanism Study Project 1981). In the terrestrial case, however, this distribution more probably represents a steady-state condition between the formation and removal of impact structures from the record. At diameters below *c.* 20 km, the cumulative size–frequency falls off, with an increasing deficit of structures at smaller diameters (Fig. 3), although Neukum & Ivanov (1994) have argued that the number of structures with diameters of *c.* 10 km is still representative. This drop-off appears to be an inherent property of the record, as it has remained through the addition of new structures to the known record (Fig. 3). The deficit of small craters is due to a combination of atmospheric crushing of weaker impacting bodies by the atmosphere (Melosh, 1981), the relative ease with which smaller (shallower) structures can be buried or eroded, and the intrinsic difficulty in recognizing smaller structures.

With the previously noted biases, it is clear that care must be exercised when estimating an average cratering rate from the terrestrial impact record. For example, to reduce the effects of the loss of older and smaller structures, Grieve (1984) restricted the sample of structures used to calculate a rate to structures ≤120 Ma and with diameters (D) ≥20 km. Detailed arguments as to why these and other restrictions to the sample used to estimate a terrestrial cratering rate are necessary have been given in Grieve (1984). The net result is that the estimated average cratering rate for the last *c.* 100 Ma is $(5.6 \pm 2.8) \times 10^{-15}\,km^{-2}\,a^{-1}$ for $D \geq 20$ km (Grieve & Shoemaker 1994). The relatively high (±50%) uncertainty attached to this estimate reflects concerns of small number of statistics and completeness of search for existing impact structures. Although it is generally believed that the bulk of the larger impact structures have been recognized on the better searched areas of the Earth, e.g. the North American craton (Fig. 2a), this estimated average cratering rate illustrates just how poorly the record is known in other areas (Fig. 2a). For example, the average cratering rate suggests that *c.* 17 ± 8 structures with $D \geq 20$ km should have been formed in an area the size of Africa (*c.* $30 \times 10^6\,km^2$) in 100 Ma. The known impact record of Africa indicates no known impact structures of the appropriate size and younger than 100 Ma.

The terrestrial cratering rate estimated from impact structures is comparable with rate estimates based on astronomical observations of Earth-crossing bodies (Shoemaker *et al.* 1990; Rabinowitz *et al.* 1994). These rates, however, are higher than those estimated from crater counts at the Apollo 12 and 15 maria sites on the

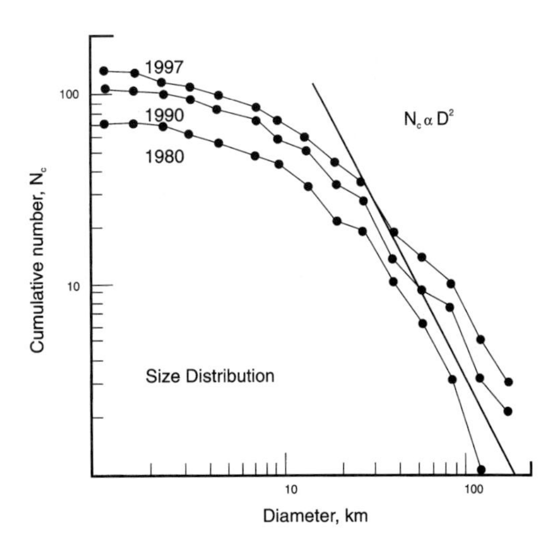

Fig. 3. Size–frequency distribution of known terrestrial impact structures in 1980, 1990 and at present. (Note similar shapes of the distributions over time, with a fall-off from a power-law approximation at *c.* 20 km diameter. See text for more details.)

Moon (Basaltic Volcanism Study Project 1981), which represent an integration of the cratering rate in the Earth–moon system over *c.* 3.2 Ga. This has prompted some debate as to whether the cratering rate has remained constant over the last *c.* 3.2 Ga (Grieve & Shoemaker 1994; Neukum & Ivanov 1994). Given the uncertainties associated with the observations and rescaling lunar impacts to terrestrial conditions, which have been discussed in some detail in Neukum & Ivanov (1994), it is prudent to be equivocal as to whether this discrepancy is apparent or real. Moreover, it is difficult to derive an astrophysical mechanism to increase the cratering rate in more recent geological time, although it has been suggested that it could be due to an increase in the flux of long-period comets as a result of oscillations within the Galactic plane (Shoemaker *et al.* 1990).

Approximately 40% of known terrestrial impact structures have been dated isotopically, most commonly from the analysis of impact melt rocks. Although impact events are highly energetic, involving the conversion of considerable kinetic energy in the impacting body to waste heat in the target rocks, they are also highly transient events in which pressure–temperature decay from the point of impact is very rapid (Ahrens & O'Keefe 1977; Melosh 1989). As a result, by far the bulk of the target rocks (*c.* 90%) affected by an impact event are subjected to insufficient temperature–time integrals to disturb significantly their isotopic systems (Deutsch & Schärer 1994). The bulk of the isotopic 'ages' for terrestrial impact structures are K–Ar or more recent ^{40}Ar–^{39}Ar dates. Fine-grained, generally clast-rich, impact melt rocks are not easy to date isotopically, because of inherited argon in incompletely degassed clasts. As a result, stepwise ^{40}Ar–^{39}Ar dates are preferable to whole-rock K–Ar or ^{40}Ar–^{39}Ar dates (Bottomley *et al.* 1990). In some cases, reproducible whole-rock K–Ar dates can be in considerable error with respect to dating the actual age of the impact event (see Currie 1971; Mak *et al.* 1976). Errors in assigning an age can also occur with ^{40}Ar–^{39}Ar dates, particularly if the materials analysed are not completely melted (see Kunk *et al.* 1989; Izett *et al.* 1993). There are only a few cases where impact melt rocks are of sufficient grain size or have sufficient compositional variation that Rb–Sr isochrons are of use (Deutsch & Schärer 1994). Generally, the most precise ages come from U–Pb dating of highly shocked zircons (Krogh *et al.* 1993, 1996) and new zircons crystallized in impact melt lithologies (Hodych & Dunning 1992). Such cases, however, are currently very few in number.

The majority of known terrestrial impact structures have biostratigraphic dates. In most cases they are minimum estimates, given that the time span between impact and infilling of the topographic depression of the craterform is not known. In other cases, the time of impact is only known to be less than the age of the target rock; for example, Eagle Butte (Table 1) is formed in Cretaceous rocks and given an age of <65 Ma. In some such cases, the degree of erosion can serve to provide a relative crude constraint but is often open to interpretation and debate, for example, at the Slate Islands (Table 1; see Grieve & Robertson 1976; Sharpton *et al.* 1996). For those structures with a poorly constrained age, there is an additional human bias. Their ages tend to be listed as some numerical value ending in 5 or 10 Ma. That is, an uncertain age is more likely to be estimated at, for example, 100 ± 20 Ma than 102 or 98 ± 20 Ma. This introduces yet another set of biases for any detailed attempts to interpret the significance of currently known impact ages.

Recognition of terrestrial impact events

'Craterforms'

By far the bulk of known terrestrial impact events have been identified through the occurrence of an impact structure. The basic forms of impact structures on the terrestrial planets are relatively fixed. Smaller structures are known collectively as simple structures and are characterized by a bowl-shaped depression surrounded by a structurally uplifted rim area (Figs 4 and 5). The canonical terrestrial example is Barringer or Meteor Crater (Fig. 4; Table 1; Barringer 1906). At larger diameters, impact structures have a complex form (Figs 6 and 7), which is characterized by an uplifted central area, shallow annulus and structurally modified rim area. This transition in craterform occurs over a relatively restricted range of diameters, as exemplified on the Moon, which provides much of the knowledge base on the relatively pristine morphology and morphometry of impact structures (e.g. Pike 1977). In the terrestrial case, the transition from simple to complex forms occurs at 2 km and 4 km for sedimentary and crystalline (igneous, metamorphic) target rocks, respectively. Similar to the lunar case, there does, however, appear to be some overlap in form near the transition diameters. Within complex impact structures a general progression in form has been recognized from those with a central peak, a central peak

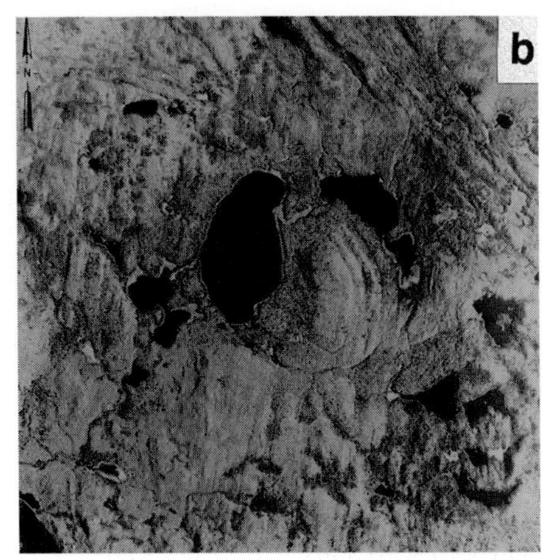

Fig. 4. (**a**) Oblique aerial photograph of canonical example of a terrestrial simple impact structure. Barringer or Meteor Crater (Table 1). (**b**) Vertical aerial photograph of the highly degraded (rim removed, interior filled with sediments) Brent simple impact structure (Table 1).

and interior ring, an interior ring to, finally, multi-ring basins. Again, lunar impact structures provide the major basis for this subdivision but this progression is recognized throughout the terrestrial planets (e.g. Wood & Head 1976).

In most cases, terrestrial impact structures are severely modified by erosion (Fig. 4), and in some cases, the craterform has been completely removed. In these cases, the presence of an impact structure is recognized by the structural and geological effects of impact in the target rocks rather than the presence of a characteristic craterform. For example, Gosses Bluff (Table 1) has a positive topographical form consisting of an annular ring of hills *c.* 5 km in diameter (Fig. 6). They consist of erosionally resistant beds from within the original central uplifted area of a complex impact structure (Fig. 7). The original craterform, which has an estimated diameter of *c.* 22 km, has been removed by erosion. There are several other impact structures, which have some form of rings, e.g. Manicouagan (Table 1; Floran & Dence 1976), Haughton (Table 1; Robertson & Sweeney 1983) but it is not clear whether these are primary forms or secondary erosional artefacts

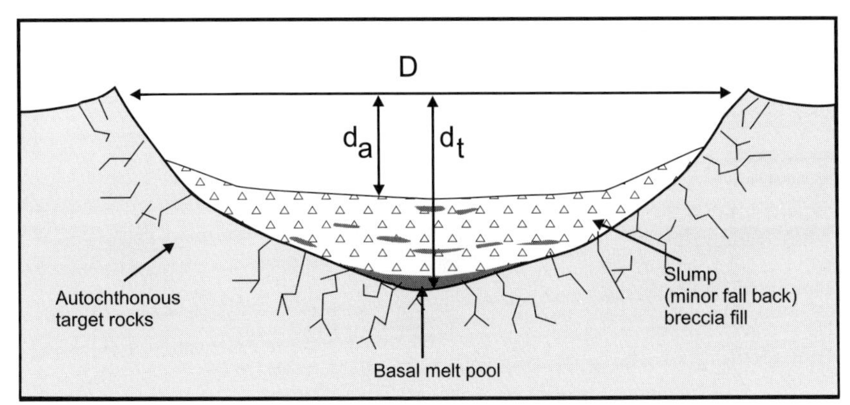

Fig 5. Schematic cross-section of a terrestrial simple impact structure, indicating various lithological morphometric and structural attributes. *D*, diameter; d_a, apparent depth; d_t, true depth. (See text for morphometric relations.)

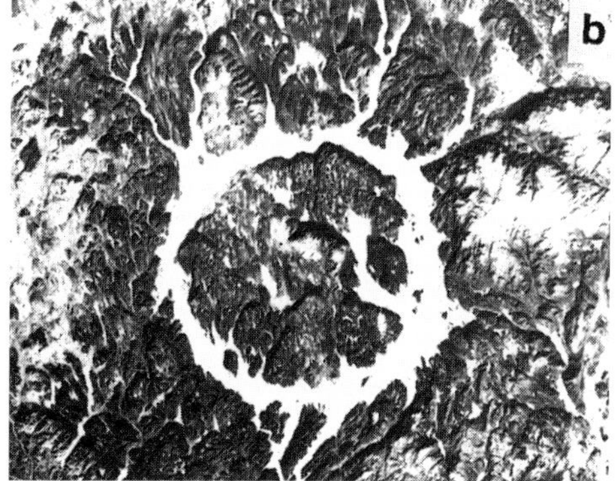

Fig. 6. (**a**) Oblique aerial photograph of the highly degraded Gosses Bluff complex impact structure (Table 1). (See text for more detail.) (**b**) LANDSAT image of the Manicouagan complex impact structure (Table 1). Annular lake is, in part, man-made and has a diameter of *c.* 65 km.

(Grieve & Head 1983). There have been claims that the largest known terrestrial impact structures have multi-ring forms, e.g. Chicxulub (Sharpton *et al.* 1993), Sudbury (Table 1; Stöffler *et al.* 1994; Spray & Thompson 1995) and Vredefort (Therriault *et al.* 1997). Although certain of their geological and geophysical attributes form annuli, it is not clear that these correspond, or are related in origin, to the obvious topographical rings observed in lunar multi-ring basins (Spudis 1993). In addition, Sudbury is also highly deformed by tectonic activity. There are a number of other cases, e.g. Gardnos in Norway (Table 1; French *et al.* 1977) and Beaverhead in the USA (Table 1; Hargraves *et al.* 1994), of tectonically deformed terrestrial impact structures.

There are also other subtleties to the character of craterforms in the terrestrial record that do not appear on the other terrestrial planets. A number of relatively young, and, therefore, only slightly eroded, complex impact structures (e.g. Haughton, Ries, Zhamanshin (Table 1)) do not have an emergent central peak or other interior topographical expression of a central uplift (Garvin & Schnetzler 1994). These structures are in mixed targets of platform sediments overlying basement. Although there are no known comparably young complex structures entirely in crystalline targets, the buried and well-preserved Boltysh structure (Table 1), which is of comparable size, has an emergent central peak, similar to the appearance of lunar central peak craters. This difference in

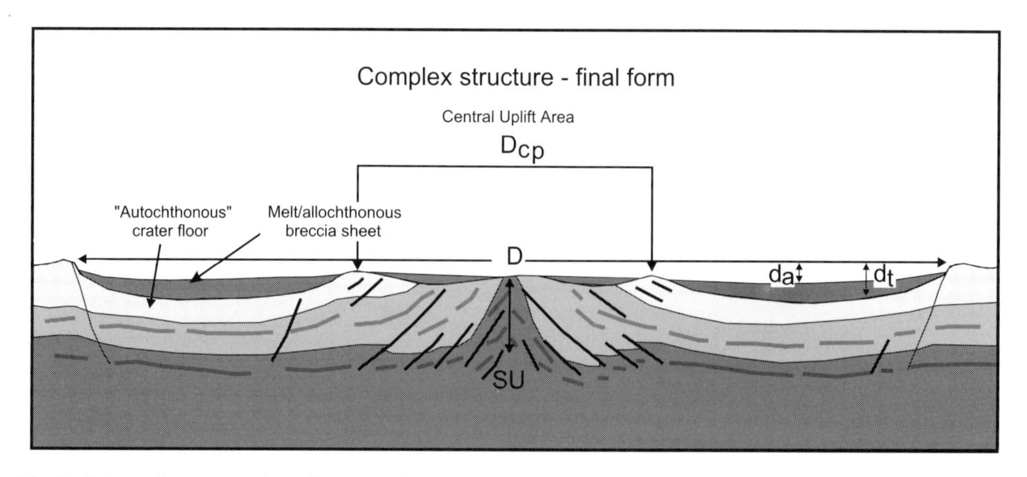

Fig. 7. Schematic cross-section of a terrestrial complex impact structure, indicating various lithological morphometric and structural attributes. D, diameter; d_a, apparent depth; d_t, true depth; D_{cp}, diameter central uplift; SU, stratigraphic uplift. (See text for morphometric relations.)

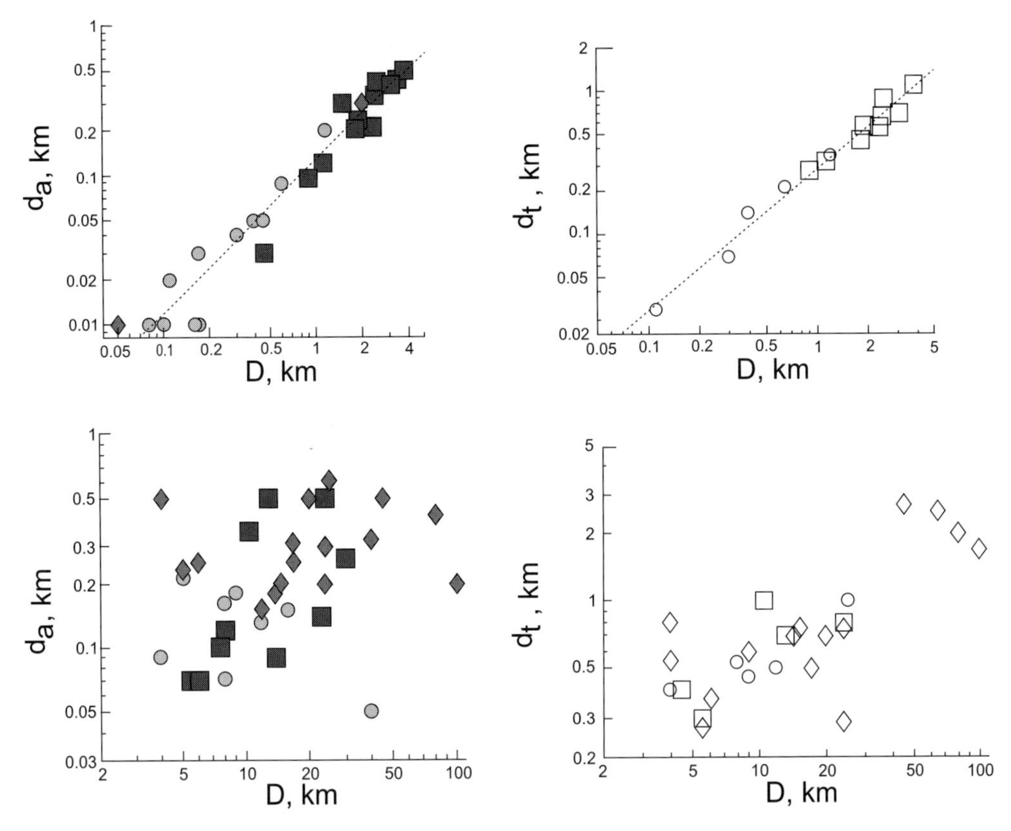

Fig. 8. Apparent (d_a, closed symbols) and true depth (d_t, open symbols) relations with rim diameter (D) for terrestrial impact structures in sedimentary (circles), mixed (diamonds) and crystalline (squares) targets. Top, simple structures. Bottom, complex structures. (See text for more details and significance of the data.)

form is probably a target rock effect but it has not been studied in detail.

Previous attempts to define morphometric relations, in particular depth–diameter relations, were based on terrestrial impact structures with little modification and, therefore, better quality data (Grieve *et al.* 1989; Grieve & Pesonen 1992). This, however, severely limited the number of data. The Geological Survey of Canada maintains a database on terrestrial impact structures, and the approach used here was to query the database and use all available data to define depth–diameter relations (see Figs 5 and 7 for definitions of the variables). The results are shown in Fig. 8. The relations for simple structures (Fig. 5) are d_a (apparent depth) $= 0.12D^{1.02}$ ($n = 22$, $r = 0.95$) and d_t (true depth) $= 0.28D^{0.98}$ ($n = 14$, $r = 0.98$), where D is diameter, n is number of structures used to define the relation and r is the correlation coefficient. These are very similar to previously defined relations (Grieve *et al.* 1989). No meaningful depth–diameter relations are forthcoming from the complete database for complex impact structures (Fig. 8). Previous relations defined $d_a = 0.12D^{0.30}$ and $d_a = 0.15D^{0.43}$ for sedimentary and crystalline targets, respectively (Grieve & Pesonen 1992). These indicate that the depths of structures in sedimentary targets are shallower than for equivalent-sized structures in crystalline targets. This would appear to be confirmed by the more complete dataset (Fig. 8). It probably reflects a target rock effect but warrants better documentation and further study, as it pertains to variable rock strength in cratering mechanics. Less severe strength-related target rock effects on the morphology of complex structures have been noted on the Moon (Cintala *et al.* 1977).

Unlike depth, stratigraphic uplift (SU, Fig. 7) variation with diameter at complex impact structures is fairly well constrained ($n = 24$) with $SU = 0.86D^{1.03}$ (Grieve & Pilkington 1996).

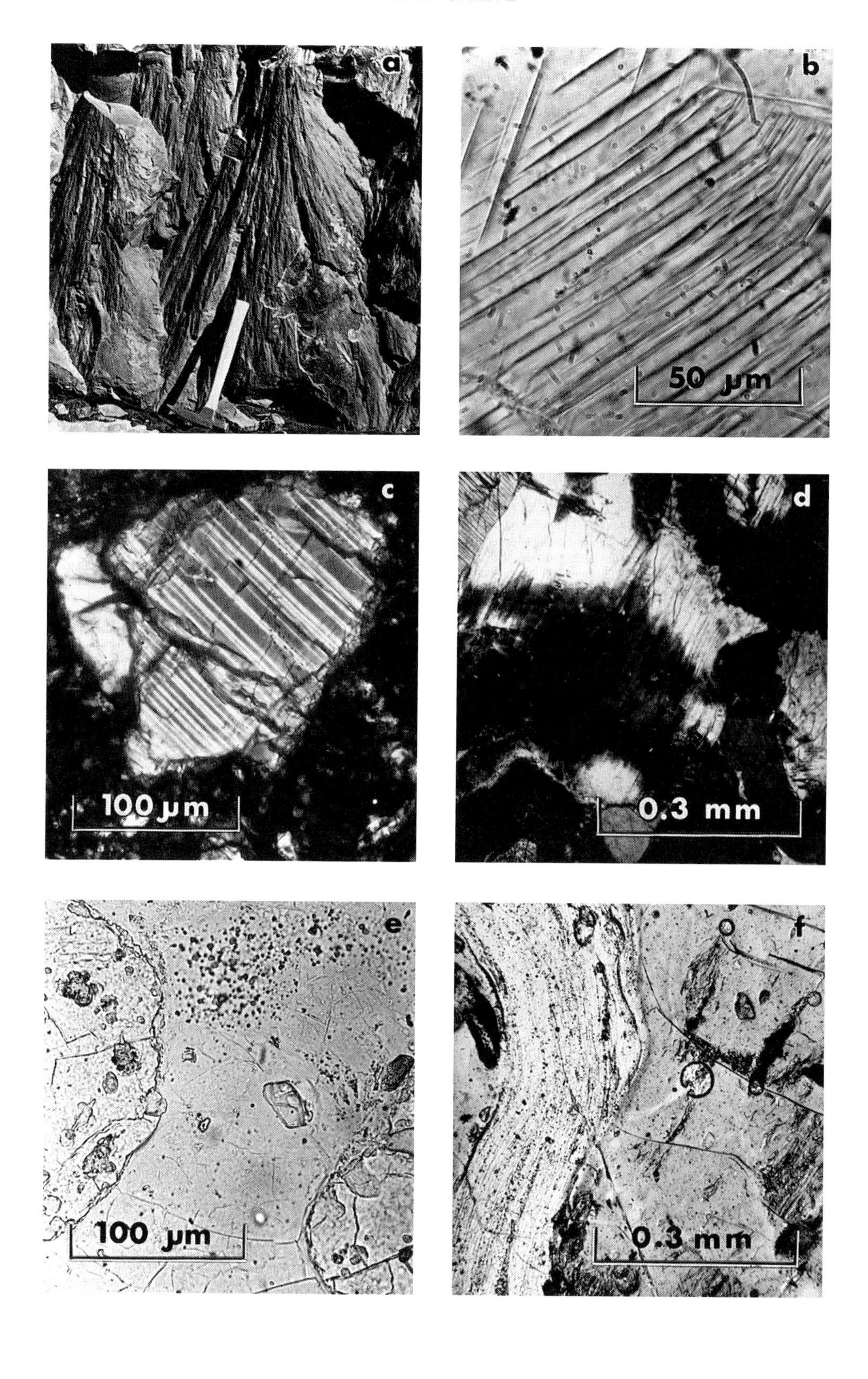
a
b
50 μm
c
100 μm
d
0.3 mm
e
100 μm
f
0.3 mm

Similarly, the diameter of the central uplift area (D_{cp}, Fig. 7) at its maximum radial expressions constrained ($n=44$) to $D_{cp}=0.31D^{1.02}$ (Therriault *et al.* 1997).

Shock metamorphism

Given the deleterious effects of the terrestrial environment on impact craterforms, reducing many to vague circular or quasi-circular topographical, geological and/or geophysical anomalies, the principal criterion for the recognition of terrestrial impact structures is the occurrence of shock metamorphic effects in the target rocks. The overall characteristics of shock metamorphic effects have been elaborated in some detail in French & Short (1968) and Stöffler (1971, 1972, 1974). Similarly, the spatial distribution of shock effects relative to various structural attributes of terrestrial impact structures has been illustrated in Grieve & Pilkington (1996).

On impact, a portion of the impacting body's kinetic energy is partitioned into kinetic energy, which results in the formation of a craterform in the target rocks. The bulk of the remainder of the kinetic energy is partitioned into increasing the internal energy of the target rocks. The target rocks are compressed and experience extremely high transient pressures and temperatures during the passage of the shock wave; for example, the peak pressures experienced in crystalline target rocks by the impact of a stony body, such as a chondritic asteroid, at $25\,km\,s^{-1}$ is *c.* 100 GPa (10 mbar) (Ahrens & O'Keefe 1977). During shock compression, considerable pressure–volume work is done on the target rocks. On decompression, not all this mechanical work is recovered. Decompression occurs along a release adiabat and some of the work is trapped in the target rocks as post-shock waste heat. This leads to the heating, melting and even vaporization of part of the target rocks. In the case of the impacting body, the peak pressure and, therefore, the post-shock temperatures are sufficient to lead generally to melting and vaporization.

The combined effect of this compression and heating is the production of a series of irreversible changes, which are manifested in individual minerals and the target rocks as a whole. These changes are known collectively as shock metamorphic effects. Shock metamorphic effects are produced at pressures and temperatures well beyond those encountered in terrestrial metamorphism caused by igneous and tectonic activity. The physical conditions upon impact are a function of the initial impact parameters: projectile and target rock type and impact velocity, which determine peak pressures at impact, and projectile size, which determines the absolute radial distance at which particular shock metamorphic effects occur. Shock metamorphic effects are also produced on vastly different time-scales from endogenic metamorphic effects and disequilibrium is the rule not the exception.

The only known diagnostic megascopic indicator of shock is the occurrence of shatter cones (Dietz 1968). These conical striated fracture surfaces are best developed in fine-grained, homogeneous lithologies, such as carbonates and quartzites (Fig. 9). They have, however, been noted in coarser-grained crystalline rocks, such as granites, but are rare and generally poorly developed (Dietz 1972). Shatter cones develop at relatively low shock pressures of 2–6 GPa (Roddy & Davis 1977) but have been found in rocks subjected to *c.* 25 GPa (Milton 1977). As a result, they are generally developed in the parautochthonous rocks flooring impact structures and are most often exposed in eroded central uplift structures of complex impact structures. Several studies (e.g. Manton 1965; Milton 1977; Stesky & Halls 1983) have indicated that, when the beds containing shatter cones at such impact structures are rotated back to their pre-impact position, the majority of the apices of the shatter cones point to a central location representing the point of impact or origin of the shock wave. Their early formation is also attested to by the occurrence of shatter cone fragments in breccia deposits, including dyke breccias in the floor of some impact structures (e.g. Halls & Grieve 1976).

The most common shock metamorphic effect is the occurrence of so-called planar deformation features (PDFs) in tectosilicates (Fig. 9; Hörz 1968). When fresh, these microscopic features

Fig. 9. Various shock metamorphic effects. (**a**) Shatter cones in quartzite, Sudbury (Table 1). (**b**) Sets of planar features with $\{10\bar{1}3\}$ orientations in quartz. Hardhat nuclear explosion, plane-polarized light. (**c**) Deformation bands in clinopyroxene, cross-polarized light. (**d**) Partial transformation of plagioclase (An_{54}) to maskelynite. Manicouagan (Table 1), cross-polarized light. (**e**) Small clusters of high refractive index coesite in two grains of diaplectic silica glass (left and lower right), separated by maskelynite containing a small apatite crystal. Ries (Table 1), plane-polarized light. (**f**) Vesicular feldspar glass showing flow lines (left) in a matrix of homogenized melt containing many small inclusions. West Clearwater (Table 1), plane-polarized light.

are filled with glass (Fig. 9; von Engelhardt & Bertsch 1969). At older impact structures, they are more commonly annealed and are manifested as linear chains of inclusions. Such features are called decorated PDFs (Robertson *et al.* 1968). Most studies of PDFs have focused on quartz (Fig. 9) because of its generally inert nature, its relatively simple crystal structure and its widespread occurrence in terrestrial rocks. Planar deformation features in quartz develop at shock pressures *c.* 5–10 GPa and higher. The nature of PDFs and other manifestations of the shock metamorphism in quartz have been discussed in considerable detail in Stöffler & Langenhorst (1994) and Grieve *et al.* (1996) and are not repeated here.

The discovery of PDFs in quartz was the first physical evidence of impact in K–T boundary samples (Bohor *et al.* 1984; Bohor 1990). Although this generated some debate from non-impact proponents for the origin of the K–T boundary (e.g. Carter *et al.* 1990), the characteristics of shock-produced PDFs are well established and distinct from those of other lamellar micro-structures produced in other endogenic high-strain environments. Some of the confusion in the literature regarding whether endogenic geological processes can produce characteristic shock metamorphic effects can be traced to the usage by certain workers of the term 'shocked' to describe minerals with optical mosaicism, cleavage and kinking (e.g. Carter *et al.* 1986). Although these features are characteristic of lowgrade shocked minerals, they are not diagnostic shock metamorphic effects (Grieve *et al.* 1996) and can occur at relatively high strain rates generated by some endogenic geological processes.

At pressures in excess of *c.* 30 GPa and *c.*40 GPa, shock-induced disorder is sufficient to render feldspar and quartz, respectively, to a glass (Fig. 9). These are solid-state glasses with properties distinct from fusion glasses (Stöffler & Hornemann 1972). They are referred to as diaplectic or thetomorphic glasses. The term 'maskelynite' is reserved for shock-induced solid-state plagioclase glass (Bunch *et al.* 1967), which was first observed in the meteorite Shergotty (Tschermak 1872). In addition to progressive structural disorder, shock can also generate metastable high-pressure mineral polymorphs, e.g. stishovite and coesite from quartz (Fig. 9; Chao *et al.* 1961, 1962), diamond from graphite (Masaitis *et al.* 1972) and others. It has been suggested that diamond, and SiC, can also be produced in impact events by chemical vapour deposition from the vaporized ejecta plume (Hough *et al.* 1995).

By 50–60 GPa, there is sufficient post-shock waste heat trapped after decompression that individual minerals, mineral mixtures (Fig. 9) and finally whole rocks begin to melt (Fig. 10). Whole-rock melting leads to the formation of so-called impact melt rocks. They occur as glass bombs in ejecta (von Engelhardt 1990), as glassy to crystalline lenses within the breccia fill of simple and complex structures, or as annular sheets surrounding the central uplift of complex crater structures (Figs. 6 and 7; Grieve *et al.* 1977). When crystallized, impact melt sheets have igneous textures but tend to be heavily charged with clastic debris towards their lower and upper contacts. They may, therefore, have a textural resemblance to endogenic igneous rocks. Both represent crystallized silicate melts. In some cases, impact melt rocks were identified initially as volcanic lithologies, e.g. Dellen (Table 1; Svenonius 1895; Lundquist 1963); Jänisjärvi (Table 1; Eskola 1921); Manicouagan (Rose 1955; Bérard 1962) and others. Impact melt rocks, however, can have an unusual chemistry compared with endogenic volcanic rocks, as their composition reflects the melting of a mixture of target rocks, as opposed to partial melting and/or fractional crystallization relationships of some progenitor. Isotopic analyses indicate that such parameters as $^{87}Sr/^{86}Sr$ and $^{143}Nd/^{144}Nd$ initial ratios reflect those of the precursor lithologies, whereas isochron dating methods indicate much younger crystallization ages, which are related to the impact event (e.g. Jahn *et al.* 1978; Faggart *et al.* 1985).

Enrichments above target rock levels in siderophile elements, and sometimes Cr, have been identified in some impact melt rocks (Palme 1982). These are due to an admixture of up to a few per cent of meteoritic impacting body material. In some cases the relative abundances of the various siderophiles have constrained the impacting body to the level of meteorite class, e.g. East Clearwater (Table 1) was formed by a Cl chondrite (Palme *et al.* 1979). In other cases, no siderophile anomaly has been identified. This may be due to the inhomogeneous distribution of meteoritic material within the impact melt rocks and sampling variations (Palme *et al.* 1981) or to differentiated, non-siderophile enriched impacting bodies, such as basaltic achondrites (Wolf *et al.* 1980).

Estimates of impacting body types, taken from the literature without qualification, are dominated by those with chondritic compositions (table II in Grieve & Pesonen (1996)). This may be misleading. Impacting body identification at many of the impact structures in the former USSR is generally based on relatively

Fig. 10. Outcrop of massive coherent impact melt rocks at Manicouagan (top, Table 1) and Popigai (bottom, Table 1). (Note trees for scale and mega-breccia underlying melt rocks at Popigai.) Bottom photograph courtesy V. L. Masaitis (Karpinsky Institute, St Petersburg).

few elements and inter-elemental ratios. Also, the geochemical signature of chondritic impacting bodies is the most easily identifiable, through their relatively high abundances of both siderophile elements and Cr. Identification of non-chondritic impacting bodies, e.g. Mistastin (Table 1), can be more of a negative result that the body was not a chondrite, rather than the firm indication that it was, as suggested, an iron body. Similar arguments apply to the identification of basaltic achondrites, which have compositions most similar to terrestrial rocks and tend to represent a negative result in terms of meteoritic contamination in impact rocks.

There has been recent interest in using Re–Os isotope systematics to detect a meteoritic component in impact melt rocks. This approach has been used to identify a meteoritic component in K–T boundary sediments (Luck & Turekian 1983). New, more sensitive, analytical techniques have extended the utility of Re–Os isotopes; particularly, where the target rocks are already relatively enriched in siderophile or platinum group elements (Koeberl *et al.* 1994). It is possible that the increased sensitivity of Re–Os isotopic techniques may help detect additional cases of faint meteoritic signatures in impact lithologies. For example, the technique has been used most recently at Vredefort, where 0.2% of meteoritic material was detected in several samples of the so-called Vredefort Granophyre (Koeberl *et al.* 1996*a*), generally believed to represent impact melt rock (e.g. Therriault *et al.* 1996). Unfortunately, Re–Os systematics are, in themselves, not an effective discriminator between meteorite classes.

Stratigraphic record

Known occurrences of impact-related materials in the stratigraphic record are very limited (Fig. 11). They have been detailed in Grieve (1997), and are only summarized here. Before the interpretation that geochemical anomalies in K–T boundary sediments were due to impact, the known record of impact in the stratigraphic column was limited to the occurrence of the Australasian, Ivory Coast and North American and moldavite tektite and microtektite strewn fields. The Ivory Coast microtektites are associated with Bosumtwi (Table 1; Schnetzler *et al.* 1967). The association of moldavite tektites with the Ries structure has also been recognized for some time (von Engelhardt & Hörz 1965). Most recently, the Chesapeake Bay structure (Table 1) appears to be the best candidate for the source of the North American strewn field (Koeberl *et al.* 1996*b*). The source of the Australasian field is currently unknown.

As a result of the interest generated by the K–T discoveries and possible connection between

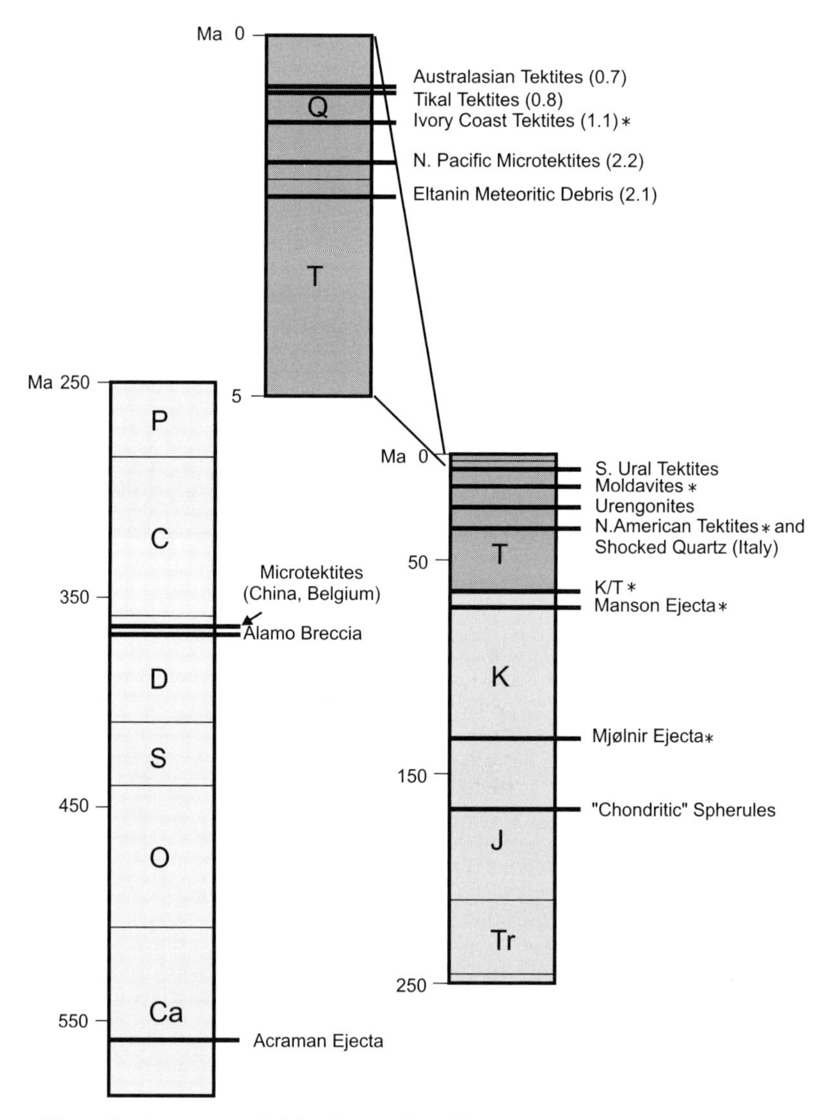

Fig. 11. Known evidence for impact recorded in the stratigraphic column. (See text and Grieve (1997) for more details.)

impact events and other short-term events in the geological record, there have been searches for siderophile element (mostly Ir) anomalies at other major stratigraphic boundaries. This work has had mixed success, with positive (e.g. Xu *et al.* 1985, 1989), negative (e.g. Orth *et al.* 1990) and conflicting results (see Hsu *et al.* 1985; Orth *et al.* 1990). In some cases, weak anomalies have been reported but, at this time, there is no compelling reason, or confirmatory evidence, to ascribe them (e.g. see table II in Rampino & Haggerty (1996)) to impact-related processes. By far the majority of known impact events recorded in the stratigraphic column (Fig. 11) were recognized initially through the occurrence of physical, not geochemical, evidence of impact. Unlike the K–T event, however, none is global in scale. For example, ejecta from Acraman (Table 1) have been recognized up to *c.* 500 km from the structure, as a distinct clastic layer in shales of the Bunyeroo Formation of the Adelaide Geosyncline (Williams 1986, 1994). There is an associated siderophile anomaly, although there has been post-impact, low-temperature remobilization (Wallace *et al.* 1990). In the case of the underwater Mjølnir structure in the Barents Sea confirmation of an impact origin for the structure comes, in fact, from shocked materials and a *c.* 1 ng g^{-1} Ir anomaly in ejecta encountered in a drill hole 30 km NNE of the structure (Dypvik *et al.* 1996).

The relatively sparse evidence for impact-related material and geochemical anomalies in the stratigraphic record is not as contradictory as it appears. Approximately 30% of the recognized impact materials occur in the most recent 2 Ma of the sedimentary record (Fig. 11). The best documented events are recorded in material recovered from the sedimentary record of the present ocean basins. This is the segment of the stratigraphic record that has the highest signal-to-noise ratio, with respect to the possible detection of anomalous impact-related horizons. Detection of similar events in lithified materials from the palaeo-ocean or palaeocontinental environment will be more challenging.

The impact signal of the K–T event is recog nizable globally, because large impact events have the capacity to produce atmospheric blow-out (Melosh 1989). In essence, a hole is blown out in the atmosphere above the impact site. This permits some impact-related materials, entrained in the impact fireball, to rise above the atmosphere and, thus, encircle the globe in relatively short time periods. Model calculations indicate that it does not require a Chicxulub-sized event to result in atmospheric blow-out. Impact events resulting in impact structures in the *c.* 20 km size-range can produce atmospheric blow-out (Melosh 1989). Given the estimated terrestrial cratering rate, there are obviously many more traces of impact events to be detected in the stratigraphic column and this would appear to be a 'growth industry' for the future.

Kyte *et al.* (1993) analysed *c.* 25 m of piston core LL44-GPC3 from the central North Pacific. The time period covered was the last *c.* 70 Ma. They noted only one detectable Ir peak and that was at the K–T boundary (Kyte & Wasson 1986). Away from the peak, the Ir accumulation rate varied by *c.* 30% of the mean (Kyte *et al.* 1993). It has been argued, therefore, that the lack of other identifiable Ir peaks in this oceanic piston core may be related to the geochemical signal-to-noise ratio of an individual impact event against the background variation because of the daily influx of cosmic dust (Grieve 1997). Given, therefore, the likelihood that even relatively large impact events result in a relatively weak geochemical signal, depending on the sedimentary environment, it would seem that it is necessary to conduct a detailed physical examination for impact-related materials of any suggested impact horizons in the stratigraphic column, in concert with geochemical studies.

Impact and Earth evolution

As the frequency of occurrence of impacts has varied throughout geological time (Fig. 12), the potential for impact to act as an agent for change has varied. Many of the effects of impact on Earth evolution have been masked or overtaken by terrestrial processes. Given the relatively poor terrestrial record, the occurrence of many of the effects is based on inferences from other planetary bodies and model considerations.

The earliest (unknown) Earth

The basic working hypotheses for the formation of the terrestrial planets in the solar nebula is the accretion of small bodies by collision and the subsequent growth to larger bodies. Numerical simulations of planetary accretion envisage the collision and growth of 10^8–10^{12} bodies with diameters of *c.* 10 km (Ohtsuki *et al.* 1988; Wetherill & Stewart 1989). These asteroidal-sized bodies grow rapidly (in 10^4–10^5 a) by impact to *c.* 20–30 embryonic planets with masses of the order of 5×10^{26} g (10% of the mass of the present Earth). This early stage of planetary accretion is believed to be followed by a stage involving truly giant impacts between the

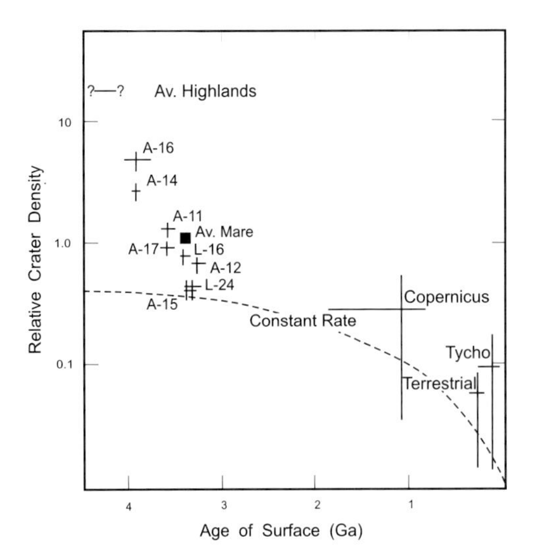

Fig. 12. Relative variation in crater density with time in the Earth-Moon system. (Note rapid increase in cratering rate before *c.* 3.5 Ga.) Data from Basaltic Volcanism Study Project (1981).

embryos. During this stage, encounter velocities are pumped up to 1–10 km s^{-1} and a few planetary bodies with masses of *c.* 10^{27} g are formed in 10^6–10^8 a (Wetherill 1990). In the context of planetary formation, impact is the most fundamental process.

In the case of the Earth, the last embryonic planetary collision may have resulted in the formation of the Earth's Moon (Hartmann *et al.* 1986). The hypothesis that the Moon formed by the collision of a large body with Earth, resulting in the ejection of iron-deficient crustal and mantle materials, was proposed originally by Hartmann & Davis (1975). Numerical simulations of a glancing impact of a Mars-sized body with the Earth result in the formation of an Earth-orbiting accretionary disk from which a Moon-like body can form (Benz *et al.* 1986; Kipp & Melosh 1986; Cameron & Benz 1991). The key is that the bulk of the material that forms the accretionary disc is originally jetted terrestrial or impacting body mantle material. Condensation and reaccretion result in volatile- and siderophile-depleted material, which goes to form the Moon. Although the collision hypothesis for the origin of the Moon is consistent with the constraints of the dynamics of the Earth–Moon system and the geochemical nature of the Moon, it remains a model. The consequences of such a giant impact for the proto-Earth, however, would have been severe. They would have included massive remelting of the proto-Earth (Tonks & Melosh 1993) and the loss of the original atmosphere (Cameron 1983). The latter could explain the differences between the noble gases contents in the atmospheres of Earth and Venus (Donahue & Pollack 1983).

After planetary formation, the subsequent high rate of bombardment by the remaining tail of accretionary debris is recorded on the Moon (Fig. 12) and other terrestrial planets that have preserved portions of their earliest crust. It is not recorded on the Earth, because of the high level of endogenic geological activity and the resultant relatively young surface. The lunar surface is the source of most interpretations of the effects of such an early, high impact flux on geological evolution. In the case of the Moon, a minimum of 6000 craters with $D > 20$ km are known to have been formed during this early period (Wilhelms 1987). In addition, there are *c.* 45 known impact basins ranging in diameter from Bailly at 300 km to the South Pole Aitken basin at 2600 km (Spudis 1993).

Throughout geological time, the Earth has received more impacts than the Moon. Not only is the Earth physically larger but it has a much larger gravitational cross-section. From the lunar cratering rate, and assuming relatively low approach velocities of 6–10 km s^{-1} at this time in Solar System evolution, it is estimated that *c.* 200 impact basins >1000 km may have been formed on the Earth from *c.* 4.6 to 3.8 Ga (Grieve *et al.* 1990). There has been considerable speculation as to the potential effects of these basin-sized impacts on the Earth. These include agents for the formation of proto-ocean basins (Frey 1980), proto-continental nuclei (Goodwin 1976; Grieve 1980), establishing the initial freeboard (Hargraves 1976), episodic crustal evolution (Glikson 1996), as well as potential genetic relations between basin-sized impacts and concentric geological and geophysical patterns in the older areas of the Earth's crust (Weiblen & Schultz 1978; Polosukhin 1981). By analogy with the lunar case, it is likely that few terrestrial surface rocks would have survived intact through this period of heavy bombardment.

Previous speculations have, to a first order, assumed that the effects of basin-sized impacts on the early Earth were similar to those on the Moon, modified by terrestrial environmental conditions, e.g. thinner lithosphere, free water, etc. Recent advances in understanding large-scale impact processes now suggest that this may not be the case. Planetary gravity is not a variable in determining the volume of impact melt produced in a specific impact event. Planetary gravity, however, is a variable in limiting the dimensions of the resultant craterform. As a result, the volume of impact melt produced in a specific impact event relative to the diameter of the impact structure increases with increasing planetary gravity or event size. This is termed differential scaling (Grieve & Cintala 1992; Cintala & Grieve 1994). Model calculations, comparing the Moon and Venus, indicate that melt volumes in events capable of producing 1000 km sized basins on Venus were such that the resultant craterform developed in impact melt. As melt has no strength, multi-ring basins, as are manifested by structures such as Orientale on the Moon, would not develop (Grieve & Cintala 1995). Terrestrial surface gravity is even greater than that on Venus (*c.* 980 v. 890 $cm\,s^{-2}$, respectively) and a similar situation would occur in impacts capable of producing Orientale-sized (*c.* 900 km) basins on the early Earth. On these grounds, therefore, analogies with lunar multi-ring basins appear unwarranted and reasoned speculations on the effects of such large impact events on the early terrestrial geosphere need to be rethought.

Basin-sized impacts on the early Earth will have affected the existing atmosphere, hydrosphere and the potential development of the biosphere. The vapour plume created by high-velocity ejecta in sufficiently large impacts can blow off the atmospheric mass above the tangent plane through the point of impact. Melosh & Vickery (1989) estimated that the primordial atmosphere on the early Earth could have been reduced in density through such a process. Model calculations suggest that the thermal energy released in the atmosphere by an impact on Earth of a body in the 500 km size-range, similar to the present-day asteroids Pallas and Vesta, is sufficient to vaporize the present world's oceans and would have effectively sterilized the surface of the early Earth (Sleep *et al.* 1989). In such an event, the Earth would have been enveloped by a hot rock vapour atmosphere that would radiate heat downward onto the surface, with a resultant surface temperature of 2000 K. Sleep *et al.* (1989) estimated that it would take 3 ka for the water-saturated atmosphere to rain out and re-form the oceans. They estimated that smaller impacts by bodies in the 200 km size-range would be sufficient to evaporate the photic zone of the oceans. Maher & Stevenson (1988) have examined this potential for the impact frustration of life from a temporal point of view. They concluded that life could survive in a deep marine setting as early as 4.2–4.0 Ga. Smaller impacts, however, would continue to make the surface inhospitable until *c.* 4.0–3.8 Ga. Similarly, photosynthesis would have been difficult before *c.* 3.8 Ga, because of relatively frequent impact-induced major climatic excursions. With the recent discovery of evidence for terrestrial life at *c.* 3.85 Ga (Mojzsis *et al.* 1996), it appears that these model predictions may be slightly severe or that primitive life was more tenacious than previously thought. Whatever the case, major impact events early in Earth history were undoubtedly a negative forcing function for the development and evolution of early life.

The later (known) Earth

The only potential evidence of large-scale impact events in the Archaean is the occurrence of extensive (>3000 km^2) spherule beds up to 2 m thick in 3.2–3.5 Ga Archaean greenstone belts in South Africa and Western Australia (Lowe & Byerly 1986; Lowe *et al.* 1989). These spherule beds, which are enriched in siderophile elements (e.g. up to 160 $ng\,g^{-1}$ Ir) in roughly chondritic proportions, have been interpreted as the recondensed distal debris from Archaean impacts in which the impacting body had a diameter in the range 20–50 km (Lowe *et al.* 1989). More recently, this interpretation has been severely challenged (Koeberl & Reimold 1995) and the issue of their origin is an open question. The oldest known terrestrial impact structures are Sudbury and Vredefort (Table 1). They have reconstructed original diameters of *c.* 250 km (Deutsch & Grieve 1994) and *c.* 300 km (Therriault *et al.* 1997), respectively. Although it is expected that such large events would have had a deleterious effect on the climate and biosphere, no direct evidence is known at present. These structures, however, also affected the local geology in a manner that is to human benefit. Both impact structures are the sites of world-class ore deposits (Grieve & Masaitis 1994).

An extensive literature exists concerning the evidence for a major impact event at the K–T boundary and its association with a mass extinction in the terrestrial biosphere 65 Ma ago. It is not the intention to review the evidence and interpretations here. The unequivocal physical evidence for impact contained in K–T boundary deposits consists of the occurrence of: PDFs in quartz, feldspar, zircon and chromite (Bohor *et al.* 1984; Izett 1987; Bohor 1990), stishovite (McHone *et al.* 1989), diamonds (Gilmour *et al.* 1992), high-temperature magnesioferrite spinels, believed to be vapour condensates (Kyte & Smit 1986; Kyte & Bostwick 1995), and various spherules, generally altered (Smit & Klaver 1981; Montanari *et al.* 1983) but including the tektite-like glass spherules in Haiti and other

Caribbean sites (Sigurdsson *et al.* 1991). The chemical evidence consists primarily of a global siderophile anomaly in K–T boundary deposits, indicative of an admixture of meteoritic material (e.g. Alvarez *et al.* 1980; Ganapathy 1980; Kyte *et al.* 1980; Smit & Hertogen 1980).

The original hypothesis by Alvarez *et al.* (1980) suggested global darkening and cessation of photosynthesis because of ejecta in the atmosphere as the killing mechanism for the K–T mass extinction. Early atmospheric evolution models involving dust loadings of 5×10^{18} g, the amount expected from a K–T-sized impact event, indicated optical depths in the atmosphere below that required for photosynthesis for about 3 months and below freezing temperatures on land for about 6 months (Toon *et al.* 1982). They also indicated similar effects for dust loadings down to two orders of magnitude lower (Gerstl & Zardecki 1982; Toon *et al.* 1982). Such dust loadings would be achieved during the formation of impact structures in the 50–100 km size range, i.e. corresponding in size to several known Phanerozoic-aged terrestrial impact structures that are not known to be associated with global mass extinctions. This suggests that the original 'lights-out' scenario of Alvarez *et al.* (1980) requires modification.

Soot has also been identified in K–T boundary deposits and its origin has been ascribed to global wild-fires (Wolbach *et al.* 1985). The calculated global abundance of this soot is 7×10^{16} g, sufficient to produce optical depths far below those required for photosynthesis (Wolbach *et al.* 1990). The soot, therefore, may enhance or even overwhelm the effects produced by global dust clouds. During the developing study of the K–T event, increasing emphasis has been placed on understanding the effects of vaporized and melted ejecta on the atmosphere. It is this hot expanding vapour cloud that provides a mechanism for igniting global wild-fires. Models of the thermal radiation produced by the re-entry of ejecta condensed from the vapour plume of a K–T-sized impact indicate a thermal radiation pulse on the Earth's surface equivalent to *c.* 10 kW m^{-2} for up to several hours (Melosh *et al.* 1990). This thermal radiation may be directly responsible, or indirectly responsible through the global killing and drying of vegetation that is later ignited by lightning strikes, for the global wild-fires. The calculations of Melosh *et al.* (1990) also indicate that the thermal radiation produced by the K–T event is at the lower end of that required to ignite living plant materials. Other model calculations that indicate the prompt and direct ignition of wild-fires would have been continental not global in scale (Hildebrand 1993).

Although originally contentious, there is now little doubt that the Chicxulub structure (Hildebrand *et al.* 1991) in the Yucatan peninsula is the K–T impact structure. Chicxulub is buried by about 1 km of Tertiary sediments but evidence for impact, in the form of PDFs in quartz and feldspar in drill core from deposits interior and exterior to the structure, as well as impact melt rocks, has been reported (e.g. Hildebrand *et al.* 1991; Kring & Boynton 1992; Sharpton *et al.* 1992). Variations in the concentration and size of shocked quartz grains and the thickness of K–T boundary deposits (Bohor *et al.* 1987; Izett 1987) point towards a source for these materials in the Americas. In addition, the geochemistry of K–T tektite-like glasses from Haiti matches the mixture of lithologies found at the Chicxulub site (Sigurdsson *et al.* 1991). Isotopic ages for the impact melt rocks at Chicxulub of 64.98 ± 0.05 Ma are indistinguishable from the K–T tektites at 65.07 ± 1.00 Ma (Swisher *et al.* 1992).

A considerable thickness of anhydrite ($CaSO_4$) occurs in the target rocks at Chicxulub. Impact heating would produce *c.* 10^{17}–10^{18} g of sulphur aerosols in the atmosphere (Brett 1992; Pope *et al.* 1994). It is not clear, however, what percentage of these aerosols would rapidly recombine with solids in the immediate area of the impact. Atmospheric modelling suggests that these sulphur aerosols would reduce light levels below those needed for photosynthesis for 6–9 months. In addition, if most of the aerosols were in the form of SO_2, solar transmission would drop to 10–20% of normal (a cloudy day) for up to an additional *c.* 13 years (Pope *et al.* 1994). It may be, therefore, that the devastating effects of the K–T impact compared with other known large impacts are due to the unusual composition of the target rocks at Chicxulub. Whatever the case, the temporal association of an extremely large impact crater with a world-wide ejecta layer and a global mass-extinction event is well established. The cause–effect relationship is less well established and is the subject of much debate and research. The modelling of the atmosphere–biosphere effects of large impact events is in its infancy and details of potential killing mechanisms require further resolution.

Considerable interest in the potential for impact to disrupt the biological balance on Earth followed the evidence for the involvement of large-scale impact at the K–T boundary. When Raup & Sepkoski (1984, 1986) reported evidence for a periodicity to the marine extinction record over the past 250 Ma, a number of works followed claiming a similar periodicity in the terrestrial cratering record (e.g. Alvarez

& Müller 1984; Davis *et al.* 1984; Rampino & Stothers 1984). They suggested that this was the result of periodic cometary showers caused by a variety of astronomical mechanisms. These periodic cometary showers were linked statistically to terrestrial mass extinction events and to a variety of other geological and geophysical phenomena. The claims were based on time-series analyses of subsets of the terrestrial cratering record. There were a variety of selection criteria but they conformed generally to 'large' ($\geq$5 km), 'young' ($\leq$250 Ma) and 'well-dated' ($\pm$20 Ma) impact structures.

Grieve *et al.* (1988) have argued against these conclusions, noting the biases in the terrestrial impact record, as exemplified here, and that, if the uncertainties in crater age estimates are taken into account, periodicities in the cratering record are questionable. They also argued that, to have confidence in the reality of any period, the uncertainties attached to individual age estimates had to be $<$10% of the period in question, which, in this case, was *c.* 30 Ma. This is not a general property of the terrestrial cratering record (Table 1). Heisler & Tremaine (1989) reached a similar conclusion based on different statistical arguments, and Baksi (1990) detected no periodicity, if the selected impact structures were restricted to those with age estimates of sufficient accuracy and precision. Weissman (1990) also found no evidence for periodic cometary showers and challenged the proposed mechanisms for producing periodic cometary showers. He did not, however, rule out random cometary showers.

Despite these arguments, periodic cometary showers, as defined by time-series analysis of the terrestrial cratering record, are still featured (e.g. Yabushita 1994, 1996) and suggested as a causative agent for various geological phenomena on Earth (e.g. Stothers & Rampino 1990; Stothers 1993; Rampino & Haggerty 1994, 1996). Most recently, Grieve & Pesonen (1996) reanalysed the terrestrial impact record, taking into account the uncertainty in age estimates for specific impact structures and failed to detect a statistically significant periodicity. The level of detection of periodicity was no more than expected from random data. They also emphasized a point alluded to earlier. Namely, not only is the precision of many age estimates for terrestrial impact structures poor but so is the accuracy, in some cases (Deutsch & Schärer 1994).

There is a fundamental difference between obtaining a 'date' and an actual 'age' for a particular event. A 'date' is a numerical result produced from analytical procedures. An 'age' is when that 'date' can be specifically and unequivocally related to a geological event. It is only in a few cases, where multiple dating methods have been applied, and have produced consistent results, that enough confidence can be placed on the estimated age of an impact structure or event in the stratigraphic column for it to be used in time-series analysis. Similarly, there are currently insufficient high-quality data to comment upon whether or not the cratering record contains clusters of impacts of similar age that could reflect large-scale events in the asteroidal belt and relatively rapid resonant dynamics. Central to arguments concerning some of the sweeping statements regarding the effects of impact on the biosphere and geosphere in more recent geological history (e.g. Rampino & Haggerty 1996) is that even an exact temporal equivalence does not prove a cause–effect relationship. Caution is warranted. To quote D. Rea in Kerr (1997) referring to the apparent temporal coincidence between global cooling in the Pliocene and increased volcanic activity, 'the geologist's most serious disease is assigning a cause-effect to things that occur at the same time when they may not have anything to do with each other'.

There are numerous incidents of rapid biological and/or climatic change, as indicated by stable isotopes, in the terrestrial geological record. The Earth is subjected to large-scale impacts, which, given the K–T observations and the results of model calculations, have the potential to produce transient changes to the atmosphere and, therefore, the climate. It would appear, therefore, to be a matter of time until some unequivocal links are made. This will require, however, more than the apparent temporal coincidence with an impact event. It will require the documentation of chemical or physical evidence of impact at the appropriate position in the stratigraphic column. Geologists have been making observations in and around Sudbury for over 100 years. Shatter cones are ubiquitous to the area but were not recognized until Dietz (1964) predicted their occurrence. Given this experience with the nature of 'objective' observation by humans and the current level of awareness of the nature of impact-related chemical and physical products, it is likely that such evidence will be forthcoming.

The terrestrial cratering rate indicates the frequency of K–T-sized events on Earth is of the order of one every *c.* 100 $\pm$ 50 Ma. Smaller, but still significant, impact events occur on shorter time-scales (Fig. 13). The model calculations of Gerstl & Zardecki (1982) suggest that dust loadings from the formation of impact craters as small as 20 km could produce light reductions and temperature disruptions. Such impacts occur on Earth with a frequency of two or three every million years (Fig. 13). Most recently, Kring *et al.* (1996) have suggested that the addition of vaporized S, originally contained in the impacting body, to the atmosphere is sufficient in such events to also result in short-term (few years) cooling. The youngest known structure in this size range is Zhamanshin (Fig. 13). Impacts of this scale are not likely to have a recognizable effect upon the biosphere, as the potential light and temperature disruptions are of limited severity and time. The most fragile component of the present environment, however, is human civilization, which is now highly dependent on an organized and technologically complex infrastructure for its survival. Although we seldom think of human civilization in terms of million of years, there is little doubt that if civilization lasts long enough it will suffer severely or may even be destroyed by an impact event. The impact hazard to human civilization has been the subject of a number of recent studies. From a review of Gehrels (1994), it would appear that the greatest threat (the product of probability and the expected death and destruction) is from ocean-wide impact-induced tsumanis, which occur on time-scales of tens of thousands of years (Morrison *et al.* 1994). For example, model calculations indicate that the impact of even a relatively small body (*c.* 200 m in diameter) would result in a wave in the open ocean that would still be 10 m high 1000 km from the point of impact (Hills *et al.* 1994).

Although we take some collective comfort in the fact that world civilization-threatening impact events occur on average only once every half million years or so (Fig. 13), it must be remembered that impact is a random process not only in space but also in time. To emphasize this point, in March 1989 an asteroidal body, named 1989 FC, passed within 700 000 km of the Earth (Chapman *et al.* 1989). Although 700 000 km is a considerable distance, it translates to a miss of the Earth by only a few hours, when orbital velocities are considered. This Earth-crossing body was not discovered until it had passed the Earth. It was estimated to be in the 0.5 km size range. If it had impacted the Earth, it would have released the equivalent of *c.* 1000 megatons TNT (*c.* 4.2×10^{18} J) of energy, sufficient to produce an impact structure only slightly smaller than Zhamanshin (Fig. 13), and possible global consequences (Chapman & Morrison 1989).

Impact events do occur on human time-scales. For example, the Tunguska event on 30 June 1908 was due to the atmospheric explosion of a relatively small, <100 m, body at an altitude ≤10 km. The energy released, based on the

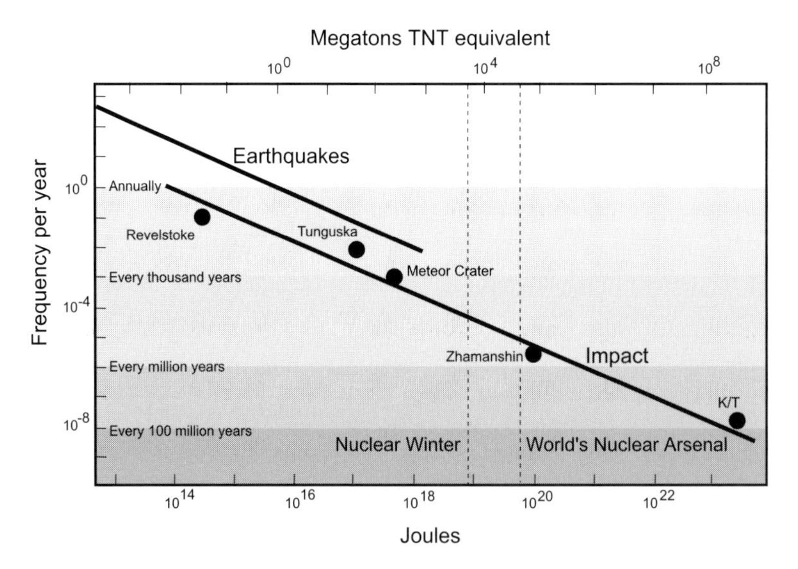

Fig. 13. Frequency of impacts with specific energies (in joules and megatons TNT equivalent). Also shown is comparable relation for earthquakes and some specific impact events. The Revelstoke and Tunguska events were in the atmosphere. Also indicated are the energies of the world's nuclear arsenal (*c.* 10^{20} J) and that believed required to induce a nuclear winter (*c.* 10^{19} J).

energy required to produce the observed seismic disturbances, has been estimated equivalent to the explosion of >10–20 megatons TNT (>(4.2–8.4) × 10^{16} J) (Chyba *et al.* 1993). It resulted in the devastation of *c.* 2000 km^2 of Siberian forest. Events such as Tunguska occur on a time-scale of hundreds of years (Fig. 13).

Concluding remarks

The detailed study of impact events on Earth is a relatively recent addition to the spectrum of studies engaged in by the geological sciences. More than anything, it was preparations for and, ultimately, the results of the lunar and the planetary exploration programme that provided the impetus and rationale for their study. This review has summarized some of the basic properties of the terrestrial record of impact cratering, which is characterized by inherent biases that result from the active geological environment of Earth. The terrestrial record of impact has made important contributions to our understanding of impact processes. Answers to many questions are known to, at least, the first order. Many details require clarification, however, and some problems, for instance the characteristics of superficial deposits, such as ejecta, are difficult to address in the active geological environment of Earth.

It is also apparent that impact can no longer be considered a process of interest only to the planetary community. It is a process that has fundamentally affected terrestrial evolution. For example, without the K–T impact the present-day biosphere may not have been dominated by mammals, and, in particular, humans. Similarly, without a Mars-sized impact forming the Moon, and the resulting tidal forces on the Earth, one can only speculate how the littoral zone, the most important area in the terrestrial ecosystem, would have evolved and been populated. Impact is the most catastrophic geological process known and is fundamental to the nature of the Solar System. Impact events have happened on Earth throughout geological time and will happen again. Although the occurrence of a large impact event has a low probability on the time-scale of the human lifespan, the consequences of its occurrence will be globally disastrous. Given the will and the appropriate support, it lies within our power, at least, to identify the potential threat to our civilization through systematic astronomical observations.

M. Ford and J. Smith assisted in manuscript preparation and J. Rupert and A. Therriault produced Figs 2 and 8, respectively. They are thanked for their contribution. A. Therriault also reviewed an earlier version of the manuscript. Formal reviews by A. Deutsch and R. Hutchison are appreciated. This paper is Geological Survey of Canada Contribution 1996480.

References

AHRENS, T. J. & O'KEEFE, J. D. 1977. Equations of state and impact-induced shock-wave attenuation on the Moon. *In*: RODDY, D. J., PEPIN, R. O. & MERRILL, R. B. (eds) *Impact and Explosion Cratering*. Pergamon, New York, 639–656.

ALVAREZ, L. W., ALVAREZ, W., ASARO, F. & MICHEL, H. V. 1980. Extraterrestrial cause for the Cretaceous–Tertiary extinction. *Science*, **208**, 1095–1108.

ALVAREZ, W. & MÜLLER, R. 1984. Evidence from crater ages for periodic impact on the Earth. *Nature*, **308**, 718–720.

BAKSI, A. K. 1990. Search for periodicity in global events in the geologic record: Quo Vadimus? *Geology*, **18**, 983–986.

BARRINGER, D. M. 1906. Coon Mountain and its crater. *Proceedings of the Academy of Natural Sciences of Philadelphia*, **57**, 861–886.

BASALTIC VOLCANISM STUDY PROJECT 1981. *Basaltic Volcanism on the Terrestrial Planets*. Pergamon, New York.

BENZ, W., SLATTERY, W. L. & CAMERON, A. G. W. 1986. Snapshots from a three-dimensional modelling of a giant impact. *In*: HARTMANN, W. K., PHILLIPS, R. J. & TAYLOR, G. J. (eds) *Origin of the Moon*. Lunar and Planetary Institute, Houston, 617–620.

BÉRARD, J. 1962. *Summary of geological investigations in the area bordering Manicouagan and Mushalagan Lakes, Saguenay County*. Quebec Department of National Resources, Preliminary Report **489**, 1–14.

BOHOR, B. F. 1990. Shock-induced microdeformations in quartz and other mineralogical indications of an impact event at the Cretaceous–Tertiary boundary. *Tectonophysics*, **171**, 359–372.

——, FOORD, E. E., MODRESKI, P. J. & TRIPLEHORN, D. M. 1984. Mineralogic evidence for an impact event at the Cretaceous–Tertiary boundary. *Science*, **224**, 867–869.

——, MODRESKI, P. J. & FOORD, E. E. 1987. Shocked quartz in the Cretaceous–Tertiary boundary clays: evidence for a global distribution. *Science*, **236**, 705–709.

BOON, J. D. & ALBRITTON, C. C. 1938. Established and supposed examples of meteoritic craters and structures. *Field and Laboratory*, **6**, 44–56.

BOTTOMLEY, R. J., YORK, D. & GRIEVE, R. A. F. 1990. ^{40}Argon–^{39}Argon dating of impact craters. *Proceedings of the 20th Lunar and Planetary Science Conference*, 421–431.

BRETT, R. 1992. The Cretaceous–Tertiary extinction: a lethal mechanism involving anhydrite target rocks. *Geochimica et Cosmochimica Acta*, **56**, 3603–3606.

BUCHER, W. H. 1936. Cryptovolcanic structures in the United States (with discussion). *16th International Geological Congress, Report*, **2**, 1055–1084.
BUNCH, T. E., DENCE, M. R. & COHEN, A. J. 1967. Natural terrestrial maskelynite. *American Mineralogist*, **52**, 244–253.
CAMERON, A. G. W. 1983. Origin of the atmospheres of the terrestrial planets. *Icarus*, **56**, 195–201.
—— & BENZ, W. 1991. The origin of the Moon and the single impact hypothesis IV. *Icarus*, **92**, 204–216.
CARTER, N. L., OFFICER, C. B., CHESNER, C. A. & ROSE, W. I. 1986. Dynamic deformation of volcanic ejecta from the Toba caldera: possible relevance to the Cretaceous/Tertiary boundary phenomena. *Geology*, **14**, 380–383.
——, —— & DRAKE, C. L. 1990. Dynamic deformation of quartz and feldspar: clues to causes of some natural crises. *Tectonophysics*, **171**, 373–391.
CHAO, E. C. T., FAHEY, J. J. & LITTLER, J. 1961. Coesite from Wabar crater, near Al Hadida, Arabia. *Science*, **133**, 882–883.
——, ——, ——, & MILTON, D. J. 1962. Stishovite, SiO_2, a very high pressure new mineral from Meteor Crater, Arizona. *Journal of Geophysical Research*, **67**, 419–421.
CHAPMAN, C. R. & MORRISON, D. 1989. *Cosmic Catastrophes*. Plenum, New York.
——, —— & BOWELL, E. 1989. Hazards from Earth-approachers: implications of 1989 FC's 'near miss' (abstract). *Meteoritics*, **24**, 258.
CHRISTENSEN, L., FREGERSLEV, S., SIMONSEN, A. & THIEDE, J. 1973. Sedimentology and depositional environment of the lower Danian fish clay from Stevns Klint, Denmark. *Bulletin of the Geological Society of Denmark*, **22**, 193–216.
CHYBA, C. F., THOMAS, P. J. & ZAHNLE, K. J. 1993. The 1908 Tunguska explosion: atmospheric disruption of a stony asteroid. *Nature*, **361**, 40–44.
CINTALA, M. J. & GRIEVE, R. A. F. 1994. The effects of differential scaling of impact melt and crater dimensions on lunar and terrestrial craters. *In*: DRESSLER, B. O., GRIEVE, R. A. F. & SHARPTON, V. L. (eds) *Large Meteorite Impacts and Planetary Evolution*. Geological Society of America, Special Paper, **293**, 51–59.
——, WOOD, C. A. & HEAD, J. W. 1977. The effects of target characteristics on fresh crater morphology: preliminary results for the moon and Mercury. *Proceeding of the 8th Lunar Science Conference*, 3409–3426.
CORNER, B., REIMOLD, W. U., BRANDT, D. & KOEBERL, C. 1997. Morokweng impact structure, Northwest Province, South Africa: geophysical imaging and shock petrographic studies. *Earth and Planetary Science Letters*, **146**, 351–364.
CURRIE, K. L. 1971. Geology of the resurgent cryptoexplosion crater at Mistastin Lake, Labrador, Canada. *Geological Survey Bulletin*, **207**.
DAVIS, M., HUT, P. & MÜLLER, R. A. 1984. Extinction of species by periodic comet showers. *Nature*, **308**, 715–717.
DE LAUBENFELS, M. W. 1956. Dinosaur extinction: one more hypothesis. *Journal of Paleontology*, **30**, 207–218.
DEUTSCH, A. & GRIEVE, R. A. F. 1994. The Sudbury Structure: constraints on its genesis from Lithoprobe results. *Geophysical Research Letters*, **21**, 963–966.
—— & SCHÄRER, U. 1994. Dating terrestrial impact events. *Meteoritics*, **29**, 301–322.
DIETZ, R. S. 1947. Meteorite impact suggested by the orientation of shatter-cones at the Kentland, Indiana, disturbance. *Science*, **105**, 42–43.
——1964. Sudbury Structure as an astrobleme. *Journal of Geology*, **72**, 412–434.
——1968. Shatter cones in cryptoexplosion structures. *In*: FRENCH, B. M. & SHORT, N. M. (eds) *Shock Metamorphism of Natural Materials*. Mono, Baltimore, 267–285.
——1972. Shatter cones (shock fractures) in astroblemes. *Proceedings of the 24th International Geological Congress, Montreal*, 112–118.
DONAHUE, T. M. & POLLACK, J. B. 1983. Origin and evolution of the atmosphere of Venus. *In*: HUNTER, D. M., COLIN, L., DONAHUE, T. M. & MOROZ, V. I. (eds) *Venus*. University of Arizona Press, Tucson, 1003–1036.
DYPVIK, H., GUDLAUGSSON, S. T., TSIKALAS, F. *et al.* 1996. Mjølnir structure: an impact crater in the Barents Sea. *Geology*, **24**, 779 782.
ESKOLA, P. 1921. On volcanic necks in Lake Jänisjärvi. *Bulletin de la Commission Géologique de Finlande*, **55**, 3–13.
FAGGART, B. E., BASU, A. R. & TATSUMOTO, M. 1985. Origin of the Sudbury complex by meteoritic impact: neodymium isotopic evidence. *Science*, **230**, 436–439.
FLORAN, R. J. & DENCE, M. R. 1976. Morphology of the Manicouagan ring-structure, Quebec, and some comparisons with lunar basins and craters. *Proceedings of the 7th Lunar Science Conference*, 2845–2865.
FRENCH, B. M. 1990. 25 years of the impact-volcanic controversy: Is there anything new under the Sun or inside the Earth? *Eos Translations, American Geophysical Union*, **71**, 411–414.
—— & SHORT, N. M. (eds) 1968. *Shock Metamorphism of Natural Materials*. Mono, Baltimore.
——, KOEBERL, C., GILMOUR, I., SHIREY, S. B., DONS, J. A. & NATERSTAD, J. 1977. The Gardnos impact structure, Norway: petrology and geochemistry of target rock and impactites. *Geochimica et Cosmochimica Acta*, **61**, 873–904.
FREY, H. 1980. Crustal evolution of the early Earth: the role of major impacts. *Precambrian Research*, **10**, 195–216.
GANAPATHY, R. 1980. A major meteorite impact on the Earth 65 million years ago: evidence from the Cretaceous–Tertiary boundary clay. *Science*, **209**, 921–923.
GARVIN, J. B. & SCHNETZLER, C. C. 1994. The Zhamanshin impact feature: a new class of complex crater? *In*: DRESSLER, B. O., GRIEVE, R. A. F. & SHARPTON, V. L. (eds) *Large Meteorite Impacts and Planetary Evolution*. Geological Society of America, Special Paper, **293**, 249–257.
GEHRELS, T. (ed.) 1994. *Hazards Due to Comets and Asteroids*. University of Arizona Press, Tuscon.

GERSTL, S. A. W. & ZARDECKI, A. 1982. Reduction of photosynthetically active radiation under extreme stratospheric aerosol loads. *In*: SILVER, L. T. & SCHULTZ, P. H. (eds) *Geological Implications of Impacts of Large Asteroids and Comets on the Earth*. Geological Society of America, Special Paper, **190**, 201–210.

GILMOUR, I., RUSSELL, S. S., ARDEN, J. W., LEE, M. R., FRANCHI, I. A. & PILLINGER, C. T. 1992. Terrestrial carbon and nitrogen isotopic ratios from Cretaceous–Tertiary boundary nanodiamonds. *Science*, **258**, 1624–1626.

GLIKSON, A. Y. 1996. Mega-impacts and mantle-melting episodes: tests of possible correlations. *Australian Geological Survey Organisation Journal of Australian Geology and Geophysics*, **16**, 587–607.

GOODWIN, A. M. 1976. Giant impacting and the development of continental crust. *In*: WINDLEY, B. F. (ed.) *The Early History of the Earth*. Wiley, New York, 77–88.

GOSTIN, V. A. & THERRIAULT, A. M. 1997. Tookoonooka, a large buried Early Cretaceous impact structure in the Eromanga Basin of southwest Queensland, Australia. *Meteoritics and Planetary Science*, **32**, 593–599.

GRIEVE, R. A. F. 1980. Impact bombardment and its role in proto-continental growth on the early Earth. *Precambrian Research*, **10**, 217–247.

——1984. The impact cratering rate in recent time. *Proceedings of the 14th Lunar and Planetary Science Conference. Journal of Geophysical Research, Supplement*, **89**, B403–B408.

——1997. Extraterrestrial impact events: the record in the rocks and stratigraphic column. *Palaeogeography, Palaeoclimate, Palaeoecology*, **132**, 5–23.

—— & CINTALA, M. J. 1992. An analysis of differential impact melt-crater scaling and implications for the terrestrial impact record. *Meteoritics*, **27**, 526–538.

—— & ——1995. Impact melting on Venus: some considerations for the nature of the cratering record. *Icarus*, **114**, 68–79.

—— & HEAD, J. W. 1983. The Manicouagan impact structure: an analysis of its original dimensions and form. *Journal of Geophysical Research, Supplement*, **88**, A807–A818.

—— & MASAITIS, V. L. 1994. The economic potential of terrestrial impact craters. *International Geology Review*, **36**, 105–151.

—— & PESONEN, L. J. 1992. The terrestrial impact cratering record. *Tectonophysics*, **216**, 1–30.

—— & ——1996. Terrestrial impact craters: their spatial and temporal distribution and impacting bodies. *Earth, Moon and Planets*, **72**, 357–376.

—— & PILKINGTON, M. 1996. The signature of terrestrial impacts. *Australian Geological Survey Organisation Journal of Australian Geology and Geophysics*, **16**, 399–420.

—— & ROBERTSON, P. B. 1976. Variations in shock deformation at the Slate Islands impact structure, Lake Superior, Canada. *Contributions to Mineralogy and Petrology*, **58**, 37–49.

—— & SHOEMAKER, E. M. 1994. The record of past impact on Earth. *In:* GEHRELS, T. (ed.) *Hazards Due to Comets and Asteroids*. University of Arizona Press, Tuscon, 417–462.

——, DENCE, M. R. & ROBERTSON, P. B. 1977. Cratering processes: as interpreted from the occurrence of impact melts. *In*: RODDY, D. J., PEPIN, R. O. & MERRILL, R. B. (eds) *Impact and Explosion Cratering*. Pergamon, New York, 791–814.

——, GARVIN, J. B., CODERRE, J. M. & RUPERT, J. 1989. Test of a geometric model for the modification stage of simple impact crater development. *Meteoritics*, **24**, 83–88.

——, LANGENHORST, F. & STÖFFLER, D. 1996. Shock metamorphism of quartz in nature and experiment: II. Significance in geoscience. *Meteoritics and Planetary Science*, **31**, 6–35.

——, PILKINGTON, M. & PARMENTIER, E. 1990. Large impact basins and the early Earth (abstract). *International Workshop on Meteorite Impact on the Early Earth, Perth, Australia.* Lunar and Planetary Institute Contribution **746**, 18–19.

——, RUPERT, J., SMITH, J. & THERRIAULT, A. 1995. The record of terrestrial impact cratering. *GSA Today*, **5**(189), 194–196.

——, SHARPTON, V. L., RUPERT, J. D. & GOODACRE, A. K. 1988. Detecting a periodic signal in the terrestrial cratering record. *Proceedings of the 18th Lunar and Planetary Science Conference*, 375–382.

GUDLAUGSSON, S. T. 1993. Large impact crater in the Barents Sea. *Geology*, **21**, 291–294.

HAINES, P. W. 1996. Goyder impact structure, Arnhem Land, Northern Territory. *Australian Geological Survey Organisation Journal of Australian Geology and Geophysics*, **16**, 561–566.

HALLS, H. C. & GRIEVE, R. A. F. 1976. The Slate Islands: a probable complex meteorite impact structure in Lake Superior. *Canadian Journal of Earth Sciences*, **13**, 1301–1309.

HARGRAVES, R. B. 1976. Precambrian geologic history. *Science*, **193**, 363–371.

——, KELLOG, K. S., FISKE, P. S. & HOUGEN, S. B. 1994. Allochthonous impact-shocked rocks and superimposed deformations at the Beaverhead site in southwest Montana. *In*: DRESSLER, B. O., GRIEVE, R. A. F. & SHARPTON, V. L. (eds) *Large Meteorite Impacts and Planetary Evolution*. Geological Society of America, Special Paper, **293**, 225–235.

HARTMANN, W. K. & DAVIS, D. R. 1975. Satellite-sized planetesimals and lunar origin. *Icarus*, **24**, 504–515.

——, PHILLIPS, R. J. & TAYLOR, G. J. (eds) 1986. *Origin of the Moon*. Lunar and Planetary Institute, Houston.

HEAD, J. W. & SOLOMON, S. C. 1981. Tectonic evolution of the terrestrial planets. *Science*, **213**, 62–76.

HEISLER, J. & TREMAINE, S. 1989. How dating uncertainties affect the detection of periodicity in extinctions and craters. *Icarus*, **77**, 213–219.

HILDEBRAND, A. R. 1993. The Cretaceous/Tertiary boundary impact (or the dinosaurs didn't have a chance). *Journal of the Royal Astronomical Society of Canada*, **87**, 77–118.

——, PENFIELD, G. T., KRING, D. A., PILKINGTON, M., CAMARGO, A. Z., JACOBSEN, S. B. & BOYNTON, W. V. 1991. Chicxulub crater: a possible Cretaceous/Tertiary boundary impact crater on the Yucatan Peninsula, Mexico. *Geology*, **19**, 867–871.

HILLS, J. G., NEMCHINOV, I. V., POPOV, S. & TETEREV, A. V. 1994. Tsunami generated by small asteroidal impacts. *In*: GEHRELS, T. (ed.) *Hazards Due to Comets and Asteroids*. University of Arizona Press, Tucson, 779–789.

HODYCH, J. P. & DUNNING, G. R. 1992. Did the Manicouagan impact trigger end-of-Triassic mass extinction? *Geology*, **20**, 51–54.

HÖRZ, F. 1968. Statistical measurements of deformation structures and refractive indices in experimentally shock loaded quartz. *In*: FRENCH, B. M. & SHORT, N. M. (eds) *Shock Metamorphism of Natural Materials*. Mono, Baltimore, 243–253.

HOUGH, R. M., GILMOUR, I., PILLINGER, C. T., ARDEN, J. W., GILKES, K. W. R., YUAN, J. & MILLEDGE, H. J. 1995. Diamond and silicon carbide in impact melt rock from the Ries impact crater. *Nature*, **378**, 41–44.

HSÜ, K. J., OBERHÄNSLI, H., GAO, J. Y., SUN, S., CHEN, H. & KRÄHENBÜL, U. 1985. 'Strangelove ocean' before the Cambrian explosion. *Nature*, **316**, 809–811.

IZETT, G. A. 1987. *The Cretaceous–Tertiary (K–T) boundary interval, Raton Basin, Colorado and New Mexico, and its content of shock-metamorphosed minerals: implications concerning the K–T boundary impact–extinction theory*. US Geological Survey Open-File Report **87-606**.

——, COBBAN, W. A., OBRADOVICH, J. D. & KUNK, M. J. 1993. The Manson impact structure: $^{40}Ar/^{39}Ar$ age and its distal impact ejecta in the Pierre shale in southeastern South Dakota. *Science*, **262**, 729–732.

JAHN, B., FLORAN, R. J. & SIMONDS, C. H. 1978. Rb–Sr isochron age of the Manicouagan melt sheet, Quebec, Canada. *Journal of Geophysical Research*, **83**, 2799–2803.

JANSA, L. F. & PE-PIPER, G. 1987. Identification of an underwater extraterrestrial impact crater. *Nature*, **327**, 612–614.

KERR, R. A. 1997. Geophysicists peer into fiery core and icy ocean depths. *Science*, **275**, 160–161.

KIPP, M. E. & MELOSH, H. J. 1986. A preliminary numerical study of colliding planets. *In*: HARTMANN, W. K., PHILLIPS, R. J. & TAYLOR, G. J. (eds) *Origin of the Moon*. Lunar and Planetary Institute, Houston, 643–648.

KOEBERL, C. & REIMOLD, W. U. 1995. Early Archaean spherule beds in the Barberton Mountain Land, South Africa: no evidence for impact origin. *Precambrian Research*, **74**, 1–33.

——, POAG, C. W., REIMOLD, W. U. & BRANDT, D. 1996*b*. Impact origin of Chesapeake Bay structure and the source for the North American tektites. *Science*, **271**, 1263–1266.

——, REIMOLD, W. V. & SHIREY, S. B. 1996*a*. Re–Os isotope and geochemical study of the Vredefort Granophyre: clues to the origin of the Vredefort structure, South Africa. *Geology*, **24**, 913–916.

——, SHIREY, S. B. & REIMOLD, W. U. 1994. Re–Os isotope systematics as a diagnostic tool for the study of impact craters (abstract). *In*: *New Developments Regarding the K/T Event and Other Catastrophes in Early Earth History*. Lunar and Planetary Institute Contribution **825**, 61–63.

KRING, D. A. & BOYNTON, W. V. 1992. Petrogenesis of an augite-bearing melt rock in the Chicxulub structure and its relationship to K/T impact spherules in Haiti. *Nature*, **358**, 141–144.

——, MELOSH, H. J. & HUNTEN, D. M. 1996. Impact-induced perturbations of atmospheric sulfur. *Earth and Planetary Science Letters*, **140**, 201–212.

KROGH, T. E., KAMO, S. L. & BOHOR, B. F. 1993. Fingerprinting the K/T impact site and determining the time of impact by U–Pb dating of single shocked zircons from distal ejecta. *Earth and Planetary Science Letters*, **119**, 425–429.

——, —— & ——1996. Shock metamorphosed zircons with correlated U–Pb discordance and melt rocks with concordant protolith ages indicate an impact origin for the Sudbury Structure: earth processes: reading the isotopic code. Special American Geophysical Union Publication, Geophysical Monograph, **95**, 343–353.

KUNK, M. J., IZETT, G. A., HAUGERUD, R. A. & SUTTER, J. F. 1989. ^{40}Ar–^{39}Ar dating of the Manson impact structure: a Cretaceous–Tertiary boundary crater candidate. *Science*, **244**, 1565–1568.

KYTE, F. T. & BOSTWICK, J. A. 1995. Magnesioferrite spinel in Cretaceous/Tertiary boundary sediments of the Pacific basin: remnants of hot, early ejecta from the Chicxulub impact? *Earth and Planetary Science Letters*, **132**, 113–127.

—— & SMIT, J. 1986. Regional variations in spinel compositions: an important key to the Cretaceous/Tertiary event. *Geology*, **14**, 485–487.

—— & WASSON, J. T. 1986. Accretion rate of extraterrestrial matter: iridium deposited 33 to 67 million years ago. *Science*, **232**, 1225–1229.

——, HEATH, G. R., LEINEN, M. & ZHOU, L. 1993. Cenozoic sedimentation history of the central North Pacific: inferences from the elemental geochemistry of core LL44-GPC3. *Geochimica et Cosmochimica Acta*, **57**, 1719–1740.

——, ZHOU, Z. & WASSON, J. T. 1980. Siderophile-enriched sediments from the Cretaceous–Tertiary boundary. *Nature*, **288**, 651–656.

——, —— & ——1988. New evidence on the size and possible effects of a late Pliocene oceanic asteroid impact. *Science*, **241**, 63–65.

LEHTOVAARA, J. J. 1992. Söderfjärden: a Cambrian impact crater in western Finland (extended abstract). *Tectonophysics*, **216**, 157–161.

LOWE, D. R. & BYERLY, G. R. 1986. Early Archean silicate spherules of probable impact origin, South Africa and Western Australia. *Geology*, **14**, 83–86.

——, ——, ASARO, F. & KYTE, F. J. 1989. Geological and geochemical record of 3400-million-year-old terrestrial meteorite impacts. *Science*, **245**, 959–962.

LUCK, J. M. & TUREKIAN, K. K. 1983. Osmium-187/Osmium-186 in manganese nodules and the Cretaceous–Tertiary boundary. *Science*, **222**, 613–615

LUNDQUIST, G. 1963. *Description of the soil map of Garleborg County* (in Swedish). Sveriges Geologiska Undersokningen, Serien Ca, **42**, 60.

MAHER, K. A. & STEVENSON, D. J. 1988. Impact frustration of the origin of life. *Nature*, **331**, 612–614.

MAK, E. K., YORK, D., GRIEVE, R. A. F. & DENCE, M. R. 1976. The age of the Mistastin Lake crater, Labrador, Canada. *Earth and Planetary Science Letters*, **31**, 345–357.

MANTON, W. I. 1965. The orientation and origin of shatter cones in the Vredefort Ring. *Annals of the New York Acadamy of Sciences*, **123**, 1017–1049.

MASAITIS, V. L., FUTERGENDLER, S. I. & GNEVUSHEV, M. A. 1972. Diamonds in impactites of the Popigai meteorite crater (in Russian). *Zapiskio Vsesoyuznoge Mineralogicheskogo Obshchestva*, **101**, 108–112.

MCHONE, J. F., NIEMAN, R. A., LEWIS, C. F. & YATES, A. M. 1989. Stishovite at the Cretaceous–Tertiary boundary, Raton, New Mexico. *Science*, **243**, 1182–1184.

MELOSH, H. J. 1981. Atmospheric breakup of terrestrial impactors. *In*: SCHULTZ, P. H. & MERRILL, P. B. (eds) *Multi-Ring Basins*. Pergamon, New York, 29–35.

——1989. *Impact Cratering: a Geologic Process*. Oxford University Press, New York.

—— & VICKERY, A. M. 1989. Impact erosion of the primordial atmosphere of Mars. *Nature*, **338**, 487–489.

——, SCHNEIDER, N. M., ZAHNLE, K. J. & LATHAM, D. 1990. Ignition of global wildfires at the Cretaceous/Tertiary boundary. *Nature*, **343**, 251–254.

MILTON, D. J. 1977. Shatter cones—an outstanding problem in shock mechanics. *In*: RODDY, D. J., PEPIN, R. O. & MERRILL, R. B. (eds) *Impact and Explosion Cratering*. Pergamon, New York, 703–714.

MOJZSIS, S. J., ARRHENIUS, G., MCKEEGAN, K. D., HARRISON, T. M., NUTMAN, A. P. & FRIEND, C. R. L. 1996. Evidence for life on Earth before 3,800 million years ago. *Nature*, **384**, 55–59.

MONTANARI, A., HAY, R. L., ALVAREZ, W., ASARO, F., MICHEL, H. V. & ALVAREZ, L. W. 1983. Spheroids at the Cretaceous–Tertiary boundary are altered impact droplets of basaltic composition. *Geology*, **11**, 668–671.

MORRISON, D., CHAPMAN, C. R. & SLOVIC, P. 1994. The impact hazard. *In:* GEHRELS, T. (ed.) *Hazards Due to Comets and Asteroids*. University of Arizona Press, Tucson, 59–91.

NEUKUM, G. & IVANOV, B. A. 1994. Crater size distributions and impact probabilities on Earth from lunar, terrestrial, planet, and asteroid cratering data. *In*: GEHRELS, T. (ed.) *Hazards Due to Comets and Asteroids*. University of Arizona Press, Tucson, 359–416.

OHTSUKI, K., NAKAGAWA, Y. & NAKAZAWA, K. 1988. Growth of Earth in nebular gas. *Icarus*, **75**, 552–565.

ORTH, C. J., ATTREP, M. J. & QUINTANA, L. R. 1990. Iridium abundances across bio-event horizons in the fossil record. *In*: SHARPTON, V. L. & WARD, P. D. (eds) *Global Catastrophes in Earth History*. Geological Society of America, Special Paper, **247**, 45–60.

PAL, P. C. & CREER, K. M. 1986. Geomagnetic reversal spurts and episodes of extraterrestrial catastrophism. *Nature*, **320**, 148–150.

PALME, H. 1982. Identification of projectiles of large terrestrial impact craters and some implications for the interpretation of Ir-rich Cretaceous/Tertiary boundary layers. *In*: SILVER, L. T. & SCHULTZ, P. H. (eds) *Geological Implications of Impacts of Large Asteroids and Comets on the Earth*. Geological Society of America, Special Paper, **190**, 223–233.

——, GOEBEL, E. & GRIEVE, R. A. F. 1979. The distribution of volatile and siderophile elements in the impact melt of East Clearwater (Quebec). *Proceedings of the 10th Lunar and Planetary Science Conference*, 2465–2492.

——, GRIEVE, R. A. F. & WOLF, R. 1981. Identification of the projectile at Brent crater, and further considerations of projectile types at terrestrial craters. *Geochimica et Cosmochimica Acta*, **45**, 2417–2424.

PIKE, R. J. 1977. Size-dependence in the shape of fresh impact craters on the Moon. *In*: RODDY, D. J., PEPIN, R. O. & MERRILL, R. B. (eds) *Impact and Explosion Cratering*. Pergamon, New York, 489–509.

POLOSUKHIN, V. P. 1981. Traces of ancient intensive meteorite bombardment of the Earth (in Russian). *Doklady Akademii Nauk SSSR*, **260**, 1434–1437.

POPE, K. O., BAINES, K. H., OCAMPO, A. C. & IVANOV, B. A. 1994. Impact winter and the Cretaceous/Tertiary extinctions: results of the Chicxulub asteroid impact model. *Earth and Planetary Science Letters*, **128**, 719–725.

RABINOWITZ, D., BOWELL, E., SHOEMAKER, E. & MUINONEN, K. 1994. The population of Earth-crossing asteroids. *In*: GEHRELS, T. (ed.) *Hazards Due to Comets and Asteroids*. University of Arizona Press, Tucson, 285–312.

RAMPINO, M. R. & HAGGERTY, B. M. 1994. Extraterrestrial impacts and mass extinctions of life. *In*: GEHRELS, T. (ed.) *Hazards Due to Comets and Asteroids*. University of Arizona Press, Tucson, 827–857.

—— & ——1996. The 'Shiva hypothesis': impacts, mass extinctions and the galaxy. *Earth, Moon, and Planets*, **72**, 441–460.

—— & STOTHERS, R. B. 1984. Terrestrial mass extinction, cometary impacts and the Sun's motion perpendicular to the galactic plane. *Nature*, **308**, 709–712.

RAUP, D. M. 1990. Impact as a general cause of extinction: a feasibility test. *In*: SHARPTON, V. L. & WARD, P. D. (eds) *Global Catastrophes in Earth History*. Geological Society of America, Special Paper, **247**, 27–32.

—— & SEPKOSKI, J. J. 1984. Periodicity of extinctions in the geologic past. *Proceedings of the National Acadamy of Sciences of the USA*, **81**, 801–805.

—— & ——1986. Periodic extinctions of families and genera. *Science*, **231**, 833–836.

ROBERTSON, P. B. & SWEENEY, J. F. 1983. Haughton impact structure: structural and morphological aspects. *Canadian Journal of Earth Sciences*, **20**, 1134–1151.

——, DENCE, M. R. & VOS, M. A. 1968. Deformation in rock-forming minerals from Canadian craters. *In*: FRENCH, B. M. & SHORT, N. M. (eds) *Shock Metamorphism of Natural Materials*. Mono, Baltimore, 433–452.

RODDY, D. J. & DAVIS, L. K. 1977. Shatter cones formed in large-scale experimental explosion craters. *In*: RODDY, D. J., PEPIN, R. O. & MERRILL, R. B. (eds) *Impact and Explosion Cratering*. Pergamon, New York, 715–750.

——, PEPIN, R. O. & MERRILL, R. B. (eds) 1977. *Impact and Explosion Cratering*. Pergamon, New York.

ROSE, R. R. 1955. *Manicouagan Lake–Mushalagan Lake area, Quebec*. Canada Geological Survey Paper **55-2**, Map.

RYDER, G., FASTOVSKY, D. & GARTNER, S. (eds) 1996. *New Developments regarding the KT Event and Other Catastrophes in Earth History*. Geological Society of America, Special Paper, **307**.

SCHNETZLER, C. C., PHILPOTTS, J. A. & THOMAS, H. H. 1967. Rare-earth and barium abundances in Ivory Coast tektites and rocks from the Bosumtwi crater area, Ghana. *Geochimica et Cosmochimica Acta*, **31**, 1987–1993.

SHARPTON, V. L. & WARD, P. D. (eds) 1990. *Global Catastrophes in Earth History*. Geological Society of America, Special Paper, **247**.

——, BURKE, K., CAMARGO-ZANOGUERA, A. *et al.* 1993. Chicxulub multiring impact basin: size and other characteristics derived from gravity analysis. *Science*, **261**, 1564–1567.

——, DALRYMPLE, G. B., MARIN, L. E., RYDER, G., SCHURAYTZ, B. C. & URRUTIA-FUCUGAUCHI, J. 1992. New links between the Chicxulub impact structure and the Cretaceous/Tertiary boundary. *Nature*, **359**, 819–821.

——, DRESSLER, B. O., HERRICK, R. R., SCHNIEDERS, B. & SCOTT, J. 1996. New constraints on the Slate Islands impact structure, Ontario, Canada. *Geology*, **24**, 851–854.

SHOEMAKER, E. M. & SHOEMAKER, C. S. 1996. The Proterozoic impact record of Australia. *Australian Geological Survey Organisation Journal of Australian Geology and Geophysics*, **16**, 379–398.

——, WOLFE, R. F. & SHOEMAKER, C. S. 1990. Asteroid and comet flux in the neighborhood of Earth. *In*: SHARPTON, V. L. & WARD, P. D. (eds) *Global Catastrophes in Earth History*. Geological Society of America, Special Paper, **247**, 155–170.

SIGURDSSON, H., D'HONDT, S., ARTHUR, M. A., BRALOWER, T. J., ZACHOS, J. C., VAN FOSSEN, M. & CHANNELL, J. E. T. 1991. Glass from the Cretaceous/Tertiary boundary in Haiti. *Nature*, **349**, 482–487.

SILVER, L. T. & SCHULTZ, P. H. (eds) 1982. *Geological Implications of Impacts of Large Asteroids and Comets on the Earth*. Geological Society of America, Special Paper, **190**.

SLEEP, N. H., ZAHNLE, K. J., KASTING, J. F. & MOROWITZ, H. J. 1989. Annihilation of ecosystems by large asteroid impacts on the early Earth. *Nature*, **342**, 139–142.

SMIT, J. & HERTOGEN, J. 1980. An extraterrestrial event at the Cretaceous–Tertiary boundary. *Nature*, **285**, 158–200.

—— & KLAVER, G. 1981. Sanidine spherules at the Cretaceous–Tertiary boundary indicate a large impact event. *Nature*, **292**, 47–49.

SPRAY, J. G. & THOMPSON, L. M. 1995. Friction melt distribution in terrestrial multi-ring impact basins. *Nature*, **373**, 130–132.

SPUDIS, P. D. 1993. *The Geology of Multi-ring Impact Basins*. Cambridge, Cambridge University Press.

STESKY, R. M. & HALLS, H. C. 1983. Structural analysis of shatter cones from Slate Islands, northern Lake Superior. *Canadian Journal of Earth Sciences*, **20**, 1–18.

STÖFFLER, D. 1971. Progressive metamorphism and classification of shocked and brecciated crystalline rocks in impact craters. *Journal of Geophysical Research*, **76**, 5541–5551.

——1972. Deformation and transformation of rock-forming minerals by natural and experimental shock processes. I. Behavior of minerals under shock compression. *Fortschritte der Mineralogie*, **49**, 50–113.

——1974. Deformation and transformation of rock-forming minerals by natural and experimental shock processes. II. Physical properties of shocked minerals. *Fortschritte der Mineralogie*, **51**, 256–289.

—— & HORNEMANN, U. 1972. Quartz and feldspar glasses produced by natural and experimental shock. *Meteoritics*, **7**, 371–394.

—— & LANGENHORST, F. 1994. Shock metamorphism of quartz in nature and experiment: 1. Basic observation and theory. *Meteoritics*, **29**, 155–181.

——, DEUTSCH, A., AVERMANN, M., BISCHOFF, L., BROCKMEYER, P., BUHL, D., LAKOMY, R. & MÜLLER-MOHR, V. 1994. The formation of the Sudbury Structure, Canada: towards a unified impact model. *In*: DRESSLER, B. O., GRIEVE, R. A. F. & SHARPTON, V. L. (eds) *Large Meteorite Impacts and Planetary Evolution*. Geological Society of America, Special Paper, **293**, 303–318.

STOTHERS, R. B. 1993. Impact cratering at geologic stage boundaries. *Geophysical Research Letters*, **20**, 887–890.

—— & RAMPINO, M. R. 1990. Periodicity in flood basalts, mass extinctions and impacts; a statistical view and a model. *In*: SHARPTON, V. L. & WARD, P. D. (eds) *Global Catastrophes in Earth History*. Geological Society of America, Special Paper, **247**, 9–18.

SVENONIUS, F. 1895. *Geological investigations in Garleborg County* (in Swedish). Sveriges Geologiska Undersokningen, Serien C, **152**, 54–59.

SWISHER, C. C., GRAJALES-NISHIMURA, J. M., MONTANARI, A. *et al.* 1992. Coeval $^{40}Ar/^{39}Ar$ ages of 65.0 million years ago from Chicxulub crater melt rock and Cretaceous–Tertiary boundary tektites. *Science*, **257**, 954–958.

THERRIAULT, A. M., GRIEVE, R. A. F. & REIMOLD, W. U. 1997. Original size of the Vredefort Structure: implications for the geological evolution of the Witwatersrand Basin. *Meteoritics and Planetary Science*, **32**, 71–77.

——, REIMOLD, W. U. & REID, A. M. 1996. Field relations and petrography of the Vredefort Granophyre. *South African Journal of Geology*, **1**, 1–21.

TONKS, W. B. & MELOSH, H. J. 1993. Magma ocean formation due to giant impacts. *Journal of Geophysical Research*, **98**, 5319–5333.

TOON, O. B., POLLACK, J. B., ACKERMAN, T. P., TURCO, R. P., MCKAY, C. P. & LIU, M. S. 1982. Evolution of an impact-generated dust cloud and its effects on the atmosphere. *In*: SILVER, L. T. & SCHULTZ, P. H. (eds) *Geological Implications of Impacts of Large Asteroids and Comets on the Earth.* Geological Society of America, Special Paper, **190**, 187–200.

TSCHERMAK, G. 1872. The Shergotty and Gopalpur meteorites (in German). *Akademie der Wissenschaften in Wien, Mathematisch-Naturwissenschaftliche Klasse, Teil 1*, **65**, 122–145.

UREY, H. 1973. Comet collisions and geological periods. *Nature*, **242**, 32–33.

VON ENGELHARDT, W. 1990. Distribution, petrography and shock metamorphism of the ejecta of the Ries crater in Germany – a review. *Tectonophysics*, **171**, 259–273.

—— & BERTSCH, W. 1969. Shock induced planar deformation structures in quartz from the Ries Crater, Germany. *Contributions to Mineralogy and Petrology*, **20**, 203–234.

—— & HÖRZ, F. 1965. Ries glasses and moldavite (in German). *Geochimica et Cosmochimica Acta*, **29**, 609–620.

WALLACE, M. W., GOSTIN, V. A. & KEAYS, R. R. 1990. Spherules and shard-like clasts from the late Proterozoic Acraman impact ejecta horizon, South Australia. *Meteoritics*, **25**, 161–165.

WEIBLEN, P. W. & SCHULTZ, K. J. 1978. Is there any record of meteorite impact in the Archean rocks of North America? *Proceedings of the 9th Lunar and Planetary Science Conference*, 2749–2773.

WEISSMAN, P. R. 1990. The cometary impactor flux at the Earth. *In*: SHARPTON, V. L. & WARD, P. D. (eds) *Global Catastrophes in Earth History.* Geological Society of America, Special Paper, **247**, 171–180.

WETHERILL, G. W. 1990. Calculation of mass and velocity distributions of terrestrial and lunar impactors by use of theory of planetary accumulation (abstract). *International Workshop on Meteorite Impact on the Early Earth, Perth, Australia.* Lunar and Planetary Institute Contribution, **746**, 54–55.

—— & STEWART, G. R. 1989. Accumulation of a swarm of small planetesimals. *Icarus*, **77**, 330–357.

WILHELMS, D. E. 1987. The geologic history of the Moon. US Geological Survey Professional Paper, **1348**, 302.

WILLIAMS, G. E. 1986. The Acraman impact structures: source of ejecta in late Precambrian shales, South Australia. *Science*, **233**, 200–203.

——1994. Acraman: a major impact structure from the Neoproterozoic of Australia. *In*: DRESSLER, B. O., GRIEVE, R. A. F. & SHARPTON, V. L. (eds) *Large Meteorite Impacts and Planetary Evolution.* Geological Society of America, Special Paper, **293**, 209–224.

WOLBACH, W. S., GILMOUR, I. & ANDERS, E. 1990. Major wildfires at the Cretaceous/Tertiary boundary. *In*: SHARPTON, V. L. & WARD, P. D. (eds) *Global Catastrophes in Earth History.* Geological Society of America, Special Paper, **247**, 391–400.

——, LEWIS, R. S. & ANDERS, E. 1985. Cretaceous extinctions: evidence for wildfires and search for meteoritic material. *Science*, **230**, 167–170.

WOLF, R., WOODROW, A. B. & GRIEVE, R. A. F. 1980. Meteoritic material at four Canadian impact craters. *Geochimica et Cosmochimica Acta*, **44**, 1015–1022.

WOOD, C. A. & HEAD, J. W. 1976. Comparison of impact basins on Mercury, Mars and the Moon. *Proceedings of the 7th Lunar Science Conference*, 3629–3651.

XU, D. Y., MA, S. L., CHAI, Z. F., MAO, X. Y., SUN, Y. Y., ZHANG, Q. W. & YANG, Z. Z. 1985. Abundance variation of iridium and trace elements at the Permian/Triassic boundary at Shangsi in China. *Nature*, **314**, 154–156.

——, SUN, Y.-Y., ZHANG, Q.-W., YAN, Z., HE, J.-W. & CHAI, Z.-F. 1989. *Astrogeological Events in China.* Van Nostrand Reinhold, New York.

YABUSHITA, S. 1994. Are periodicities in crater formations and mass extinctions related? *Earth, Moon and Planets*, **64**, 207–216.

——1996. Are cratering and probably related geological records periodic? *Earth, Moon and Planets*, **72**, 343–356.

Identification of meteoritic components in impactites

CHRISTIAN KOEBERL

Institute of Geochemistry, University of Vienna, Althanstrasse 14, A-1090 Vienna, Austria (e-mail: christian.koeberl@univie.ac.at)

Abstract: The verification of an extraterrestrial component in impact-derived melt rocks or breccias can be of diagnostic value to provide confirming evidence for an impact origin of a geological structure. Geochemical methods are used to determine the presence of the minor traces of such a component. This is because even under ideal conditions, meteoritic fragments are preserved only at very young and small craters, but are destroyed in larger and older structures as a result of erosion and/or vaporization. The absence of meteorite fragments can, therefore, not be used as evidence against an impact origin of a crater structure. The determination of such an extraterrestrial component can be done by studying the concentrations and interelement ratios of siderophile elements, especially the platinum group elements, which are several orders of magnitude more abundant in meteorites than in terrestrial upper-crustal rocks. More recently, the Re–Os isotopic system has been used for identifying the presence of a meteoritic component in a number of impact melt rocks and breccias. This method avoids the ambiguities that may affect the usage of elemental abundances and ratios; for example, if the target rocks have high abundances of siderophile elements, or if the siderophile element concentrations in the impactites are very low. The presence of a meteoritic component has been established for about 41 impact structures (out of about 150 known on Earth), which reflects also the detail in which these structures were studied. In some of the smaller of these craters, fragments of the meteorite (the 'projectile') are preserved, but in most others, only a minor extraterrestrial component (mostly <1 wt %) is mixed in with impact melt rocks or breccias. The identification of the projectile type (i.e. meteorite type) is not straightforward, as vapour fractionation during the impact, as well as post-impact fractionation in the melt or hydrothermal mobilization, may lead to changes in the interelement ratios.

Impact cratering is one of the most important surface-shaping and modifying processes for the terrestrial planets. Whereas such crater structures are easily recognized on planets and satellites with negligible or no atmosphere, such as the Moon, Mercury, or Mars, impact craters are much less obvious on the surface of the Earth. A variety of terrestrial processes, such as volcanism, tectonics and erosion, conspire to obliterate the impact cratering record on the Earth. This makes the detection of impact craters on the Earth difficult. About 150 impact structures are known on Earth (e.g. Grieve *et al.* 1995). Details of the formation of impact structures, the variety of impact lithologies, and the use of shock metamorphism to identify them, are described in the literature (see, e.g., French & Short 1968; Stöffler 1972; Melosh 1989; Koeberl 1994*a*; Stöffler and Langenhorst 1994).

The detection of meteoritic components in impact-derived rocks, either at or in the crater (e.g. in autochthonous and allochthonous impact breccias and melt rocks), or in distal ejecta, is of great diagnostic value in confirming an impact origin for a certain geological structure or an ejecta layer. The discovery of a possible extraterrestrial component in the Cretaceous–Tertiary (K–T) boundary clay led Alvarez *et al.* (1980) to their now well-known proposal that a large asteroid or comet impact was responsible for the environmental catastrophe 65 Ma ago that led to one of the largest mass extinctions in Earth history. Their discovery, high levels of siderophile element abundances, especially the platinum group elements (PGEs), in the K–T boundary clay, was confirmed in later studies at numerous locations around the world. Although opponents of the impact hypothesis proposed terrestrial sources to explain the high PGE abundances, the clearly chondritic interelement ratios, and abundances that are two to five orders of magnitude above those of most terrestrial rocks made any such sources highly unlikely. Finally, more than 10 years after the original discovery by Alvarez *et al.* (1980), the source crater was identified: the large (possibly up to 280 km diameter) Chicxulub impact structure in Mexico (e.g. Hildebrand *et al.* 1991; Sharpton *et al.* 1992; Swisher *et al.* 1992). Thus, the detection of a meteoritic component in distal ejecta led to the identification of a large, previously unknown impact structure. However, a more common problem is to determine if a geological structure

KOEBERL, C. 1998. Identification of meteoritic components in impactites. *In*: GRADY, M. M., HUTCHISON, R., MCCALL, G. J. H. & ROTHERY, D. A. (eds) *Meteorites: Flux with Time and Impact Effects*. Geological Society, London, Special Publications, **140**, 133–153.

of unknown or controversial origin may have formed by impact.

The evidence for an impact origin is not well established for a number of terrestrial structures. Although various criteria have been proposed for the identification of impact craters (including structural, geophysical, petrological, and geochemical parameters), only one such criterion is commonly accepted as providing unambiguous evidence for an impact origin. This is the presence of shock metamorphic effects, as no such effects related to non-impact processes have ever been observed in nature (see, e.g. Sharpton & Grieve (1990), Stöffler & Langenhorst (1994) and Grieve *et al.* (1996) for reviews of this topic). However, many craters are deeply eroded, obscuring any shock record, or, in the case of ejecta layers, shocked rocks and minerals are very rare, making their discovery unlikely. Thus, the detection of a clear extraterrestrial component can provide decisive evidence for the impact origin of a geological structure. Such a component could be preserved in impact melt that is injected into the basement, which would lead to its preservation even in a deeply eroded crater. Various aspects and methods regarding a meteoritic component in impactites will be discussed in this paper.

Meteoritic material in craters and impactites

The observation that meteorites are rarely present at most meteorite impact craters on Earth may seem like a contradiction, but is the result of the extreme temperature conditions during an impact event, which, in general, lead to vaporization of the meteoritic impactor. For smaller craters (less than about 1.5 km in diameter) a small fraction of the initial mass of the meteorite may survive, because of spallation during entry into the atmosphere or lower impact velocity resulting from atmospheric drag. This spallation effect is probably rare for large impactors; however, the age of an impact structure may provide a bias as well. Meteoritic material is easily eroded within a few thousand years. Small craters (smaller than a few kilometres in diameter) form fairly often on Earth, once every few thousand years. However, large craters form with a much lower frequency, and therefore even if spalled meteorite fragments survive such an impact, erosion will have destroyed any remains over the millions of years since crater formation. Thus, even under ideal conditions, meteoritic fragments are likely to be present only at very young and small craters. The absence of meteorite fragments cannot be used as evidence against an impact origin of a crater structure.

The detection of traces of the meteoritic projectile in melted or brecciated target rocks is a rather generally applicable impact-diagnostic method. Such a detection allows the establishment of the impact origin for a crater structure. The meteoritic projectile undergoes vaporization in the early phases of crater formation. A small amount of the meteoritic vapour is incorporated with the much larger quantity of target rock vapour and melt, which later forms impact melt rocks, melt breccias, or glass. Most often the contribution of meteoritic matter to these impactite lithologies is very small (commonly $\ll 1\%$), leading to only minor chemical changes in the resulting impactites. Distal ejecta may also contain a small meteoritic contribution, as the famous K–T boundary example demonstrates (for further examples, see later discussion).

Determination of meteoritic component by siderophile element analyses

The differentiation of such a minute amount of meteoritic contamination from the compositional signature of the normal terrestrial target rocks is extremely difficult. Only elements that have high abundances in meteorites, but low abundances in terrestrial crustal rocks, such as the siderophile elements, can be used to detect such a meteoritic component. Another complication is the existence of a variety of meteorite groups and types (the three main groups are stony meteorites, iron meteorites, and stony-iron meteorites, in order of decreasing abundance), which have widely varying siderophile element compositions. Distinctly higher siderophile element contents in impact melts, compared with target rock abundances, can be indicative of the presence of either a chondritic or an iron meteoritic component. Achondritic projectiles are much more difficult to discern, because they have significantly lower abundances of the key siderophile elements.

Siderophile (and related) elements that have often been used are Ni, Co, and Cr. In addition, the interelement ratios of these elements are an effective discriminator. If the meteoritic contribution is $>0.1\%$, it is possible to distinguish between stony meteorites (chondrites) and iron meteorites, because chondrites have high abundances of Cr (typically about 0.26 wt%), whereas iron meteorites have Cr abundances that are much more variable, but typically about 100 times lower than those of chondrites, and low Ni/Cr or Co/Cr ratios. Therefore, the Cr abundance and evaluation of the Ni/Cr or Co/Cr ratios in impact melts can be used to distinguish between iron and chondritic projectiles (e.g. Palme *et al.* 1978; Evans *et al.* 1993). Unfortunately, Co, Cr, and Ni do not show particularly low abundances in terrestrial rocks. Thus, there may be a relatively high (and variable) indigenous component that is derived from the target rocks. If any conclusions are to be drawn from the abundances of elements such as Cr, Ni, and Co, a detailed study of their abundances in the target rocks needs to be done. From mixing calculations it is possible to determine the relative proportions of the various target rock types that make up impact melt rocks or breccias. It is then possible to subtract the siderophile element contribution that is derived from the target rocks from the breccia or melt rock composition and arrive at a corrected meteoritic component.

Studies done in the wake of analyses of lunar rocks have shown that the platinum group elements (PGEs: Ru, Rh, Pd, Os, Ir, Pt) and Au are better suited for identifying a meteoritic component (e.g. Morgan *et al.* 1975; Palme *et al.* 1978, 1979; Morgan & Wandless 1983; Evans *et al.*, 1993). The abundances of the PGEs in chondrites and most iron meteorites are several orders of magnitude higher than those in terrestrial crustal rocks, as illustrated for Ir in Fig. 1. The range of Ir and Os abundances in chondrites is about 400–800 ppb, whereas iron meteorites show a much wider range (see Fig. 1). In contrast, continental crustal rocks contain on the order of 0.02 ppb Ir or Os (e.g. Taylor & McLennan 1985; see Table 1, Fig. 1). Thus, the signal-to-background ratio is very high for the PGEs (i.e. low indigenous concentrations; high concentrations in the 'contaminating' meteorite).

The admixture of minor quantities of meteoritic material (chondrite, iron meteorite) to crustal target rocks yields significantly elevated abundances of the PGEs in the resulting impact melt rocks or breccias. For example, if only 1 wt % of a chondritic meteorite is mixed into terrestrial crustal target rocks, the resulting breccia or melt will contain about 4 ppb of Ir. Even 0.1 wt % of a chondritic component still results in 0.4 ppb of Ir (or Os) in the breccia, which is at least one order of magnitude higher than average background values. Achondrites have a much wider range of Ir and Os contents, and, in general, lower abundances than for chondrites. Ranges of contents of about 0.1–760 ppb Ir and 0.02–5 ppb Os have been reported (Mason 1979). It is not clear if the apparent differences between the Ir and Os contents are real or the result of analytical problems, because Gros *et al.* (1976) reported comparable

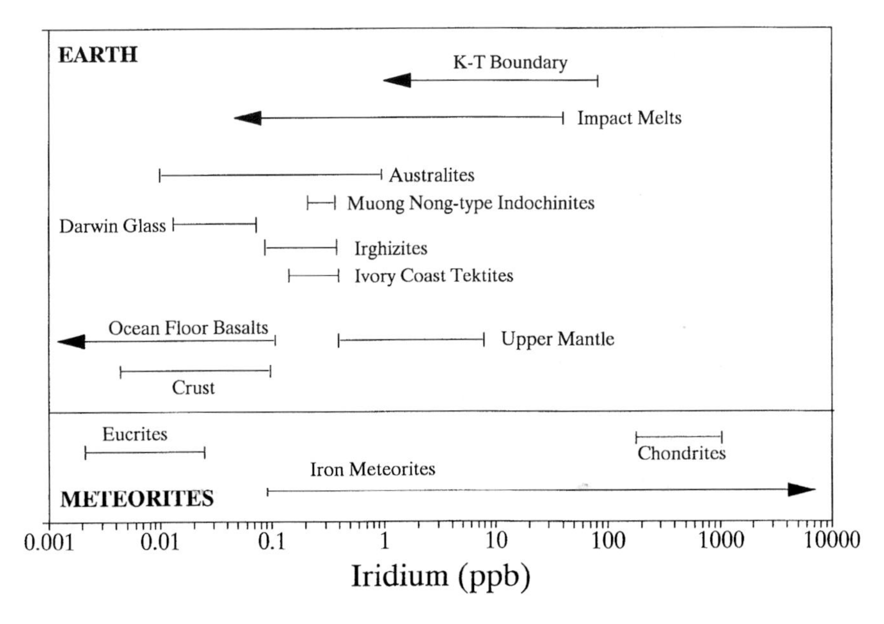

Fig. 1. Schematic diagram of range of Ir abundances in some meteorite types, terrestrial rocks, impactites, and distal ejecta. Typical ranges are shown. Data from Morgan (1978), Palme *et al.* (1981), Palme (1982) and Meisel & Koeberl (1990).

Table 1. *Abundances of Re and Os in various terrestrial and extraterrestrial rocks*

Rock	Re (ppb)	Os (ppb)	Re/Os	References
C1 chondrites	36.5	486	0.075	Anders & Grevesse (1989)
Allende (CV3 chondrite)	68.6	851	0.081	Walker & Morgan (1989)
Ordinary chondrites	57	660	0.086	Allègre and Luck (1980)
IIAB iron meteorites	1.39–4799	12.7–65270	0.073–0.11	Morgan *et al.* (1992)
IIIAB iron meteorites	2.83–1444	17.1–18430	0.078–0.202	Morgan *et al.* (1992)
Moore County eucrite	0.06	0.44	0.136	Mason (1979)
Juvinas eucrite	0.01	0.018	0.55	Mason (1979)
Achondrites	0.06–38.6	0.02–5	–	Mason (1979)
Tholeiite basalt	0.84	0.03	28	Allègre & Luck (1980)
Amphibolite	0.44	0.011	40	Walker *et al.* (1991)
Peridotite	0.43	5.3	0.081	Allègre & Luck (1980)
Gneiss	0.049	0.011	4.45	Walker *et al.* (1991)
Granite (USA)	0.077	0.033	2.33	Walker *et al.* (1991)
Granite (Australia)	0.22	0.007	31.4	Walker *et al.* (1991)
Sandstone	0.033	0.019	1.74	Koeberl *et al.* (1994*c*)
Av. upper continental crust	0.5	0.02	25	Taylor & McLennan (1985)

abundances of Ir and Os in the Juvinas eucrite (0.028 and 0.018 ppb, respectively).

Confirmed enrichments of Ir and other PGEs in impactite-derived rocks provides good evidence for the presence of a meteoritic component (see, e.g. Morgan *et al.* 1975; Palme *et al.* 1978; Palme 1982). In a number of these studies, PGE abundances and interelement ratios were used in an attempt to resolve the type or class of meteorite (e.g. Morgan *et al.* 1975; Palme *et al.* 1978, 1979; Evans *et al.* 1993).

Using siderophile element abundances and interelement ratios, the projectile type was determined for a number of terrestrial craters. A list of such projectile identifications for 41 impact structures is given in Table 2. Some comments on this table are in order, however. First, only impact craters or structures with diameters larger than 100 m are listed. Several smaller impact craters with known projectile types exist, and in all these cases iron meteorites have been found as remnants of the impactor; for such small craters, the change from a simple excavation hole to an actual hypervelocity impact crater is not well defined. As a general rule, the larger the crater, the more likely it is that the projectile has been vaporized. Some of the smaller structures listed in Table 2 are in fact crater fields consisting of several more or less circular structures, such as Wabar, Henbury and Odessa. Also, the Rio Cuarto crater field in Argentina is unusual because it consists of at least ten rimmed, oblong structures ranging in size between about 0.1×0.3 and $4.5 \times 1.1\,km^2$, which formed by a very low angle (grazing) impact of a chondritic body, with the top of the impactor being decapitated and producing the down-range crater structures (e.g. Schultz *et al.* 1994).

In Table 2, a few points are worth noticing. Almost all of the smaller impact craters (less than about 1.2 km in diameter) were formed by iron projectiles of known composition, with the exception of the Saltpan crater, which was formed by the impact of a chondritic body. The types of most of these iron meteorites are known, because fragments of the meteorites were still present at the crater sites. All the craters with meteorite fragments are relatively young (less than *c.* 0.1 Ma), whereas at older craters (e.g. Aouelloul at 3.1 Ma) no remnants of the actual impactor were recovered. Although the Saltpan crater is only 0.22 Ma old, no remnants of the impacting chondrite were found (probably because of erosion of any remaining fragments). In contrast, a small mass of the chondritic impactor (of H4/5 type) that formed the much larger Rio Cuarto crater field is still preserved, as this structure is young (probably <4000 years).

It should also be pointed out that conflicting identifications have been made for a number of impact structures, and that many identifications are highly uncertain. For example, both iron and chondritic projectiles were proposed for the Bosumtwi (Ghana), Zhamanshin (Kazakhstan), and Rochechouart (France) impact structures. For others, the question mark indicates uncertain identifications (for example, those that are based only on the study of Ni, Co, and Cr, without determination of the PGEs). The reasons for some of the associated problems are discussed in the following section.

Problems with meteoritic component identification

Apart from the seemingly straightforward cases in which meteorite fragments are found at a crater, inferences about the projectile type have to be made from the abundances and interelement ratios of various siderophile elements in impactites. However, several complications are introduced by complex fractionation processes that seem to take place during the formation of impact glasses and melts.

In recent studies of impact glasses from some small craters for which the meteorite has been partly preserved (e.g. Meteor Crater, Wabar, and some Australian craters), it was found that the siderophile elements show strong (and highly variable!) fractionation in the inter-element ratios compared with the initial ratios in the impacting meteorite. The changes do not seem to be correlated in a straightforward way with any physical or chemical properties of the elements (Attrep *et al.* 1991; Mittlefehldt *et al.* 1992*a*, *b*). Mittlefehldt *et al.* (1992*b*) ruled out simple vapour fractionation of the pure elements, or post-depositional fractionation. Instead, they proposed that the siderophile elements may have been fractionated from each other during the early phases of the impact, while the projectile was undergoing decompression and before mixing with the target materials, although they were unable to explain the fractionation with a specific model.

In general, though, there is a correlation between the amount of melt and vapour with increasing impact velocity (e.g. Pierazzo *et al.* 1997), which may, thus, be the most important factor controlling the incorporation of a meteoritic component (the amount of which is known to vary widely between craters of similar size, and also within impact breccias and melt rocks from a single crater). However, this energy scaling relationship would not explain the fractionation within a single crater. This observation does not bode well for any attempts to directly infer projectile types of small craters from siderophile element ratios in impactites. And indeed, the element ratios at, for example, Aouelloul or Brent, do not readily conform to those of any known meteorite types, but can be interpreted in various ways.

To allow the identification of a meteoritic component in impactites, the abundances of PGEs and other siderophile elements in the impactites have to be compared with those in the target rocks. Ideally, these indigenous concentrations should be determined and subtracted from the abundances found in the impact melt rocks, and yield the 'pure' meteoritic abundance ratios (e.g. Morgan & Wandless 1983). Thus, it is necessary to analyse all rock types that are known or suspected to be present in the pre-impact target area (in some cases, clasts within breccias can be used). Mixing calculations (see, e.g. French *et al.* 1997) can be used to establish the relative proportions of the target rock types that were mixed to form impact melt rocks or breccias, which then allows one to subtract a proper indigenous component from the melt or breccia composition. This procedure may yield good results only if the target stratigraphy is simple (i.e. not involving more than one or two rock types), and if the composition of the target rocks is uniform. In reality, though, the exact target rocks that were involved in forming impact melt rocks or breccia are not always well known; for example, because of later erosion of the target stratigraphy, or because of very low or highly variable indigenous PGE concentrations (see, e.g. Schmidt & Pernicka 1994; Pernicka *et al.* 1996).

The Au/Ir ratio has also been used to distinguish between cosmic and terrestrial signatures (Palme 1982). However, Au is much more mobile than Ir under a number of terrestrial conditions (e.g. weathering, diagenesis, metamorphism), which may lead to non-chondritic Ir/Au ratios even if a meteoritic component is present in impact-derived rocks. Specifically, Au often shows high indigenous abundances in some terrestrial rocks, yielding low (non-chondritic) Ir/Au ratios. Matters are further complicated by the fact that PGEs are enriched in certain terrestrial rock types (see below in the section on Ivory Coast tektites). The PGE patterns of the mantle and in some mantle-derived rocks may be similar to those of chondrites, and the PGE abundances in mantle rocks are also higher than those in the crust by factors of about 10^2–10^3, making a distinction between an exposed mantle section or a component in the impactites derived from an ultramafic precursor rock and a meteoritic component difficult, if not impossible.

A case in point is the discussion regarding the possible presence of a meteoritic signature in the 3–3.4 Ma old Barberton Mountain (South Africa) spherule layers. Lowe *et al.* (1989) and Kyte *et al.* (1992) interpreted the presence of high concentrations of iridium and other PGEs in some of the spherule samples as evidence for an extraterrestrial origin of these spherule beds. Koeberl *et al.* (1993) and Koeberl & Reimold (1995) also found high contents of, e.g. Ir (up to 2700 ppb), Ni (0.96 wt %), and Cr (1.6 wt %). However, abundances of these

Table 2. *Terrestrial impact structures with inferred impactor types*

Name	Country	Diameter (km)	Location	Bolide type	Evidence	References
Wabar	Saudia Arabia	0.10*	21°30′N, 50°28′E	IIIAB iron	M, S	Morgan *et al.* (1975), Mittlefehldt *et al.* (1992*b*)
Kaalijärvi	Estonia	0.11*	58°24′N, 22°40′E	IAB	M	Buchwald (1975)
Henbury	Australia	0.16*	24°35′S, 133°09′E	IIIAB	M, S	Taylor (1967)
Odessa	USA	0.17*	31°45′N, 102°29′W	IIIAB	M	Buchwald (1975)
Boxhole	Australia	0.17	22°37′S, 135°12′E	IIIAB	M	Buchwald (1975)
Macha	Russia	0.3*	59°59′N, 118°00′E	Iron	MS	Gurov (1996)
Aouelloul	Mauritania	0.4	20°15′N, 12°41′W	Iron (IIIB, IIID?)	S, Os	Morgan *et al.* (1975), Koeberl *et al.* (1998)
Monturaqui	Chile	0.46	23°56′S, 68°17′W	IAB	M, S	Bunch & Cassidy (1972), Buchwald (1975)
Wolfe Creek	Australia	0.9	19°18′S, 127°46′E	IIIAB	M, S	Attrepp *et al.* (1991)
Meteor (Barringer)	USA	1.2	35°02′N, 11101′W	IAB	M, S	Morgan *et al.* (1975), Mittlefehldt *et al.* (1992*a*)
Saltpan	South Africa	1.2	25°24′S, 28°05′E	Chondrite	S, Os	Koeberl *et al.* (1994*a*)
New Quebec	Canada	3.4	61°17′N, 73°40′W	Chondrite (L?)	S	Grieve (1991), Evans *et al.* (1993)
Brent	Canada	3.8	46°05′N, 78°29′W	Chondrite	S	Palme *et al.* (1981), Evans *et al.* (1993)
Gow Lake	Canada	4	56°27′N, 104°29′W	Iron?	(S)	Wolf *et al.* (1980)
Rio Cuarto	Argentina	4.5*	30°52′S, 64°14′W	Chondrite (H)	M, S, Os	Schultz *et al.* (1994), Koeberl *et al.* (in prep.)
Ilyinets	Ukraine	4.5	49°06′N, 29°12′E	Iron?	S	Grieve & Shoemaker (1994)
Sääksjärvi	Finland	5	61°24′N, 22°24′E	Stony-iron?	S	Palme (1982)
Gardnos	Norway	6	60°40′N, 09°00′E	Chondrite	S, Os	French *et al.* (1997)
Wanapitei	Canada	7.5	46°45′N, 80°45′W	Chondrite	S	Wolf *et al.* (1980), Evans *et al.* (1993)
Mien	Sweden	9	56°25′N, 14°52′E	Stone?	S	Palme *et al.* (1980)
Bosumtwi	Ghana	11	06°30′N, 01°25′W	Chondrite? Iron?	S, Os	Palme *et al.* (1978), Koeberl and Shirey (1993)

Ternovka	Ukraine	12	48°01′N, 33°05′E	Chondrite?	S	Grieve & Shoemaker (1994)
Nicholson Lake	Canada	12.5	62°40′N, 10°241′W	Achondrite	S	Wolf *et al.* (1980)
Zhamanshin	Kazakhstan	13.5	48°20′N, 60°58′E	Chondrite (Iron?)	S	Glass *et al.* (1983) (Palme *et al.* 1978)
Dellen	Sweden	15	61°55′N, 16°39′E	Stone?	S	Palme *et al.* (1980)
Obolon	Ukraine	15	49°30′N, 32°55′E	Iron?	S	Grieve & Shoemaker (1994)
Lappajärvi	Finland	17	63°12′N, 23°42′E	Chondrite	S	Göbel *et al.* (1980)
Elgygytgyn	Russia	18	67°30′N, 172°00′E	Achondrite?	S	Grieve and Shoemaker (1994)
Clearwater East	Canada	22	56°05′N, 74°07′W	Chondrite	S	Palme *et al.* (1979)
Rochechouart	France	23	45°50′N, 00°56′E	Chondrite? Iron?	S	Wolf *et al.* (1980), Lambert (1982)
Ries	Germany	24	48°53′N, 10°37′E	Achondrite?	S	Morgan *et al.* (1979), Schmidt and Pernicka (1994)
Boltysh	Ukraine	25	48°45′N, 32°10′E	Chondrite?	S	Grieve & Shoemaker (1994)
Strangways	Australia	25	15°12′S, 133°35′E	Achondrite	S	Morgan & Wandless (1983)
Mistastin	Canada	28	55°53′N, 63°18′W	Iron?	S	Wolf *et al.* (1980)
Manson	USA	38	42°35′N, 94°33′W	Chondrite	S, Os	Pernicka *et al.* (1996), Koeberl & Shirey (1996)
Acraman	Australia	90	32°01′S, 135°27′E	Chondrite	Es	Gostin *et al.* (1989), Wallace *et al.* (1990)
Kara	Russia	65–100	69°12′N, 65°00′E	Chondrite?	S	Nazarov *et al.* (1989), (1990)
Popigai	Russia	100	71°30′N, 111°00′E	Chondrite	S	Masaitis & Raikhlin (1985), Masaitis (1992)
Chicxulub	Mexico	180–280	21°20′N, 89°30′W	Chondrite	S, Os, Es	Koeberl *et al.* (1994*d*), Schuraytz *et al.* (1996)
Morokweng	South Africa	200	26°31′S, 23°32′E	Chondrite	S	Koeberl *et al.* (1997*b*)
Vredefort	South Africa	300	27°00′S, 27°30′E	Chondrite	S, Os	Koeberl *et al.* (1996*b*)

* Crater field; largest dimension of largest structure is given.
Evidence: S, siderophile element enrichment and/or pattern; Os, Os isotopic ratio; M, meteorite fragments; MS, metallic spherules; Es, siderophile element enrichment in ejecta.
Only craters larger than 0.1 km in diameter are listed.

elements in chondritic meteorites are approximately 600 ppb, 1.4 wt %, and 0.35 wt %, respectively. This would indicate respective meteoritic components in the Barberton samples of 450%, 70%, and 460%. Thus, Koeberl & Reimold (1995) concluded that these extraordinarily high concentrations cannot be primary meteoritic signatures. The high abundances of the siderophile elements in some Barberton samples and their enrichment in secondary minerals (they are present almost exclusively in various sulphides) indicate that these elements have been remobilized and reconcentrated. The PGE interelement ratios have changed during remobilization as well. Thus, the PGE abundance patterns and ratios are not primary and cannot be used as an argument in favour of an impact origin.

A further complication is the possibility of siderophile element fractionation in the impact melt while it is still molten. This effect may be significant in larger craters, because there the melt can stay hot for many thousands of years. Different mineral phases, such as sulphides or oxides (e.g. magnetite, chromite), may take up various proportions of the PGEs or other siderophile elements, leading to an irregular distribution of these elements and possibly fractionated interelement ratios and patterns. Such irregular distribution of siderophiles is known from, for example, the East and West Clearwater impact structures (Palme *et al.* 1979), or the Chicxulub impact structure (Koeberl *et al.* 1994*d*; see also below). Hydrothermal processes associated with the hot impact melt may also change PGE abundances. In contrast to a widely held opinion, actual data show that meteoritic components are often inhomogeneously distributed in impact melt rocks and breccias (e.g. data in Palme *et al.* 1979; Koeberl *et al.* 1994*a, c, d*; Schuraytz *et al.* 1994).

A variety of fractionation effects have also been documented for distal ejecta from the K–T boundary impact at various localities around the world (e.g. Evans *et al.* 1993). For example, high PGE abundances were discovered in impact ejecta from the Acraman structure in Australia (see Table 2), but show deviations from chondritic patterns as a result of low-temperature hydrothermal alteration (e.g. Gostin *et al.* 1989; Wallace *et al.* 1990; see also Colodner *et al.* 1992). However, space reasons prevent a more detailed discussion of distal impact ejecta in this paper. All these problems make it difficult to properly identify a meteoritic component, and, especially, a specific projectile type, and can yield fortuitous results. Thus, a more selective method for the identification of a meteoritic component would be very valuable.

Re–Os isotope analyses

In contrast to using PGE abundances, the Re–Os isotopic system has numerous advantages. This method is superior with respect to detection limit and selectivity, as discussed by Koeberl & Shirey (1993, 1996) and Koeberl *et al.* (1994*a, c, d*). In principle, the abundances of Re and Os and the $^{188}Os/^{187}Os$ isotopic ratios, which are measured by very sensitive mass spectrometric techniques, allow one to distinguish the isotopic signatures of meteoritic and terrestrial Os. Meteorites (and the terrestrial mantle) have much higher (by factors of 10^4–10^5) PGE contents than terrestrial crustal rocks. In addition, meteorites have relatively low Re and high Os abundances (Table 1), resulting in Re/Os ratios ≤ 0.1, whereas the Re/Os ratio of terrestrial crustal rocks is usually no less than ten. The $^{188}Os/^{187}Os$ isotopic ratios for meteorites and terrestrial crustal rocks are significantly different.

^{187}Os is formed from the β-decay of ^{187}Re, leading to a high $^{187}Os/^{188}Os$ ratio in crustal rocks (Fig. 2). Osmium is much more abundant in meteorites than Re, leading to only small changes in the meteoritic $^{187}Os/^{188}Os$ ratio with time (Fig. 2). Thus, meteorites have low $^{187}Os/^{188}Os$ ratios of about 0.11–0.18. Because of the high Os abundances in meteorites, the addition of even a very small meteoritical contribution to the crustal target rocks leads to an almost complete change of the Os isotopic signature of the resulting impact melt or breccia.

In addition to elemental abundance enrichments, the Re–Os isotopic system adds the Os isotopic composition as a distinguishing factor between terrestrial and meteoritic components in impact craters and ejecta. Although high abundances of Os could be due to crustal enrichment processes and incorporation of ore minerals, noncrustal Os isotopic compositions unambiguously indicate the presence of a meteoritic or mantle component. When mantle sources can be excluded on the basis of other evidence (see below), noncrustal Os isotopic ratios provide very good evidence for a meteoritic component in impact-derived rocks. This concept was first suggested by Turekian (1982) for distinguishing between a cosmic and a terrestrial component in Cretaceous–Tertiary (K–T) boundary clays. Later, only a few Re–Os studies of K–T boundary clays were done (e.g. Luck & Turekian 1983; Lichte *et al.* 1986, Krähenbühl *et al.* 1988; Esser & Turekian 1989; Meisel *et al.* 1995). The first attempt at using this system for impact crater studies was made by Fehn *et al.* (1986), but met with mixed success because of low analytical sensitivity. Only the development of the sensitive negative thermal ionization mass spectrometry (NTIMS) method (e.g. Creaser *et al.* 1991; Völkening *et al.* 1991) and selective chemical procedures (e.g. Shirey & Walker 1995) made impact crater studies feasible (Koeberl & Shirey 1993).

Even though the Os isotopic composition adds a new level of discrimination between terrestrial and extraterrestrial compositions, careful petrological studies, as well as detailed major and trace element analyses, and, if possible, supplementary Rb–Sr and/or Sm–Nd data, are essential for complete sample characterization, as components derived from ultramafic precursor rocks may mimic the presence of a meteoritic contribution, and their presence or absence must be established. However, common mantle-derived rocks have PGE abundances that are about 100 times lower than those in chondrites (see Table 1). Thus, a 100-times higher ultramafic contribution (compared with meteorites) is needed to explain elevated Os abundances and lower $^{187}Os/^{188}Os$ ratios; and such large contributions are easily distinguished by petrological and geochemical means. If properly used, Re–Os isotopic measurements of target rocks and impactites may provide confirming evidence for an impact origin of a geological structure, with a diagnostic value similar to that of shock metamorphism.

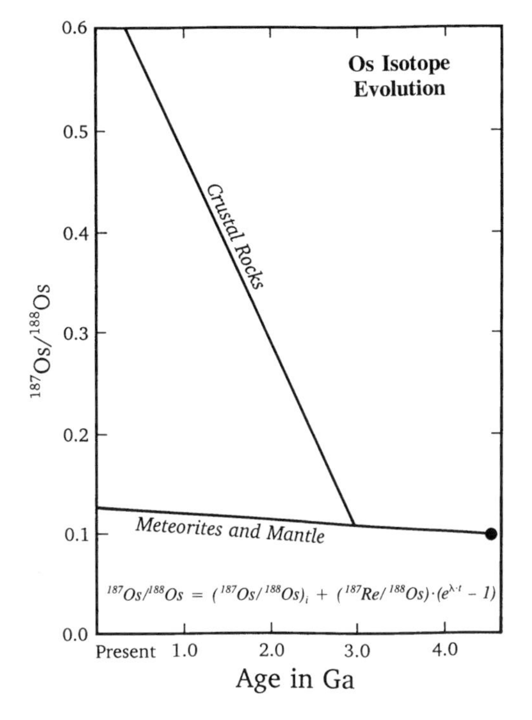

Fig. 2. Schematic evolution of the $^{187}Os/^{188}Os$ ratio of meteorites and the Earth's mantle and crust, after Faure (1986). The solid dot on the right side of the diagram marks the estimates of the isotopic starting composition ($[^{187}Os/^{188}Os]_{4.56\,Ga} = 0.0965$; Walker & Morgan, 1989) of the Earth's mantle and meteorites 4.56 Ga ago. Because of the relatively low Re/Os ratio of <0.1, the $^{187}Os/^{188}Os$ ratio increases only slightly with time, as indicated by the 'Meteorites and Mantle' line. The 'Crust' line follows a hypothetical crustal extraction event 3 Ga ago. During most magmatic processes, including crustal extraction, Os is strongly retained in the mantle, whereas Re partitions somewhat into the melt, leading to Re/Os ratios $\gg 1$ in many crustal rocks. Thus, the $^{187}Os/^{188}Os$ ratio of crustal rocks increases rapidly with time as a result of the production of radiogenic ^{187}Os by decay of ^{187}Re, as indicated in the diagram ($^{187}Re/^{188}Os \approx 12$). Therefore, many old crustal rocks have $^{187}Os/^{188}Os$ ratios (and absolute Re and Os abundances) that are distinctly different from those of mantle material and meteorites.

Selected case studies

The following paragraphs provide a selection of case studies, which were chosen to represent a variety of different types of craters, of different approaches to determine the meteoritic component, or showing complications in the identification. Hardly any of these cases poses identical problems, demonstrating the necessity of detailed studies.

Darwin crater and impact glass

Darwin glass occurs in the vicinity of Mount Darwin, about 20 km south of Queenstown, Tasmania, Australia, and has been interpreted as either of tektite origin or an impact glass formed by fusion of silicate sediments by meteorite impact. Geochemical studies of Darwin glass showed that it is not of volcanic origin, that the chemical composition of the glass resembles terrestrial sediments, that at least two chemical groups of Darwin glass can be distinguished, and that the geochemistry of the glass is consistent with a terrestrial origin by meteorite impact (see Meisel *et al.* (1990) and references therein, for a discussion). Previously, the absence of an impact crater associated with the glass provided problems for the impact theory, but in 1972 R. J. Ford found a crater-like structure near Mount Darwin. The structure is situated 26 km SSE of Queenstown, at the eastern boundary of the strewn field, which has been estimated to extend over 400 km^2. The area is densely vegetated and outcrops of country rocks are rare, making a detailed geological investigation difficult. The structure is situated within lightly metamorphosed Silurian and Devonian slates, argillites, and faulted and disrupted quartzites.

Meisel *et al.* (1990) performed a detailed geochemical study of the composition of Darwin glass, and identified three distinct groups: A (low Fe, Al (LFe,Al), or average Darwin glass), B (HFe,Al), and C (HMg,Na). The glasses of group C also show anomalous enrichments of several elements, e.g. Cr, Mn, Co, and Ni. These enrichments in group C glasses cannot be explained by contributions from ultrabasic rocks (which are not known in the area anyway). However, mixtures with meteoritic material also do not provide a satisfactory fit. Darwin glass does not show any significant Ir enrichments. Both Ir and Os abundances in Darwin glass are very low (Meisel & Koeberl 1990), as shown in a chondrite-normalized plot of siderophile element abundances in Fig. 3. Admixture of material from iron meteorites gives too high Fe, Co and Ni, and too low Cr and Mn contents. Chondritic contaminations would yield Ir abundances in the glass that are several orders of magnitude above the observed levels. Better fits are obtained for an achondritic contamination, but again give excess Ir. An ultrabasic contribution gives better results, except for higher Mg, but no such rocks are known from the target area. The siderophile element pattern of Darwin glass is actually very similar to that of average upper crust, indicating the necessity of a significant correction for indigenous values. As no detailed geochemical studies of target rocks at the Darwin crater are available, the indigenous component is not known. Thus, at the present time it is not possible to explain the enrichments of Cr, Mn, Co, and Ni in glasses of group C in a satisfactory way.

Australasian tektites

Tektites are natural glasses occurring on Earth in four distinct strewn fields: Australasian, Ivory Coast, Central European, and North American. Geochemical arguments have shown that tektites have been derived by hypervelocity impact melting from terrestrial upper-crustal rocks (see reviews by Koeberl (1986, 1994*b*)). Tektites occur in three different forms on land (Muong Nong-type, splash-form, and ablation shaped), and as microtektites predominantly obtained from deep-sea cores. Tektites are distal ejecta, which do not occur directly at a source crater, in contrast to impact glasses, which are found directly in or at the respective source crater. This has made the identification of the source crater somewhat difficult (Koeberl 1994*b*). Nevertheless, at least two of the four Cenozoic tektite strewn fields have been associated with known impact craters: the Ries crater in southern Germany and the Central European field, and the Bosumtwi crater in Ghana and the Ivory Coast field are rather firmly linked. In addition, the newly discovered large Chesapeake Bay

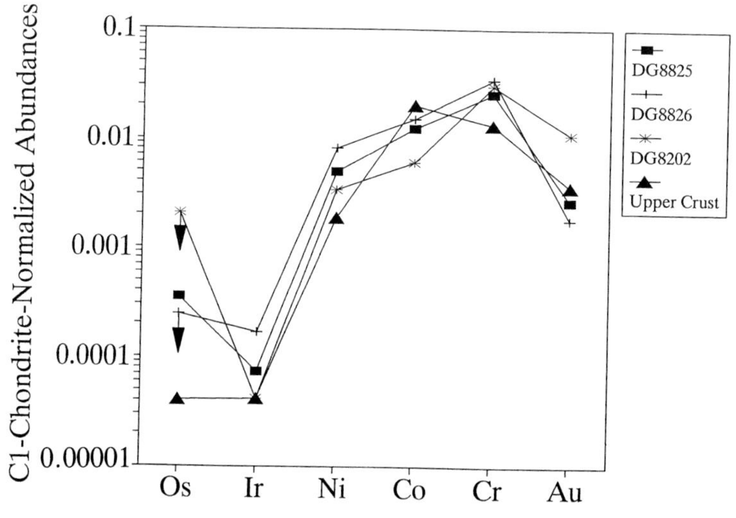

Fig. 3. Chondrite (Orgueil) normalized siderophile element abundances in three samples of Darwin glass (Meisel & Koeberl 1990; Meisel *et al.* 1990) compared with average upper-crustal concentrations (Taylor & McLennan 1985).

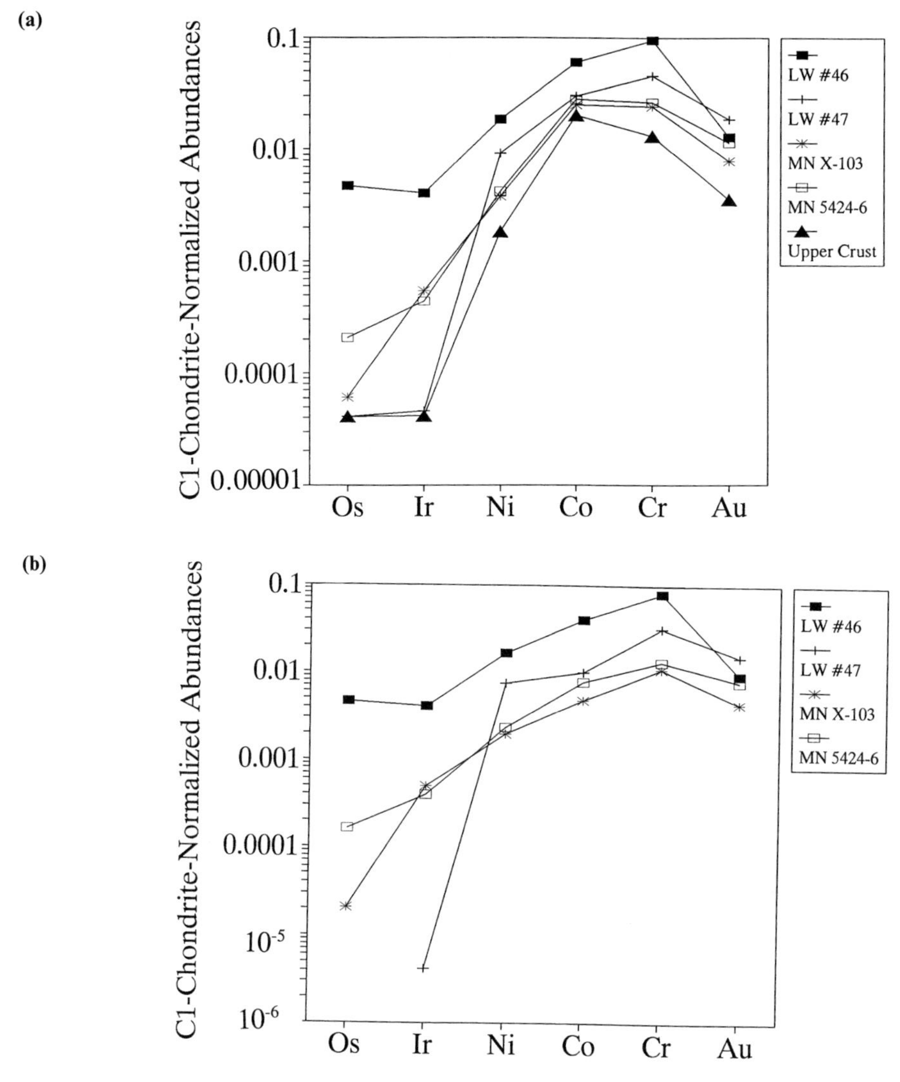

Fig. 4. (**a**) Chondrite (Orgueil) normalized siderophile element abundances in two HMg australite samples (Morgan 1978) and two Muong Nong type indochinites (Meisel & Koeberl 1990; C. Koeberl, unpublished data), compared with average upper-crustal concentrations (Taylor & McLennan 1985). (**b**) Same samples as in (**a**), but corrected for average upper-crustal (indigenous) concentrations.

impact structure (Poag *et al.* 1994; Koeberl *et al.* 1996*a*) is a likely source crater for the North American tektites. This leaves the Australasian tektites as the only strewn field without a clear choice for a source crater.

Attempts to determine a meteoritic component in Australasian tektites have not yielded unambiguous results. Chao *et al.* (1964) reported on Ni–Fe-rich spherules in some philippinites and suggested that these spherules are the remnant of meteoritic matter, but Ganapathy & Larimer (1983) concluded from Ir, W, As, Co, and Au abundances in the spherules that they more probably formed by *in situ* reduction from target material. Morgan (1978) analysed six HMg (high Mg content) australites by radiochemical neutron activation analysis (RNAA) for a selection of volatile and siderophile element concentrations. Only one of these samples showed a distinct enrichment in siderophile elements, whereas the other five do not indicate such an enrichment. Two of these analyses are plotted in Fig. 4a, normalized to C1 chondritic abundances. Sample Lake Wilson

No. 46 has a clear enrichment in all siderophile elements compared with the other samples and average continental crust.

Morgan (1978) also determined several other PGEs, which yielded a relatively flat pattern similar to about $4 \times 10^{-3} \times$ C1 values; however, the Cr, Co, and Ni concentrations seem high even if average crustal indigenous concentrations are subtracted (Fig. 4b). This indicates that either the projectile was not chondritic or the exact target rocks (which are currently not known) had higher than crustal Ni, Co, and Cr contents. The first possibility is supported by the lack of Ir enrichments in most of the australites (Morgan 1978) and also in the few other tektites from the Australasian strewn field analysed to date. For example, Muong Nong-type indochinites (Meisel & Koeberl 1990; C. Koeberl, unpublished data) have near-crustal Ir and Os contents (Fig. 4a,b) and Re–Os isotopic characteristics that are closer to crustal values than to meteoritic values.

A minor but discernible enrichment of Ir compared with adjacent sediment layers was found in deep-sea core samples of an ejecta layer associated with Australasian microtektites (Koeberl 1993), presenting the first evidence supporting the observation of Morgan (1978) of a cosmic component associated with Australasian tektites.

Bosumtwi crater (Ghana) and Ivory Coast tektites

The Bosumtwi crater in Ghana, which has a diameter of about 11 km, was inferred to be the Ivory Coast tektite source crater, based on geographical proximity, because these tektites and the crater have the same age (Gentner *et al.* 1969; Koeberl *et al.* 1989, 1997*a*), similar chemical composition (Schnetzler *et al.* 1967; Jones 1985), as well as Rb/Sr (Schnetzler *et al.* 1966; Kolbe *et al.* 1967) and oxygen isotopic characteristics (Taylor & Epstein 1966; Chamberlain *et al.* 1993). Palme *et al.* (1978, 1981) analysed two Ivory Coast tektites by RNAA and found Ir and Os abundances of 0.24 and 0.33 ppb and 0.099 and 0.199 ppb, respectively. These values are higher than those of average crustal rocks. This result led Palme *et al.* (1981) to suspect that an iron projectile might have been responsible for the Bosumtwi crater. However, this conclusion was rejected by Jones (1985), who, in turn, suggested that the high siderophile element abundances in the tektites could be derived from the target rocks, because the Bosumtwi crater is in an area of known gold mineralization.

Thus, the Ivory Coast tektites were a good candidate for trying out the potential of the Re–Os isotopic method to constrain the meteoritic component. Koeberl & Shirey (1993) determined the abundances and isotopic ratios of Os and Re in Ivory Coast tektites, Bosumtwi impact glass samples, and target rocks from the Bosumtwi crater. They found high Os abundances in the target rocks (Fig. 5), which seems to confirm the suspicion of Jones (1985) that the PGEs and Au in the tektites could very well be derived from the Bosumtwi country rocks, which have much higher Os contents than normal crustal rocks. Thus, the chemical abundance data do not allow any unambiguous conclusions regarding the presence of a cosmic component in Ivory Coast tektites.

However, the determination of the $^{187}Os/^{188}Os$ ratios in these samples revealed significant differences. The $^{187}Os/^{188}Os$ ratios of the tektites range from 0.153 to 0.209; these values overlap the range for carbonaceous chondrites and iron meteorites, but are inconsistent with the origin of the Os from crustal rocks. On the other hand, Bosumtwi crater target rocks were found to have $^{187}Os/^{188}Os$ ratios ranging from 1.48 to 4.98 (Koeberl & Shirey 1993), which are characteristic values for old continental crust, and clearly different from those of the tektites. The Bosumtwi impact glass falls between the target rock and tektite values. The variable and relatively high Os abundances in the target rocks indicate that some of the Os in the tektites is derived from the country rocks. The large difference in isotopic ratios, though, indicates

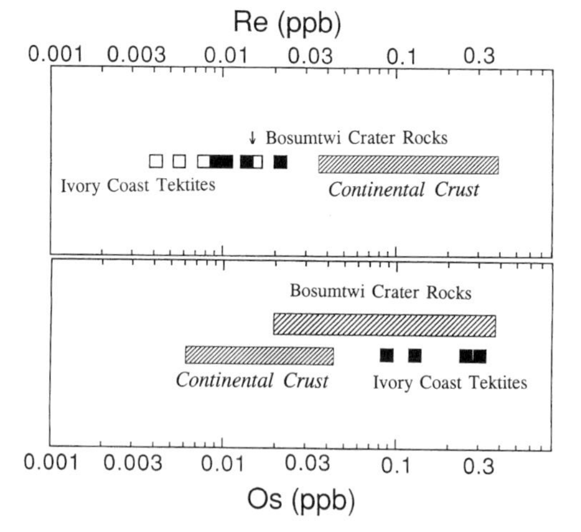

Fig. 5. Range of target rock compositions for Re and Os from the Lake Bosumtwi impact structure, Ghana, compared with concentrations measured in Ivory Coast tektites (data from Koeberl & Shirey (1993)).

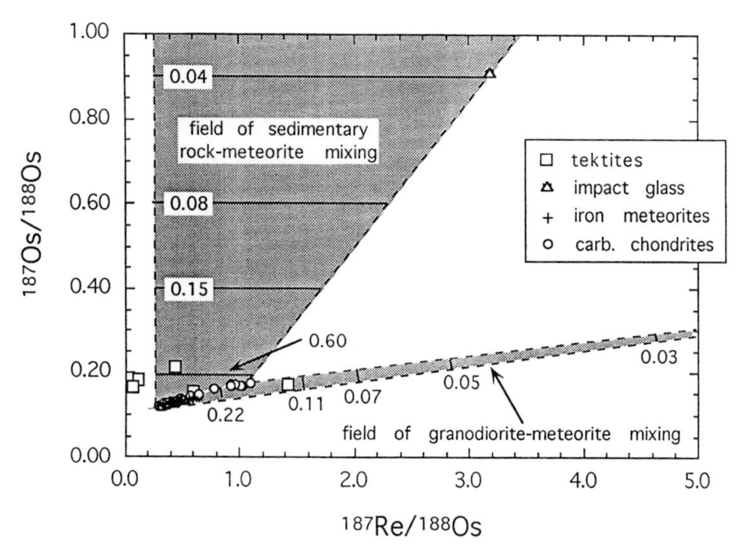

Fig. 6. $^{187}Os/^{188}Os$ v. $^{187}Re/^{188}Os$ diagram showing the mixing relationships between two types of target rock from the Bosumtwi crater, compared with Ivory Coast tektite data, indicating admixture of a minor meteoritic component to predominantly sedimentary target rocks (data from Koeberl & Shirey (1993)).

that this fraction does not exceed 10–20% of the total Os in the tektites, because otherwise the isotopic values would not remain close to meteoritic ratios. Figure 6 also provides an explanation for the initially somewhat puzzling observation that most tektites plot to the left of the meteorite data array, as the isotopic composition of the tektites results from mixing of Os from country rocks with Os from a meteorite. From the isotopic ratios, and based on chondritic abundances, Koeberl & Shirey (1993) estimated a meteoritic contribution to the tektite composition not exceeding 0.05–0.1 wt %.

Saltpan crater (South Africa)

The 1.2 km diameter Saltpan crater is located at 25°24′30″S and 28°04′59″E, about 40 km NNW of Pretoria, South Africa. The origin of the crater, either by explosive volcanism or meteorite impact, has been discussed by various researchers earlier this century (e.g. Wagner 1920; Rohleder 1933; Milton & Naeser 1971). The crater formed in 2.05 Ga old Nebo granite of the Bushveld Complex, and at the crater rim, various intrusive rocks are exposed. These rocks were earlier interpreted as evidence of a volcanic origin of the crater, but are now known to be a feature of the regional geology and not related to the local cratering event (e.g. Reimold *et al.* 1992; Koeberl *et al.* 1994*a,b*; Brandt and Reimold 1995). In 1988–1989, a drill core was obtained from the centre of the crater, to a depth of 200 m. The drill core penetrated 90 m of lacustrine sediments that are underlain by 53 m of unconsolidated suevitic breccia. The discovery of shock metamorphic effects in various minerals from the suevitic breccia provided confirming evidence for an impact origin of the Saltpan crater (Reimold *et al.* 1992). The age of the crater-forming event was determined by fission track measurements on impact glasses that were isolated from the suevite, showing that the crater was formed about 220 ka ago (Koeberl *et al.* 1994*b*).

Koeberl *et al.* (1994*a*) found that the compositions of the suevites are indistinguishable from those of the granites. Impact glasses isolated from the suevites also show predominantly granitic compositions, with the exception of significant Mg and Fe enrichments. The Mg and Fe abundances of most of the impact glasses show an excellent positive correlation (Fig. 7), with a slight Fe excess compared with C1 abundances, but an enstatite chondrite (with higher relative Fe than Mg contents) would fit such a correlation perfectly. In addition, enrichments in Mn, Cr, Co, Ni, and Ir are obvious in the handpicked glasses (Fig. 8), but less evident for the bulk suevite. The Ir concentration in the impact glasses is higher than the respective enrichment of other siderophiles, such as Co,

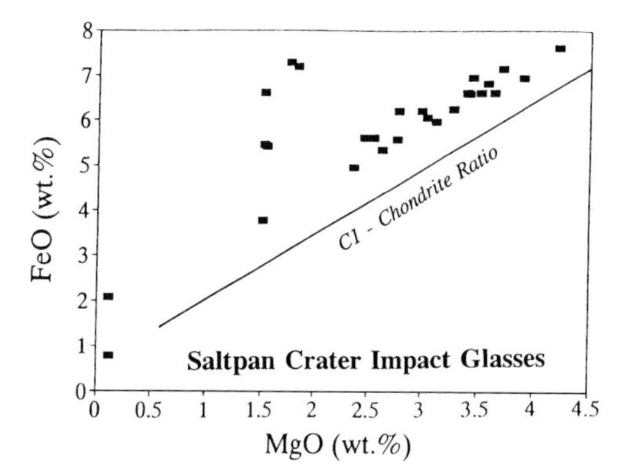

Fig. 7. Correlation of Mg and Fe concentrations in impact glasses from the Saltpan crater, South Africa (Koeberl *et al.* 1994*a*).

Cr, and Ni (Fig. 8). Koeberl *et al.* (1994*a*) concluded that these enrichments are the result of a significant meteoritic contribution to the impact glass composition. About 10% of a chondritic component is necessary to explain the elevated abundances of these elements in the impact glasses. An iron meteorite is an unlikely alternative because it cannot explain the enrichments in Mg and Cr.

The Re–Os isotope characteristics of basement granites and suevitic breccias from the Saltpan crater were analysed by Koeberl *et al.* (1994*a*). The Os contents of the two granitic target rocks are 6.3 and 7.0 ppt, respectively, which is consistent with values expected for granite. The $^{187}Os/^{188}Os$ ratios of those granites are high, with values of 0.713 and 0.736, respectively (Fig. 9). Such high values are within the range expected for old crustal rocks (e.g. Esser & Turekian 1993). In contrast, the Os abundances in the bulk suevitic breccias are more than ten times higher, at 81.4 and 75.5 ppt, respectively. They have low $^{187}Os/^{188}Os$ ratios of 0.205 and 0.206, respectively. The $^{187}Os/^{188}Os$ v. $^{187}Re/^{188}Os$ diagram (Fig. 9) shows that isotopic ratios of the breccias are consistent with derivation by mixing of a meteoritic component with basement granites. There is no indication from petrographical studies and major and trace element chemical analyses for the presence of any peridotitic, ultramafic, or any other mantle component in the breccias. After correction of the average Os abundance in the breccias for the indigenous Os content of the granites, an excess of about 0.072 ppb Os is obtained. Assuming an average chondritic Os abundance of 486 ppb (Anders & Grevesse 1989) results in about 0.015 wt % of a chondritic component in the bulk breccia.

The estimate of 10% chondritic component in the glass fragments can be compared with the result derived from the Re–Os studies for the bulk breccia. Impact glasses make up ≤ 1 vol % of the suevite, and the handpicked glasses have been selected mainly because of their brown or green colour and make up a subset of about

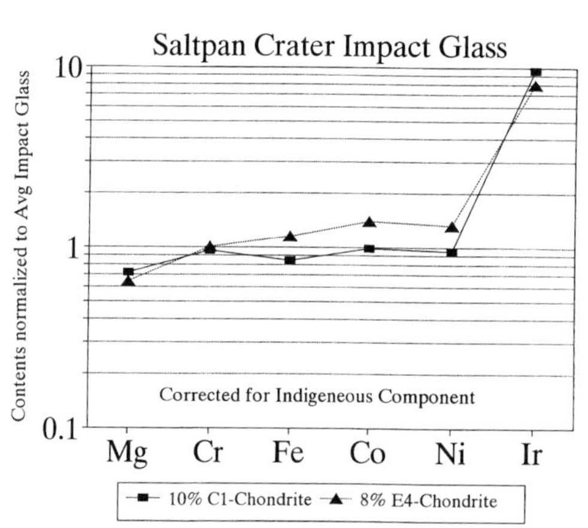

Fig. 8. Saltpan crater (South Africa); model siderophile element abundances calculated for 10% of a C1 chondrite and 8% of an enstatite chondrite, normalized to the average abundances measured in the impact glass, indicating a deficiency of the measured impact glasses in Ir (i.e. the model provides too much Ir). After Koeberl *et al.* (1994*a*).

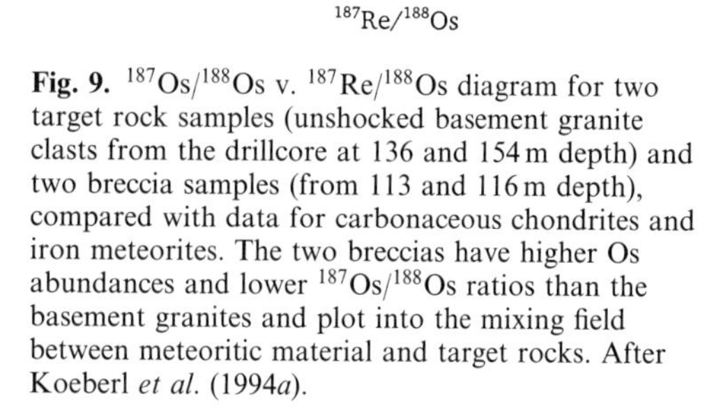

Fig. 9. $^{187}Os/^{188}Os$ v. $^{187}Re/^{188}Os$ diagram for two target rock samples (unshocked basement granite clasts from the drillcore at 136 and 154 m depth) and two breccia samples (from 113 and 116 m depth), compared with data for carbonaceous chondrites and iron meteorites. The two breccias have higher Os abundances and lower $^{187}Os/^{188}Os$ ratios than the basement granites and plot into the mixing field between meteoritic material and target rocks. After Koeberl *et al.* (1994*a*).

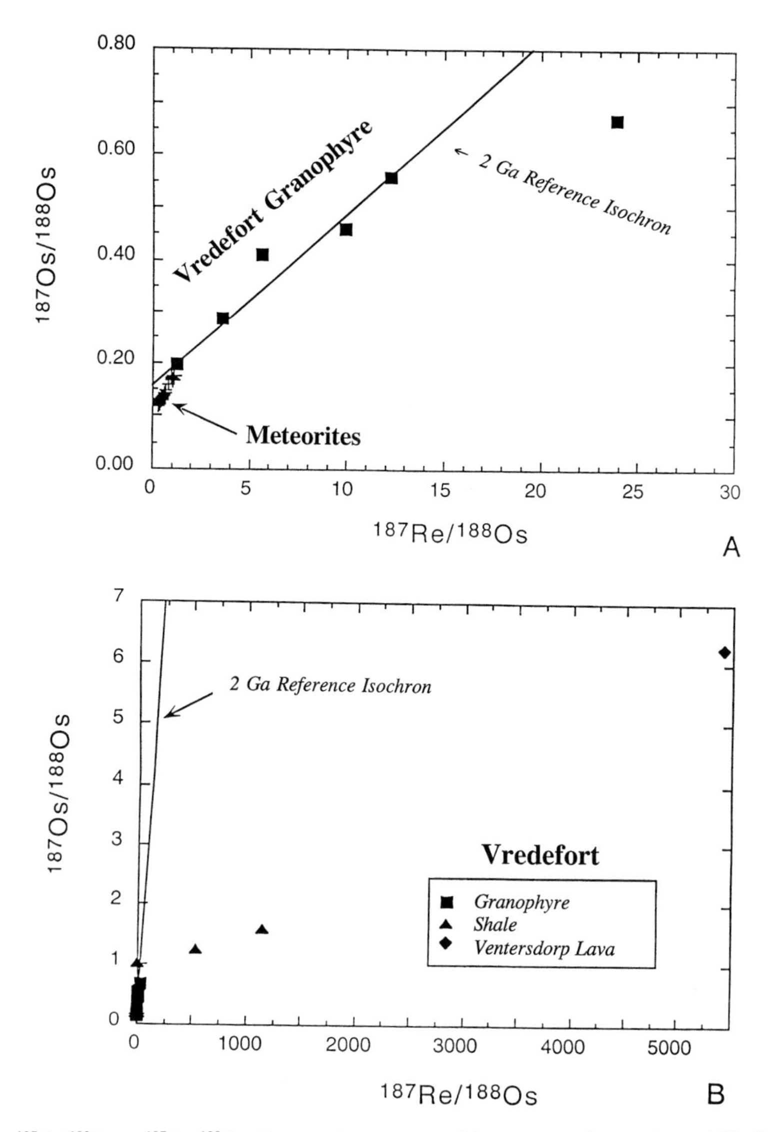

Fig. 10. $^{187}Os/^{188}Os$ v. $^{187}Re/^{188}Os$ diagram for some possible source rock samples and Vredefort Granophyre samples, compared with data for carbonaceous chondrites and iron meteorites. (**a**) Granophyre samples (all but one) are close to the 2 Ga reference isochron, which intersects the meteorite data array. Thus, the Granophyre samples have a narrow range of initial $^{187}Os/^{188}Os$ (2 Ga) ratios that overlap with the range of meteoritic initial ratios, which indicates the presence of a meteoritic component in all Granophyre samples. (**b**) Target rocks show significantly different compositions. After Koeberl *et al.* (1996*b*).

one-tenth of the total number of impact glasses. This rough estimate yields about 0.01 wt % of a chondritic component in the bulk breccia, which agrees well with the 0.015 wt % obtained from the Re–Os studies.

The meteoritic contribution found for the Saltpan impact glasses is rather high compared with glasses or melt rocks at other craters (e.g. Morgan 1978; Palme *et al.* 1978; Palme 1982). An exception is impact melt from the West Clearwater crater with up to about 10% meteoritic contribution (Palme *et al.* 1979). In addition, the Saltpan crater is the only small crater known to have been made by a chondritic meteorite, in contrast to other small craters (less than about 3 km in diameter), for which the projectile type is known, which were formed by iron meteorite projectiles (Grieve 1991).

Vredefort (South Africa)

The Vredefort structure is located about 120 km southwest of Johannesburg, with a current diameter of about 100 km. The formation of the Vredefort structure has been dated at 2024 ± 5 Ma from U–Pb ages of zircons as well as (bulk) laser ^{40}Ar–^{39}Ar ages, both from pseudotachylitic breccia (Kamo *et al.* 1995; Spray *et al.* 1995). Thus, the Vredefort event occurred about 30 Ma. after the formation of the Bushveld Complex. The origin of the structure, which may have initially been as large as 300 km in diameter, has been controversial (see, e.g. Reimold 1993). Only recently, impact-characteristic shock metamorphic effects were found in the form of basal Brazil twins in quartz in Vredefort rocks (Leroux *et al.* 1994) and planar deformation features (PDFs) in zircon (Kamo *et al.* 1995). (For a detailed review of Vredefort, see Reimold & Gibson (1996).)

Granophyric rock dykes with an age of 2 Ga are exposed in the basement core of the structure and along the boundary between the core and the supracrustal rocks of the so-called collar. In the internal model for the origin of Vredefort, it was suggested that the granophyre represents an igneous intrusion (e.g. Bisschoff 1972). In contrast, in the impact model it was proposed that the dykes represent impact melt that was injected into fractures in the floor of the impact structure (French *et al.* 1989; French & Nielsen 1990). The ambiguity of earlier determinations of the Ir content in the granophyre made this structure another ideal target for an Re–Os isotopic study.

Recently, Koeberl *et al.* (1996*b*) found that most Vredefort Granophyre samples have considerably higher Os contents than the country rocks from which the granophyre is likely to have been formed by mixing. The ^{187}Re/^{188}Os and ^{187}Os/^{188}Os ratios of the Vredefort Granophyre scatter about a 2 Ga isochron, with the majority of the initial ^{187}Os/^{188}Os ratios (at 2 Ga) ranging from 0.13 to 0.22 (Fig. 10a). These values overlap the meteoritic data range and indicate that all the granophyre samples contain some meteoritic Os. In addition, the Re–Os isotopic composition of the granophyre is significantly different from that any of the target rocks (Koeberl *et al.* 1996*b*; Fig. 10b). Isotopic composition and Os abundance suggest that the Vredefort Granophyre contains up to 0.2% of a chondritic component, confirming that the Vredefort Granophyre represents an impact melt rock.

Chicxulub (Mexico)

The 65 Ma old Chicxulub impact structure is by now almost unequivocally accepted as being related to the K–T boundary (e.g. Hildebrand *et al.* 1991), and it was, of course, the discovery of PGE enrichments in K–T boundary layers that started the search for this impact structure. In some of the melt rocks from the Chicxulub drill cores, Ir contents of up to 13.5 ppb were found, indicating the presence of an extraterrestrial component (Sharpton *et al.* 1992; Koeberl *et al.* 1994*d*; Schuraytz *et al.* 1994). However, the Ir distribution (and that of other siderophile elements) was found to be very inhomogeneous. Schuraytz *et al.* (1994) found pyrite crystals with high and variable contents of Ni and Co, and some opaque mineral grains also contained high Ir, which they interpreted as evidence for extensive post-impact hydrothermal activity that may have led to a redistribution of the siderophile elements. Some of the inhomogeneity may also be due to inhomogeneous distribution within the impact melt body.

In an Re–Os isotopic study, Koeberl *et al.* (1994*d*) found 25 ppb Os, and very low ^{187}Os/^{188}Os ratio of 0.113 in a melt rock sample. These values are inconsistent with derivation from old continental crust, but close to the meteoritic data array and indicate the presence of a meteoritic component. However, as mentioned before, the similarity between mantle and meteorite Os isotopic compositions requires supporting studies to exclude the presence of any mantle components. In the case of Chicxulub, the large size of the structure may suggest that mantle material could have been excavated. However, impact models indicate that the Chicxulub-forming impact event excavated to a depth of 17–20 km, which is within the upper part of the crust. The depth of the transient cavity (about 45–60 km) includes excavation plus downward displacement, and thus, it is unlikely that the crater-forming event could have mixed mantle material into the Chicxulub melt rocks. Also, trace element, Rb–Sr, and Sm–Nd isotopic characteristics of the samples are typical of rocks from the continental crust (see, e.g. Kring (1993) and Koeberl (1996) for reviews). Mid-ocean ridge basalt (MORB) and related basalts contain only sub-ppb amounts of Os and ^{187}Os/^{188}Os ratios that are slightly higher than that observed in the Chicxulub melt rock (Koeberl *et al.* 1994*d*). Depleted lithospheric mantle xenoliths are the only terrestrial rocks known with subchondritic ^{187}Os/^{188}Os ratios, but Os abundances in xenoliths are too low (2–3 ppb) to explain the high abundances observed in the Chicxulub melt rock. In addition, no basalts or ultramafic bodies have been observed in the Chicxulub target area or in impact breccias (Sharpton *et al.* 1992). Consequently, the Os abundance and isotopic data suggest that the Chicxulub melt rocks contain up to 3 wt % of a chondritic component, which is within the range of meteoritic components reported for large craters (see, e.g. Palme, 1982).

Schuraytz *et al.* (1996) found two micrometre-sized particles in Chicxulub melt rock that consist of almost pure Ir. One particle seems to be pure Ir (99 wt %), and the other one contains a few weight per cent of other PGEs (e.g. Os; B. Schuraytz, pers. comm. 1996). It is not clear at this time if these metal nuggets represent primary remains of the impactor or later differentiation products from the hot impact melt. In addition, Kyte (1996) described an unusual inclusion in K–T boundary sediments from Deep Sea Drilling Project (DSDP) drill core 576 (western North Pacific). This altered fragment has almost chondritic Ir, Fe, and Cr abundances (within a factor of <2) and was interpreted by Kyte (1996) as a possible fragment of the K–T boundary impactor. However, the explanation of this fragment may be much more complicated, as other important elemental abundances either have not yet been determined or have non-chondritic abundances or ratios. For example, Au has an abundance of 1000 times the chondritic concentration (on the order of >100 ppm(!); Kyte 1996), which makes it unlikely that this fragment is a meteorite.

Conclusions

The determination of an extraterrestrial component in melt rocks and breccias can be of value for assessing a possible impact origin of a geological structure from which these rocks are derived. For example, small impact structures (less than *c.* 0.5 km in diameter) often do not show well-developed shock metamorphic effects (such as planar deformation lamellae in rock-forming minerals), and the presence of a meteoritic component in impactites can be the only definitive criterion for an impact origin. Geochemical methods are used to determine the presence of the minor traces (usually <1%) of such a component. The determination of such an extraterrestrial component can be done by studying the concentrations and interelement ratios of siderophile elements, especially the platinum group elements, which are several orders of magnitude more abundant in meteorites than in terrestrial upper-crustal rocks.

However, elemental analyses can lead to ambiguous results if target rocks have high abundances of siderophile elements, or for very low siderophile element concentrations in the impactites. The mechanism by which the extraterrestrial component is mixed into the target material is not well understood. It was proposed that the impact velocity plays a major role in defining the amount of meteoritic matter that ends up in the impactites. High impact velocities are supposed to yield very low meteoritic abundances in the impactites, whereas low impact velocities should produce large extraterrestrial components in melt rocks and breccias (see, e.g. Palme 1982). However, in such a scenario it is difficult to understand why rocks of the East Clearwater impact structures contain up to about 10% of a chondritic component, whereas a cosmic component seems to be absent in rocks from the supposedly contemporaneous West Clearwater impact structure.

In some cases, even the high amounts of siderophile elements present at some structures, such as at Sudbury, may not be of extraterrestrial origin. Walker *et al.* (1991) and Dickin *et al.* (1992) found high initial $^{187}Os/^{188}Os$ ratios ranging from 0.56 to 0.91, which indicate that the Os in the sulphide ores was derived from ancient crust, with only a minor possible contribution from mantle rocks. However, a preliminary study by Cohen *et al.* (1997) yielded low $^{187}Os/^{188}Os$ ratios in some sulphur-poor inclusions from the Sudbury sublayer, which could indicate a mantle component, or, in the absence of a clear Nd isotope mantle signature in these samples, could conceivably represent a meteoritic component.

Using the Re–Os isotopic system for identifying the presence of a meteoritic component in impact melt rocks and breccias avoids the ambiguities that may affect the usage of elemental abundances and ratios. The identification of the projectile type, however, is not straightforward, as vapour fractionation during the impact, as well as post-impact fractionation in the melt or hydrothermal mobilization, may lead to changes in the interelement ratios, and the Re–Os isotope data do not easily allow distinction between individual projectile types.

Acknowledgements: I am grateful to R. Hutchison (Natural History Museum, London) for the invitation to contribute to this symposium. I also appreciate the help of D. Jalufka in preparing some of the figures, as well as various contributions by W. U. Reimold (University of the Witwatersrand, Johannesburg), S. B. Shirey (DTM, Carnegie Institution, Washington), and T. Meisel (Mining University,

Leoben) to the studies of several of the discussed impact structures. Perceptive reviews by I. Gilmour and H. Palme are gratefully acknowledged. This work was supported in part by the Austrian Fonds zur Förderung der wissenschaftlichen Forschung, Project Y-58-GEO.

References

ALLÈGRE, C. J. & LUCK, J. M. 1980. Osmium isotopes as petrogenetic and geological tracers. *Earth and Planetary Science Letters*, **48**, 148–154.

ALVAREZ, L. W., ALVAREZ, W., ASARO, F. & MICHEL, H. V. 1980. Extraterrestrial cause for the Cretaceous–Tertiary extinction. *Science*, **208**, 1095–1108.

ANDERS, E., & GREVESSE, N. 1989. Abundances of the elements: meteoritic and solar. *Geochimica et Cosmochimica Acta*, **53**, 197–214.

ATTREP, M., ORTH, C. J., QUINTANA, L. R., SHOEMAKER, C. S., SHOEMAKER, E. M. & TAYLOR, S. R. 1991. Chemical fractionation of siderophile elements in impactites from Australian meteorite craters. *Lunar and Planetary Science*, **XXII**, 39–40.

BISSCHOFF, A. A. 1972. The dioritic rocks at the Vredefort Dome. *Transactions, Geological Society of South Africa*, **75**, 31–45.

BRANDT, D., & REIMOLD, W. U. 1995. The geology of the Pretoria Saltpan crater and the surrounding area. *South African Journal of Geology*, **98**, 287–303.

BUCHWALD, V. F. 1975. *Handbook of Iron Meteorites*. University of California Press, Berkeley.

BUNCH, T. E., & CASSIDY, W. A. 1972. Petrographic and electron microprobe study of the Monturaqui impactite. *Contributions to Mineralogy and Petrology*, **36**, 95–112.

CHAMBERLAIN, C. P., BLUM, J. D. & KOEBERL, C. 1993. Oxygen isotopes as tracers of tektite source rocks: an example from the Ivory Coast tektites and Lake Bosumtwi crater. *Lunar and Planetary Science*, **XXIV**, 267–268.

CHAO, E. C. T., DWORNIK, E. J. & LITTLER, J. 1964. New data on the nickel–iron spherules from South-East Asian tektites and their implications. *Geochimica et Cosmochimica Acta*, **28**, 971–980.

COHEN, A. S., BURNHAM, O. M., HAWKESWORTH, C. J., LIGHTFOOT, P. C. & COOPER, M. 1997. Os isotope study of ultramafic inclusions and separated sulphides in the Sublayer, Sudbury Igneous Complex, Ontario. *Abstracts, The Origin and Fractionation of Highly Siderophile Elements in the Earth's Mantle, Workshop, Max Planck Institute, Mainz*, p. 29–30.

COLODNER, D. C., BOYLE, E. A., EDMOND, J. M. & THOMSON, J. 1992. Post-depositional mobility of platinum, iridium and rhenium in marine sediments. *Nature*, **358**, 402–404.

CREASER, R. A., PAPANASTASSIOU, D. A. & WASSERBURG, G. J. 1991. Negative thermal ion mass spectrometry of osmium, rhenium and iridium. *Geochimica et Cosmochimica Acta*, **55**, 397–401.

DICKIN, A. P., RICHARDSON, J. M., CROCKET, J. H., MCNUTT, R. H. & PEREDERY, W. V. 1992. Osmium isotope evidence for a crustal origin of platinum group elements in the Sudbury nickel ore, Ontario, Canada. *Geochimica et Cosmochimica Acta*, **56**, 3531–3537.

ESSER, B. K. & TUREKIAN, K. K. 1989. Osmium isotopic composition of the Raton Basin Cretaceous-Tertiary boundary interval. EOS Transactions, American Geophysical Union, **70**, 717.

—— & ——1993. The osmium isotopic composition of the continental crust. *Geochimica et Cosmochimica Acta*, **57**, 3093–3104.

EVANS, N. J., GREGOIRE, D. C., GRIEVE, R. A. F., GOODFELLOW, W. D. & VEIZER, J. 1993. Use of platinum-group elements for impactor identification: Terrestrial impact craters and Cretaceous–Tertiary boundary. *Geochimica et Cosmochimica Acta*, **57**, 3737–3748.

FAURE, G. 1986. *Principles of Isotope Geology*, 2nd edn. Wiley, New York, 341–361.

FEHN, U., TENG, R., ELMORE, D. & KUBIK, P. W. 1986. Isotopic composition of osmium in terrestrial samples determined by accelerator mass spectrometry. *Nature*, **323**, 707–710.

FRENCH, B. M., & NIELSEN, R. L. 1990. Vredefort bronzite granophyre: chemical evidence for an origin as meteorite impact melt. *Tectonophysics*, **171**, 119–138.

—— & SHORT, N. M. (eds) 1968. *Shock Metamorphism of Natural Materials*. Mono, Baltimore.

——, KOEBERL, C., GILMOUR, I., SHIREY, S. B., DONS, J. A. & NATERSTAD, J. 1997. The Gardnos impact structure, Norway: petrology and geochemistry of target rocks and impactites. *Geochimica et Cosmochimica Acta*, **61**, 873–904.

——, ORTH, C. J. & QUINTANA, C. R. 1989. Iridium in the Vredefort bronzite granophyre: impact melting and limits on a possible extraterrestrial component. *In*: *Proceedings 19th Lunar and Planetary Science Conference*. Cambridge University Press, New York, 733–744.

GANAPATHY, R. & LARIMER, J. W. 1983. Nickel–iron spherules in tektites: non-meteoritic in origin. *Earth and Planetary Science Letters*, **65**, 225–228.

GENTNER, W., STORZER, D. & WAGNER, G. A. 1969. New fission track ages of tektites and related glasses. *Geochimica et Cosmochimica Acta*, **33**, 1075–1081.

GLASS, B. P., FREDRIKSSON, K. & FLORENSKY, P. V. 1983. Microirghizites recovered from a sediment sample from the Zhamanshin impact structure. *Proceedings of the 14th Lunar and Planetary Science Conference, Journal of Geophysical Research*, **88**, B319–B330.

GÖBEL, E., REIMOLD, W. U., BADDENHAUSEN, H. & PALME, H. 1980. The projectile of the Lapajärvi impact crater. *Zeitschrift für Naturforschung*, **35a**, 197–203.

GOSTIN, V. A., KEAYS, R. R. & WALLACE, M. W. 1989. Iridium anomaly from the Acraman impact ejecta horizon: impacts can produce sedimentary iridium peaks. *Nature*, **340**, 542–544.

GRIEVE, R. A. F. 1991. Terrestrial impact: the record in the rocks. *Meteoritics*, **26**, 175–194.

—— & SHOEMAKER, E. M. 1994. The record of past impacts on Earth. *In*: GEHRELS, T. (ed.) *Hazards due to Comets and Asteroids*. University of Arizona Press, Tucson, 417–462.

——, LANGENHORST, F. & STÖFFLER, D. 1996. Shock metamorphism in nature and experiment: II. Significance in geoscience. *Meteoritics and Planetary Science*, **31**, 6–35.

——, RUPERT, J., SMITH, J. & THERRIAULT, A. 1995. The record of terrestrial impact cratering. *GSA Today*, **5**(10), 193–196.

GROS, J., TAKAHASHI, H., HERTOGEN, J., MORGAN, J. W. & ANDERS, E. 1976. Composition of the projectiles that bombarded the lunar highlands. *Proceedings of the 7th Lunar Science Conference*, 2403–2425.

GUROV, E. P. 1996. The group of Macha craters in Western Yakutia. *Lunar and Planetary Science*, **XXVII**, 473–474.

HILDEBRAND, A. R., PENFIELD, G. T., KRING, D. A., PILKINGTON, M., CAMARGO, Z. A., JACOBSEN, S. B. & BOYNTON, W. V. 1991. Chicxulub crater: a possible Cretaceous–Tertiary boundary impact crater on the Yucatan Peninsula, Mexico. *Geology*, **19**, 867–871.

JONES, W. B. 1985. Chemical analyses of Bosumtwi crater target rocks compared with Ivory Coast tektites. *Geochimica et Cosmochimica Acta*, **49**, 2569–2576.

KAMO, S. L., REIMOLD, W. U., KROGH, T. E. & COLLISTON, W. P. 1995, Shocked zircons in Vredefort pseudotachylite and the U–Pb zircon age of the Vredefort impact event. *In*: *Centennial Geocongress, Johannesburg, Geological Society of South Africa*, 566–569.

KOEBERL, C. 1986. Geochemistry of tektites and impact glasses. *Annual Review of Earth and Planetary Science*, **14**, 323–350.

——1993. Extraterrestrial component associated with Australasian microtektites in a core from ODP Site 758B. *Earth and Planetary Science Letters*, **119**, 453–458.

——1994*a*. African meteorite impact craters: characteristics and geological importance. *Journal of African Earth Sciences*, **18**, 263–295.

——1994*b*. Tektite origin by hypervelocity asteroidal or cometary impact: target rocks, source craters, and mechanisms. *In*: DRESSLER, B. O., GRIEVE, R. A. F. & SHARPTON, V. L. (eds) *Large Meteorite Impacts and Planetary Evolution, Sudbury 1992*. Geological Society of America, Special Paper, **293**, 133–151.

——1996. Chicxulub – the K–T boundary impact crater: a review of the evidence, and an introduction to impact crater studies. *Abhandlungen der Geologischen Bundesanstalt, Wien*, **53**, 23–50.

—— & REIMOLD, W. U. 1995. Early Archean spherule beds in the Barberton Mountain Land, South Africa: no evidence for impact origin. *Precambrian Research*, **74**, 1–33.

—— & SHIREY, S. B. 1993. Detection of a meteoritic component in Ivory Coast tektites with rhenium–osmium isotopes. *Science*, **261**, 595–598.

—— & ——1996. Re–Os isotope study of rocks from the Manson impact structure. *In*: KOEBERL, C. & ANDERSON, R. R. (eds) *The Manson Impact Structure, Iowa: Anatomy of an Impact Crater*. Geological Society of America, Special Paper **302**, 311–339.

——, ARMSTRONG, R. A. & REIMOLD, W. U. 1997*b*. Morokweng, South Africa: a large impact structure of Jurassic–Cretaceous boundary age. *Geology*, **25**, 731–734.

——, BOTTOMLEY, R., GLASS, B. P., & STORZER, D. 1997*a*. Geochemistry and age of Ivory Coast tektites and microtektites. *Geochimica et Cosmochimica Acta*, **61**, 1745–1772.

——, ——, ——, —— & YORK, D. 1989. Geochemistry and age of Ivory Coast tektites. *Meteoritics*, **24**, 287.

——, POAG, C. W., REIMOLD, W. U. & BRANDT, D. 1996*a*. Impact origin of Chesapeake Bay structure and the source of North American tektites. *Science*, **271**, 1263–1266.

——, REIMOLD, W. U. & BOER, R. H. 1993. Geochemistry and mineralogy of Early Archean spherule beds, Barberton Mountain Land, South Africa: evidence for origin by impact doubtful. *Earth and Planetary Science Letters*, **119**, 441–452.

——, —— & SHIREY, S. B. 1994*a*. Saltpan impact crater, South Africa: geochemistry of target rocks, breccias, and impact glasses, and osmium isotope systematics. *Geochimica et Cosmochimica Acta*, **58**, 2893–2910.

——, —— & ——1996*b*. A Re–Os isotope and geochemical study of the Vredefort Granophyre: clues to the origin of the Vredefort structure, South Africa. *Geology*, **24**, 913–916.

——, —— & ——1998. The Aouelloul Crater, Mauritania: On the Problem of Confirming the Impact Origin of a Small Crater. *Meteoritics and Planetary Science*, **33**, in press.

——, ——, —— & LE ROUX, F. G. 1994*c*. Kalkkop crater, Cape Province, South Africa: confirmation of impact origin using osmium isotope systematics. *Geochimica et Cosmochimica Acta*, **58**, 1229–1234.

——, SHARPTON, V. L., SCHURAYTZ, B. C., SHIREY, S. B., BLUM, J. D. & MARIN, L. E. 1994*d*. Evidence for a meteoritic component in impact melt rock from the Chicxulub structure. *Geochimica et Cosmochimica Acta*, **58**, 1679–1684.

——, STORZER, D. & REIMOLD, W. U. 1994*b*. The age of the Saltpan impact crater, South Africa. *Meteoritics*, **29**, 374–379.

KOLBE, P., PINSON, W. H., SAUL, J. M. & MILLER, E. W. 1967. Rb–Sr study on country rocks of the Bosumtwi crater, Ghana. *Geochimica et Cosmochimica Acta*, **31**, 869–875.

KRÄHENBÜHL, U., GEISSBÜHLER, M., BÜHLER, F. & EBERHARDT, P. 1988. The measurement of osmium isotopes in samples from a Cretaceous/Tertiary (K/T) section of the Raton Basin, USA. *Meteoritics*, **23**, 282.

KRING, D. A. 1993. The Chicxulub impact event and possible causes of K/T boundary extinctions. *In*: BOAZ, D. & DORNAN, M. (eds) *Proceedings of the*

First Annual Symposium of Fossils of Arizona: Mesa (Arizona), Mesa Southwest Museum and Southwest Paleontological Society, 63–79.

KYTE, F. T. 1996. A piece of the KT bolide? *Lunar and Planetary Science*, **XXVII**, 717–718.

——, ZHOU, L. & LOWE, D. R. 1992. Noble metal abundances in an Early Archean impact deposit. *Geochimica et Cosmochimica Acta*, **56**, 1365–1372.

LAMBERT, P. 1982. Anomalies within the system: Rochechouart target rock meteorite. *In*: SILVER, L. T. & SCHULTZ, P. H. (eds), *Geological Implications of Impacts of Large Asteroids and Comets on the Earth*. Geological Society of America, Special Paper, **190**, 243–249.

LEROUX, H., REIMOLD, W. U. & DOUKHAN, J. C. 1994. A T.E.M. investigation of shock metamorphism in quartz from the Vredefort dome, South Africa. *Tectonophysics*, **230**, 223–239.

LICHTE, F. E., WILSON, S. M., BROOKS, R. R., REEVES, R. D., HOLZBECHER, J. & RYAN, D. E. 1986. New method for the measurement of osmium isotopes applied to a New Zealand Cretaceous/Tertiary boundary shale. *Nature*, **322**, 816–817.

LOWE, D. R., BYERLY, G. R., ASARO, F. & KYTE, F. T. 1989. Geological and geochemical record of 3400-million-year-old terrestrial meteorite impacts. *Science*, **245**, 959–962.

LUCK, J. M. & TUREKIAN, K. K. 1983. Osmium-187/Osmium-186 in manganese nodules and the Cretaceous–Tertiary boundary. *Science*, **222**, 613–615.

MASAITIS, V. L. 1992. Impactites from Popigai crater. *In*: DRESSLER, B. O., GRIEVE, R. A. F. & SHARPTON, V. L. (eds) *Large Meteorite Impacts and Planetary Evolution*. Geological Society of America, Special Paper, **293**, 153–162.

—— & RAIKHLIN, A. J. 1985. The Popigai crater formed by the impact of an ordinary chondrite (in Russian). *Doklady Akademii Nauk SSSR*, **286**, 1476–1478.

MASON, B. 1979. *Meteorites. Data of Geochemistry*. US Geological Survey Professional Paper, **440-B-1**, Chapter B, Part 1.

MEISEL, T. & KOEBERL, C. 1990. Siderophile elements in selected impact glasses and melts and the possibility of determining the composition of the impactor. *Meteoritics*, **25**, 385.

——, —— & FORD, R. J. 1990. Geochemistry of Darwin impact glass and target rocks. *Geochimica et Cosmochimica Acta*, **54**, 1463–1474.

——, KRÄHENBÜHL, U. & NAZAROV, M. A. 1995. Combined osmium and strontium isotopic study of the Cretaceous–Tertiary boundary at Sumbar, Turkmenistan: a test for impact vs. volcanic hypothesis. *Geology*, **23**, 313–316.

MELOSH, H. J. 1989. *Impact Cratering: a Geologic Process*. Oxford University Press. New York.

MILTON, D. J. & NAESER, C. W. 1971. Evidence for an impact origin of the Pretoria Salt Pan, South Africa. *Nature, Physical Science*, **299**, 211–212.

MITTLEFEHLDT, D. W., SEE, T. H. & HÖRZ, F. 1992*a*. Projectile dissemination in impact melts from Meteor crater, Arizona. *Lunar and Planetary Science*, **XXIII**, 919–920.

——, —— & ——1992*b*. Dissemination and fractionation of projectile materials in the impact melts from Wabar crater, Saudi Arabia. *Meteoritics*, **27**, 361–370.

MORGAN, J. W. 1978. Lunar crater glasses and high-magnesium australites: trace element volatilization and meteoritic contamination. *Proceedings of the 9th Lunar and Planetary Science Conference*, 2713–2730.

—— & WANDLESS, G. A. 1983. Strangways Crater, Northern Territory, Australia: siderophile element enrichment and lithophile element fractionation. *Journal of Geophysical Research*, **88**, A819-A829.

——, HIGUCHI, H., GANAPATHY, R. & ANDERS, E. 1975. Meteoritic material in four terrestrial meteorite craters. *Proceedings of the 6th Lunar Science Conference*, 1609–1623.

——, JANSSENS, M.-J., HERTOGEN, H., GROS, J. & TAKAHSHI, H. 1979. Ries impact crater: search for meteoritic material. *Geochimica et Cosmochimica Acta*, **43**, 803–815.

——, WALKER, R. J. & GROSSMAN, J. N. 1992. Rhenium–osmium isotope systematics in meteorites I: Magmatic iron meteorite groups IIAB and IIIAB. *Earth and Planetary Science Letters*, **108**, 191–202.

NAZAROV, M. A., BARSUKOVA, L. D., BADJUKOV, D. D., KOLESOV, G. M. & NIZHEGORODOVA, I. V. 1990. The Kara impact structure: iridium abundances in the crater rocks. *Lunar and Planetary Science*, **XXI**, 849–850.

——, ——, ——, ——, —— & ALEKSEEV, A. S. 1989. Geology and chemistry of the Kara and Ust-Kara impact craters. *Lunar and Planetary Science*, **XX**, 764–765.

PALME, H. 1982. Identification of projectiles of large terrestrial impact craters and some implications for the interpretation of Ir-rich Cretaceous/Tertiary boundary layers. *In*: SILVER, L. T. & SCHULTZ, P. H. (eds) *Geological Implications of Impacts of Large Asteroids and Comets on Earth*. Geological Society of America, Special Paper, **190**, 223–233.

——, GÖBEL, E. & GRIEVE, R. A. F. 1979. The distribution of volatile and siderophile elements in the impact melt of East Clearwater (Quebec). *Proceedings of the 10th Lunar and Planetary Science Conference*, 2465–2492.

——, GRIEVE, R. A. F. & WOLF, R. 1981. Identification of the projectile at the Brent crater, and further considerations of projectile types at terrestrial craters. *Geochimica et Cosmochimica Acta*, **45**, 2417–2424.

——, JANSSENS, M.-J., TAKAHASI, H., ANDERS, E. & HERTOGEN, J. 1978. Meteorite material at five large impact craters. *Geochimica et Cosmochimica Acta*, **42**, 313–323.

——, RAMMENSEE, W. & REIMOLD, W. U. 1980. The meteoritic component of impact melts from European impact craters. *Lunar and Planetary Science*, **XI**, 848–851.

PERNICKA, E., KAETHER, D. & KOEBERL, C. 1996. Siderophile element concentrations in drill core samples from the Manson crater. *In*: KOEBERL, C. & ANDERSON, R. R. (eds) *The Manson Impact Structure, Iowa: Anatomy of an Impact Crater*. Geological Society of America, Special Paper, **302**, 325–330.

PIERAZZO, E., VICKERY, A. M. & MELOSH, H. J. 1997. A reevaluation of impact melt production. *Icarus*, **127**, 408–423.

POAG, C. W., POWARS, POPPE, L. J. & MIXON, R. B. 1994. Meteoroid mayhem in Ole Virginny: source of the North American tektite strewn field. *Geology*, **22**, 691–694.

REIMOLD, W. U. 1993. A review of the geology of and deformation related to the Vredefort Structure, South Africa. *Journal of Geological Education*, **41**, 106–117.

—— & GIBSON, R. L. 1996. Geology and evolution of the Vredefort impact structure, South Africa. *Journal of African Earth Sciences*, **23**, 125–162.

——, KOEBERL, C., PARTRIDGE, T. C. & KERR, S. J. 1992. Pretoria Saltpan crater: impact origin confirmed. *Geology*, **20**, 1079–1082.

ROHLEDER, H. P. T. 1933. The Steinheim basin and the Pretoria Salt Pan: volcanic or meteoritic origin? *Geological Magazine*, **70**, 489–498.

SCHMIDT, G. & PERNICKA, E. 1994. The determination of platinum group elements (PGE) in target rocks and fall-back material of the Nördlinger Ries impact crater (Germany). *Geochimica et Cosmochimica Acta*, **58**, 5083–5090.

SCHNETZLER, C. C., PHILPOTTS, J. A., & THOMAS, H. H. 1967. Rare earth and barium abundances in Ivory Coast tektites and rocks from the Bosumtwi crater area, Ghana. *Geochimica et Cosmochimica Acta*, **31**, 1987–1993.

——, PINSON, W. H. & HURLEY, P. M. 1966. Rubidium–strontium age of the Bosumtwi crater area, Ghana, compared with the age of the Ivory Coast tektites. *Science*, **151**, 817–819.

SCHULTZ, P. H., KOEBERL, C., BUNCH, T., GRANT, J. & COLLINS, W. 1994. Ground truth for oblique impact processes: new insight from the Rio Cuarto, Argentina, crater field. *Geology*, **22**, 889–892.

SCHURAYTZ, B. C., LINDSTROM, D. J., MARÍN, L. E., MARTINEZ, R. R., MITTLEFEHLDT, D. W., SHARPTON, V. L. & WENTWORTH, S. J. 1996. Iridium metal in Chicxulub impact melt: forensic chemistry on the K-T smoking gun. *Science*, **271**, 1573–1576.

——, SHARPTON, V. L. & MARIN, L. E. 1994. Petrology of impact-melt rocks at the Chicxulub multiring basin, Yucatán, Mexico. *Geology*, **22**, 868–872.

SHARPTON, V. L. & GRIEVE, R. A. F. 1990. Meteorite impact, cryptoexplosion, and shock metamorphism: A perspective on the evidence at the K/T boundary. *In*: SHARPTON, V. L. & WARD, P. D. (eds) *Global Catastrophes in Earth History*. Geological Society of America, Special Paper, **247**, 301–318.

——, DALRYMPLE, G. B., MARIN, L. E., RYDER, G., SCHURAYTZ, B. C., & URRUTIA-FUCUGAUCHI, J. 1992. New links between the Chicxulub impact structure and the Cretaceous/Tertiary boundary. *Nature*, **359**, 819–821.

SHIREY, S. B., & WALKER, R. J. 1995. Carius tube digestion for low-blank rhenium–osmium analysis. *Analytical Chemistry*, **67**, 2136–2141.

SPRAY, J. G., KELLEY, S. P. & REIMOLD, W. U. 1995. Laser probe argon-40/argon-39 dating of coesite and stishovite-bearing pseudotachylytes and the age of the Vredefort impact event. *Meteoritics*, **30**, 335–343.

STÖFFLER, D. 1972. Deformation and transformation of rock-forming minerals by natural and experimental shock processes: 1. Behaviour of minerals under shock compression. *Fortschritte der Mineralogie*, **49**, 50–113.

—— & LANGENHORST, F. 1994. Shock metamorphism of quartz in nature and experiment: I. Basic observations and theory. *Meteoritics*, **29**, 155–181.

SWISHER, C. C., GRAJALES-NISHIMURA, J. M., MONTANARI, A. *et al.* 1992. Coeval $^{40}Ar/^{39}Ar$ ages of 65.0 million years ago from Chicxulub crater melt rock and Cretaceous–Tertiary boundary tektites. *Science*, **257**, 954–958.

TAYLOR, H. P. & EPSTEIN, S. E. 1966. Oxygen isotope studies of Ivory Coast tektites and impactite glass from the Bosumtwi crater, Ghana. *Science*, **153**, 173–175.

TAYLOR, S. R. 1967. Composition of meteorite impact glass across the Henbury strewn field. *Geochimica et Cosmochimica Acta*, **31**, 961–968.

—— & MCLENNAN, S. M. 1985. *The Continental Crust: its Composition and Evolution*. Blackwell Scientific, Oxford.

TUREKIAN, K. K. 1982. Potential of $^{187}Os/^{186}Os$ as a cosmic versus terrestrial indicator in high iridium layers of sedimentary strata. *In*: SILVER, L. T. & SCHULTZ, P. H. (eds) *Geological Implications of Impacts of Large Asteroids and Comets on the Earth*. Geological Society of America, Special Paper, **190**, 243–249.

VÖLKENING, J., WALCZYK, T. & HEUMANN, K. G. 1991. Osmium isotope ratio determinations by negative thermal ionization mass spectrometry. *International Journal of Mass Spectrometry and Ion Processes*, **105**, 147–159.

WAGNER, P. A. 1920. Note on the volcanic origin of the Salt Pan on the farm Zoutpan, No. 467. *Transactions, Geological Society of South Africa*, **23**, 52–58.

WALKER, R. J., & MORGAN, J. W. 1989. Rhenium–osmium isotope systematics of carbonaceous chondrites. *Science*, **243**, 519–522.

——, NALDRETT, A. J., LI, C., & FASSETT, J. D. 1991. Re–Os isotope systematics of Ni–Cu sulfide ores, Sudbury Igneous Complex, Ontario: evidence for a major crustal component. *Earth and Planetary Science Letters*, **105**, 416–429.

WALLACE, M. W., GOSTIN, V. A., & KEAYS, R. R. 1990. Acraman impact ejecta and host shales: evidence for low-temperature mobilization of iridium and other platinoids. *Geology*, **18**, 132–135.

WOLF, R., WOODROW, A. & GRIEVE, R. A. F. 1980. Meteoritic material at four Canadian impact craters. *Geochimica et Cosmochimica Acta*, **44**, 1015–1022.

Mapping Chicxulub crater structure with gravity and seismic reflection data

A. R. HILDEBRAND[1], M. PILKINGTON[1], C. ORTIZ-ALEMAN[2], R. E. CHAVEZ[2], J. URRUTIA-FUCUGAUCHI[2], M. CONNORS[3], E. GRANIEL-CASTRO[4], A. CAMARA-ZI[4], J. F. HALPENNY[5] & D. NIEHAUS[6]

[1] *Geological Survey of Canada, 615 Booth Street, Ottawa, Ontario, Canada K1A 0E9*
[2] *Instituto de Geofisica, UNAM, Ciudad Universitaria, Codigo 04510, México, D. F., México*
[3] *Athabasca University, 1, University Drive, Athabasca, Alberta, Canada T9S 3A3*
[4] *Facultad de Ingeniería, Universidad Autónoma de Yucatán, Apdo. Postal No. 150, Cordemex 97111, Mérida, Yucatán, México*
[5] *Geomatics Canada, 615 Booth Street, Ottawa, Ontario, Canada K1A 0E9*
[6] *PCI Enterprises, 50 West Wilmot Street, Richmond Hill, Ontario, Canada, L4B 1M5*

Abstract: Aside from its significance in establishing the impact–mass extinction paradigm, the Chicxulub crater will probably come to exemplify the structure of large complex craters. Much of Chicxulub's structure may be 'mapped' by tying its gravity expression to seismic-reflection profiles revealing an ~180 km diameter for the now-buried crater. The distribution of karst topography aids in outlining the peripheral crater structure as also revealed by the horizontal gradient of the gravity anomaly. The fracturing inferred to control groundwater flow is apparently related to subsidence of the crater fill. Modelling the crater's gravity expression based on a schematic structural model reveals that the crater fill is also responsible for the majority of the negative anomaly. The crater's melt sheet and central structural uplift are the other significant contributors to its gravity expression. The Chicxulub impact released $\sim 1.2 \times 10^{31}$ ergs based on the observed collapsed disruption cavity of ~86 km diameter reconstructed to an apparent disruption cavity (D_{ad}) of ~94 km diameter (equivalent to the excavation cavity) and an apparent transient cavity (D_{at}) of ~80 km diameter. This impact energy, together with the observed $\sim 2 \times 10^{11}$ g global Ir fluence in the Cretaceous–Tertiary (K–T) fireball layer indicates that the impactor was a comet estimated as massing $\sim 1.8 \times 10^{18}$ g of ~16.5 km diameter assuming a 0.6 g cm^{-3} density. Dust-induced darkness and cold, wind, giant waves, thermal pulses from the impact fireball and re-entering ejecta, acid rain, ozone-layer depletion, cooling from stratospheric aerosols, H_2O greenhouse, CO_2 greenhouse, poisons and mutagens, and oscillatory climate have been proposed as deleterious environmental effects of the Chicxulub impact with durations ranging from a few minutes to a million years. This succession of effects defines a temperature curve that is characteristic of large impacts. Although some patterns may be recognized in the K–T extinctions, and the survivorship rules changed across the boundary, relating specific environmental effects to species' extinctions is not yet possible. Geochemical records across the boundary support the occurrence a prompt thermal pulse, acid rain and a ~5000 year-long greenhouse. The period of extinctions seems to extend into the earliest Tertiary.

Aside from its role in determining Earth history by ending the Cretaceous Period (Alvarez *et al.* 1980; Hildebrand *et al.* 1991), the Chicxulub crater will probably come to exemplify the structure of large complex craters due to its relatively good preservation. Also, the Chicxulub impact constrains ejecta dispersion from a large impact, as its ballistic ejecta are known at dozens of localities, and its impact fireball deposit is the only example known as a discrete layer (e.g. Bohor *et al.* 1987; Izett 1990; Hildebrand & Boynton 1990; Hildebrand 1993; Smit 1994; Smit *et al.* 1996; Kyte *et al.* 1996).

Although the Chicxulub crater lies buried on the Yucatán Peninsula, understanding its structure is aided by extensive regional geophysical surveys acquired by Petróleos Méxicanos in the course of petroleum exploration (e.g. Villagomez 1953), and the overlying half-water, half-land flat terrain which allows application of a variety of geophysical (and drilling) techniques (e.g. Camargo & Suárez 1994; Pilkington *et al.* 1994; Hildebrand *et al.* 1995; Ward *et al.* 1995; Urrutia-Fucugauchi *et al.* 1996; Morgan *et al.* 1997; Christeson *et al.* 1997; Campos-Enriquez *et al.* 1997).

HILDEBRAND, A. R. *et al.* 1998. Mapping Chicxulub crater structure with gravity and seismic reflection data. *In*: GRADY, M. M., HUTCHISON, R., MCCALL, G. J. H. & ROTHERY, D. A. (eds) *Meteorites: Flux with Time and Impact Effects*. Geological Society, London, Special Publications, **140**, 155–176.

The 'mapping' of buried craters is hindered by the covering sediments, although crater formation in a shallow water depositional environment can result in relatively complete preservation. Drilling offers a means of testing models derived from other techniques (and supplies necessary material for many sample-based studies), but any mapping effort using drilling is generally prohibitively expensive with techniques currently available (which are also depth limited). Therefore geophysical techniques are generally applied; such efforts are aided by impact-generated lithologies and structure having distinct seismic and potential field signatures (e.g. Pilkington & Grieve 1992), and the understanding of crater structure as gained from exposed examples. Conventional 2-D reflection seismic techniques can produce useful cross sections, but can only map a crater if many lines are run (e.g. Dypvik *et al.* 1996). The recently developed 3-D reflection seismic method can provide a more complete picture of buried craters, although at considerable expense (e.g. Isaac & Stewart 1993). Nevertheless, seismic reflection studies are hampered by progressive loss of resolution towards crater centres as is the case at Chicxulub (e.g. Camargo & Suárez 1994) and the general lack of reflectors in craters formed in crystalline rocks. Seismic velocity studies, however, can resolve crater structure in zones where coherency has been lost (e.g. Christeson *et al.* 1997)

Potential field studies are conducted economically (relative to seismic efforts) so that magnetic- and gravity-field anomalies associated with buried craters have been delineated in many cases. Useful data can be obtained for craters in both sedimentary and crystalline terranes (e.g. Pilkington & Grieve 1992). Mapping crater structure using potential fields is limited unless extensive petrophysical information is available from the crater lithologies, otherwise quantitative interpretations are not well constrained (although the extent of the anomalies in plan may be determined). If sufficient samples are not available to provide petrophysical data, crater structure may be mapped by tying observed potential field anomalies to crater structural features as revealed by 2-D seismic profiles through the crater, thus constraining structure in three dimensions. In this paper a mapping and interpretation of Chicxulub crater structure is presented based on tying seismic reflection data to the crater's gravity signature, coupled with gravity-field modelling.

Mapping Chicxulub crater structure

Figure 1 shows the two main data sets which are used to map Chicxulub crater structure: horizontal gravity gradients and two offshore seismic lines. Also shown is the distribution of karst topography which provides additional constraints.

The gravity data set largely comprises the Bouguer gravity anomaly onshore and the free air anomaly offshore resulting in not strictly comparable data; the free air anomaly will be lower than the Bouguer anomaly in proportion to the water depth (0.069 mGal/m using 2.67 g cm^{-3} for reconstituted rock). However, as the sea floor is remarkably flat on the shelf, and slopes very gradually from the coast, no local anomalies (that would create features in a horizontal gradient plot) are introduced by the juxtapositioning of the two data types. The magnitude of the difference is ~4 mGals in the deep water at the northern edge of the map. The onshore data are from surveys conducted by Petróleos Méxicanos beginning in 1947 (e.g. Villagomez 1953), and recent surveys conducted since recognition of the crater (Hildebrand *et al.* 1995; Kinsland *et al.* 1995; Connors *et al.* 1996). The latter surveys have closed some data gaps and added ~1000 useful stations (precision usually exceeds 0.1 mGals) to the land-based data set. Integration of the data sets is not without problems as the earlier data are mislocated by up to 2 km (due to lack of accurate maps at the time of the survey) and differences in survey values of up to 2 mGals between the old and new data sets. In case of conflict the newer values have been used. The offshore data were largely acquired by Ness *et al.* (1991) using a shipborne dynamic meter with a mean cross-over precision of 1.5 mGals (internal disagreements range up to ~5 mGals). These data sets have been augmented by Petróleos Méxicanos aerogravity data collected over the sea and coastal mangrove swamps in the western part of the map area, which have been previously largely unsurveyed. These data have had their base level adjusted to conform to the land and shipborne data sets. Although filtered, the airborne data contain numerous artefacts parallel to the survey boundaries oriented NNW and WSW.

The horizontal gradient of the gravity anomaly data was computed by root-sum-squaring the gradient in two orthogonal directions in a 3.75 × 3.75 km^2 cell (Fig. 1). A series of concentric gradient features roughly centred at 21.29°N, 89.52°W is revealed comparable to similar maps of Hildebrand *et al.* (1994 1995) and Connors *et al.* (1996), although the gravity field is better constrained to the west in this version. Only the largest gradients are traced offshore, partly due to survey gaps and partly due to the greater imprecision of offshore surveys. The outermost feature occurs at 80 to 95 km radius, indicating

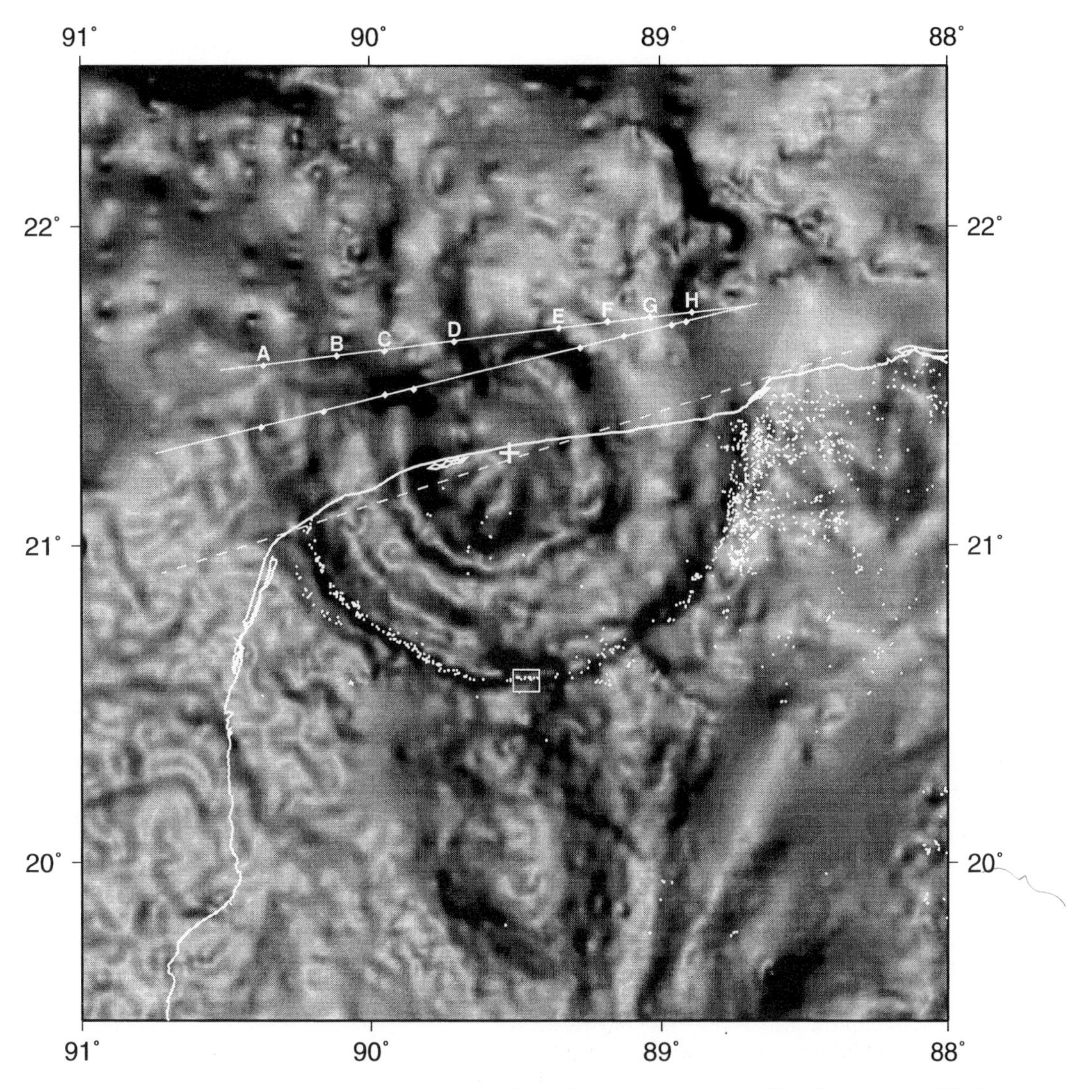

Fig. 1. Horizontal gravity anomaly gradient over the Chicxulub crater, northwestern Yucatán peninsula, Mexico. Darker shading indicates steeper gradients. The regions that appear blurred on the plot are areas lacking any measurements to constrain the gravity field. Fewer data exist offshore and the available ship tracks appear as generally N–S oriented artefacts. Much of the western offshore area is covered by aerogravity data with data artefacts oriented NNW and ENE. The coastline is indicated by a white line, the centre of the crater by a cross, and the small white rectangle indicates the location of Fig. 2. The white dots on land indicate the positions of water-filled karst features known as cenotes by the Yucatecos; the straight lines offshore indicate the positions of Petróleos Méxicanos seismic reflection profiles which roughly parallel the shore: D92RP001 and 002, north to south, respectively. The letters locate structural features as detailed in Table 1: A, outermost slump fault; B, C, F, slump faults with >1 km displacement. The H position is a western limit for the eastern basin margin, and G only approximately locates a fault with >1 km displacement as seismic resolution is lost to the east. Positions D & E mark the easterly and westerly inner margins of the resolved stratigraphy of the innermost slumped block, and mark upper limits for the edges of the collapsed disruption cavity. See Fig. 3 for a perspective view of these features on line D92RP002. The letters are annotated to only the northern profile; white diamonds indicate the corresponding positions on the southern profile. The white dashed line indicates the gravity profile modelled in Fig. 8.

the margin of the buried impact structure. No sign of concentric structure in the gravity field is revealed at greater radii as advocated by Sharpton *et al.* (1993; 1996; see Hildebrand (1997) for a discussion.) The centre of the central gradient features is located slightly to the southwest of the centre of the peripheral features and radii to the east and west are slightly greater than radii to the south. Until the northern portion of the crater is mapped, the latter observation is uncertain as the crater centre may be mislocated.

The karst topography data set used to map crater structure is illustrated in Fig. 1. Pope *et al.* (1993, 1996), Hildebrand *et al.* (1994, 1995), Perry *et al.* (1995) and Connors *et al.* (1996) have described the subcircular distribution of flooded sinkholes, locally known as cenotes, that are the most prominent surface expression of the buried crater. Their distribution in Fig. 1 was mapped from 1: 50 000 topographic maps covering all the northern Yucatán Peninsula (e.g. Connors *et al.* 1996) and from 1 : 75 000 aerial photographs on the western side of the crater. Not all the cenotes are indicated on the topographic maps, but the lack is most significant to the west. The main cenote ring is 160–165 km in diameter and comprises cenotes 30 to 300 m across. Figure 2 illustrates a portion of the main ring in the south. The cenotes are generally near circular reflecting their origin as sinkholes, although elongated

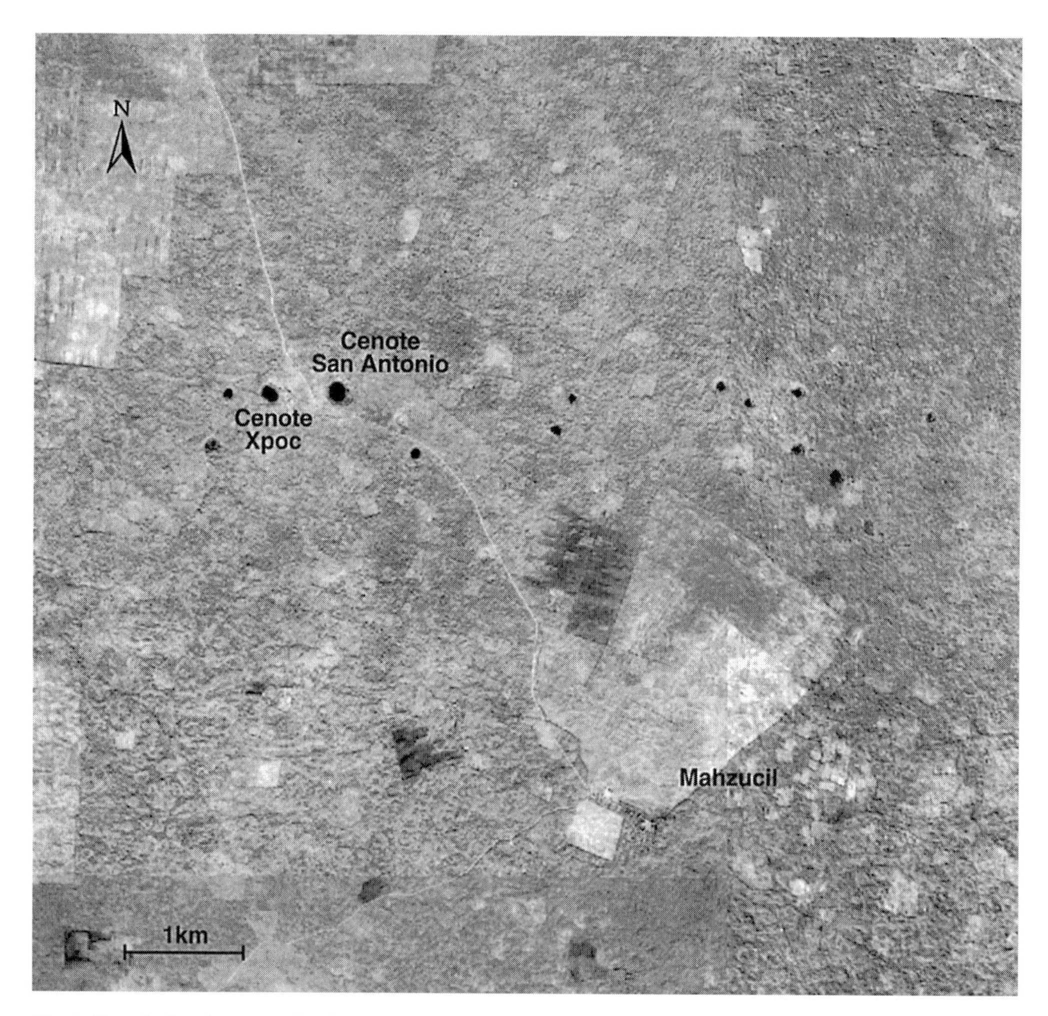

Fig. 2. Detail of main cenote ring from a geographically corrected mosaic constructed from ~1: 75 000 black and white aerial photographs available from the Instituto Nacional Estadistica Geografia e Informatica of Mexico (see Fig. 1 for location). Mahzucil is a rural village. As no surface drainage occurs on the northwestern Yucatán, the cenotes have been used as water sources since time immemorial.

'doublet' examples occur. The width of the ring ranges up to ~5 km. Partial rings occur exterior to the main ring in the southwest and east. The cenotes of the ring are generally larger than the numerous cenotes located to the east of the buried crater and are thereby distinguishable on aerial photographs and detailed maps. The cenotes' occurrence corresponds to peripheral gravity gradient features of the buried impact structure (e.g. Hildebrand *et al.* 1995), and their locations are presumed to reflect preferential groundwater flow/dissolution along concentric fracturing in the overlying Tertiary sediments. (The cenote ring is developed in the uppermost 100 m of the ~500 m of post-impact sediments that bury the eroded crater rim.) This fracturing is thought to have been induced by long-term and continuing subsidence of the crater lithologies and the crater fill. Such post-burial, long duration subsidence is recorded at other craters smaller than Chicxulub (e.g. Dypvik *et al.* 1996), and ~100 m of total subsidence is indicated by greater thicknesses of the uppermost Tertiary sediments burying the crater relative to the remainder of the peninsula (e.g. Lopez Ramos 1975).

Figure 1 locates two seismic-reflection profiles, the third data set used to map Chicxulub crater structure. Camargo & Suárez (1994) presented and described these two seismic lines which were shot in 1992 to study the buried crater. Figure 3 shows their interpretation of line D92RP002, the line closest to the coast (located at least 27 km offshore due to required minimum water depths of ~20 m for a reflection survey of this type) which cuts a chord ~30 km north of the crater's centre. The line therefore records crater structure somewhat tangentially. The vertical scale is valid only for the shallowest and lowest velocity sediments; the lower parts of the section will have vertical scales compressed by up to ~3×. The main crater features revealed are the Tertiary crater fill, the

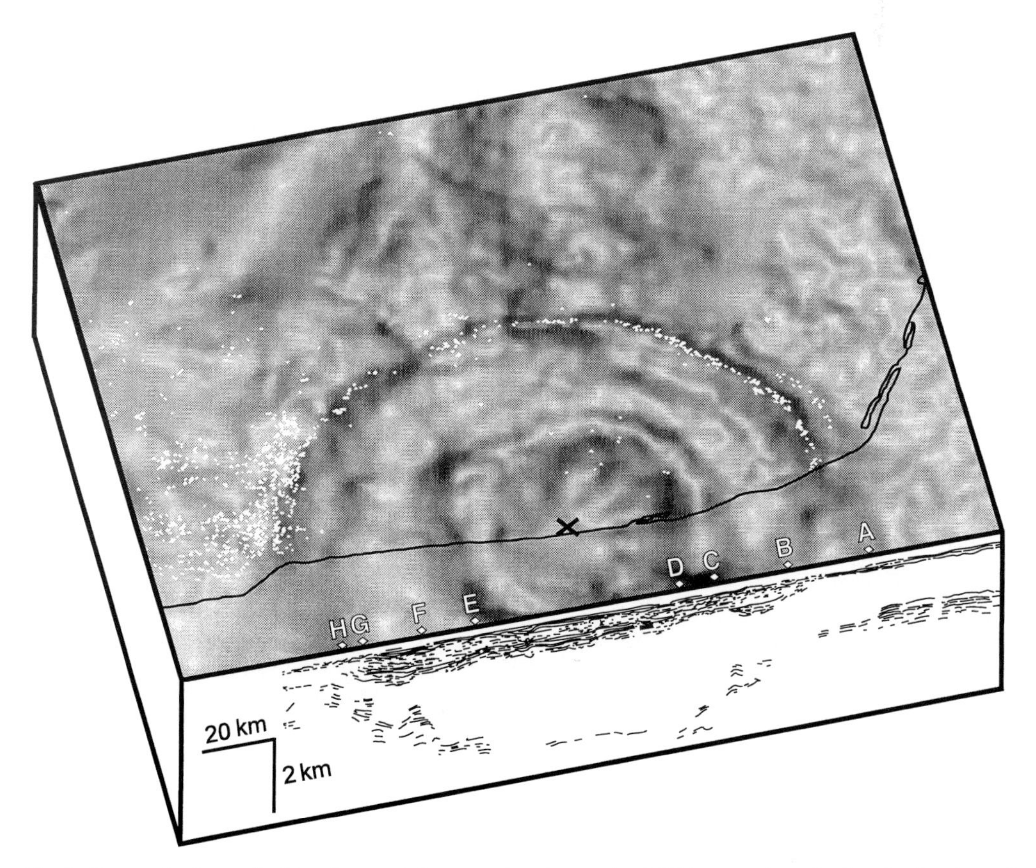

Fig. 3. Perspective figure (looking due south from an elevation angle of 50° juxtaposing seismic reflection data of line D92RP002 with horizontal gradient of gravity anomaly data as shown in Fig. 1. Letters indicate crater structural features as described in Table 1 and plotted in Figs 1 & 5. Cenotes are indicated by white dots; coastline with a black line; centre of crater with a black cross. Vertical exaggeration of ~10×.

peak ring, and the zone of slumping. Poorly developed/imaged rim uplifts may also be seen. The slump blocks are fortuitously imaged in part, despite sometimes severe distortion, due to the presence of strong reflectors in the lower part of the 3–4 km-thick sedimentary sequence on the platform. These reflectors are assumed to represent impedance contrasts of the Lower Cretaceous interbedded carbonates and evaporites recorded in nearby drill holes (e.g. Lopez Ramos 1975; Ward *et al.* 1975). The crater features annotated in Fig. 3 are listed (and located) in Table 1 for both seismic lines. The feature picks were made from the uninterpreted profiles as presented by Camargo & Suárez (1994) rather than from their interpretations. However, agreement results in most cases with the most easterly slump fault location being the most discrepant. We picked the westerly limit for this fault (indicated by 'H') as shown in Fig. 3; Camargo and Suárez interpret additional reflectors that might delimit a fault-bounded rim uplift ~5 km east of this position.

Mapping the buried crater follows by tying the gravity gradient and karst features to the crater structures as revealed by the seismic reflection profiles. The horizontal gradient features are 'mapped' by a computer routine as shown in Fig. 4. This technique is used in interests of objectivity. The cenote distributions are used to map structure normally associated with the peripheral gravity features where no gravity stations are available or are sparsely distributed. Figure 5 shows the results of this 'mapping' together with some correlations to the seismic profiles. The correlations are tentative (and are limited to the strongest gradient features) due to the ~20 km-wide coastal gap in gravity survey coverage closed only in the west by the relatively imprecise airborne data. The provisional correlations include the peripheral slump fault, the edges of the Tertiary crater fill, and the innermost edges of the slumped blocks. In the west a splitting occurs between the peripheral slump fault and the large peripheral gradient that correlates to the edge of the Tertiary crater fill (assumed to represent preferential infilling of the crater on the northwest side of the crater by prompt backwash). The main cenote ring is therefore interpreted to correlate with the edge of the crater fill rather than necessarily to a slump fault (in some places the edge of the Tertiary fill and peripheral slump faults will still correlate). Subsidence of the crater fill is assumed to result in the concentric fracturing that led to cenote ring formation, which explains why the cenote ring has a width of up to ~5 km; this width is assumed to be influenced by the width of basin slope which is 7 to 10 km at the seismic lines. Thus, this origin replaces the fault-based subsidence advocated by Hildebrand *et al.* (1995) as the cause of the fracturing localizing the main cenote ring. However, the outer partial rings may still reflect late stage movement on the peripheral slump faults, and both types of subsidence may contribute where the basin margin coincides with slumping. Less well defined gradient features between ~50 and 80 km radius that are presumed to correlate to structure in the zone of slumping may not be traced to the nearest offshore seismic line. The peak ring is not apparently correlated

Table 1. *Locations of Chicxulub crater structural features on two Petróleos Méxicanos seismic lines based on interpretations of Camargo & Suárez (1994). Shot point interval is ~25 m*

		Line D92PR001			Line D92PR002		
		Shot point	Latitude	Longitude	Shot point	Latitude	Longitude
	Western end	904	21.55°	90.52°	1000	21.29°	90.745°
	Rim uplift	~920–1460			1800–2550		
A	Perimeter slump	1550	21.565°	90.37°	2550	21.37°	90.38°
B	Outer >1 km slump	2600	21.595°	90.115°	3500	21.42°	90.16°
	Crater basin slope	2550–2800			3400–3850		
C	Second >1 km slump	3300	21.61°	89.95°	4400	21.47°	89.95°
D	Innermost slump block	4300	21.64°	89.71°	4800	21.49°	89.85°
	Peak ring	4750–5200			4350–4900		
	Peak ring	5800–7050			6600–7700		
E	Innermost slump block	5800	21.68°	89.35°	7250	21.62°	89.275°
F	Second >1 km slump	6500	21.70°	89.18°	7900	21.655°	89.125°
G	?Outer >1 km slump?	7500?	21.715°	89.035°	8700?	21.695°	88.94°
H	Western limit for perimeter slump and crater basin edge	7700	21.73°	88.89°	~8800	21.70°	88.91°
	Eastern end	8600	21.75°	88.695°	9874	21.755°	88.67°

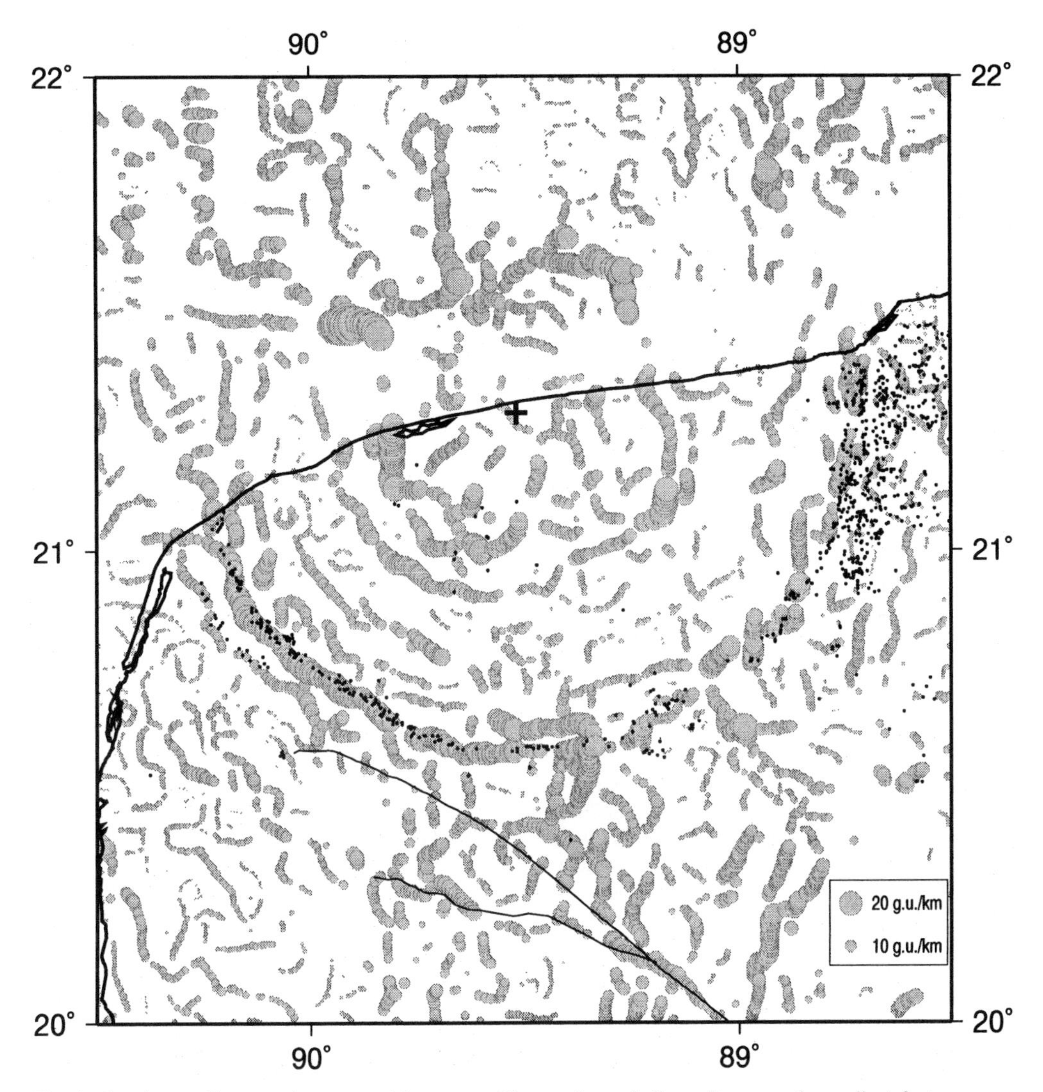

Fig. 4. Gravity gradient maxima map with cenotes. The maxima of all gravity anomaly gradient features >0.01 mGal/km are plotted as gray dots which typically form irregular lines. Dot sizes correspond to gradient magnitude (in legend 10 g.u. equals 1 mGal). The maxima routine was modified by W. Roest from that described by Blakely & Simpson (1986). The coastline is indicated by a black line, cenotes as solid black dots, Ticul fault splays as thin dark lines and the crater centre by a cross.

to any particular gradient feature (contrary to Hildebrand *et al.* 1995), and does not correlate to the gravity low at ~35 km radius (e.g. Pilkington *et al.* 1994) or the gravity high at ~50 km radius (e.g. Sharpton *et al.* 1993) as previously proposed, lying at ~45 km radius.

The rim-to-rim diameter of the Chicxulub crater that existed before erosion may be inferred from this mapping of the buried structure. We define a crater rim as 'the topographic high formed by the top of the ejecta blanket immediately exterior to the outermost, inward-facing, slump-fault scarp' in accord with several previous definitions (e.g. see Croft 1985). The peripheral slump features (located immediately interior to rim uplifts) seen on the seismic lines (with associated gradient features) have a maximum radius of ~93 km at ~3 km depth. Using a dip of 60° for the marginal normal fault adds ~2 km to the pre-erosion rim position. Thus the maximum inferred crater radius is 95 km. To the south the outermost gradient features

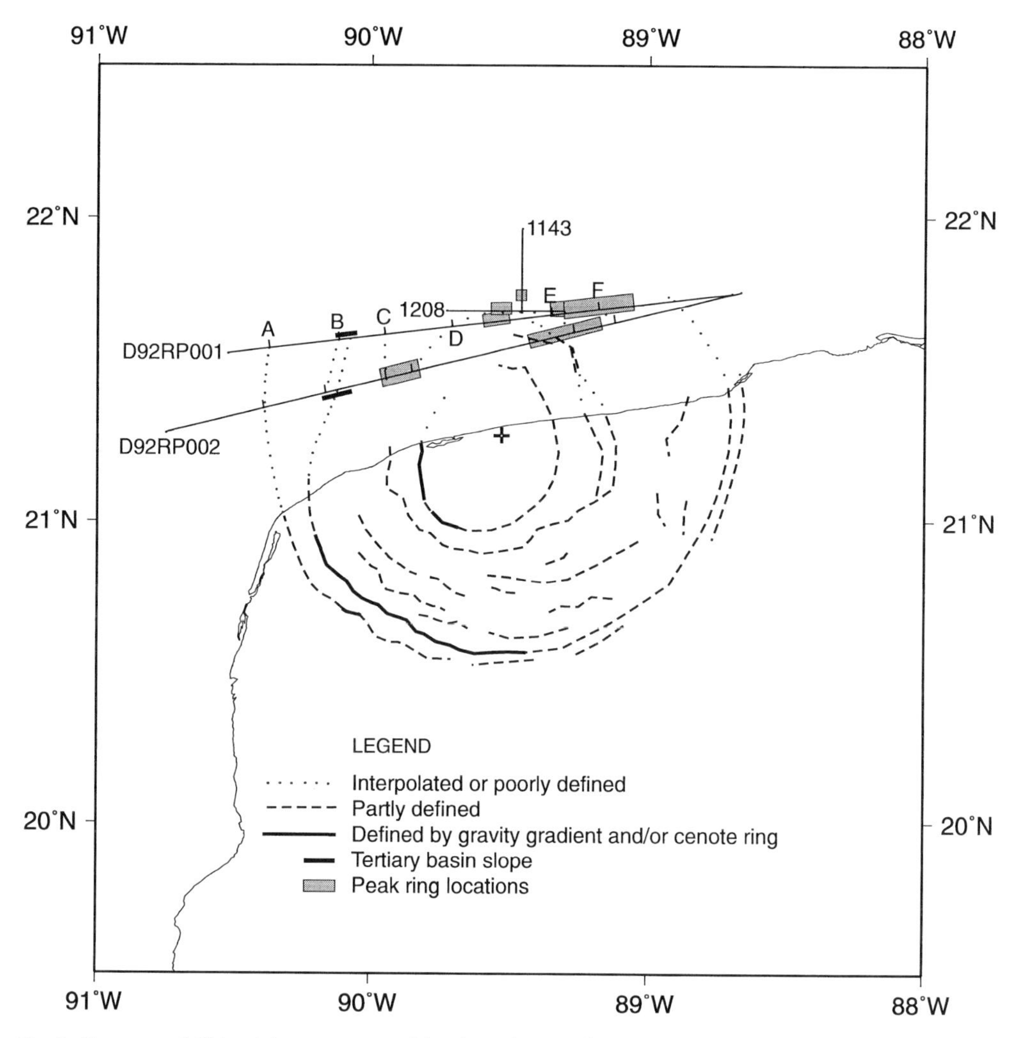

Fig. 5. Structure of Chicxulub crater mapped in plan using gravity anomaly gradients and cenote distributions (see Figs 1 & 4). Crater features as revealed by offshore seismic reflection profiles are also indicated (see Fig. 3 and Table 1). Note that the poorly known peak ring is broader to the northeast and that the zone of slumping appears more ordered to the southwest. The Yucatán coastline and the crater centre are indicated by a black line and a black cross, respectively. The shot point locations of the Tertiary basin slopes and peak rings are listed in Table 1.

show radii as low as ~80 km, however, with the remainder of the peripheral structures in between, indicating a crater size between these bounds. The originally advocated crater diameter of ~180 km remains reasonable for the buried impact structure, but rim reconstruction would add ~5 km to this figure with rim radius variability of 5 to 10 km. Chicxulub's diameter will need minor revision as the remainder of the crater is mapped and our understanding of the subsurface structure represented by the gradients and karst topography improves. Figure 5 shows two seismic-line segments (profiles 1143 and 1208) from the grid of pre-discovery seismic surveys that Petróleos Méxicanos collected over the Yucatán Platform. These data plus the seismic data recently described by Morgan *et al.* (1997) when completely processed will provide enough coverage to map the poorly expressed half of the crater that lies under the sea.

Transient cavity size and impact energy

Energy scaling relations for large impact craters are extrapolations from smaller artificial impact and explosion craters. Melosh (1989) discusses development of these relations and views the relevant parameter for comparison/calculation in a natural crater to be the diameter of the transient 'crater' at the level of the pre-impact surface (D_{at} of Fig. 6) from the way experimental craters are measured (e.g. Schmidt & Housen 1987). The term 'transient crater' has been used to describe two slightly different forms during crater growth/relaxation, leading to some ambiguity in its meaning and application. Natural impact craters and sufficiently large artificial craters are lined with breccia and melt (and vapour). Melosh (1989) defined the transient 'crater' as the cavity lined with breccia and melt which collapses into the bowl-shaped crater thus enlarging the resulting crater; see page 129 of Melosh (1989) for illustrative Fig. 8.3 and the quantitative relation $D_t = 0.84D$ where D_t and D are transient 'crater' and final simple crater diameters, respectively. Melosh advocates use of this crater element to estimate impact energy of craters (Melosh 1989; pers. comm.). In contrast, Dence *et al.* (1977) and Croft (1980) defined the 'transient crater' as bounded by the interface between the breccias (and melt) and the impacted rocks at its maximum dimension and prior to the slumping that occurs in sufficiently large craters to produce complex crater morphology. The term 'transient cavity' was applied to this same bounding interface during its growth phase; the separate term was used to distinguish intermediate stages from the end stage. During discussion of Chicxulub's impact energy (and most large craters), the width of the zone of disrupted target rocks (that subsequently collapsed through formation of slumped blocks) has been described as a transient crater (or cavity) following Dence *et al.* (e.g. Hildebrand *et al.* 1991; Sigurdsson *et al.* 1992; Sharpton *et al.* 1993; Pilkington *et al.* 1994; Hildebrand *et al.* 1995; Morgan *et al.* 1997). This use of the term and the related 'collapsed transient cavity' refer to a cavity of disrupted target rocks rather than the 'transient crater' of Melosh, and apparently lead to invalid impact energy estimates.

Figure 6 illustrates the application of two terms to clarify the discussion of these crater elements. Reflecting the sentiment that these crater elements exist transiently (if at all, sometimes only being devices to discuss evolution of impact craters during their enlargement and collapse), it seems useful to refer to them as 'cavities' to distinguish them from the post-collapse structures which are craters. The term 'transient cavity' is applied to the cavity lined with breccias and melt, the 'transient crater' of Melosh (1989). D_t and D_{at} define the maximum extent of this cavity from rim to rim and at the pre-impact level of the target surface, respectively. The term 'disruption cavity' is applied to the cavity bounded by the interface between the breccias and undisrupted target, the 'transient crater' of Dence *et al.* (1977) and Croft (1980). D_d and D_{ad} analogously define the maximum extent of this cavity before marginal slumping of undisrupted target rocks begins (if the impact exceeds the transitional size for complex crater formation). The disruption cavity becomes the upper crater walls in a simple crater and defines a

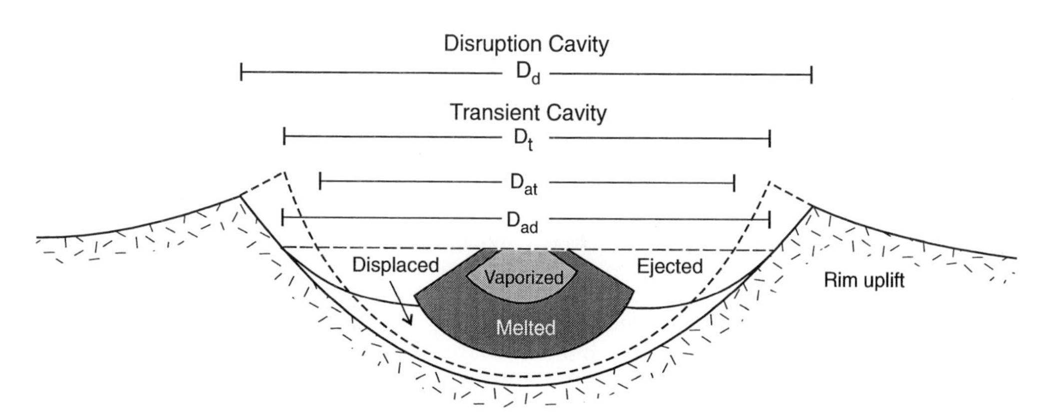

Fig. 6. Schematic distinguishing an impact crater's transient (diameter D_t) and disruption (diameter D_d) cavities. At the pre-impact ground surface these diameters are D_{at} and D_{ad} respectively. The stippled pattern under the crater and the exterior surface indicates fracturing of the target; the horizontal dashed line indicates the position of the pre-impact surface within the crater. The unit lining the disruption cavity to form the transient cavity walls is composed primarily of melted and brecciated target with a minor projectile component. In the past, these two 'cavities' have both been termed transient craters without differentiation.

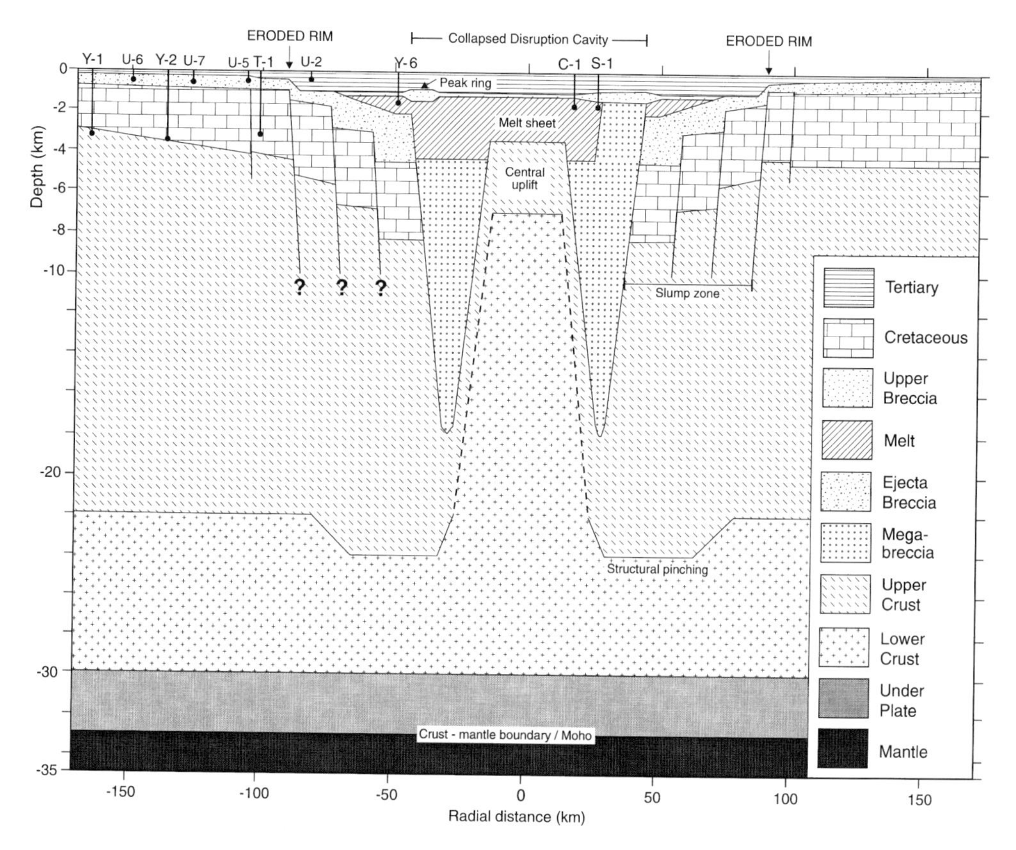

Fig. 7. Schematic whole-crustal structural model of Chicxulub crater (modified from that of Pilkington *et al.* 1994 and Hildebrand 1997). Vertical exaggeration of 7.3 times. Wells plotted are: S-1, Sacapuc-1; C-1, Chicxulub-1; Y-6, Yucatán-6; U-2, UNAM-2; T-1, Ticul-1; U-5, UNAM-5; U-7, UNAM-7; Y-2, Yucatán-2; U-6, UNAM-6; Y-1, Yucatán-1. Upper structure/well placement within the collapsed disruption cavity is roughly oriented west to east; well placement on the left side of the diagram is roughly SSE to NNW, left to right. Note that extension of the faults to the zone of structural pinching is uncertain.

simple crater's diameter excepting for the small modification to rim diameter from the ejecta resting on the rim uplift. At a complex crater the inner edge of the slumped blocks defines the collapsed disruption cavity; the collapsed transient cavity has no observable feature, being reconstructed only with knowledge of breccia and melt volumes. The disruption cavity corresponds to the limit where the impact-generated shock wave is no longer able to brecciate the target. This surface approximately corresponds to shock magnitude decreasing below the Hugoniot elastic limit of the impacted rocks with a complicating effect that pre-existing fractures 'weaken' the target allowing additional brecciation; see Asphaug *et al.* (1996) for a discussion. However, an instrumented nuclear explosion (Borg 1972) suggests that the Hugoniot elastic limit is a reasonable approximation of the brecciation/disruption limit for terrestrial craters in crystalline rocks. The 'excavation cavity' (e.g. Dence *et al.* 1977; Croft 1980) or 'crater of excavation' (e.g. McGetchin *et al.* 1973) has the same diameter as the disruption cavity (D_d); for clarity the term 'excavation cavity' seems preferable.

To calculate the energy released by the Chicxulub impact the internal structure of this complex crater must be understood well enough to constrain the size of the collapsed disruption cavity. Figure 7 shows a schematic structural cross section for the crater modified from those of Pilkington *et al.* (1994) and Hildebrand (1997). The upper 8 km of crater structure are constrained by drill-hole, seismic-reflection, and potential-field data. The primary uncertainties associated with the shallow structure are the peak ring's subsurface structure (Fig. 7 illustrates the two possibilities described by Pilkington *et al.* (1994)), and the thickness of the melt sheet (Hildebrand

1997). Chicxulub's deep structure is constrained by scaling in a self-similar (or proportional) way to the structure of well understood smaller craters, and by conserving volumes. For example, the vertical displacement and diameter of the central structural uplift can be estimated from relations established by studying other craters (e.g. Grieve *et al.* 1981; Pike 1985; Therriault *et al.* 1997). The zone of structural pinching can be similarly constrained by analogy to those at smaller craters plus, by inspection, the principle that the volume of the central uplift (V_{cu}) must equal the volume withdrawn from the zone of structural pinching (V_{sp}; assuming effects like bulking or injection of melt into the central uplift are insignificant). This volume conservation has been confirmed by observation at well constrained smaller craters (e.g. Offield & Pohn 1977). Similarly the volume of near-surface slumping (V_{ss}) must equal that of the zone of structural pinching plus the volume of contraction of the disruption cavity (V_{cd}) during slumping ($V_{ss} = V_{sp} + V_{cd}$ where inwards motion yields a volume charge defined as positive). However, the geometry of all the deep structural elements of the crater remain to be well constrained (Hildebrand 1997), although seismic studies should be able to provide much useful information (e.g. Morgan *et al.* 1997; Christeson *et al.* 1997) and magnetotelluric studies may also help (Campos-Enriquez *et al.* 1997).

The validity of this schematic model can be assessed by its utility as a starting point for forward modelling efforts. Figure 8 shows a two-dimensional gravity model along a central profile parallel to the coast as located in Fig. 1. A reasonable approximation of the observed gravity field can be made using the approximate geometry of the schematic cross section. The mass deficiencies in this model differ substantially from those previously published by Pilkington *et al.* (1994) and Hildebrand *et al.* (1995) due to density contrast revisions based on measurements of ~200 samples of Chicxulub crater lithologies and the surrounding impacted stratigraphy. The most significant mass deficiency shift is from the slumped blocks to the crater fill. Density measurements showed that the deep-water facies

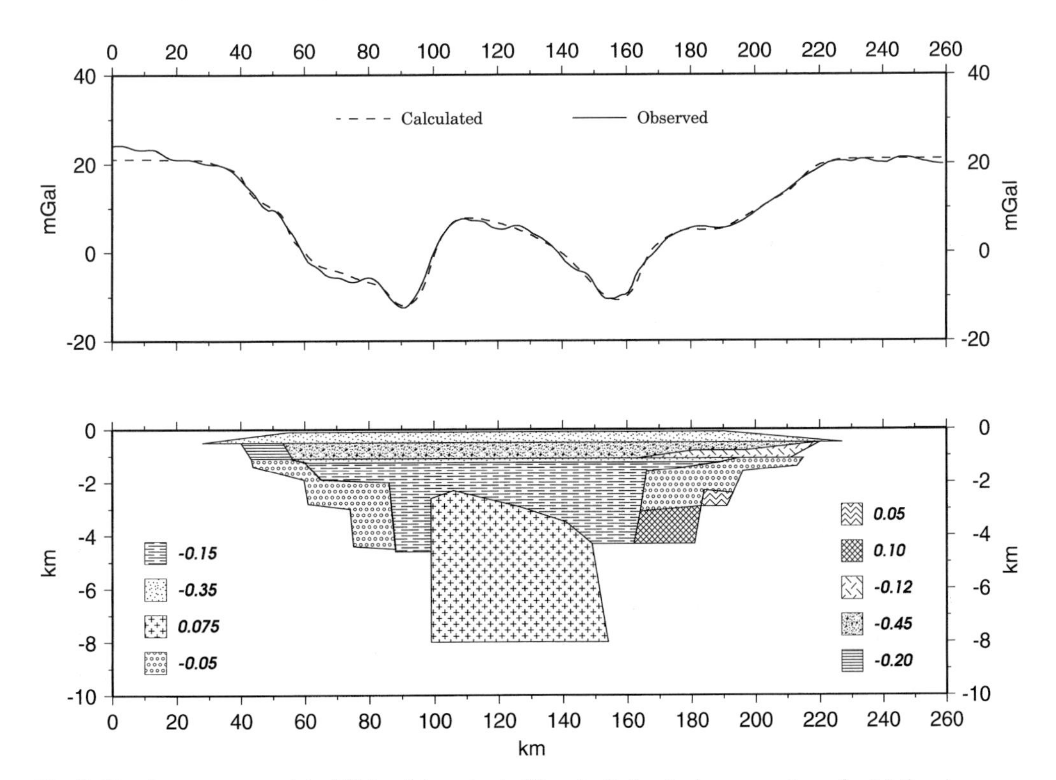

Fig. 8. Density contrast model of Chicxulub crater to 8 km depth (vertical exaggeration of ~8.5 times). Density contrasts of the outlined blocks (in g cm^{-3}) are listed and indicated by patterns. Gravity data modelled are from a near-central profile running from 20.9128°N, 90.7248°W to 21.6114°N, 88.3254°W, a distance of 260 km (see Fig. 1). A regional gradient of 0.035 mGals/km (increasing eastwards) has been removed from this profile. The observed data and model results are shown by solid and dashed lines, respectively.

of the Tertiary crater fill had densities of 2.0 to 2.3 g cm^{-3}, creating large density contrasts relative to the surrounding Upper Cretaceous rocks. This results in approximately two thirds of Chicxulub's mass deficiency occurring in the crater fill, and needing none in the slumped blocks (where little porosity is expected). The large-amplitude, short-wavelength gradient features associated with the crater are also somewhat easier to model with shallower mass deficiency. The second notable difference is that the impact melt has an increased negative density contrast based on the new measurements. With the peak ring now known to be at a greater radius and apparently density neutral, the collapsed disruption cavity filling between the central uplift and the slumped blocks is the only candidate to produce the ~35 km-radius, annular gravity low at Chicxulub. The observed density contrast of ~0.15 g cm^{-3} is sufficient to produce the observed low, although some of the mass deficiency may also lie in breccias underlying the melt sheet. Although a regional gradient has been removed across the model (and its location was partly chosen to minimize regional effects), some regional basement anomalies are still thought to be expressed in the observed profile. Small model blocks have been added to compensate for these suspected regional effects, mostly on the east side, but they may be spurious additions. One exception is the step in the large marginal slope on the west side which is assumed to reflect a prism of promptly backwashed material.

The collapsed disruption cavity as indicated by the gravity model is slightly smaller than that indicated by the seismically resolved inner edges of the slumped blocks. The positions of the latter as indicated in Figs 1, 3 & 5 have an average radius of 45 km. The positions of the inner slumped block margins in the gravity model are interior to the gradient maxima. As the gravity model lacks detailed density contrast control and is not three-dimensional, we can only provisionally estimate that the collapsed disruption cavity margin lies at ~43 km radius. Assuming a slump motion angle of 60° along normal faults and a downdrop of ~4.5 km (Camargo & Suárez 1994), reconstructing this block to its pre-impact position moves it outwards ~2.5 km. This amount of lateral translation is not unreasonable as it volumetrically corresponds to ~10 000 km^3 of contraction of the disruption cavity (assuming inwards motion along a cavity wall 15 km high). The volume on top of the slump blocks is ~40 000 km^3. As the estimated volume of the central uplift is comparable, the net inwards motion of the collapsing disruption cavity wall cannot be significantly larger without requiring a significantly smaller central uplift; gravity modelling apparently precludes the latter. Finally, since the imaged reflection is at the base of the sedimentary sequence, and assuming that the disruption cavity wall has a slope of ~60°, an additional ~1.5 km must be added to yield a ~47 km radius for the reconstructed disruption cavity at the pre-impact surface level (D_{ad} = 94 km). Using $D_{at} = 0.84D_{ad}$, slightly modified from Melosh (1989), the apparent transient cavity diameter of Chicxulub was 80 km. This implies impact energies somewhat smaller than previous estimates, as noted above.

Nature of the impactor

Sigurdsson *et al.* (1992) and Hildebrand (1992) argued that the Chicxulub impactor must have been a comet rather than an asteroid because any undifferentiated asteroid which can provide the observed global Ir fluence does not provide enough energy to excavate the Chicxulub crater. The average Ir fluence in the Cretaceous–Tertiary fireball layer is ~40 ng cm^{-2} (Orth *et al.* 1987; Hildebrand 1992) yielding ~200 000 tonnes globally. This corresponds to a carbonaceous chondrite asteroid (the largest and most massive undifferentiated asteroid possible) of 7.4 km diameter (4.2×10^{17} g), assuming a specific gravity of 2.0 g cm^{-3} and an Ir concentration of 474 ppb (e.g. Anders & Grevesse 1989). An undifferentiated impactor is indicated by the near chondritic distribution of the siderophile trace elements in the fireball layer (e.g. Kyte *et al.* 1985). Using the energy-scaling relation of Schmidt & Housen (1987) with constants as listed by Melosh (1989), this asteroid at the maximum possible terrestrial impact velocity for a prograde solar orbit of ~42 km s^{-1} will make an apparent transient cavity 71 km across with vertical incidence (assuming a target density of 2.8 g cm^{-3}); for a more typical asteroidal impact velocity of ~20 km s^{-1} (Shoemaker *et al.* 1990), a 51 km-diameter cavity results. Considering a comet as an impactor, assuming comet Halley composition (Anders & Grevesse 1989) with dust composition determined by the mass spectrometer of the Giotto spacecraft (Jessberger *et al.* 1988), and gas to dust ratios of Delsemme (1988) yields an impactor mass of 1.8×10^{18} g (~30% chondritic material) to deliver the observed Ir fluence (assuming that the cometary silicate dust fraction is similar to that of carbonaceous chondrites). The comet will have a diameter of 16.5 km using a density of 0.6 g cm^{-3} on the basis of: (1) an average density for Halley's comet based on its response to dynamic forces (Weissman

1990), and (2) the density range of 0.5–0.7 g cm^{-3} derived for comet Shoemaker–Levy 9 by Asphaug & Benz (1994) with a preferred value of ~0.6 g cm^{-3} to explain the observed variation in the comet's fragments' impacts with Jupiter. This impactor will produce an appar-ent transient cavity of 80 km diameter with an impact velocity of 33 km s^{-1} at vertical incidence. The velocity needs to increase to 37 km s^{-1} to compensate for an impact angle of ~60° (Hildebrand *et al.* 1997) based on the empirical relation observed by Gault & Wedekind (1978). These velocities are compatible with the average and most probable impact velocities of short period comets of 37.5 and 28.9 km s^{-1}, respectively (Weismann 1990). An impact energy of 1.2×10^{31} ergs results. As long-period comets can impact with velocities of up to 72 km s^{-1}, changes in the silicate fraction of the comet may be compensated by varying the impact velocity.

Chicxulub impact environmental effects

Since Alvarez *et al.* (1980) provided evidence that the terminal Cretaceous mass extinction was due to an impact, suggesting that darkness from the resulting global dust cloud was the lethal environmental change, many other impact-generated environmental stresses have been proposed. Table 2 lists these plus selected references; this table has only three additions (thermal pulse from re-entering ejecta; cooling from atmospheric aerosols; oscillatory/disrupted climate) and one deletion (impact-triggered volcanism) from the compilation published by Wolbach *et al.* (1990) suggesting that the recognition of the Chicxulub crater has not added much to our understanding of the impact's environmental effects. Indeed, many of these effects had been broadly stated more than two centuries ago by de Maupertuis (1750) while describing a cometary impact:

> 'One can't doubt that most of the animals would perish, if it happened that they were reduced to supporting the very excessive heat, or to swim in fluids very different than their own, or to breathe strange vapours. It would only be the most robust animals and possibly the lowest which would remain alive. Entire species would be destroyed; . . .'

However, Chicxulub's recognition has added a good deal of *quantification*, and as exploration of the crater (and its ejecta) proceeds, input to environmental perturbation modelling will improve further. Toon *et al.* (1997) have reviewed most of the environmental consequences of the Chicxulub impact using an energy of ~1.3×10^{31} ergs for their assessment, fortuitously in agreement with that calculated above, but while noting that the size of the impact was controversial. Suggestions of larger disruption cavity sizes (e.g. Sharpton *et al.* 1993, 1996) had created the appearance of an order of magnitude uncertainty. As the crater mapping presented above shows, the diameter (~86 km at a traceable horizon) of Chicxulub's collapsed disruption cavity is now

Table 2. *Suggested agents of environmental damage induced by the Chicxulub impact; this table is updated and modified from that produced by Wolbach* et al. *(1990); see Toon* et al. *(1997) for a similar tabulation. The modifiers 'proximal' and 'regional' indicate potential extinction agents operating within ~1000 and ~5000 km radial distance of Chicxulub, respectively*

Environmental change agent	Duration	Reference(s)
1. Dust veil (darkness and cold)	Months	i, ii, iii
2. Proximal wind	Hours	iv
3. Proximal giant waves	Hours	iv, v, vi, vii, viii
4. Proximal to regional fireball irradiance	Minutes	ix, xxviii
5. Regional thermal pulse from re-entering ejecta	Hour	x
6. Acid rain (nitrogen- and sulfur-based)	Year	xi, xii, xiii, xiv, xv, xvi
7. Stratospheric aerosols (cold)	Decades	xiv, xv, xvii, xviii
8. Ozone layer depletion (ultraviolet exposure)	Decades	xiv, xvii
9. H_2O greenhouse	Decades	iv
10. CO_2 greenhouse	Millennia	xii, xiv, xix, xx
11. Poisons and mutagens	Years to millenia?	xi, xii, xxi, xxii, xxiii, xxiv, xxv
12. Oscillatory/disrupted climate	Million years	xxvi, xxvii

References cited: (i) Alvarez *et al.* 1980; (ii) Toon *et al.* 1982; (iii) Covey *et al.* 1990; (iv) Emiliani *et al.* 1981; (v) Ahrens & O'Keefe 1983; (vi) Bourgeois *et al.* 1988; (vii) Hildebrand & Boynton 1990; (viii) Smit *et al.* 1996; (ix) Melosh *et al.* 1990; (x) Opik 1958; (xi) Lewis *et al.* 1982; (xii) Prinn & Fegley 1987; (xiii) Zahnle 1990; (xiv) Brett 1992; (xv) Sigurdsson *et al.* 1992; (xvi) D'Hondt *et al.* 1994; (xvii) McKay & Thomas 1982; (xviii) Pope *et al.* 1994; (xix) O'Keefe & Ahrens 1989; (xx) Wolbach *et al.* 1990; (xxi) Wolbach *et al.* 1985; (xxii) Wolbach *et al.* 1988; (xxiii) Gilmour *et al.* 1990; (xxiv) Venkatesan & Dahl 1989; (xxv) Hsu *et al.* 1982; (xxvi) Zachos & Arthur 1986; (xxvii) D'Hondt *et al.* 1996; (xxviii) Adushkin & Nemchinov 1994.

known to ~5% uncertainty, and future modelling efforts should therefore consider narrower ranges associated with quantities of energy and volatiles released by the impact. Additional constraints of shock regimes that caused shock devolatization at Chicxulub will be available as the volume of melt produced is determined from exploration of the crater's melt sheet and ejecta blanket, and exploration of the impacted stratigraphy establishes the impacted lithologies. The extinction potential of an impact like Chicxulub is based on the prodigious amount of energy released in a geological instant (e.g. Grieve 1982). The estimated 1.2×10^{31} ergs released (in minutes) equals the global endogenic heat output (~3.5×10^{13} W; Turcotte & Schubert 1982) of ~1000 years, and corresponds to the energy released by all subaerial volcanoes in 50 000 years. In large impacts such as Chicxulub, approximately half of the impact energy is partitioned into the crater's ejecta

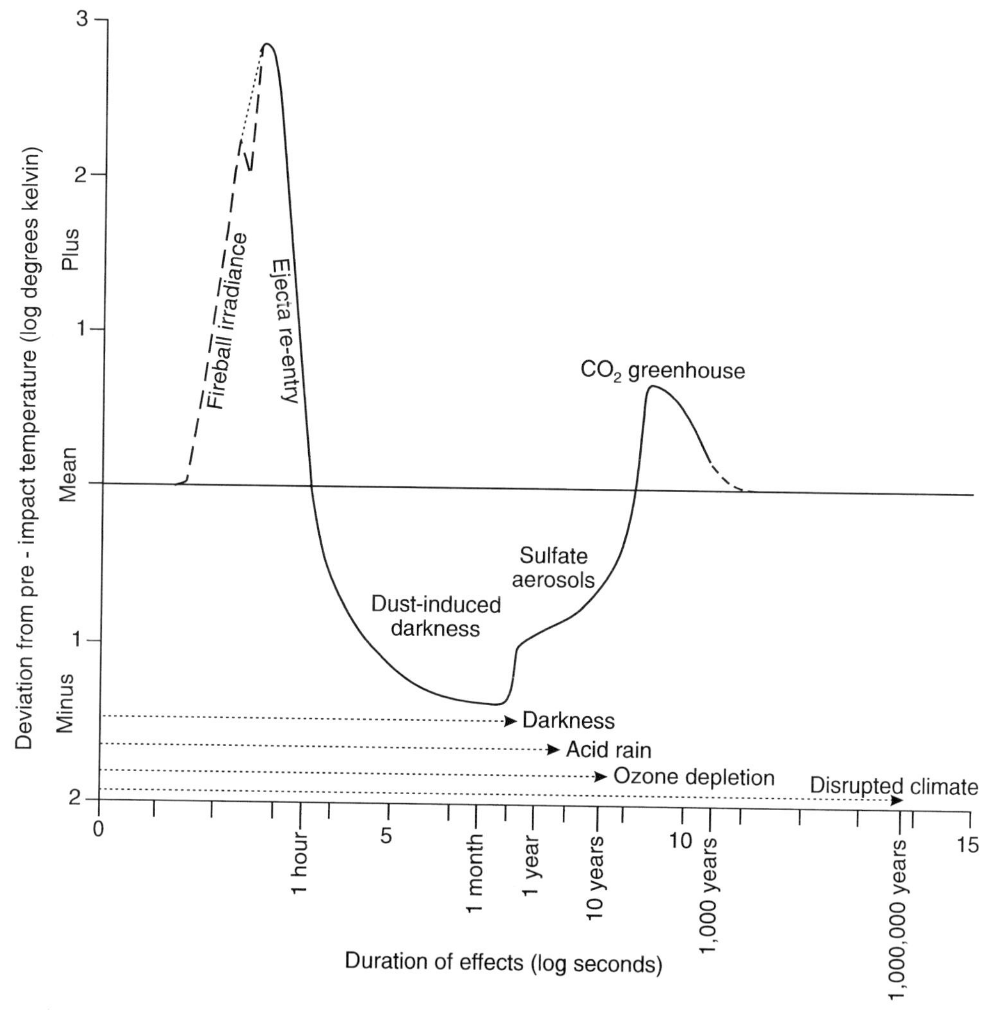

Fig. 9. Schematic illustration of temperature perturbations induced by the Chicxulub impact. Environmental effects of varying durations produce a succession of temperature changes. The dashed portions of the curve indicate greater uncertainty. The log of the temperature deviation (in Kelvin) is plotted against the log of time (in seconds). The portrayed high-temperature pulse(s) due to ejecta re-entry and fireball irradiance is (are) valid for locations ~3000 km from the crater; at distances greater than ~5000 km the thermal pulse will be relatively insignificant due to thinning of the ejecta (and increasing distance from the fireball). The thermal pulse associated with the impact fireball will also be geographically restricted, although its extent remains to be well understood. Magnitudes of temperature changes will be most applicable to continental interiors. Durations of other well evidenced environmental perturbations are also indicated by dashed lines.

(O'Keefe & Ahrens 1982), allowing relatively efficient coupling of the released energy to the terrestrial atmosphere and widespread surfaces. Aside from impacts, the only other natural events capable of delivering instantaneous energy pulses of this magnitude to the Earth are also extra-terrestrial in origin, such as changes in the Sun or nearby stars.

Figure 9 schematically illustrates the temperature perturbations caused by the Chicxulub impact together with the durations of other suggested extinction mechanisms. The first temperature excursion is caused by the radiant energy of the fireball. This geographically restricted effect remains to be well modelled for the Chicxulub impact (or any similar impact) as the calculations done to date are sketchy (Opik 1958; Adushkin & Nemchinov 1994). The Shoemaker–Levy 9 impacts on Jupiter provide empirical examples (e.g. Carlson 1995; Carlson *et al.* 1996) from which the schematic curve in Fig. 9 is estimated, but the duration, magnitude and the geographic extent of the fireball's irradiation remain to be determined. The thermal pulse from re-entering ejecta modelled by Melosh *et al.* (1990) as a global effect is likely to have been geographically restricted and primarily due to the emplacement of the ejecta layer. The fireball layer is uniform in thickness globally and consists partly of relatively fine-grained particles from its observed settling history. Smit *et al.* (1992) show the most comprehensive evidence of sizeable particles in the fireball layer which are still relatively fine grained, so it is problematic for modelling the thermal pulse with the ballistically distributed ~0.5 mm-diameter particles. The material of the fireball layer was probably distributed in significant part by hydrodynamic flow of the collapsing fireball (e.g. Colgate & Petschek 1985). The ejecta layer is composed of mm-sized particles, so its ~5000 km-radius extent probably marks the region of the thermal pulse due to re-entering ejecta (e.g. Hildebrand 1993). The cooling due to dust-induced darkness has been shown to be moderated by heat transferred from the oceans to continental interiors (Covey *et al.* 1990) so that temperature drops are not as large as once thought (e.g. Toon *et al.* 1982), although still significant. The cooling due to stratospheric aerosols (which is expected to predominate over the concurrent greenhouse effect from H_2O) was only briefly considered for the K–T impact (McKay & Thomas 1982) before Chicxulub was established as the K–T source crater. The recognition that the Chicxulub impact shock-devolatized sulphates lead to concurrent suggestions that the impact created a decade-long period of cooling due to stratospheric sulphates (e.g. Sigurdsson *et al.* 1992; Brett 1992); Pope *et al.* (1994) modelled this effect.

The subsequent CO_2-induced, millenia-scale greenhouse has been long suggested, although its magnitude and duration have been controversial. Four sources of CO_2 are available to cause this greenhouse: (1) respiration of CO_2 from the ocean after the death of organisms in the surface layer (e.g. Hsu *et al.* 1982) or acidification of the surface layer of the ocean (e.g. Lewis *et al.* 1982; D'Hondt *et al.* 1994); (2) shock devolatization of carbonate within the stratigraphy impacted at Chicxulub (e.g. O'Keefe & Ahrens 1989); (3) combustion of vegetation (e.g. Wolbach *et al.* 1990); (4) oxidation of carbon in the impacting comet in the fireball (3.2×10^{17} g of C present from Anders & Grevesse (1989) would yield 1.2×10^{18} g of CO_2). All of these mechanisms can provide CO_2 in quantities comparable to the pre-industrial atmospheric CO_2 content of $\sim 1.5 \times 10^{18}$ g. As models of atmospheric CO_2 estimate terminal Cretaceous atmospheric abundances were ~2 times greater (e.g. Berner 1994), the magnitude of the greenhouse effect would be proportionally lessened, but four different sources probably combined to produce a CO_2 increase exceeding the latest Cretaceous atmospheric abundance. The existence of a substantial perturbation to atmospheric CO_2 is evidenced by a ~ -2 per mil shift in the $^{12}C/^{13}C$ ratio recorded in the kerogen organic fraction at the non-marine K–T boundary site at Raton Pass (Hildebrand & Wolbach 1989). The duration of the perturbation is 3500 to 6000 years assuming that the corresponding ~17 cm-thick stratigraphic interval was deposited at rates of 20 000 to 35 000 yrs/m (Lerbekmo & St. Louis 1986; Shoemaker *et al.* 1987). An ~5000 year-long period of warming is supported by a initial ~ -2 per mil shift in the $^{17}O/^{18}O$ ratio recorded at marine K–T sites in Spain (e.g. Smit 1990). Smit interprets this shift as corresponding to a net warming of ~8°C in the ocean's surface layer.

The severity of acid rain produced by the Chicxulub impact has been debated. Generation of nitrogen-based acids by atmospheric oxidation was first considered (e.g. Lewis *et al.* 1982; Prinn & Fegley 1987; Zahne 1990), but the recognition of Chicxulub has added discussion of sulphur-based acids from shock-devolatized evaporites (e.g. Brett 1992; Sigurdsson *et al.* 1992; Pope *et al.* 1994; D'Hondt *et al.* 1994; Ivanov *et al.* 1996; Toon *zet al.* 1997). It remains to be determined if sufficient quantities of acid were produced to acidify the surface layer of the ocean; modelled quantities are close to that necessary to substantially lower the pH of the surface layer (e.g, D'Hondt *et al.* 1994). However, the chemical leaching of cations as predicted by Prinn & Fegley (1987) is supported by the presence of Hg anomalies at non-marine K–T boundary sites (Hildebrand & Boynton 1989).

Almost all impacts will show a similar sequence of temperature changes and other stresses as indicated for Chicxulub, with magnitude of the temperature perturbations (and other effects) primarily corresponding to impact energy. Some effects will also be sensitive to the types of impacted protolith and projectile. For example, the Chicxulub impact released carbon- and sulphur-based volatiles to an exceptional extent because it impacted a 3 to 4 km-thick sequence of carbonates and evaporites (e.g. Lopez Ramos 1975; Ward *et al.* 1995). However, even if the impacted terrane had not contained volatile-rich lithologies, all undifferentiated impactors would still have contributed some amount of sulphur to induce the sulphate-aerosol cooling (e.g. Kring *et al.* 1996; Toon *et al.* 1997), and CO_2 is released from three other sources besides the target including the impactor. The decade-long period after the dust-induced darkness is the most potentially variable part of the impact temperature curve as the cooling effect from aerosols is competing with the warming effect of the H_2O-induced greenhouse. An impact of a sulphur-deficient impactor, such as a differentiated asteroid, into the deep ocean may produce a positive temperature deviation with the temperature effect going from dust-induced cooling to an H_2O-induced greenhouse.

How the patterns of extinction at the K–T boundary match the effects of the Chicxulub impact remains to be well-understood, and the character of the mass extinction is controversial. For most workers the synchroneity of the mass extinction and the Chicxulub impact indicate causality. Indeed, the existence of the extinction level drew attention to the boundary clay layer, study of which eventually led to recognition of the crater. Chicxulub thus marks the first and only case of impact recognition by studying the biospheric record, although much work has subsequently been done studying other extinction levels (e.g. Orth *et al.* 1990) without similar success. The rules governing survival for the Chicxulub impact were different from those governing success during times of background extinction (e.g. Jablonski 1986; Gallagher 1991). From the outset some extinction patterns seemed clear. For example, 'generalist', 'opportunistic' or 'disaster' species were able to survive the event (e.g. Smit & Romein 1985; Jiang & Gartner 1986). Large creatures and those living in the surface layer of the ocean were hardest hit (e.g. Russell 1979; Hansen *et al.* 1987; Smit 1994), while detritus feeders in both marine and freshwater environments persisted (e.g. Sheehan & Hansen 1986; Hansen *et al.* 1987). Amongst land plants, species dependent upon pollination agents other than wind (e.g. insects) preferentially went extinct (e.g. Sweet 1994). Thus, simply, larger biota dependent on an ordered, elaborate ecosystem were more likely to become extinct and smaller generalist scavengers persisted. A consensus seems to be emerging that some survivorship occurred to be followed by an extended if restricted (order 1000 to 100 000 years) period of extinctions in the marine (e.g. Keller 1989; Smit 1994). Therefore, although the Chicxulub impact does seem to have caused many extinctions in a prompt sense of days to months, it seems that other environmental changes caused by the impact produced extinctions significantly after its prompt effects. Still longer-term extinctions may be more related to changing evolutionary dynamics in impact-perturbed ecosystems, rather than to direct lethality of the environmental changes such as the CO_2 greenhouse or oscillatory climate.

Conclusions

Mapping the Chicxulub crater by tying the structural cross sections provided by seismic-reflection profiles to the gravity-field plan view provides a general tracing of the seismologically resolvable elements over the southern two thirds of the structure. The correlations between the offshore seismic data and the onshore gravity data set are tentative due to the ~20 km-wide gap along most of the coast and the lower level of precision associated with marine gravity surveys. However, the outermost slump faults, the edge of the Tertiary basin fill, and the edge of the collapsed disruption cavity may all be tied to gravity gradient features in a provisional way. Preliminary data from an effort to close this data gap support these ties (e.g. Hildebrand *et al.* 1997). Modelling Chicxulub's gravity anomaly with a schematic structural model based in part on the seismic data can reproduce the observed field with density contrasts consistent with those measured to date. A majority of the mass deficiency is in the crater fill, so comparing craters' Bouguer anomalies versus crater diameter (e.g. Pilkington & Grieve 1992) will be best done by removing the fill contribution from the total anomaly observed. The diameter of the buried impact structure is ~180 km, and apparently departs from circularity by up to ~10 km although the crater centre is not precisely located. The crater size remains uncertain due to the northern third of the crater being poorly shown by publicly available data, but the diameter uncertainty is probably less than the 10% estimated by Pilkington *et al.* (1994). Reconstruction of the now-eroded crater rim would add ~5 km to this diameter. Suggestions of ~240 km or ~300 km diameters for Chicxulub (e.g. Pope *et al.* 1993, 1996; Sharpton *et al.* 1993, 1996) with collapsed disruption cavity diameters up to 170 km are precluded.

Discussion and computation of impact energies for Chicxulub have been confused by two uses of the term 'transient crater': (1) as the transient impact cavity that is lined with melt and breccia (Melosh 1989) and (2) for the surface that separates the undisrupted/unbrecciated impacted rocks from the melt and breccia lining (Dence *et al.* 1977). (The term transient crater has also been applied without distinguishing between the uncollapsed and collapsed states.) We propose that the terms 'transient cavity' and 'disruption cavity' be applied to the 'transient craters' of Melosh (1989) and Dence *et al.* (1977), respectively. Also, that 'collapsed' versions of these should be differentiated, although only the collapsed disruption cavity will be a potential observable at craters. The excavation cavity diameter corresponds to that of the apparent disruption cavity. We feel that the use of terms containing 'crater' for these ephemeral or heuristic constructs unnecessarily confuses them with the craters that result from impacts.

At ~86 km diameter Chicxulub's collapsed disruption cavity, as outlined by stratigraphy originally ~3 km below the impacted surface, is known to ~5% uncertainty. The reconstructed disruption and transient cavities of ~94 and ~80 km, respectively, correspond to an impact energy of $\sim 1.2 \times 10^{31}$ ergs using the scaling relation of Schmidt & Housen (1987). Although this collapsed disruption cavity is somewhat smaller (94 vs 110 km) than initially thought (Hildebrand *et al.* 1991), reconstructing impactors using the global K–T Ir fluence, the energy necessary to produce the Chicxulub crater still exceeds that of the largest permitted undifferentiated asteroid impact by a factor of ~20 indicating a cometary impactor. Although impact energy-scaling relations are uncertain to that extent at the scale of the Chicxulub impact (see Melosh 1989 for a discussion), this result is in accord with the terrestrial cratering population as inferred by Shoemaker *et al.* (1990) from astronomically observed populations, who found that comets are more likely than asteroids to produce large terrestrial craters, and that one cometary crater larger than 100 km diameter is expected to form in a 100 million-year period. The putative comet was probably not part of a shower as no ^{3}He anomaly occurs across the K–T boundary stratigraphy (Farley *et al.* 1997).

The size of Chicxulub's apparent disruption cavity is the starting point for assessing many of the proposed K–T mechanisms, particularly for those related to volatile release from the impacted carbonates and evaporites. Consideration of the suggested volatile-based extinction mechanisms indicates that maximal volatile release (and environmental stress) will come from cometary impacts into oceans, ice caps or carbonate/evaporite-rich terranes; a high velocity of impact will result in relatively fast, highly shocked ejecta yielding greater dispersion/perturbations. The least stressing impact will come from a differentiated silicate asteroid into an unvegetated, continental, silicate terrane resulting in minimal volatile release and relatively low-velocity ejecta. However, all sizeable impacts will produce a succession of temperature deviations (as illustrated in Fig. 9) of varying magnitudes and geographic extents. This speaks to the estimation of the recurrence interval of impact-induced mass extinctions. Impacts of Chicxulub's size are thought to occur at ~100 million-year intervals with more frequent lesser impacts, but evidence of analogous impacts has been slow in coming compared to early expectations. It remains to be understood if the Chicxulub impact was particularly deadly due to the impacted carbonate/evaporite terrane, impact velocity near the maximum possible, or that the impact magnitude was more unique than our current understanding of the flux of large impactors suggests (or an unknown factor). Ease of study of this relatively recent example is often considered as the reason that evidence of impact was discovered, but the admitted selection effect becomes questionable as more research is conducted on older extinction horizons. While many environmental stresses have been suggested for the Chicxulub impact (with evidence of some preserved in the geological record), the patterns of extinctions at the K–T boundary indicate a different order of extinction without as yet establishing specific responsible mechanisms.

We are grateful to Petróleos Méxicanos, Instituto Méxicano del Petróleo, National Imaging and Mapping Agency and G. Kinsland for provision of gravity data, and to A. Camargo for discussion of same; Petróleos Méxicanos and Camargo also generously supplied (and discussed) seismic-reflection data; J. Morgan also improved our understanding of these data. H. Melosh discussed crater collapse and generously supplied a computer program to calculate impact parameters. D. Crawford kindly discussed impact fireball scaling.

A. Rafeek provided timely drafting services. The Geophysics Commission of the Pan American Institute of Geography and History and the Meteorites and Impacts Advisory Committee to the Canadian Space Agency financially supported gravity surveys. This is Geological Survey of Canada contribution No. 1998028.

References

ADUSHKIN, V. V. & NEMCHINOV, I. I. 1994. Consequences of impacts of cosmic bodies on the surface of the Earth. *In*: GEHRELS, T. (ed.) *Hazards due to Comets and Asteroids*. University of Arizona Press, Tucson, Arizona, 721–778.

AHRENS, T. J. & O'KEEFE, J. D. 1983. Impact of an asteroid or comet in the ocean and extinction of terrestrial life. *In*: *Proceedings of the 13th Lunar and Planetary Science Conference, Part 2, Journal of Geophysical Research*, **88**, A799–A806.

ALVAREZ, L. W., ALVAREZ, W., ASARO, F. & MICHEL, H. 1980. Extraterrestrial cause for the Cretaceous/Tertiary extinction. *Science*, **208**, 1095–1108.

ANDERS, E. & GREVESSE, N. 1989. Abundances of the elements: Meteoritic and solar. *Geochimica et Cosmochimica Acta*, **53**, 2363–2380.

ASPHAUG, E. & BENZ, W. 1994. Density of comet Shoemaker-Levy 9 deduced by modelling breakup of the parent 'rubble pile'. *Nature*, **370**, 120–124.

——, MOORE, J. M., MORRISON, D., BENZ, W., NOLAN, M. C. & SULLIVAN, R. J. 1996. Mechanical and geological effects of impact cratering on Ida. *Icarus*, **120**, 158–184.

BERNER, R. A. 1994. 3GEOCARB II: A revised model of atmospheric CO_2 over Phanerozoic time. *American Journal of Science*, **294**, 56–91.

BLAKELY, R. J. & SIMPSON, R. W. 1986. Approximating edges of source bodies from magnetic or gravity anomalies. *Geophysics*, **51**, 1494–1498.

BOHOR, B. F., MODRESKI, P. J. & FOORD, E. E. 1987. Shocked quartz in the Cretaceous–Tertiary boundary clays: Evidence for a global distribution. *Science*, **236**, 705–709.

BORG, I. Y. 1972. Some shock effects in granodiorite to 270 kilobars at the Pile Driver site. *Flow and Fracture in Rocks*. American Geophysical Union Monograph, **16**, 293–311.

BOURGEOIS, J., HANSEN, T. A., WILBERG, P. L. & KAUFFMAN, E. G. 1988. A tsunami deposit at the Cretaceous–Tertiary boundary in Texas. *Science*, **241**, 567–570.

BRETT, R. 1992. The Cretaceous–Tertiary extinction: A lethal mechanism involving anhydrite target rocks. *Geochimica et Cosmochimica Acta*, **56**, 3603–3606.

CAMARGO ZANOGUERA, A. & SUÁREZ REYNOSO, G. 1994. Evidencia sismica del crater de impacto de Chicxulub. *Boletín de la Asociación Mexicana de Géofisicos de Exploracion*, **34**, 1–28.

CAMPOS-ENRIQUEZ, J. O., ARZATE, J. A., URRUTIA-FUCUGAUCHI, J. & DELGADO-RODRIGUEZ 1997. The subsurface structure of the Chicxulub crater (Yucatán, México): Preliminary results of a magnetotelluric study. *The Leading Edge*, **16**, 1774–1777.

CARLSON, R. W., WEISSMAN, P. R., SEGURA, M., HUI, J., SMYTHE, W. D., JOHNSON, T. V. & BAINES, K. H. 1995. Galileo infrared observations of the Shoemaker–Levy 9 G impact fireball: A preliminary report. *Geophysical Research Letters*, **22**, 1557–1560.

——, DROSSART, P., ENCRENAZ, TH., WEISSMAN, R., HUI, J. & SEGURA, M. 1997. Temperature, size and energy of the Shoemaker-Levy 9 G-impact fireball. *Icarus*, **128**, 251–274.

CHRISTESON, G. L., NAKAMURA, Y., BUFFLER, R. T. & CHICXULUB WORKING GROUP 1997. Structure of the Chicxulub crater from wide-angle OBS data: Evidence for a multi-ring impact basin. *Eos, Transactions, American Geophysical Union*, **78**, F399.

COLGATE, S. A. & PETSCHEK, A. G. 1984. *Cometary impacts and global distributions of resulting debris by floating*. LA-UR-84-3911, Los Alamos National Laboratory, Los Alamos, NM, 17.

CONNORS, M., HILDEBRAND, A. R., PILKINGTON, M., ORTIZ-ALEMAN, C., CHAVEZ, R. E., URRUTIA-FUCUGAUCHI, J., GRANIEL-CASTRO, E., CAMARA-ZI, A., VASQUEZ, J. & HALPENNY, J. F. 1996. Yucatán karst features and the size of Chicxulub Crater. *Geophysical Journal International*, **127**, F11–F14.

COVEY, C., GHAN, S. J., WALTON, J. J., WEISSMAN, P. R. 1990. Global environmental effects of impact-generated aerosols; Results from a general circulation model. *In*: SHARPTON, V. L. & WARD, P. D. (eds) *Global Catastrophes in Earth History; An Interdisciplinary Conference on Impacts, Volcanism, and Mass Mortality*. Geological Society of America Special Paper, **247**, 391–400.

CROFT, S. K. 1980. Cratering flow fields: Implications for the excavation and transient expansion stages of crater formation. *Proceedings of the Eleventh Lunar and Planetary Science Conference, Geochimica et Cosmochimica Acta, Supp.*, **14**, 2347–2378.

——1985. The scaling of complex craters. *Proceedings of the 15th Lunar and Planetary Science Conference, Part 2, Journal of Geophysical Research*, 90, Supplement, C828–C842.

D'HONDT, S., PILSON, M. E. Q., SIGURDSSON, H., HANSON, A. K. JR. & CAREY, S. 1994. Surface-water acidification and extinction at the Cretaceous–Tertiary boundary. *Geology*, **22**, 983–986.

——, KING, J. & GIBSON, C. 1996. Oscillatory marine response to the Cretaceous–Tertiary impact. *Geology*, **24**, 611–614.

DELSEMME, A. H. 1988. The chemistry of comets. *Philosophical Transactions of the Royal Society of London*, **A325**, 509–523.

DENCE, M. R., GRIEVE, R. A. F. & ROBERTSON, B. 1977. Terrestrial impact structures: Principal characteristics and energy considerations: *In*: RODDY, D. J., PEPIN, R. O. & MERRILL, R. B. (eds) *Impact and Explosion Cratering*. Pergamon, New York, 247–275.

DYPVIK, H., GUDLAUGSSON, S. T., TSIKALAS, F., ATTREP, M. JR., FERRELL, R. E. JR., KRINSLEY, D. H., MORK, A., FALEIDE, J. I. & NAGY, J. 1996. Mjolnir structure: An impact crater in the Barents Sea. *Geology*, **24**, 779–782.

EMILIANI, C., KRAUS, E. B. & SHOEMAKER, E. M. 1981. Sudden death at the end of the Mesozoic. *Earth and Planetary Science Letters*, **55**, 317–334.

FARLEY, K. A., MONTANARI, A., SHOEMAKER, E. M. & ROBINSON, K. 1997. Extraterrestrial ^{3}He in seafloor sediments: A comparison of the Eocene/Oligocene and Cretaceous/Tertiary boundaries. *Eos, Transactions, American Geophysical Union*, **78**, F371.

GALLAGHER, W. B. 1991. Selective extinction and survival across the Cretaceous/Tetiary boundary in the northern Atlantic Coastal Plain. *Geology*, **19**, 967–970.

GAULT, D. E. & WEDEKIND, J. A. 1978. Experimental studies of oblique impact. *Proceedings of the Ninth Lunar and Planetary Science Conference*, **3**, 3843–3875.

GILMOUR, I., WOLBACH, W. S. & ANDERS, E. 1990. Early environmental effects of the terminal Cretaceous impact. *In*: SHARPTON, L. & WARD, D. (eds) *Global Catastrophes in Earth History; An Interdisciplinary Conference on Impacts, Volcanism, and Mass Mortality*. Geological Society of America Special Paper, **247**, 391–400.

GRIEVE, R. A. F. 1982. The record of impact on Earth: Implications for a major Cretaceous/Tertiary impact event. *In*: SILVER, L. T. & SCHULTZ, P. H. (eds) *Geological Implications of Impacts of Large Asteroids and Comets on the Earth*. Geological Society of America Special Paper, **190**, 25–37.

——, ROBERTSON, P. B. & DENCE, M. R. 1981. Constraints on the formation of ring impact structures, based on terrestrial data, Multi-ring Basins. *Proceedings of Lunar and Planetetary Science*, **12A**, 37–57.

HANSEN, T., FARRAND, R. B., MONTGOMERY, H. A., BILLMAN, H. G. & BLECHSCHMIDT, G. 1987. Sedimentology and extinction patterns across the Cretaceous–Tertiary boundary internal in east Texas. *Cretaceous Research*, **8**, 229–252.

HILDEBRAND, A. R. 1992. *Geochemistry and stratigraphy of the Cretaceous/Tertiary boundary impact ejecta*. PhD dissertation, University of Arizona.

——1993. The Cretaceous/Tertiary boundary impact (or the dinosaurs didn't have a chance). *Journal of the Royal Astronomical Society of Canada*, **87**, 77–118.

——1997. Contrasting Chicxulub Crater structural models: What can seismic velocity studies differentiate? *Journal of Conference Proceedings*, **1**, 37–46.

—— & BOYNTON, W. V. 1989. Hg anomalies at the K/T boundary: Evidence for acid rain? *Meteoritics*, **24**, 277–278.

—— & ——1990. Proximal Cretaceous–Tertiary Boundary Impact Deposits in the Caribbean. *Science*, **248**, 843–847.

—— & WOLBACH, W. S. 1989. Carbon and chalcophiles at a nonmarine K/T boundary: Joint investigations of the Raton section, New Mexico, Lunar and Planetary Science, **XX**, 414–415. The Lunar and Planetary Science Institute, Houston.

——, PENFIELD, G. T., KRING, D. A., PILKINGTON, M., CAMARGO, Z. A., JACOBSEN, S. & BOYNTON, W. 1991. Chicxulub crater: A possible Cretaceous–Tertiary boundary impact crater on the Yucatán Peninsula, México. *Geology*, **19**, 867–871.

——, CONNORS, M., PILKINGTON, M., ORTIZ ALEMAN, C. & CHAVEZ, R. E. 1994. Size and structure of the Chicxulub crater. *Revista de la Sociedad Mexicana de Paleontologia*, **7**, 1, 59–68.

——, PILKINGTON, M., CONNORS, M., ORTIZ-ALEMAN, C. & CHAVEZ, R. E. 1995. Size and structure of the Chicxulub crater revealed by horizontal gravity gradients and cenotes. *Nature*, **376**, 415–417.

——, ——, HALPENNY, J. F., COOPER, R. V., CONNORS, M., ORTIZ-ALEMAN, C., CHAVEZ, R. E., URRUTIA-FUCUGAUCHI, J., GRANIEL-CASTRO, E., CAMARA-ZI, A. & BUFFLER, R. T. 1997. The Chicxulub crater as revealed by gravity and seismic surveys. *Eos, Transactions, American Geophysical Union*, **78**, F398.

HSU, K. J., HE, Q., MCKENZIE, J. A., WEISSERT, H., PERCH-NIELSEN, K., OBERHANSLI, H., KELTS, K., LABREQUE, J., TAUXE, L., KRAHENBUHL, U., PERCIVAL, S. F., WRIGHT, R., KARPOFF, A. M., PETERSEN, N., TUCKER, P., POORE, R. Z., GOMBOS, A. M., PISCIOTTO, K., CARMAN, M. F. JR. & SCHREIBER, E. 1982. Mass mortality and its environmental and evolutionary consequences. *Science*, **216**, 249–256.

ISAAC, J. H. & STEWART, R. R. 1993. 3-D seismic expression of a cryptoexplosion structure. *Canadian Journal of Exploration Geophysics*, **29**, 429–439.

IVANOV, B. A., BADUKOV, D. D., YAKOVLEV, O. I., GERASIMOV, M. V., DIKOV, YU. P., POPE, K. O. & OCAMPO, A. C. 1996. Degassing of sedimentary rocks due to Chicuxlub impact: Hydrocode and physical simulations. *In*: RYDER, G., FASTOVSKY, D. & GARTNER, S. (eds) *The Cretaceous–Tertiary Event and other Catastrophes in Earth History*. Geological Society of America Special Paper, **307**, 125–139.

IZETT, G. A. 1990. *The Cretaceous/Tertiary boundary interval, Raton Basin, Colorado and New Mexico, and its content of shock-metamorphosed minerals; Evidence relevant to the K–T boundary impact-extinction theory*. Geological Society of America Special Paper, **249**.

JABLONSKI, D. 1986. Background and mass extinctions: The alternation of macroevolutionary regimes. *Science*, **231**, 129–133.

JESSBERGER, E. K., CHRISTOFORIDIS, A. & KISSEL, J. 1988. Aspects of the major element composition of Halley's dust. *Nature*, **332**, 691–695.

JIANG, M. J. & GARTNER, S. 1986. Calcareous nonnofossil succession across the Cretaceous/Tertiary boundary in east-central Texas. *Micropaleontology*, **32**, 232–255.

KELLER, G. 1989. Extended Cretaceous/Tertiary boundary extinctions and delayed population change in planktonic foraminifera from Brazos River, Texas. *Paleooceanography*, **4**, 3, 287–332.

KINSLAND, G., HURTADO, M., CERON, A., SMYTHE, W., OCAMPO, A. & POPE, K. 1995. New gravity data over the southwestern portion of the Chicxulub impact feature, Yucatán Peninsula, México. *Lunar and Planetary Science* **XXIII**, 755–756. The Lunar and Planetary Science Institute, Houston.

KRING, D. A., MELSOH, H. J. & HUNTEN, D. M. 1996. Impact-induced perturbations of atmospheric sulfur. *Earth and Planetary Science Letters*, **140**, 201–212.

KYTE, F. T., SMIT, J. & WASSON, J. T. 1985. Siderophile interelement variations in the Cretaceous–Tertiary boundary sediments from Caravaca, Spain. *Earth and Planetary Science Letters*, **73**, 183–195.

——, BOSTWICK, J. A. & ZHOU, L. 1996. The Cretaceous–Tertiary boundary on the Pacific plate: Composition and distribution of impact debris. *In*: RYDER, G., FASTOVSKY, D. & GARTNER, S. (eds) *The Cretaceous–Tertiary Event and other Catastrophes in Earth History*. Geological Society of America Special Paper, **307**, 389–401.

LERBEKMO, J. F. & ST. LOUIS, R. M. 1986. The terminal Cretaceous iridium anomaly in the Red Deer Valley, Alberta, Canada. *Canadian Journal of Earth Science*, **23**, 120–124.

LEWIS, J. S., WATKINS, G. H., HARTMAN, H. & PRINN, R. G. 1982. Chemical consequences of major impact events on earth. *In*: SILVER, L. T. & SCHULTZ, H. (eds) *Geological Implications of Impacts of Large Asteroids and Comets on the Earth*. Geological Society of America Special Paper, **190**, 215–221.

LOPEZ RAMOS, E. 1975. Geological summary of the Yucatán Peninsula. *In*: NAIRN, A. E. M. & STEHLI, F. G. (eds) *The Ocean Basins and Margins, Vol. 3 – The Gulf of Mexico and the Caribbean*. Plenum, New York, 257–282.

DE MAUPERTUIS, L. M. 1750. Essay de Cosmologie, in Les Oeuvres de Mr. De Maupertuis, 1752, Dresden, Libraire du Roy, 1–54.

MCGETCHIN, T. R., SETTLE, M. & HEAD, J. W. 1973. Radial thickness variation in impact crater ejecta: Implications for lunar basin deposits. *Earth and Planetary Science Letters*, **20**, 226–236.

MCKAY, C. & THOMAS, G. E. 1982. Formation of noctilucent clouds by an extraterrestrial impact. *In*: SILVER, L. T. & SCHULTZ, H. (eds) *Geological Implications of Impacts of Large Asteroids and Comets on the Earth*. Geological Society of America Special Paper, **190**, 211–214.

MELOSH, H. J. 1989. *Impact Cratering: A Geologic Process*. Oxford University Press.

——, SCHNEIDER, N. M., ZAHNLE, K. J. & LATHAM, D. 1990. Ignition of global wildfires at the Cretaceous/Tertiary boundary. *Nature*, **343**, 251–254.

MORGAN, J., WARNER, M., BRITTAN, J., BUFFLER, R., CAMARGO, A., CHRISTESON, G., DENTON, P., HILDEBRAND, A., HOBBS, R., MACINTYRE, H., MACKENZIE, G., MAGUIRE, P., MARIN, L., NAKAMURA, Y., PILKINGTON, M., SHARPTON, V. L., SNYDER, D., SUAREZ, G., TREJO, A. 1997. Size and morphology of the Chicxulub impact crater. *Nature*, **390**, 472–476.

NESS, G. E., DAUPHIN, J. P., GARCIA-ABDESLEM, J. & ALVARADO-OMANA, M. E. 1991. *Bathymetry and Gravity and Magnetic Anomalies of the Yucatán Peninsula and Adjacent Areas*. Geological Society of America Map and Chart Series MCH073.

OFFIELD, T. W. & POHN, H. A. 1977. Deformation at the Decaturville impact structure, Missouri: *In*: RODDY, D. J., PEPIN, R. O. & MERRILL, R. B. (eds) *Impact and Explosion Cratering*. Pergamon, New York, 321–341.

O'KEEFE, J. D. & AHRENS, T. J. 1982. The interaction of the Cretaceous/Tertairy Extinction Bolide with the atmosphere, ocean and solid Earth. *In*: SILVER, L. T. & SCHULTZ, P. H. (eds) *Geological Implications of Impacts of Large Asteroids and Comets on the Earth*. Geological Society of America Special Paper, **190**, 103–120.

—— & ——1989. Impact production of CO_2 by the Cretaceous/Tertiary extinction bolide and the resultant heating of the Earth. *Nature*, **338**, 247–249.

OPIK, E. J. 1958. On the catastrophic effects of collisions with celestial bodies. *Irish Astronomical Journal*, **5**, 34–35.

ORTH, C. J., GILMORE, J. S. & KNIGHT, J. D. 1987. Iridium anomaly at the Cretaceous–Tertiary boundary in the Raton Basin, New Mexico Geological Society 38th Annual Field Conference Guidebook, 265–270.

——, ATTREP, M. JR & QUINTANA, L. R. 1990. Iridium abundance patterns across bio-event horizons in the fossil record. *In*: SHARPTON, V. L. & WARD, P. D. (eds) *Global Catastrophes in Earth History; An Interdisciplinary Conference on Impacts, Volcanism, and Mass Mortality*. Geological Society of America Special Paper, **247**, 45–59.

PERRY, E. C., MARIN, L., MCCLAIN, J. & VELÁQUEZ, G. 1995. Ring of Cenotes (sinkholes), northwest Yucatan, Mexico: Its hydrogeologic characteristics and possible association with the Chicxulub impact crater. *Geology*, **23**, 17–20.

PILKINGTON, M. & GRIEVE, R. A. F. 1992. The geophysical signature of impact craters. *Reviews of Geophysics*, **30**, 161–181.

——, HILDEBRAND, A. R. & ORTIZ ALEMAN, C. 1994. Gravity and magnetic field modeling and structure of the Chicxulub crater, Mexico. *Journal of Geophysical Research*, **99**, 13 147–13 162.

PIKE, R. J. 1985. Some morphologic systematics of complex impact structures. *Meteoritics*, **20**, 49–68.

POPE, K. O., OCAMPO, A. C. & DULLER, C. E. 1993. Surficial geology of the Chicxulub impact crater, Yucatán, México. *Earth, Moon and Planets*, **63**, 93–104.

——, BAINES, K. H., OCAMPO, A. C. & IVANOV, B. A. 1994. Impact winter and the Cretaceous/Tertiary extinctions. Results of a Chicxulub asteroid impact model. *Earth and Planetary Science Letters*, **128**, 719–725.

——, OCAMPO, A. C., KINSLAND, G. L., SMITH, R. 1996. Surface expression of the Chicxulub crater. *Geology*, **24**, 527–530.

PRINN, R. G. & FEGLEY, B. 1987. Bolide impacts, acid rain, and biospheric traumas at the Cretaceous–Tertiary boundary. *Earth and Planetary Science Letters*, **83**, 1–5.

RUSSELL, D. A. 1979. The enigma of the extinction of the dinosaurs. *Annual Review of Earth and Planetary Science*, **7**, 163–182.

SCHMIDT, R. M. & HOUSEN, K. R. 1987. Some recent advances in the scaling of impact and explosion cratering. *International Journal of Impact Engineering*, **5**, 543–560.

SHARPTON, V. L., BURKE, K., CAMARGO-ZANOGUERA, A., HALL, S. A., LEE, D. S., MARIN, L. E., SUÁREZ-REYNOSO, G., QUEZADA-MUZETON, J. M., SPUDIS, D. & URRUTIA-FUCUGAUCHI, J. 1993. Chicxulub multi-ring impact basin: Size and other characteristics derived from gravity analysis. *Science*, **261**, 1564–1567.

——, MARIN, L. E., CARNEY, J. L., LEE, S., RYDER, G., SCHURAYTZ, B. C., SIKORA, P., SPUDIS, P. D. 1996. A model of the Chicxulub impact basin based on evaluation of geophysical data, well logs, and drill core samples. *In*: RYDER, G., FASTOVSKY, D. & GARTNER, S. (eds) *The Cretaceous–Tertiary Event and other Catastrophes in Earth History*. Geological Society of America Special Paper, **307**, 55–74.

SHEEHAN, P. M. & HANSEN, T. A. 1986. Detritus feeding as a buffer to extinction at the end of the Cretaceous. *Geology*, **14**, 868–870.

SHOEMAKER, E. M., PILLMORE, C. L. & PEACOCK, E. W. 1987. Remanent magnetization of rocks of latest Cretaceous and earliest Tertiary age from drill core at York Canyon, New Mexico. *In*: FASSETT, J. E. & RIGBY, J. K. JR. (eds) *The Cretaceous–Tertiary Boundary in the San Juan and Raton Basins, New Mexico and Colorado*. Geological Society of America Special Paper, **209**, 131–150.

——, WOLFE, R. F. & SHOEMAKER, C. S. 1990. Asteroid and comet flux in the neighbourhood of Earth. *In*: SHARPTON, V. L. & WARD, P. D. (eds) *Global Catastrophes in Earth History; An Interdisciplinary Conference on Impacts, Volcanism, and Mass Mortality*. Geological Society of America Special Paper, **247**, 155–170.

SIGURDSSON, H., D'HONDT, S. & CAREY, S. 1992. The impact of the Cretaceous/Tertiary bolide on evaporite terrane and generation of major sulfuric acid aerosol. *Earth and Planetary Science Letters*, **109**, 543–559.

SMIT, J. 1990. Meteorite impact, extinctions and the Cretaceous–Tertiary Boundary. *Geologie en Mijnbouw*, **69**, 187–204.

——1994. Extinctions at the Cretaceous–Tertiary boundary: The link to the Chicxulub impact. *In*: GEHRELS, T. (ed.) *Hazards due to Comets and Asteroids*. University of Arizona Press, Tucson, Arizona, 859–878.

—— & ROMEIN, A. J. T. 1985. A sequence of events across the Cretaceous–Tertiary boundary. *Earth and Planetary Science Letters*, **74**, 155–170.

——, ALVAREZ, W., MONTANARI, A., SWINBOURNE, N. H. M., VAN KEMPEN, T. M., KLAVER, G. T. & LUSTENHOUWER, W. J. 1992. 'Tektites' and microcrystites at the Cretaceous Tertiary boundary: Two strewn fields, one crater? *Proceedings of Lunar and Planetary Science*, **22**, 87–100.

——, ROEP, TH. B., ALVAREZ, W., MONTANARI, A., CLAEYS, P., GRAJALES-NISHIMURA, J. M. & BERMUDEZ, J. 1996. Coarse-grained, clastic sandstone complex at the K/T boundary around the Gulf of Mexico: Deposition by tsunami waves induced by the Chicxulub impact? *In*: RYDER, G., FASTOVSKY, D. & GARTNER, S. (eds) *The Cretaceous–Tertiary Event and other Catastrophes in Earth History*. Geological Society of America Special Paper, **307**, 151–182.

SWEET, A. R. 1994. Relationships between depositional environments and changes in palynofloras across the K/T boundary interval. *In*: TRAVERSE, A. (ed.) *Sedimentation of Organic Particles*. Cambridge University Press, 461–488.

—— & BRAMAN, D. R. 1992. The K–T boundary and contiguous strata in western Canada: interactions between paleoenvironments and palynological assemblages. *Cretaceous Research*, **13**, 31–79.

THERRIAULT, A. M., GRIEVE, R. A. F. & REIMOLD, W. U. 1997. Original size of the Vredefort structure: implications for the geological evolution of the Witwatersrand Basin. *Meteoritics and Planetary Science*, **32**, 71–77.

TOON, O. B., POLLACK, J. B., ACKERMAN, T. P., TURCO, R. P., MCKAY, C. P. & LIU, M. S. 1982. Evolution of an impact-generated dust cloud and its effects on the atmosphere. *In*: SILVER, L. T. & SCHULTZ, P. H. (eds) *Geological Implications of Impacts of Large Asteroids and Comets on the Earth*. Geological Society of America Special Paper, **190**, 187–200.

——, ZAHNLE, K., MORRISON, D., TURCO, R. P. & COVEY, C. 1997. Environmental perturbations caused by the impacts of asteroids and comets. *Reviews of Geophysics*, **35**, 41–78.

TURCOTTE, D. L. & SCHUBERT, G. 1982. Geodynamics: Applications of Continuum Physics to Geological Problems. Wiley, New York.

URRUTIA-FUCUGAUCHI, J., MARIN, L. & TREJO GARCIA, A. 1996. Initial results of the UNAM scientific drilling program on the Chicxulub impact structure: Rock magnetic properties of UNAM-7 Tekax borehole. *Geofisica Internacional*, **35**, 125–133.

VENKATESAN, M. I. & DAHL, J. 1989. Further geochemical evidence for global fires at the Cretaceous–Tertiary boundary. *Nature*, **338**, 57–60.

VILLAGOMEZ, A. 1953. Un programa de exploracion para la Peninsula de Yucatan. *Boletin de la Asociacion Mexicana de Geologos Petroleros*, **5**, 77–84.

WARD, W. C., KELLER, G., STINNESBECK, W. & ADATTE, T. 1995. Yucatán subsurface stratigraphy: Implications and constraints for the Chicxulub impact. *Geology*, **23**, 873–876.

WEISSMAN, P. R. 1990. The cometary impactor flux at the Earth. *In*: SHARPTON, V. L. & WARD, P. D. (eds) *Global Catastrophes in Earth History; An Interdisciplinary Conference on Impacts, Volcanism, and Mass Mortality*. Geological Society of America Special Paper, **247**, 171–179.

WOLBACH, W. S., LEWIS, R. S. & ANDERS, E. 1985. Cretaceous extinctions: evidence for wildfires and search for meteoritic material. *Science*, **230**, 167–170.

——, GILMOUR, I., ANDERS, E., ORTH, C. J. & BROOKS, R. R. 1988. Global fire at the Cretaceous—Tertiary boundary. *Nature*, **334**, 665–669.

——, —— & ——1990. Major wildfires at the Cretaceous/Tertiary boundary: *In*: SHARPTON, V. L. & WARD, P. D. (eds) *Global Catastrophes in Earth History; An Interdisciplinary Conference on Impacts, Volcanism, and Mass Mortality*. Geological Society of America Special Paper, **247**, 391–400.

ZACHOS, J. C. & ARTHUR, M. A. 1986. Paleoceanography of the Cretaceous/Tertiary boundary event: Inferences from stable isotopic and other data. *Paleoceanography*, **1**, 5–26.

ZAHNLE, K. J. 1990. Atmospheric chemistry by large impacts. *In*: SHARPTON, V. L. & WARD, P. D. (eds) *Global Catastrophes in Earth History; An Interdisciplinary Conference on Impacts, Volcanism, and Mass Mortality*. Geological Society of America Special Paper, **247**, 271–288.

Preliminary results from a passive seismic array over the Chicxulub impact structure in Mexico

P. K. H. MAGUIRE[1], G. D. MACKENZIE[1], P. DENTON[1], A. TREJO[2], R. KIND[3] & MEMBERS* OF THE CHICXULUB WORKING GROUP

[1] *University of Leicester, University Road, Leicester LE1 7RH, UK*
[2] *Universidad Nacional Autonoma de Mexico, Mexico DF, Mexico*
[3] *GeoForschung Zentrum, Potsdam, Germany*

Abstract: A passive, 20-element, short-period (1 Hz) and broadband seismic array was deployed over the Chicxulub impact structure for *c.* 100 days in early 1996. The principal objective was to study the shear-wave anisotropy associated with the structure; in particular, to determine the presence (or absence) of radial symmetry which will allow comment on the time variance of that anisotropy. A total of 15 teleseismic, 75 regional, and 100 local events were recorded. Preliminary results from studies of the surface-wave dispersion of the local events, and a receiver function analysis of a single teleseismic event are reported here. Thirty local events have been located, a number of which originated from quarries within the array. Analysis of seismograms from three of these events demonstrates a bimodal distribution; those whose ray-paths cross the outer part of the impact structure show a strong inverse dispersion, whereas those with ray-paths crossing the centre do not. The pattern may be produced by the sedimentary depositional environment, with deeper water sedimentation in the outer part of the post-impact crater basin and shallower water sedimentation over the upraised peak-ring block at the centre. Receiver functions derived for an event originating in Peru are dominated by an efficient mode conversion, simply modelled as a P–S multiple from the base Tertiary boundary. This shows a strong correlation with distance from the centre of the impact structure and implies it has an S-wave radial symmetry. The multiple also has a variable delay probably related to the depth of the conversion boundary. Unfortunately, the Moho conversion occurs at almost exactly the same time as this surface layer sediment multiple, restricting any modelling of Moho topography and its influence on the receiver functions.

The Chicxulub structure located on the Yucatan peninsula in Mexico (Fig. 1) is widely recognized as an asteroid or comet impact site occurring at the Cretaceous–Tertiary (K–T) boundary, and possibly connected with mass extinctions at that time.

Studies of the gravity and magnetic signature (Sharpton *et al.* 1993; Pilkington & Hildebrand 1994; Espindola *et al.* 1995), the distribution of karst topography (Hildebrand *et al.* 1995; Perry *et al.* 1995; Connors *et al.* 1996), subsurface drill-hole information (Hildebrand *et al.* 1991; Sharpton *et al.* 1996) and shallow seismic reflection data (Carmargo & Suarez 1994) have been used to attempt to determine the size and geometry of the impact structure. Different groups suggest the radius of crustal deformation is between 180 km (Hildebrand *et al.* 1995) and *c.* 300 km (Sharpton *et al.* 1996) making it one of the largest impact structures formed within the inner Solar System in the past 4 Ga.

The structure lies partly offshore on the northern edge of the Yucatan peninsula, buried under *c.* 1 km of flat-lying, Tertiary rocks. The circular gravity anomalies suggest a central peak-ring or multi-ring basin morphology. Structural uplift followed the collapse of the initial transient crater, but as yet it is undecided whether the Moho was involved in this uplift. It is the only well-preserved example of a large, terrestrial, tectonically undisturbed crater and allows the possibility of measuring many of its parameters using high-resolution geophysical techniques. The resulting information will be used for constraining hypotheses concerning both planetary crater formation, and catastrophic biosphere disruption.

In early 1996 Leicester University deployed a passive seismic array over the crater to study its deep structure and to examine further the

* J. Morgan[1], M. Warner[1], J. Brittan[1], H. Macintyre[1], E. King[2], D. Snyder[3], L. Marin[4] and G. Suarez[4]
[1] Department of Geology, Imperial College, London, UK
[2] British Antarctic Survey, Cambridge, UK
[3] BIRPS, Bullard Laboratories, University of Cambridge, Cambridge, UK
[4] Universidad Nacional Autonoma de Mexico, Mexico DF, Mexico

MAGUIRE, P. K. H. *et al.* 1998. Preliminary results from a passive seismic array over the Chicxulub impact structure in Mexico. *In*: GRADY, M. M., HUTCHISON, R., MCCALL, G. J. H. & ROTHERY, D. A. (eds) *Meteorites: Flux with Time and Impact Effects*. Geological Society, London, Special Publications, **140**, 177–193.

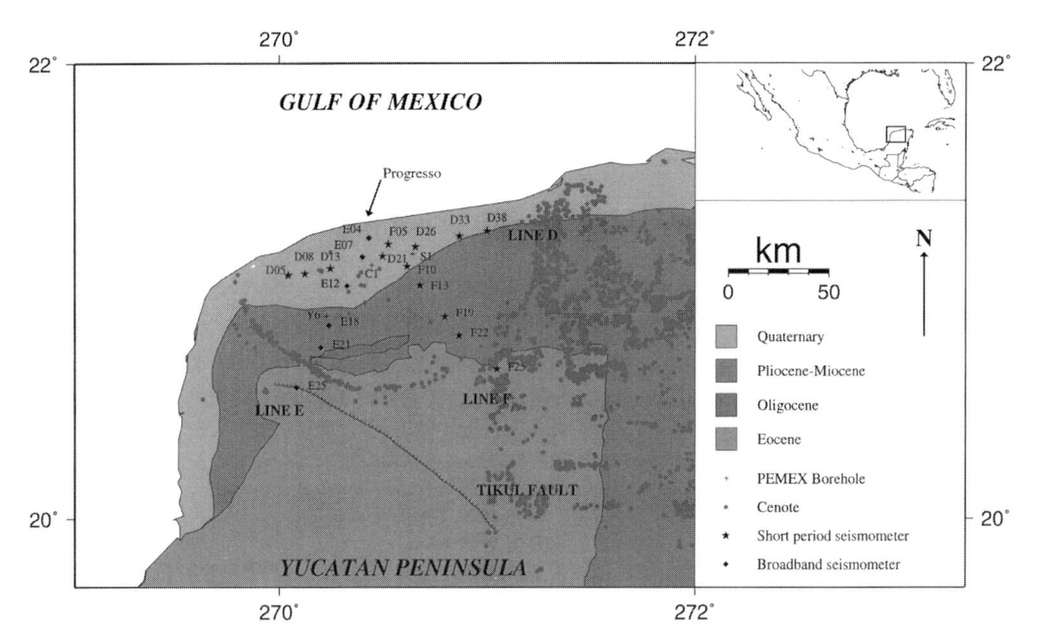

Fig. 1. Geology of the northwest Yucatan peninsula showing the location of the Leicester University Seismic Array and boreholes Y-6, S-1 and C-1 mentioned in the text. The proposed 180 km crater diameter (Hildebrand *et al.* 1995) is coincident with the ring of cenotes.

processes resulting in seismic anisotropy in the crust. In particular, the project was designed to investigate:

- the effect of the impact on crustal and upper-mantle structure, via the inversion of local, regional and teleseismic earthquake data;
- the shear-wave anisotropy associated with the structure, and its analysis in terms of: the presence of any structural, radial symmetry beneath the crater and its depth variation within the lithosphere; and/or the time variance of seismic anisotropy (i.e. is it time invariant with polarisations showing radial symmetry, or is it controlled by the present local stress alignment?).

The experiment was run in conjunction with the 1996 deep seismic reflection project carried out over the impact structure, conducted by the British Institutes Reflection Profiling Syndicate, Imperial College, London, the University of Texas in Austin, and the Universidad Nacional Autonoma de Mexico.

Data acquisition and event location

A passive seismic array, consisting of 20 sites with a nominal station spacing of 25 km, was deployed for *c.* 100 days from mid-February to mid-May 1996 along three radial arms (D, E and F) across the top of the crater, an area of *c.* $100 \times 200\,\text{km}^2$ (Fig. 1). All of the stations bar two were deployed within a radius of 90 km from the approximate centre of the crater at the village of Chicxulub Puerto close to the port of Progresso, and thus within the impact structure. It had been intended that the remaining two stations (E25 and F25) be deployed at a large distance from the structure on 'undisturbed' crust. However, budget and logistic constraints necessitated their siting at a distance of no more than 120 km from Chicxulub Puerto, outside the smaller estimate for the crater radius (Hildebrand *et al* 1995), but within the larger estimate (Sharpton *et al.* 1996).

The equipment consisted of 15 three-component sets of Teledyne Geotech S13 (1 Hz) seismometers, five Guralp CMG-40T (30 s) and one Guralp CMG-3T (120 s) three-component, broadband seismometers recording continuously to Teledyne PDAS recorders. These wrote data to 540 Mbyte external disks which were replaced every 8 days. An external global positioning satellite (GPS) receiver was connected to each PDAS to control timing and record site locations every 2 h. The broadband stations were deployed along Line E, the short-period stations along Lines D and F. Station E07 recorded from both broadband and short-period instruments for signal comparison.

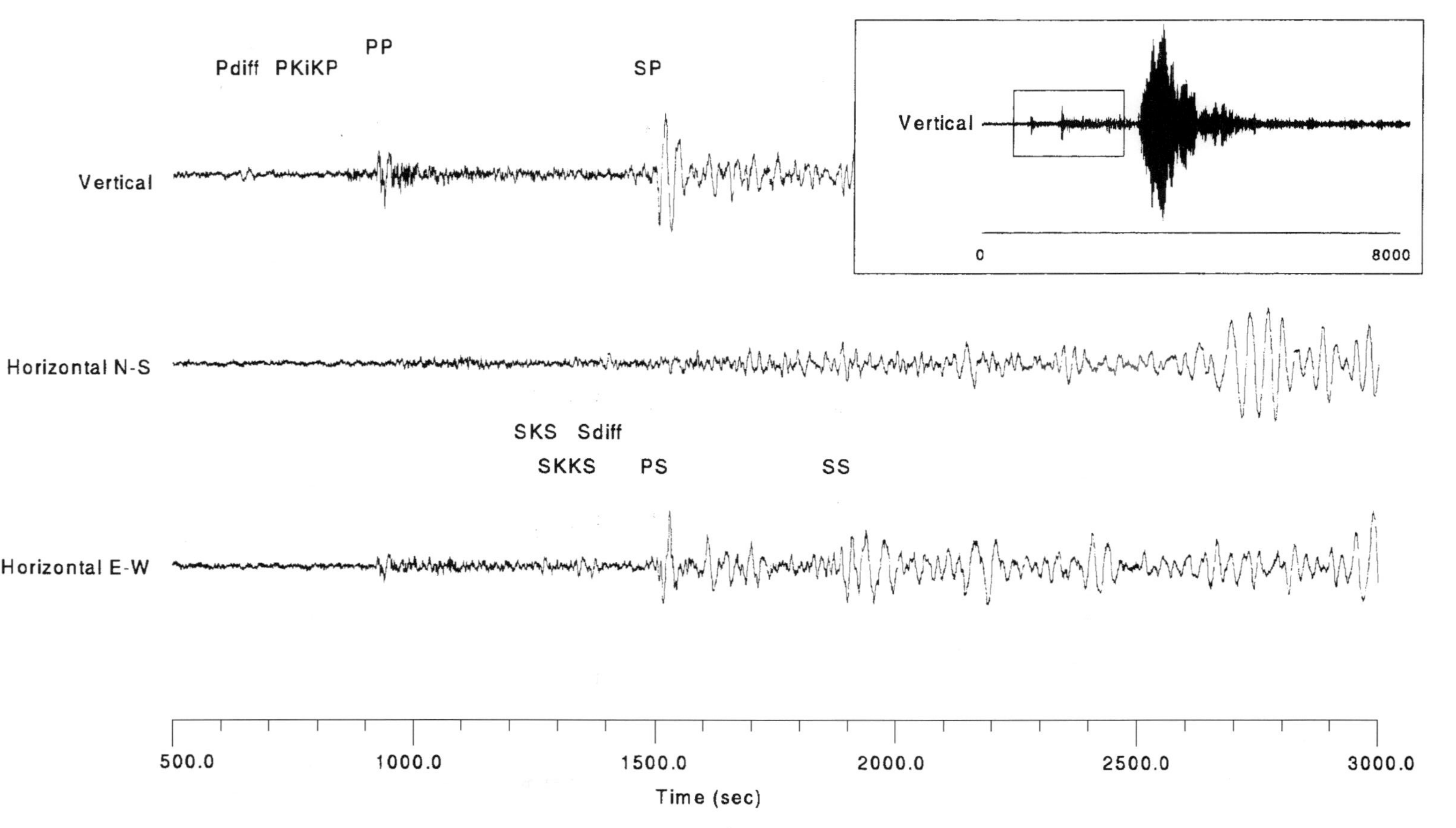

Fig. 2. Broadband record of the 29 April 1996 Solomon Islands earthquake (M_S 7.5; focal depth 44 km; epicentral distance 161°). Inset shows a complete vertical-component record.

The data recorded were downloaded from disk to tape and quality checked at the field headquarters in Chicxulub Puerto. The recordings were of high quality (Fig. 2 and 4), Line E resulting in the noisiest data owing to the proximity of the main Progresso–Merida road and its south–westward continuation towards Campeche. A total of 15 teleseismic and 75 regional events with $M_b > 4$ reported by the US Geological Survey (Fig. 3), and *c.* 100 local events were recorded.

P- and S-wave arrival times have been picked for 30 local events which have subsequently been located using HYPOINVERSE (Klein 1989). Several of these events have been identified as originating from quarries within the array. Many of the local events examined to date show an inverse surface wave dispersion (Fig. 4) for some event–station paths. A preliminary study to examine this has been carried out, and is reported below. Of the teleseismic events, nine had $M_S > 6$ and epicentral distances within the range 30–100°, making receiver function analysis possible. A further preliminary study, reported here, has attempted to identify crustal thickness variations beneath the seismic array via a receiver function study.

Local event Rayleigh-wave dispersion study

The depth variation of the displacement eigenfunctions for various crustal models shows that fundamental mode Rayleigh waves (Rg) are strong for shallow sources compared with their counterpart higher modes constituting the Lg waves (Saikia 1992). The local events recorded by the array generally include a characteristic Rg phase consistent with their originating from quarries. Saikia *et al.* (1990) investigated the lateral variation of Rg wave dispersion in southern New England to examine the sharply contrasting geology between neighbouring provinces. They showed that tectonic boundaries could be demarcated both on the surface and at depth. Significant differences in the dispersion characteristics of the Rg phase from different events into different stations of the Chicxulub

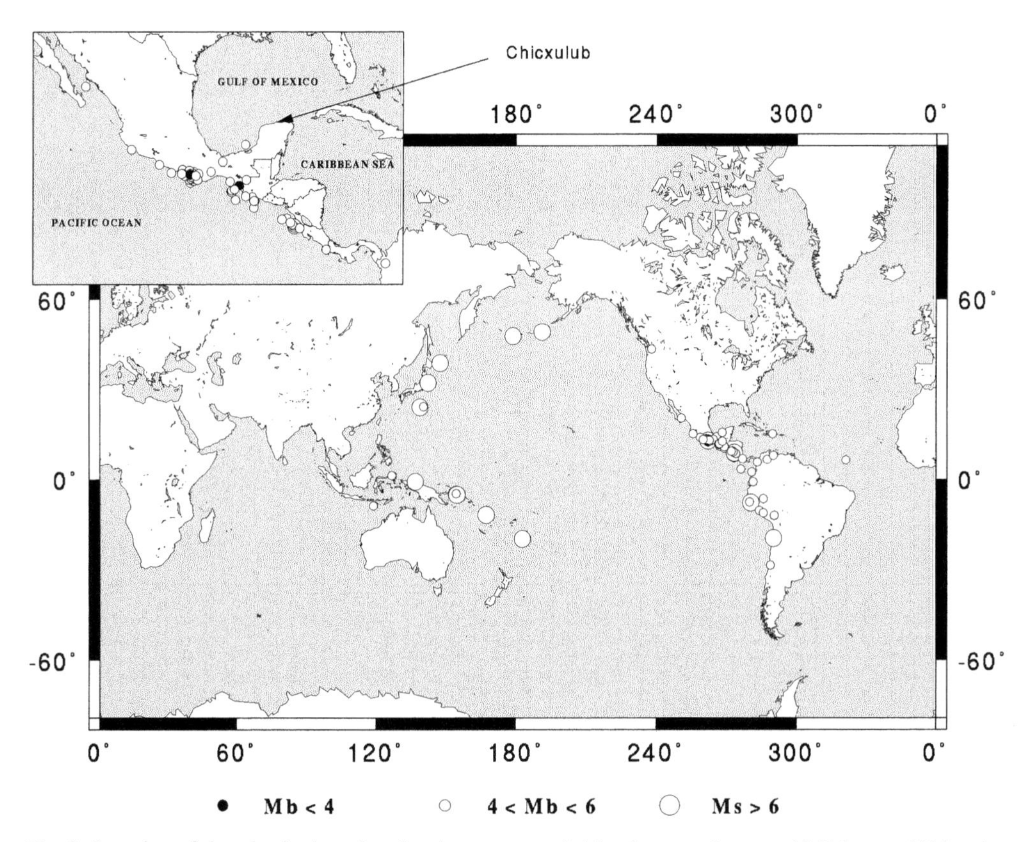

Fig. 3. Location of the teleseismic and regional events recorded by the array between 10 February 1996 and 17 May 1996.

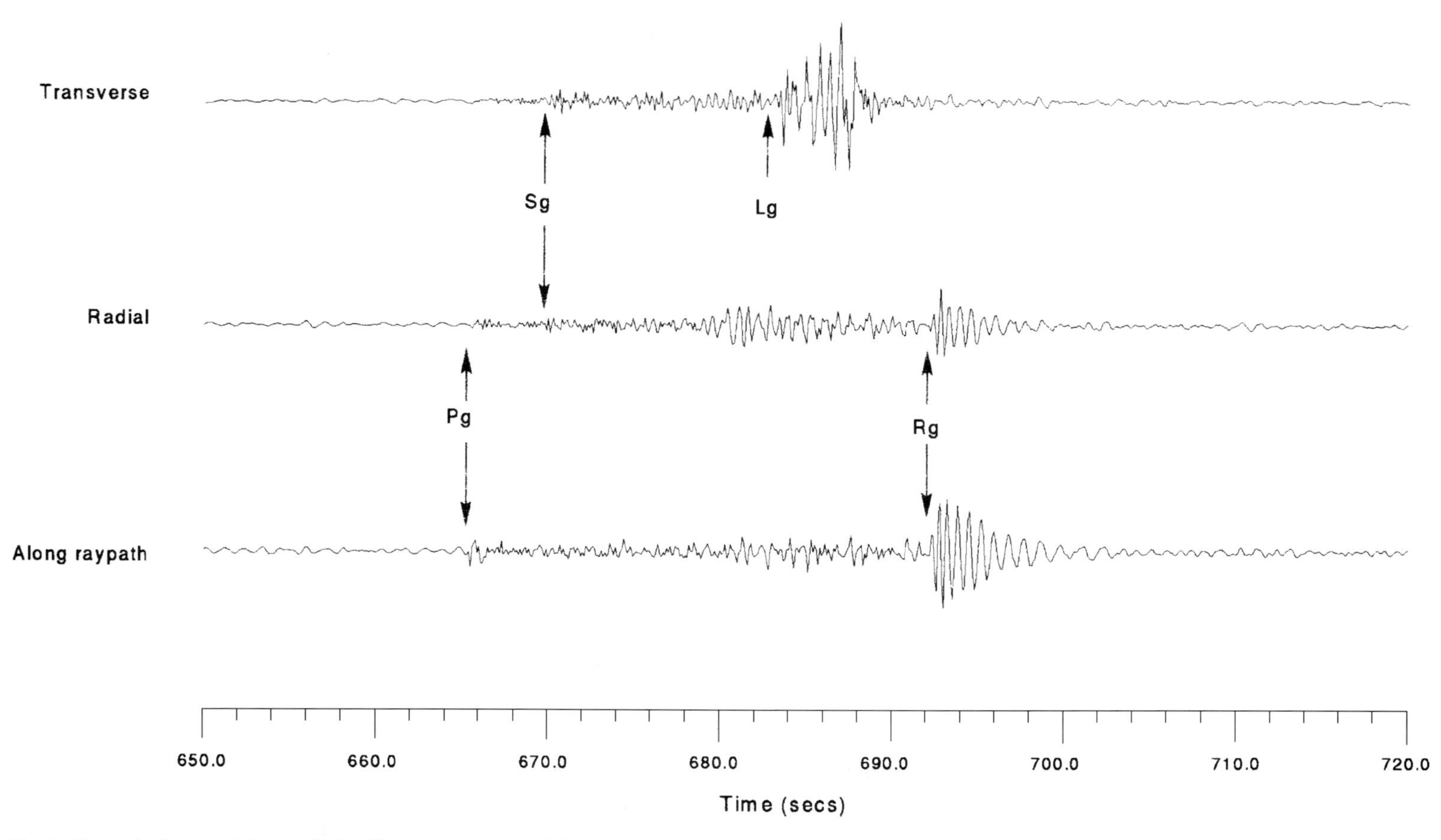

Fig. 4. Rotated, short-period record of a Flamboyanes quarry blast recorded at station D08 (distance 29 km) on 16 March 1996.

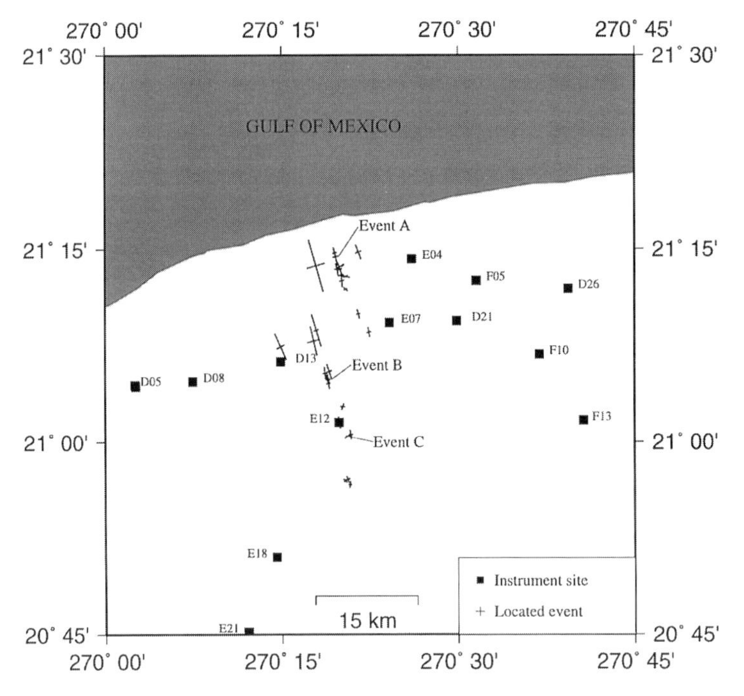

Fig. 5. Location map of local events identifying Events A, B and C used in the surface-wave dispersion study.

array suggested that an equivalent study, detailing a spatial variation in the geology beneath the array, would be productive.

The data

Of the 30 located local events (Fig. 5), three have been used in the preliminary study reported here. Two are thought to have originated from the known quarries at Flamboyanes (Event A) and Dzitya (Event B). Event C is thought to have originated from a third quarry, which was not visited. None of the three events were timed or precisely surveyed. Saikia (1992) examined how the recorded Rg phase was affected by the ripple firing technique. He suggested that it produced a signal merely slightly reduced in high-frequency content over that originating from a point source, provided the spatial extent of the entire experiment is small. His analysis incorporated a distance range of 50 km, similar to that involved for the recordings here.

Seismograms recorded on the short-period instruments at Station D08 have been rotated into the L, Q and T domain (Fig. 6) and are shown in Fig. 4. A clear, fundamental Rg phase is seen on the vertical- and radial-component seismograms arriving *c.* 26 s after the P-wave onset and, unusually, shows inverse dispersion (the short period arriving before the longer-period energy). The Rayleigh wave may be preceded by a possible higher mode.

A clearly defined Love wave, identified on the transverse component arriving *c.* 18 s after the P-wave onset, is a more complex phase possibly resulting from higher mode interference.

A combined record section (ignoring raypath azimuth variation) of the vertical-component seismograms from Event C (Fig. 7),

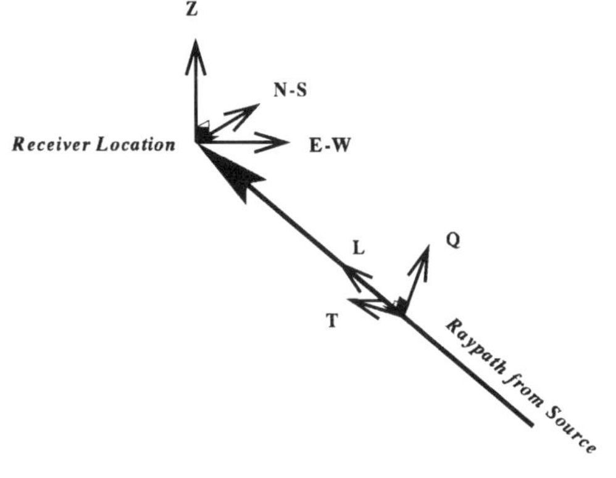

Fig. 6. Representation of the L, Q, T domain,

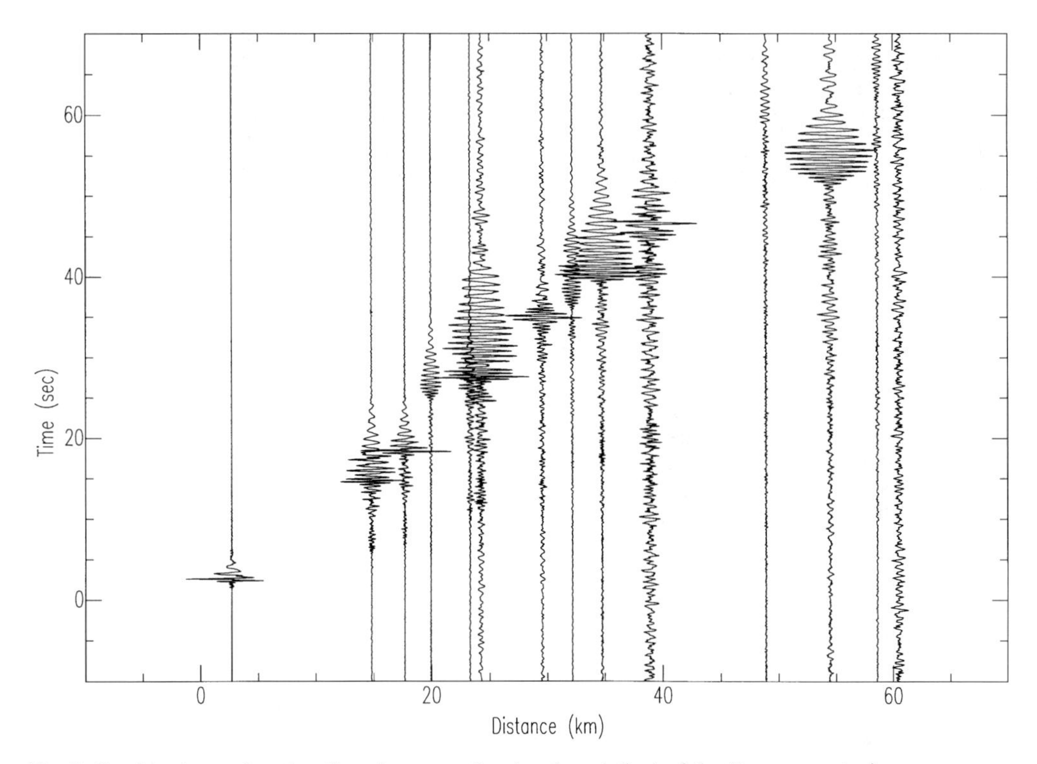

Fig. 7. Combined record section (ignoring ray-path azimuth variation) of the *Z*-component seismograms from Event C.

displayed at their calculated distance from the source, shows the Rayleigh wave propagating across the network to at least 55 km. There is an obvious variation in signal characteristic across the record section which results from the different recording azimuths. The actual distribution of the three event–station paths (Fig. 8) allows some identification of the variation in geology within the impact structure, identified from the surface-wave dispersion characteristics along those paths. It is this variation that is examined below.

The P-wave onsets have been picked for Event C to a distance of 35 km to provide a first arrival apparent velocity of *c.* 5 km s^{-1}. This almost certainly originates from material beneath the Tertiary post-impact fill, which may include melt breccia towards the centre of the crater and pre-impact platform rocks towards the margin as shown by the drill-core and well-log data (e.g. Hildebrand *et al.* 1991; Sharpton *et al.* 1996).

Data analysis

The interpretation of group velocity dispersion curves from single stations is improved if the precise location and time of the event is known, and the initial phase of each spectral component is defined at the source. In this work neither is known precisely, and before obtaining a refined velocity function that would allow optimum location of the events, no quantitative interpretation of the group velocity curves has yet been attempted. However, the variation in dispersion character over the array, derived from the three events studied, is significant and is discussed here.

Each of the vertical-component seismograms was deconvolved to remove the recording instrument effect. The short-period seismometers had been set, before deployment, to a 1 s natural period and this, together with a damping constant of 0.7, was assumed for the derivation of the deconvolution operator. Pole-zero descriptions of the broadband instruments, provided by the manufacturer, were used to derive their deconvolution operators.

Narrow-band filtered envelopes of the Rg phase, derived from the deconvolved seismograms windowed in velocity from 0.5 to 1.5 km s^{-1}, were produced to estimate the group velocities at specified periods using the technique of Dziewonski *et al.* (1969) via Burton & Blamey's

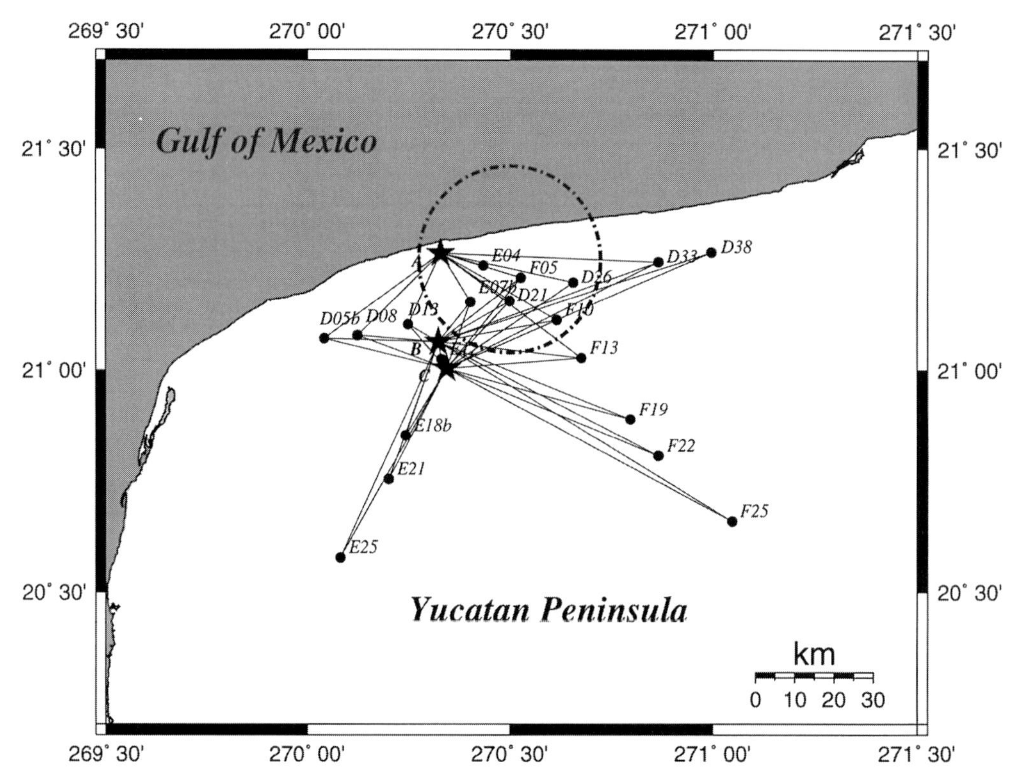

Fig. 8. Event–station connections for the three events used in the surface-wave dispersion analysis. The dashed line shows the approximate location of the boundary separating the dispersion curves into two types (see text for explanation).

(1972) TSAP code. Energy contour maps were produced in each case alongside a non-linear time plot of the associated seismogram, showing the dispersion characteristics of the Rg phase. Two examples demonstrate the two principal types of plot obtained (Fig. 9). The first type shows no clear decrease in group velocity with increasing period, whereas the second does, with velocity decreasing from initial values of *c.* 0.85 to 1.00 km s^{-1}. Two further examples of the second type are shown in Fig. 10a and b.

It is possible that the second type of curve originates from a high-velocity layer overlying a low-velocity layer, or it may simply highlight the minimum in group velocity with period displayed by Rayleigh waves for a simple two-layer elastic model with a higher-velocity lower layer (e.g. Ewing *et al.* 1957). Phase velocities are not associated with such a minimum. Therefore, to comment on the velocity–depth function, it is convenient to examine the interstation phase velocity dispersion curves. These also reduce the need for very accurate estimates of the location and timing of the event, especially, as in this study, when the resulting S-wave velocity–depth function between two stations is merely being semi-quantitatively determined, as shown below. To avoid azimuthal variations in the initial phase (Knopoff & Schwabb 1968) and to minimize or avoid lateral refraction effects (Evernden 1953, 1954) it is necessary that the source lies within a small azimuthal range of the interstation path. This preliminary analysis of the three events so far studied has revealed that only one event (Event C), recorded at two stations (E18 and E25) has provided a suitably dispersed Rayleigh wave with a suitable source–station geometry. The group velocity dispersion curves for these two stations are shown in Fig. 10a and b. The relevant seismograms were used to derive the interstation phase velocities at different periods using the cross-multiplication technique (Bloch & Hales 1968; Stuart *et al.* 1976) and incorporated in studies elsewhere (e.g. Clark & Stuart 1981). The resultant phase velocity dispersion curve is shown in Fig. 10c.

To obtain an estimate of the associated S-wave velocity depth–function, first the depth

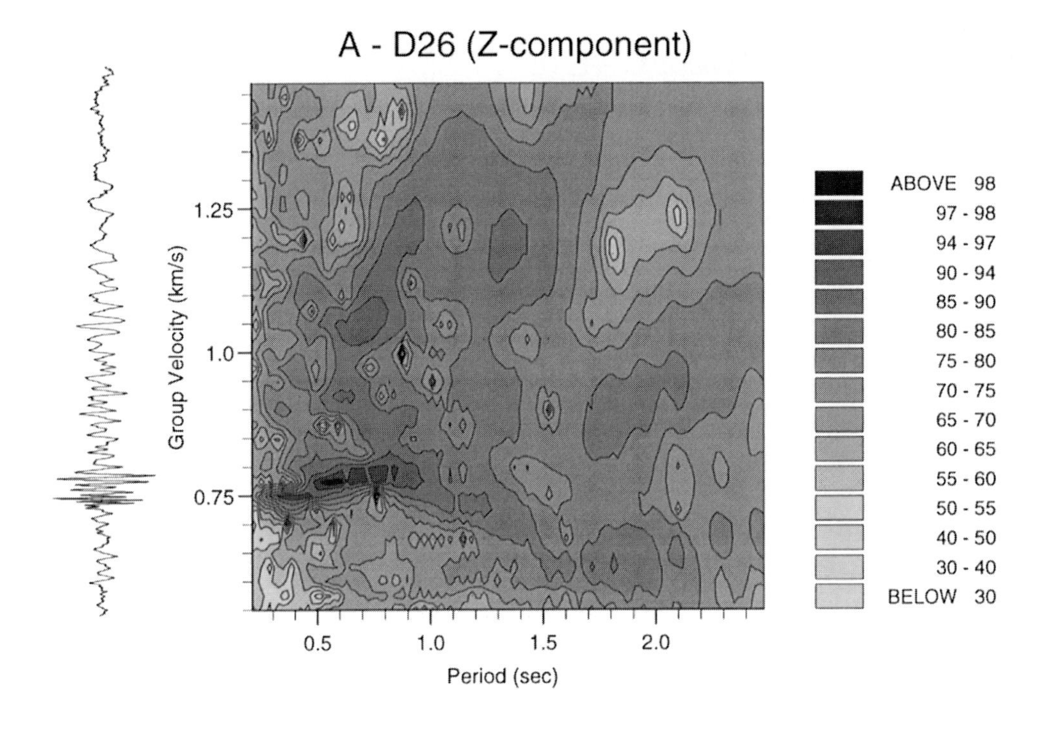

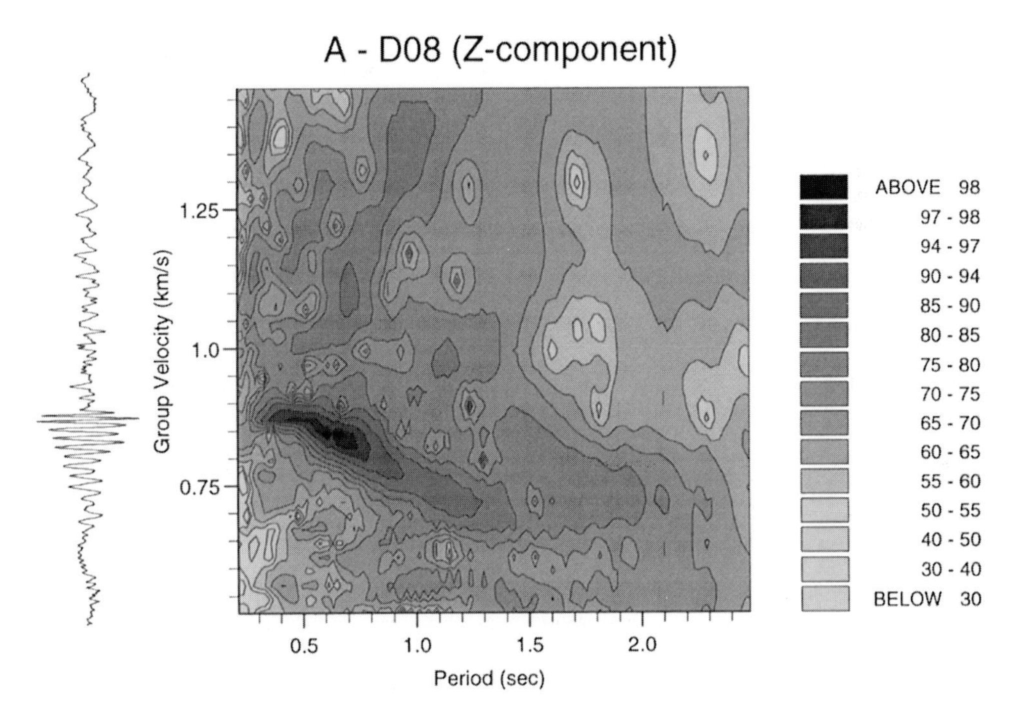

Fig. 9. Group velocity dispersion curves for Event A recorded at two stations (D08, D26). The non-linear plot of the seismogram is shown on the left. The contour shading shows the percentage energy in the narrow-band filtered seismograms at a range of periods normalized to the maximum energy in the plot.

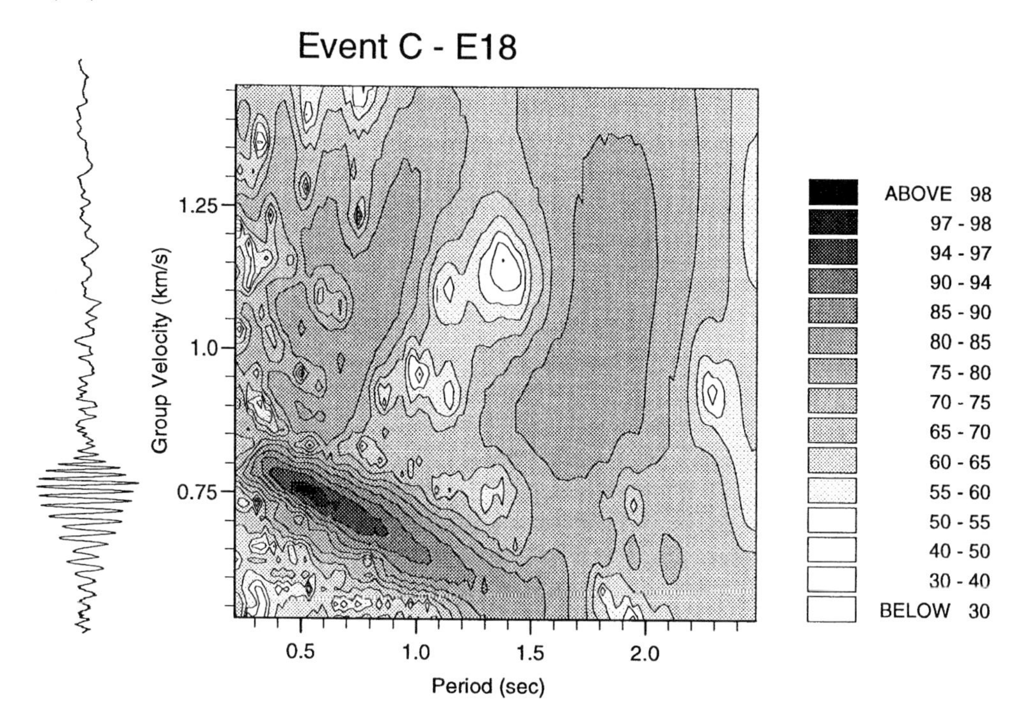

Event C - E18
Group Velocity (km/s)
1.25
1.0
0.75
0.5
1.0
1.5
2.0
Period (sec)
ABOVE 98
97 - 98
94 - 97
90 - 94
85 - 90
80 - 85
75 - 80
70 - 75
65 - 70
60 - 65
55 - 60
50 - 55
40 - 50
30 - 40
BELOW 30

(b)

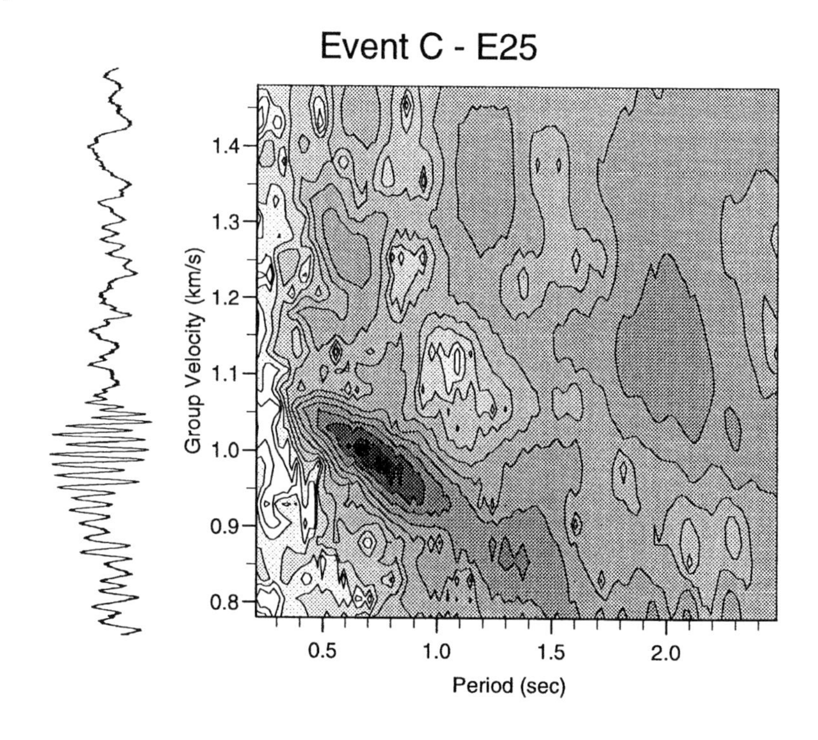

Event C - E25
Group Velocity (km/s)
1.4
1.3
1.2
1.1
1.0
0.9
0.8
0.5
1.0
1.5
2.0
Period (sec)

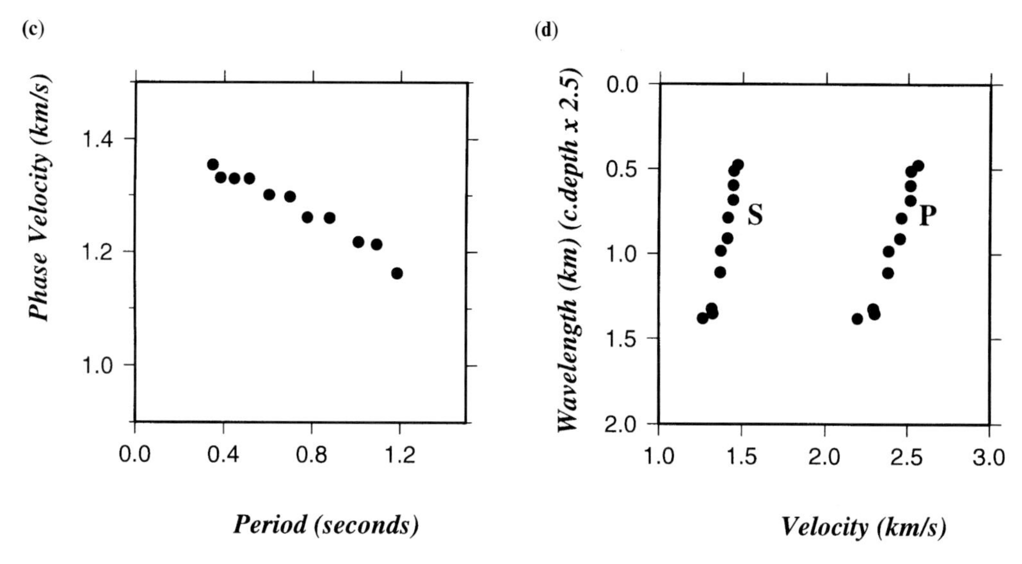

Fig. 10. (**a, b**) Group velocity dispersion curves for Event C into Stations E18 and E25. (See Fig. 9 for explanation.) (**c**) Phase velocity dispersion curve for Event C recorded at the same two stations. (**d**) Estimate of wavelength (c. 2.5 × depth) v. velocity derived for the material between Stations E18 and E25.

of the Rayleigh-wave nodal plane for a homogeneous solid (Poisson's ratio, $\sigma = 0.25$) is c. 0.2 times the signal wavelength, λ (Grant & West 1965), and an estimate of the penetration depth sampled by a particular wavelength may be taken as c. 0.4λ (Knopoff 1972). Second, the relationship between Rayleigh-wave and S-wave velocity has been assumed to be that for a homogeneous half-space (i.e. $V_R = 0.92 \times V_S$) (Fig. 10d). The P-wave velocity is shown assuming $V_P = 1.74 \times V_S$, which is considered reasonable for the Tertiary cover infilling the crater.

Discussion

From such a preliminary analysis it is not possible to define precisely a laterally varying velocity–depth function within the near-surface layers of the impact structure. However, two comments are relevant. First, from the coarse depth estimates determined and indicated in Fig. 10, the lithologies sampled by the Rayleigh waves almost certainly lie within the Tertiary section. Second, there is a bimodal distribution of velocity–depth function, with a low-velocity layer identified beneath the surface lithologies towards the margin, and no such low-velocity layer identified beneath the centre of the crater. The approximate demarcation between the two zones is shown in Fig. 8.

The Tertiary section derived from well data across the crater consists of Paleocene to Pleistocene limestones and marls beneath occasionally dolomitized and partly oolitic limestones (Lopez-Ramos 1973; Sharpton 1996). The division of the region into the two velocity–depth zones discussed previously may relate to the depositional environment, which itself will have been controlled by the final morphology of the crater basin shortly after the impact. The crater included a central peak ring block surrounded by a deeper floor, likely to have shallowed towards the crater rim. The velocity inversion towards the margin of the crater, derived from the surface-wave dispersion study, may result from pelagic sediments beneath shallow water limestones deposited sequentially as the crater basin was infilled. Towards the centre, the lack of a velocity inversion could be due to shallower water deposition on top of the central peak ring block resulting in a less well-defined low-velocity layer. This would be consistent with the thicker Tertiary sediments seen in wells Y-6 and S-1 (Fig. 1) compared with those in C-1 over the centre of the crater.

Teleseismic receiver function analysis

Teleseismic body waves have been used intensively for a number of years to investigate crustal structure using the crustal transfer method (e.g. Phinney 1964). Improvements have been made to the technique, which is now commonly referred to as the receiver function method. A teleseismic event occurring in Peru on 21 February 1996 (epicentral distance 32°; magnitude $7.3M_L$) was recorded at 14 stations within the array. Receiver functions, derived from this event, have been calculated and analysed as a first stage in the complete study.

The method

Receiver functions have been generated following the method of Kind *et al.* (1995).

- The theoretical 1 Hz seismometer response is deconvolved from the data (Fig. 11a) producing band limited (20 s; 50 Hz) displacement seismograms (Fig. 11b). This restitution is carried out in the time domain, and comparison with the broadband seismograms of the same event recorded on the 30 s Guralp CMG-40T instruments shows good correlation.
- The back-azimuth and angle of incidence of the direct P phase are determined by diagonalizing the coherence matrix of the *Z*, N–S and E–W component traces, windowed to 30 s around the direct P arrival. The displacement seismograms are then rotated into the L, Q, T coordinate system (Fig. 11c).
- The L component is assumed to consist of a compressional wave with a coda caused by source and intermediate path effects. This is deconvolved from the Q and T components (Fig. 11d). The deconvolution is carried out in the time domain with an inverse filter generated by minimizing the least-squares difference between the deconvolved L component and a desired delta-like spike (Berkhout 1977). This method preserves the original amplitude ratios in the data.

The deconvolved Q component is assumed to represent P–SV conversions at impedance discontinuities in the lithosphere and is referred to as the 'receiver function'. The T component provides an indication of anisotropy or lateral heterogeneity, which for a simple 1D velocity

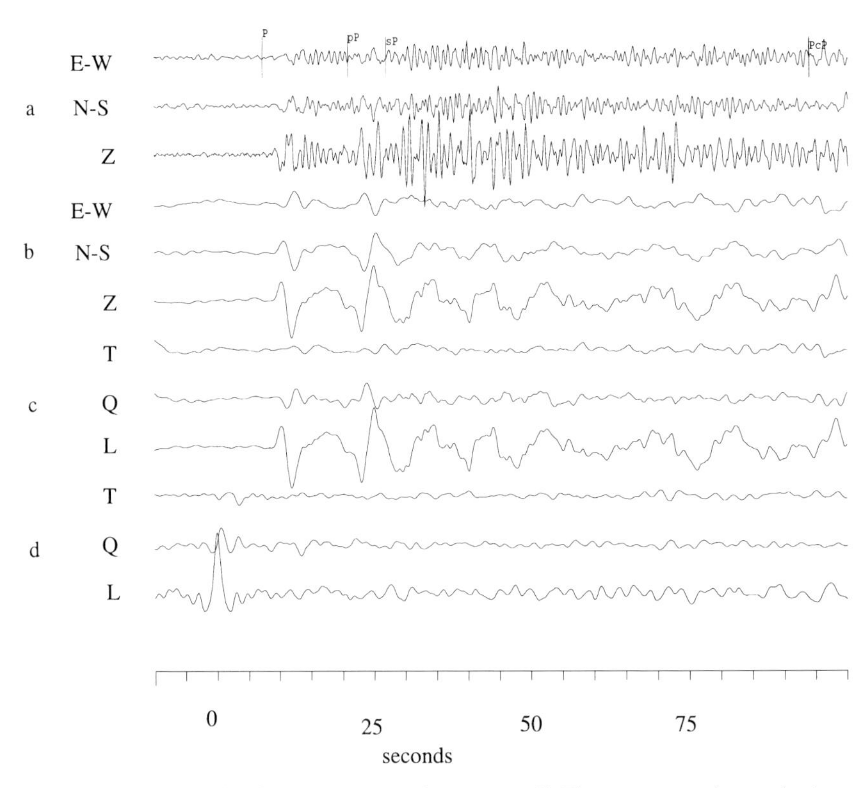

Fig. 11. (**a**) Short-period velocity three-component seismograms. (**b**) The same traces deconvolved to remove the instrument response and converted to displacement seismograms. (**c**) The displacement traces rotated into the L, Q, T domain. (**d**) The rotated displacement traces with the L component deconvolved.

structure would be zero. In situations where receiver functions have been calculated from a wide range of azimuths and distances, it is possible to attempt modelling of lateral heterogeneities. However, owing to the short recording period of this study and resulting small dataset, no such attempt can be made.

The measured receiver function can be compared with synthetic data generated from a simple, 1D velocity structure. Synthetic seismograms are converted to synthetic receiver functions by the same process of rotation and deconvolution as for the measured data. Any suitable method can be used for generating the synthetic seismograms. However, Kind *et al.* (1995) have shown that a simple plane wave method gives results comparable with those of more complete and computationally more intensive, full waveform, reflectivity methods. An inversion method can subsequently be applied, producing a best-fit theoretical model derived from the observed data. As with many geophysical problems this is a non-unique process, the final solution being strongly dependent on the *a priori* derived starting model; the numerical procedure in effect is used merely to improve the fit of the data. In this first study, a linearized inversion method described by Kind *et al.* (1995) is attempted, in which a stack of variable velocity, fixed thickness layers is inverted to produce an optimum velocity structure. The starting model consisted of a low-velocity sedimentary layer above a stack of uniform velocity layers. The quality of the fit in this preliminary work is shown graphically, together with the starting and final 1D velocity–depth functions.

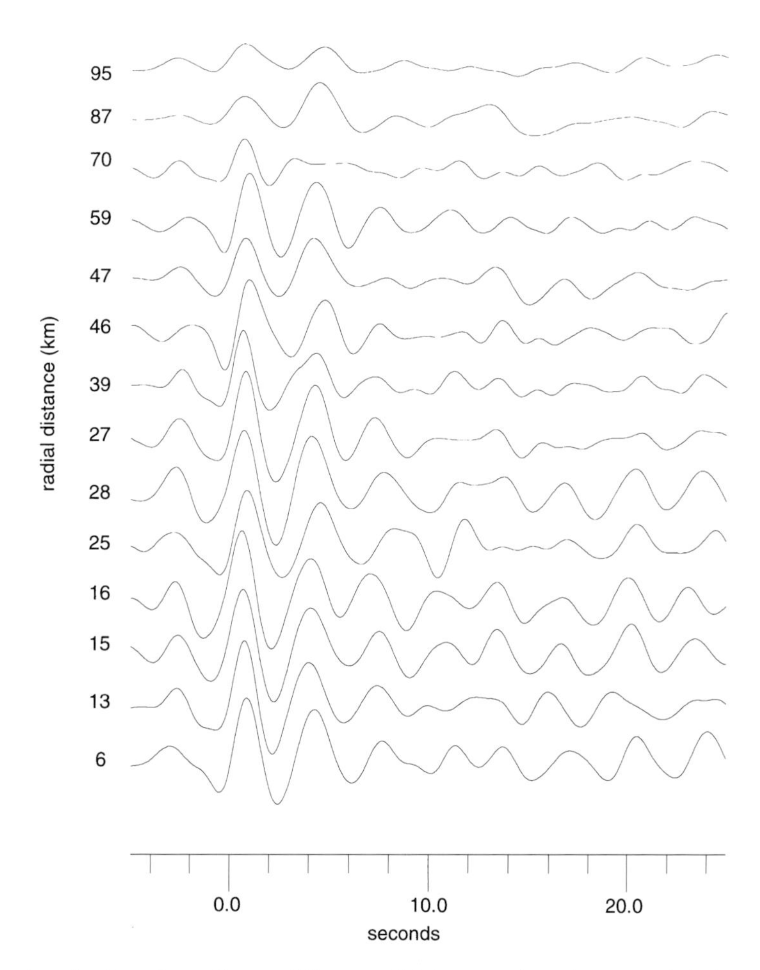

Fig. 12. Receiver functions for stations at different radial distances from the centre of the impact structure.

Results

The generated receiver functions show a dominant P–SV conversion close to the surface, identified in the high-amplitude phase occurring *c.* 1 s after the first arrival (Fig. 12). This is presumed to arise from the base Tertiary discontinuity. Figure 12 shows that there is a strong correlation between the form of the receiver function and distance (independent of azimuth) from the centre of the impact structure, implying that the structure has some radial symmetry. The data appear to fall into three groups, 0–40 km, 40–80 km and 80–100 km, the initial sediment conversion being stronger towards the centre. Also, there appears to be a slight delay in the base Tertiary conversion arrival time away from Chicxulub Puerto consistent with the possible deepening of the crater floor away from the impact structure's centre.

The results from the inversion study simply highlight the features seen qualitatively in the receiver functions (Fig. 13). The sedimentary layer conversion and multiples within the sedimentary layer are dominant, their maximum energy occurring towards the centre of the structure. This suggests a higher impedance contrast for this region as indicated by the velocity peak at 2–10 km depth for the 0–40 km radial distance model (Fig. 13). There is an indication of a Moho conversion at about 30 km depth at stations out to 80 km radial distance.

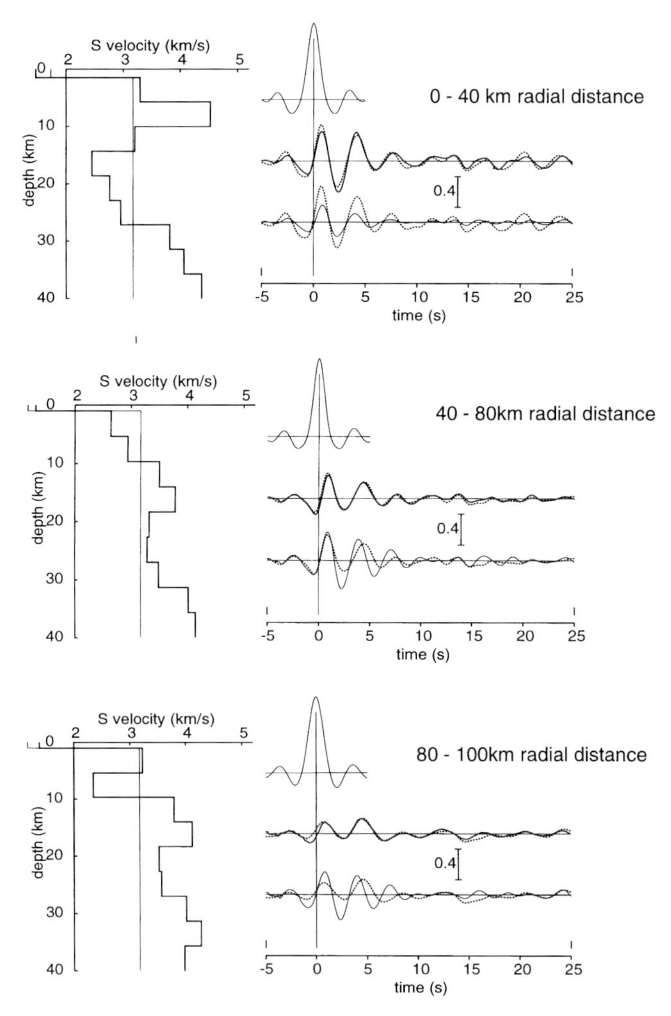

Fig. 13. Receiver function modelling. The fine line represents the starting model. The bold line represents the optimized model. The dashed line represents the 'observed' receiver function.

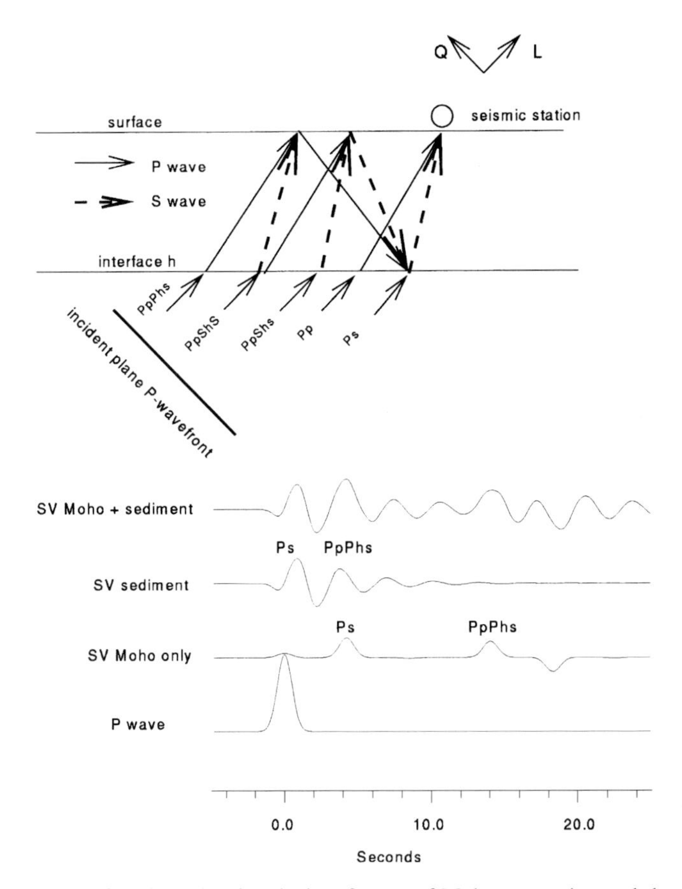

Fig. 14. Synthetic receiver functions showing the interference of Moho conversion and the sediment multiple.

However, synthetic studies show that for a reasonable crustal velocity function, the conversion from a 30 km depth Moho coincides approximately with a PpPhs multiple from the base of a 1 km sedimentary layer (Fig. 14). Thus for any realistic Earth model, the Moho depth beneath the impact structure is, at present, poorly resolved using this method. It is intended that an upper-crustal velocity structure derived from the normal incidence reflection profiling carried out as part of the Chicxulub seismic experiment will better constrain the starting models used in further studies.

Conclusions

This research is currently at a very early stage. However, the following results can be reported here:

- A total of 15 teleseismic, 75 regional and *c.* 100 local events were recorded on the Chicxulub passive seismic array, which consisted of 20 short-period and broadband stations deployed over the impact structure in the Yucatan for a period of *c.* 100 days.
- Location of 30 local events shows that many appear to have originated from a number of local quarries within the array.
- Analysis of the quarry blast Rg phase demonstrates a bimodal velocity distribution within the Tertiary sediments: a low-velocity layer is identified beneath the surface lithology towards the margin of the crater, whereas no such low-velocity layer is identified towards the impact structure's centre. This may relate to varying depositional environments within the basin after the impact, resulting from the variable depth to crater floor.
- Derivation of teleseismic receiver functions across the array from an event originating in Peru shows that there is a strong correlation between the form of the function and distance (independent of azimuth) from the

centre of the structure, demonstrating a radial symmetry beneath the array.

- The first Moho conversion seen in the receiver functions occurs at almost exactly the same time as the surface layer sediment multiple, thus apparently restricting possible modelling of Moho topography and its influence on the receiver functions.
- It is expected that analytical modelling of the surface wave dispersion curves, together with interpretation of wide-angle data recorded on the deep seismic reflection onshore array, will provide a refined upper-crustal velocity model beneath the Chicxulub array. This may then be used in the receiver function analysis to model the velocity structure at depth with greater confidence.

Future research will also focus on the upper-crustal anisotropy of the region through a study of the shear-wave polarizations across the structure. Current theories suggest that the principal cause of shear-wave splitting in the upper crust may be extensive dilatancy anisotropy (EDA) where the anisotropy is caused by the alignment of intergranular micro-cracks and pore spaces (Crampin & Lovell 1991). This alignment is a result of present-day stress and is very compliant to even small changes in local conditions. Therefore if EDA is the primary cause, a near-uniform polarization orientation across the structure might be expected. There is a possibility, however, that mineralogical orientations, fractures and shear fabrics associated with the impact will still influence the anisotropy, producing a radial pattern of polarization directions across the crater.

This research was funded by a fieldwork grant from the Royal Society of London, and supported by the NERC grant to BIRPS and Imperial College for the Chicxulub seismic project. One of us (G.D.M) was financed by NERC research studentship GT4/95/156/E. We thank all members of the Chicxulub Working Group and, in particular, A. Brisbourne of Leeds University, for their significant contribution to this research. We also thank the NERC Geophysical Equipment Pool, GeoForschung Zentrum Potsdam and the British Antarctic Survey for the loan of equipment.

References

BERKHOUT, A. J. 1977. Least square inverse filtering and wavelet deconvolution. *Geophysics*, **42**, 1369–1383.

BLOCH, S. & HALES, A. L. 1968. New techniques for the determination of surface wave phase velocities. *Bulletin of the Seismological Society of America*, **58**, 1021–1034.

BURTON, P. W. & BLAMEY, C. 1972. *A computer program to determine the spectrum and a dispersion characteristic of a transient seismic signal.* UKAEA AWRE Report **0/48/72**.

CAMARGO, Z. A. & SUAREZ, R. G. 1994. Evidencia sismica del crater de impacto de Chicxulub. *Boletin de la Asociacion Mexicana de Geofisicos de Exploracion*, **34**, 1–28.

CLARK, R. A. & STUART, G. W. 1981. Upper mantle structure of the British Isles from Rayleigh wave dispersion. *Geophysical Journal of the Royal Astronomical Society*, **67**(1), 59–75.

CONNORS, M., HILDEBRAND, A. R., PILKINGTON, M., *et al.* 1996. Yucatan karst features and the size of Chicxulub crater. *Geophysical Journal International*, **127**, F11–F14.

CRAMPIN, S. & LOVELL, J. H. 1991. A decade of shear-wave splitting in the Earth's crust: What does it mean? What use can we make of it? And what should we do next? *Geophysical Journal International*, **107**, 387–407.

DZIEWONSKI, A. M., BLOCK, S. & LANDISMAN, M. 1969. A technique for the analysis of transient seismic signals. *Bulletin of the Seismological Society of America*, **59**, 427–444.

ESPINDOLA, J. M., MENA, M., DE LA FUENTE, M. & CAMPOS-ENRIQUEZ, J. O. 1995. A model of the Chicxulub impact structure (Yucatan, Mexico) based on its gravity and magnetic signatures. *Physics of the Earth and Planetary Interiors*, **92**, 271–278.

EVERNDEN, J. E. 1953. Direction of approach of Rayleigh waves and related problems, part 1. *Bulletin of the Seismological Society of America*, **43**, 225–274.

——1954. Direction of approach of Rayleigh waves and related problems, part 2. *Bulletin of the Seismological Society of America*, **44**, 159–184.

EWING, W. M., JARDETZKY, W. S. & PRESS, F. 1957. *Elastic Waves in Layered Media.* McGraw-Hill, New York.

GRANT, F. S. & WEST, G. F. 1965. *Interpretation Theory in Applied Geophysics*. McGraw-Hill, New York.

HILDEBRAND, A. R., PENFIELD, G. T., KRING, D. A., PILKINGTON, M., CAMARGO, Z. A., JACOBSEN, S. B. & BOYNTON, W. V. 1991. Chicxulub crater: a possible Cretaceous/Tertiary boundary impact crater on the Yucatan peninsula, Mexico. *Geology*, **19**, 867–871.

——, PILKINGTON, M., CONNORS, M., ORTIZ-ALEMAN, C. & CHAVEZ, R. E. 1995. Size and structure of the Chicxulub crater revealed by horizontal gravity gradients and cenotes. *Nature*, **376**, 415–417.

KIND, R., KOSAREV, G. L. & PETERSON, N. V. 1995. Receiver functions at the stations of the German Regional Seismic Network (GRSN). *Geophysical Journal International*, **121**, 191–202.

KLEIN, F. W. 1989. *User's guide to HYPOINVERSE, a program for VAX computers to solve for earthquake locations and magnitudes.* US Geological Survey Open File Report **89-314**.

KNOPOFF, L. 1972. Observations and inversion of surface wave dispersion. *In*: RITSEMA, A. R. (ed.) *The Upper Mantle. Tectonophysics*, **13**(1–4), 497–519.

—— & SCHWAB, F. A. 1968. Apparent initial phase of a source of Rayleigh waves. *Journal of Geophysical Research*, **73**, 755–760.

LOPEZ-RAMOS, E. 1975. Geological summary of the Yucatan peninsula. *In*: NAIRN, A. E. M. & STEHLI, F. G. (eds) *The Ocean Basins and Margins, Volume 3: the Gulf of Mexico and the Caribbean*, 257–282.

PERRY, E., MARIN, L., MCCLAIN, J. & VELAZQUEZ, G. 1995. Ring of cenotes (sinkholes), northwest Yucatan, Mexico: its hydrogeologic characteristics and possible association with the Chicxulub impact crater. *Geology*, **23**, 17–20.

PHINNEY, R. A. 1964. Structure of the Earth's crust from spectral behavior of long-period body waves. *Journal of Geophysical Research*, **69**, 2997–3017.

PILKINGTON, M. & HILDEBRAND, A. R. 1994. Gravity and magnetic field modeling and structure of the Chicxulub crater, Mexico. *Journal of Geophysical Research*, **99**(E6), 13147–13162.

SAIKIA, C. K. 1992. Numerical study of quarry generated Rg as a discriminant for earthquakes and explosions: modeling of Rg in southwestern New England. *Journal of Geophysical Research*, **97 (B7)**, 11 057–11 072.

——, KAFKA, A. L., GNEWUCH, S. C. & MCTIGUE, J. W. 1990. Shear velocity and intrinsic Q structure of the shallow crust in southeastern New England from Rg wave dispersion. *Journal of Geophysical Research*, **95**, 8527–8541.

SHARPTON, V. L., MARIN, L. E., CARNEY, C., *et al.* 1996. A model of the Chicxulub impact basin based on evaluation of geophysical data, well logs, and drill core samples. *In*: RYDER, G., FASTOVSKY, D. & GARTNER, S. (eds) *The Cretaceous–Tertiary Event and other Catastrophes in Earth History*. Geological Society of America, Special Paper, **307**, 55–74.

STUART, G. W., DOUGLAS, A. & BLAMEY, C. 1976. *A computer program for the determination of the phase velocity of seismic surface waves between pairs of stations*. AWRE Report **8/76**.

Localized shock- and friction-induced melting in response to hypervelocity impact

JOHN G. SPRAY

Department of Geology, University of New Brunswick, Bailey Drive, Fredericton, N.B. E3B 5A3, Canada

Abstract: The distribution of shock veins, friction melts and cataclastic rocks in complex impact craters reflects the response of target lithologies to varying rates of strain. Within the Sudbury impact structure, thin (<2 mm), anastomosing veins, which can define shatter cone surfaces, permeate the target rocks in a *c.* 15 km wide zone around the Sudbury igneous complex (SIC). A similar relationship exists within the Vredefort impact structure, with the additional association of the high-pressure SiO_2 polymorphs coesite and stishovite. These S- (shock-dominant) type pseudotachylytes are comparable with shock veins developed in meteorites. Both are considered to form primarily by shock compression–decompression during the contact and compression stage of the collision process. Larger, thicker (up to 1 km wide) friction melt bodies constitute pseudotachylytes formed by the extreme comminution of fault walls during rebound and gravitational collapse of the transient cavity. They also appear to define the concentric fault systems of multi-ring impact basins. These E- (endogenic-) type pseudotachylytes form during the modification stage of the cratering process and post-date S-type pseudotachylytes. At Sudbury, they occur up to 80 km beyond the SIC. E-type pseudotachylytes are formed by the same mechanism as pseudotachylytes in non-impact-related fault systems: frictional melting of fault walls during seismogenic slip. Those in impact structures can be large because the extreme displacements (several kilometres) facilitated by superfaults generate massive volumes of friction melt.

Geological materials can undergo fusion and vaporization as a result of high rates of deformation. The main scenarios for attaining these conditions by natural processes are frictional melting during high-velocity slip (e.g. coseismic faulting), and melting and vaporization during hypervelocity collision (e.g. meteorite impact). Evidence for the attainment of these states comes from theoretical considerations of energy transformation (e.g. Zel'dovich & Raizer 1967; Melosh 1989) and by examination of the geological products via field observations and microscopy.

Geologists are most familiar with processes occurring at strain rates $\ll 10^{-8}\,s^{-1}$ within the isothermal regime, for example, steady-state aseismic creep within the mantle (about $10^{-14}\,s^{-1}$) and localized shear within a mylonite zone (about $10^{-11}\,s^{-1}$). However, for the purposes of understanding shock melting and frictional melting, we must look to strain rates $>10^{-2}\,s^{-1}$ within the adiabatic regime (e.g. coseismic slip at *c.* 10^{-2} to $1\,s^{-1}$, and meteorite impact at $>10^{6}\,s^{-1}$). Figure 1 shows a simplified depiction of strain rate regimes and products for strain rates of 10^{-4}–10^{14} and above. The highest strain rates equate with bulk target melting, vaporization and plasma generation. This can lead to the formation of impact melt sheets, locally developed fallback breccias (suevites) and globally dispersed ejecta layers. This work focuses on more localized shock- and friction-induced melting processes in the sub-melt sheet autochthonous to parautochthonous target material of impact structures.

Pseudotachylytes

The term pseudotachylyte was first used by Shand (1916) to describe the dark, aphanitic to typically fine-grained, dyke-like bodies seen in the Vredefort Dome of South Africa. Vredefort is now accepted as one of Earth's largest impact structures (Dietz 1961; Martini 1978; Spray *et al.* 1995; Koeberl *et al.* 1996; Therriault *et al.* 1997). Since Shand's pioneering work, pseudotachylytes have been described from myriad fault systems of endogenic origin which are demonstrably unrelated to the impact process. The association of pseudotachylytes with endogenic fault systems is commonplace, and geologists can be forgiven for believing that this affiliation, rather than with impact structures, is the type setting for the rock. Significantly, pseudotachylyte remains one of the few lithologies indicative of seismogenic activity. Under normal circumstances, and excluding deep earthquakes generated in Wadati–Benioff zones, endogenic pseudotachylytes are generated via seismogenic faulting in the brittle, upper 10–15 km of the crust. However, they can be developed at deeper levels if high enough strain rates are imposed on what would otherwise be a

SPRAY, J. G. 1998. Localized shock- and friction-induced melting in response to hypervelocity impact. *In*: GRADY, M. M., HUTCHISON, R., MCCALL, G. J. H. & ROTHERY, D. A. (eds) *Meteorites: Flux with Time and Impact Effects*. Geological Society, London, Special Publications, **140**, 195–204.

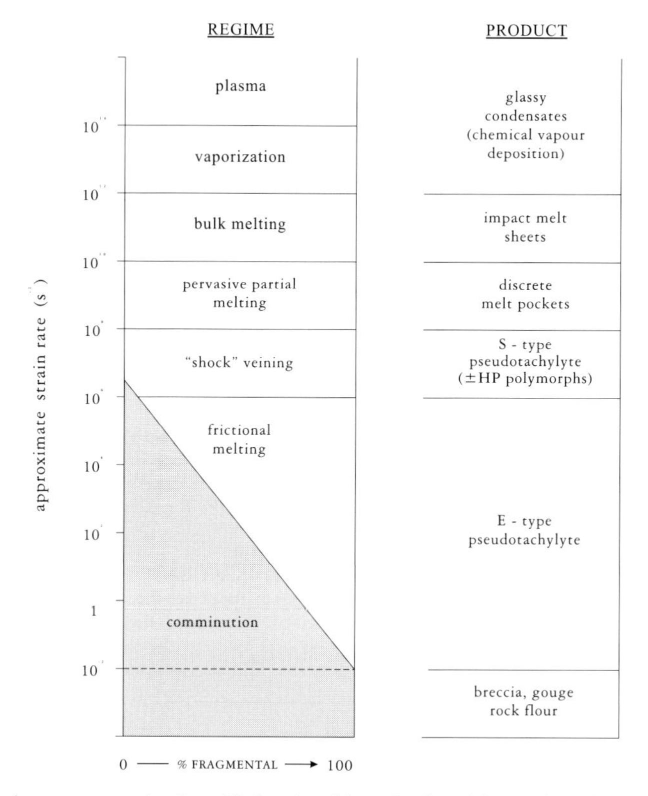

Fig. 1. Relations between comminution, frictional melting, shock melting and strain rate for high to ultrahigh rates of strain. Strain rate = v/α (per s), where v is velocity and α is slip-zone thickness.

ductile regime (Hobbs *et al.* 1986; Austrheim & Boundy 1994; White 1996).

It is in the impact setting that the more spectacular examples of pseudotachylyte are developed. Pseudotachylyte dykes commonly reach 5 m thickness at Vredefort. The largest known pseudotachylyte occurs in the Sudbury impact structure of Canada (Scott *et al.* 1996). This single, giant pseudotachylyte body (referred to locally as the Frood, or South Range, Breccia Belt) is *c.* 45 km long and up to 1 km wide. The pseudotachylyte is host to one of the world's largest Cu–Ni sulphide deposits (the Frood–Stobie ore body). The size of the larger impact-related pseudotachylytes is in marked contrast to the centimetre-wide pseudotachylytes typical of crustal fault zones of non-impact origin.

Much of the field-based literature on impact structures classifies the different forms of damaged rock as 'breccia', and pseudotachylytes have been included in this scheme (e.g. Reimold 1995). In most cases, use of the term 'breccia' for pseudotachylyte is inappropriate because the inclusions (mineral and/or rock fragments) are typically subrounded and not angular. Sibson (1975) suggested the term 'quasiconglomerate', but even this has sedimentary connotations. True breccias may occur in the footwall immediately below the melt sheet floor in many impact structures (e.g. Lakomy 1990) and, most characteristically, as fallback where they constitute suevites (e.g. von Engelhardt *et al.* 1995). Locally, footwall breccias may even extend some distance into basement lithologies via fracture and fault systems. The term pseudotachylyte should not be confused with these forms of breccia.

There has been a tendency in the literature to consider endogenic and impact-related pseudotachylytes as distinct, with the former supposedly being of frictional origin and the latter being shock generated (e.g. French *et al.* 1989; Reimold 1995). Consequently, some confusion exists as to the definition and origin of the rock type. To clarify the terminology it is important to understand how pseudotachylyte forms.

Pseudotachylytes consist of two main components:

- inherited rock and mineral clasts derived from the wall rocks. In intermediate to acid, quartzo-feldspathic lithologies typical of crystalline basement, the clasts commonly consist of quartz and feldspar, or rock fragments;

- a more basic matrix, which has undergone preferential comminution and which has probably undergone frictional melting. This matrix may have been supercooled to a glass or, more commonly, it will have crystallized to yield a microcrystalline to fine-grained igneous texture.

A given pseudotachylyte therefore possesses a bulk composition made up of both clasts and matrix which, when taken together, are of a composition usually close to that of the host rocks. It should be borne in mind, however, that, in the case of impact structures, there is evidence for the preferential focusing of fracturing, brecciation and faulting at lithological contacts, especially between rocks with contrasting physical properties (e.g. basic dykes within quartzo-feldpathic gneisses). Density contrasts are particularly important because shock-wave velocity is dependent on the density of the medium through which it passes (e.g. Boslough & Asay, 1993). Sudden changes in shock-wave velocity may induce localized fracturing or faulting if the wave encounters the contact at anything other than right angles. This syn-impact damage may subsequently facilitate post-impact faulting during crater collapse along the preformed planes of weakness. In such cases, the bulk pseudotachylyte composition may represent a blend of contrasting lithologies, as has been found for certain pseudotachylytes at Sudbury (Thompson & Spray 1996).

A contentious aspect of pseudotachylytes has been whether they are essentially cataclastic or melt derived (e.g. Wenk 1978; Maddock 1983). This controversy is unfounded because comminution ultimately leads to melting, as has been demonstrated in the laboratory during high-speed slip experiments using friction welding apparatus (Spray 1993, 1995). Whether a given pseudotachylyte is essentially cataclastic or is derived by frictional melting depends on the prevailing strain rate and the amount of displacement on the slip surface. A continuum exists between these end-members (Fig. 1). Because of the typically fine-grained to cryptocrystalline nature of the matrix to pseudotachylytes, it is usually difficult to establish in the field whether the matrix was once molten or has remained cataclastic. In the absence of a microscope, if not an electron microscope, such a distinction cannot be made at the macroscopic scale. This difficulty is usually compounded by the effects of alteration on the matrix (e.g. devitrification, metamorphic recrystallization). Figure 1 shows the relationship between comminution and friction melting and, ultimately, melting by shock. There exist certain lower-pressure features associated with shock that would occupy a field between friction-dominant pseudotachylytes and shock veins as depicted in Fig. 1. These include, under increasing shock pressures, planar microstructures (planar fractures and planar deformation features) and mosaicism (Stöffler & Langenhorst 1994; Grieve *et al.* 1996). Distinctions can be made between comminution (cataclasis), friction melting and shock melting in terms of the mechanisms at work, and these will be discussed in turn.

Comminution

The progressive comminution of a rock during slip between two abrading surfaces normally leads to a power-law particle size distribution possessing self-similarity (Turcotte 1986). It is possible to test this at the microscopic scale by measuring grain-size distribution. This has been done for a natural fault gouge, which was indeed found to possess a fractal grain-size distribution (e.g. Sammis *et al.* 1986). However, in the case of pseudotachylyte examined from the Musgrave Range of central Australia (Shimamoto & Nagahama 1992), it was revealed that the estimated amount of ultrafine clasts was smaller than the true area of the fine-grained matrix by about one order of magnitude. This was attributed to matrix melting and the subsequent formation of relatively coarser-grained igneous crystallites from the melt phase. Thus, frictional melting destroys a power-law size distribution in the smaller size fraction of comminuted rock, as a result of the smaller grains being preferentially melted. Careful analysis of pseudotachylyte microtextures should allow for discrimination between purely cataclastic and frictionally melted matrices. Scanning electron microscopy may well be required to achieve this. In the absence of frictional melting, true cataclastic rocks prevail. Within the Sudbury impact structure, truly cataclastic veins and dykes are not common. Where present, they occur towards the periphery of the structure. If fluids are available to transport comminuted material, then tuffisites may form. In the type area of the Schwäbische Alb of southern Germany (Cloos 1941), the propellant for the basaltic tuffaceous medium was volcanic gas. At Sudbury there appears little evidence for the existence of tuffisites.

Studies of pseudotachylytes examined from the Vredefort and Sudbury impact structures, as well as non-impact-related pseudotachylytes from fault zones, indicate that their matrices were derived by some degree of melting (Sibson 1975; Maddock 1983; Magloughlin 1992; Spray 1993). This is not to say that comminution and melting are mutually exclusive. Rather, at higher strain rates and/or large displacements, the rate of comminution within the bulk layer decreases with the onset of shear localization (Marone & Scholz 1989) and frictional melting. Comminution is thus a prerequisite for melting by friction (Spray 1995).

Frictional melting mechanisms for rocks and minerals

A key advance in understanding how frictional melting proceeds in geological materials has been made through the microanalysis of pseudotachylytes. Before the advent and use of the electron microprobe, and especially the analytical scanning electron microscope, it proved impossible to analyse matrix and clast compositions separately in pseudotachylytes. Attempts to physically separate clasts from matrices have proved unsuccessful, although the removal of visible clasts by hand from crushed samples does yield separates closer to the true matrix composition than to the bulk rock composition (Sibson 1975; Dressler 1984). Conversely, defocused-beam and especially rastering techniques using the electron microprobe allow for the distinction to be made between clasts and matrix *in situ* and at high resolution. The results show that the matrices of pseudotachylytes are typically more basic (less siliceous) than their bulk compositions (Schwarzman *et al.* 1983; Reimold 1991; Spray 1993; Killick 1994; Thompson & Spray 1996). This can be explained by the preferential incorporation of mafic mineral phases into the matrix during the comminution process.

Although it may seem contrary that the more basic components of a given rock undergo preferential comminution and melt first, the pathway to frictional melting is distinct from normal igneous melting. Under the high strain rate conditions of seismic slip, the controlling parameter in comminution and frictional melting is fracture toughness. This determines which minerals are more easily comminuted to produce an ultrafine matrix and which minerals survive as the larger, intact clasts. Fracture toughness data are limited, but there exists a hierarchy of susceptibility to comminution and, ultimately, frictional melting (Spray 1992). The sequence, in order of decreasing melting susceptibility, is: phyllosilicates > inosilicates (amphiboles > pyroxenes) > tectosilicates (orthoclase > plagioclase > quartz) > orthosilicates. This succession broadly follows Mohs' hardness scale M, which can be related to the indentation hardness, or true hardness H, and hence yield strength σ_y and shear yield strength k through

$$(nM)^3 \approx H \approx 3\sigma_y \approx 6k$$

where n ranges between 1.3 and 1.6, except for the extremes of the scale (Tabor 1954). Microscopic observations confirm this sequence, with quartz characteristically forming the resistant clasts in many pseudotachylytes. This sequence is compatible with a mechanism of frictional melting controlled by the mechanical behaviour of minerals during comminution. The mechanically weakest minerals thus become the energy sinks during deformation. Ultimately, the mechanically weaker minerals undergo the greatest comminution and so heat up to their melting points, as a result of the focusing of elastic, plastic and brittle deformation processes. This is corroborated by an evaluation of the efficiency of industrial ball mill comminution made by Beke (1964). It that study it was found that virtually all the energy consumed in grinding was converted to heat (*c.* 97%), with only a small fraction (<1%) actually contributing to fracture and comminution. In rocks, many of the less tough minerals (e.g. micas and amphiboles) are also hydrous. Release of their chemically bound water during comminution results in a lowering of melting points and melt viscosities within the matrix (Spray 1993). This can render pseudotachylytes very effective fault lubricants, even for acid to intermediate wall rocks whose bulk molten equivalents would normally be considered highly viscous.

Fig. 2. Photomicrographs of shock veins (S-type pseudotachylytes) in meteorites and from the Vredefort impact structure. (**a**) Shock vein in the Chantonnay olivine–hypersthene L6 chondrite. (Note deformation of olivine blades and apparent offset either side of vein.) Field of view 2.5 mm, cross-polars. (**b**) Ringwoodite (lighter phases surrounded by black phase) in shock vein from the Catherwood L6 chondrite. Field of view 0.6 mm, cross-polars. (**c**) Shock vein in Kimberley–Elsburg quartzite, Central Rand Group from the Vredefort impact structure of South Africa. The shock vein contains the high-pressure polymorphs coesite and stishovite (Martini 1978) and has been dated at 2018 ± 14 Ma by the laser fusion $^{40}Ar/^{39}Ar$ technique by Spray *et al.* (1995). Field of view 0.95 mm, cross-polars.

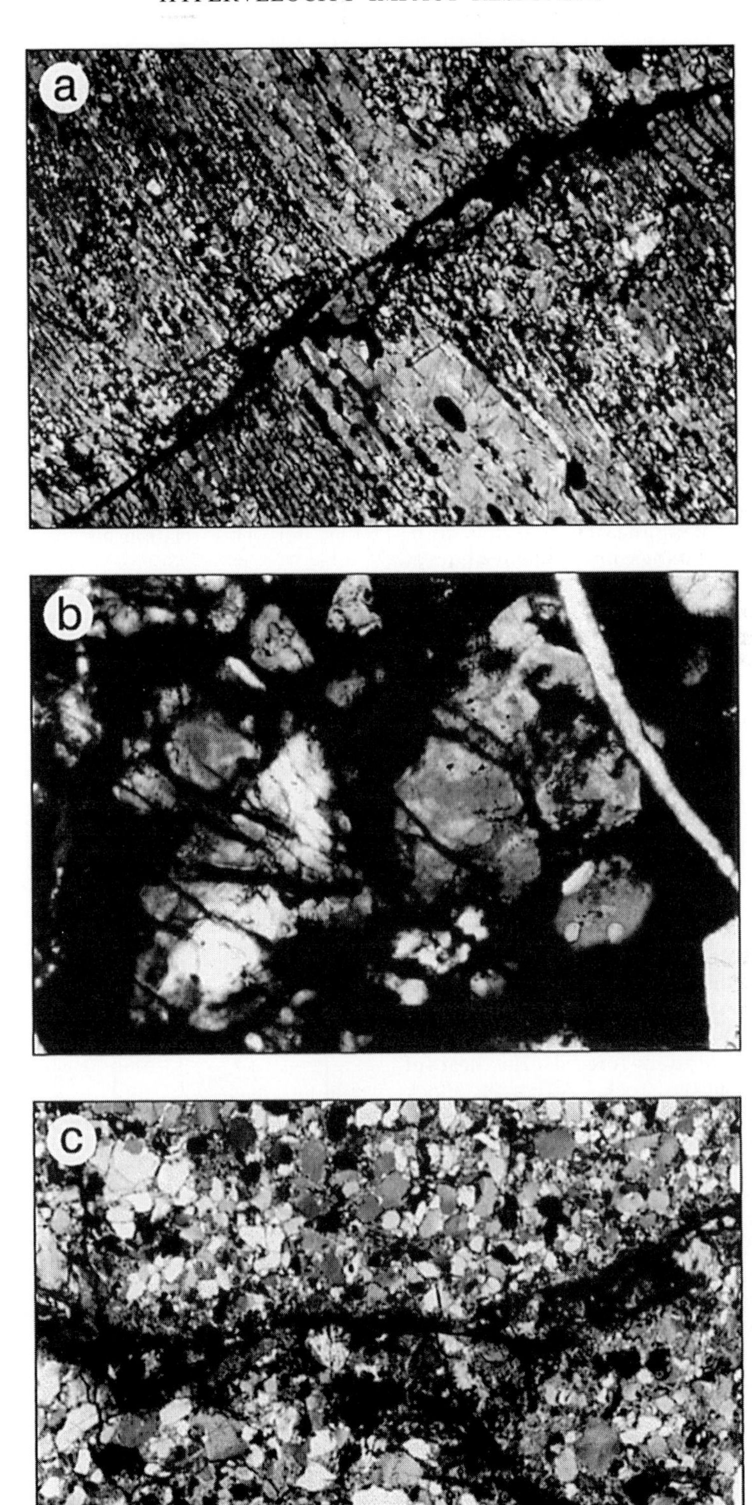
a
b
c

An additional consideration regarding viscosity determinations is that friction melts do not consist exclusively of melt, but are polyphase suspensions (as are most magmas): they comprise a liquid matrix with entrained mineral and rock fragments and, in some cases, bubbles. Depending on the geometry, size and relative volume of fragments (clasts) and bubbles, viscosities can be reduced in suspensions by as much as an order of magnitude at high shear rates. This is due to the effects of shear thinning (Spray 1993). This implies that pseudotachylytes can be highly mobile and so travel considerable distances within fracture–fault systems. This also explains how pseudotachylytes can contain exotic rock fragments or boulders: these inclusions may have been ripped off fault walls some distance away (possibly as much as kilometres) and transported along the flow trajectory to be juxtaposed with lithologically distinct wall rocks and locally derived fragments.

Shock veins

Dark, fine-grained to glassy veins, generally <2 mm wide, are a common feature in meteorites (Fig. 2a and b). Stöffler *et al.* (1991) reported *c.* 45% of the chondrites they studied as being vein bearing. The development within these veins of planar deformation features, mosaicism, diaplectic glass (e.g. maskelynite) and high-pressure polymorphs (e.g. ringwoodite, majorite) indicates formation pressures of 10–80 GPa (Bischoff & Stöffler 1992). It is implicit that these veins have formed by shock-wave compression and especially decompression, where the energy released is equal to or exceeds the heat of vaporization of the rock. Although shock remains the dominant driving force for their formation, most so-called shock veins do exhibit micro-offset (e.g. Begemann & Wlotza 1969; Axon & Stefle-Perkins 1975; Spray 1994; Osborn *et al.* 1997), thus a frictional contribution to their energy budget and formation mechanism cannot be neglected. Critically, these so-called shock veins show many similarities to certain pseudotachylytes developed in terrestrial impact structures, as first alluded to by Barnes (1939).

In the Sudbury impact structure, thin (<2 mm) pseudotachylyte fracture–vein systems appear to be restricted to basement rocks <15 km beyond the Sudbury igneous complex, a distance that roughly coincides with the occurrence of shatter cones. These veins are pervasive within this inner damage zone and it is difficult to stand on any rock outcrop without observing them. They typically form anastomosing fracture–vein systems that show little evidence of offset at the macroscopic scale (Fig. 3). At Sudbury, this type of pseudotachylyte may be found on shatter cone surfaces (Gibson & Spray 1998), as has also been recognized for the shatter cones at Vredefort (Martini 1978). This highly pervasive fracturing, combined with micro-displacement, may facilitate gross (megascopic) hydrodynamic behaviour within the innermost damage zone. Otherwise, there is no evidence at the microscopic scale for the homogeneous fluidization of target rocks as suggested by Baldwin (1972) and further discussed by Melosh (1989).

The fine, fracture–vein systems are cross-cut (post-dated) by more discrete, thicker (centimetre to tens of metres wide), planar-bodied, dyke-like pseudotachylytes which exhibit offset (Fig. 3c). This relationship between different

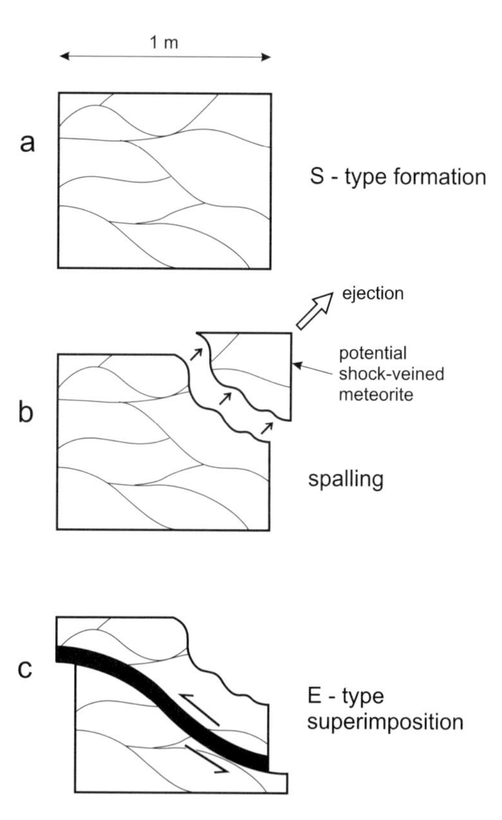

Fig. 3. (**a**) Cross-section of a block of target rock pervaded by thin (≪1 cm) shock-related pseudotachylyte veins. (**b**) Decompression may lead to near-surface spalling and ejection of veined fragment. (**c**) Subsequent modification processes may result in friction-dominant pseudotachylytes exploiting the pre-existing vein system.

types of pseudotachylyte has been observed in other impact structures. Based largely on his investigation of the Rochechouart impact structure, Lambert (1981) presented evidence for the existence of two distinct varieties of so-called 'breccia dikes' within footwall rocks: Type A, equated with the passage of the shock wave (compression stage), and Type B, generated during pressure release (subtype B_1) or subsequent crater modification (subtype B_2). It was also noted that the Type A breccia veins were generally thinner than the Type B variety. Martini (1978, 1991) adopted a similar classification for the Vredefort pseudotachylytes. He noted that his A-type occurs as thin veins which can contain high-pressure polymorphs, such as coesite and stishovite in the appropriate lithology (Fig. 2c), whereas his B-type post-dated the A-type, was generally larger, and was formed during the tensional period immediately following the passage of the shock wave.

So-called Type A pseudotachylytes have been found within rock fragments deposited in fallback breccias of the Onaping Formation at Sudbury, as have shatter cone fragments and clasts containing shock features (Muir & Peredery 1984), but Type B pseudotachylytes have not. This supports the contention that the so-called Type A variety is formed as a result of the passage of the shock front (the contact and compression stage) such that vein-bearing fragments nearer the surface could be ejected during the crater excavation stage (Fig. 3b) and be incorporated into the fallback (or reach escape velocity to potentially yield shock-veined meteorites), whereas the Type B pseudotachylytes are formed after jetting and spalling. Field evidence from the Sudbury impact structure indicates that the Type B variety is generated during transient cavity collapse and subsequent gravity-driven crustal adjustments of the crater modification stage (Spray & Thompson 1995; Thompson & Spray 1996), which may include certain rebounds effects. Furthermore, the Type B variety appears more closely allied to slip rather than shock, and in this respect is similar to the endogenic pseudotachylytes developed in non-impact-related crustal fault zones. On this basis, it is suggested that pseudotachylytes in impact structures are divided into two subtypes: S-type (for shock related) and E-type (for endogenic related). Unlike the Type B variety of Lambert (1981) and Martini (1991), this new scheme divorces the E-type from any association with decompression or tension immediately following shock compression, which in any case remains an inherent part of the shock event (Zel'dovich & Raizer 1967). In this scheme, the E-type

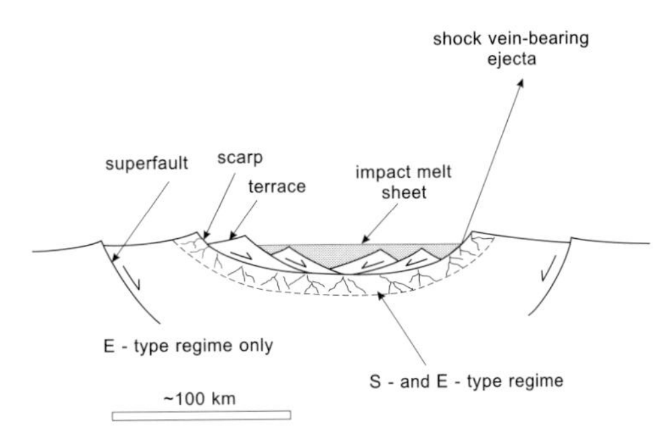

Fig 4. Cross-section of a large impact structure (e.g. Sudbury, Vredefort) showing impact melt sheet, collapsed transient cavity walls, superfaults, terraces and superfault scarps. S-type pseudotachylytes are developed in the central part of the structure. Those adjacent to the impact melt sheet may be annealed and destroyed. S-type pseudotachylytes may evolve in time to the E-type variant within the collapsed transient cavity during the modification stage, with only E-type pseudotachylytes forming beyond the collapse zone. Fallback removed for clarity.

pseudotachylytes are no different from pseudotachylytes developed in non-impact settings. However, because of the large displacements associated with transient cavity collapse leading to the formation of terraces (Fig. 4) and rebound, E-type pseudotachylytes can reach considerable thicknesses (tens to hundreds of metres as opposed to centimetres).

Figure 5 shows a simplified plan view of a large complex impact structure. S-type pseudotachylytes are generated in the central zone up to the collapsed transient cavity perimeter. The development of a pervasive S-type fracture–vein system may actually facilitate the collapse and modification of the transient cavity via subsequent E-type pseudotachylyte development. The larger E-type pseudotachylytes are generated along superfaults which form terraces and scarps. Much of this inner zone evidence may be obscured or totally destroyed by the adjacent superheated melt sheet through secondary melting (assimilation) and annealing processes. The melt sheet may egress the footwall target rocks via radial and concentric fracture–fault systems. E-type pseudotachylytes are developed out to the periphery of the impact structure, where the most distal morphometric feature is defined by a fault scarp or monocline. Interaction between radial and concentric fracture–fault systems is likely to lead to the inward movement of the trapeziform segments via the development of thrusts and zones of transpression and transtension (not shown).

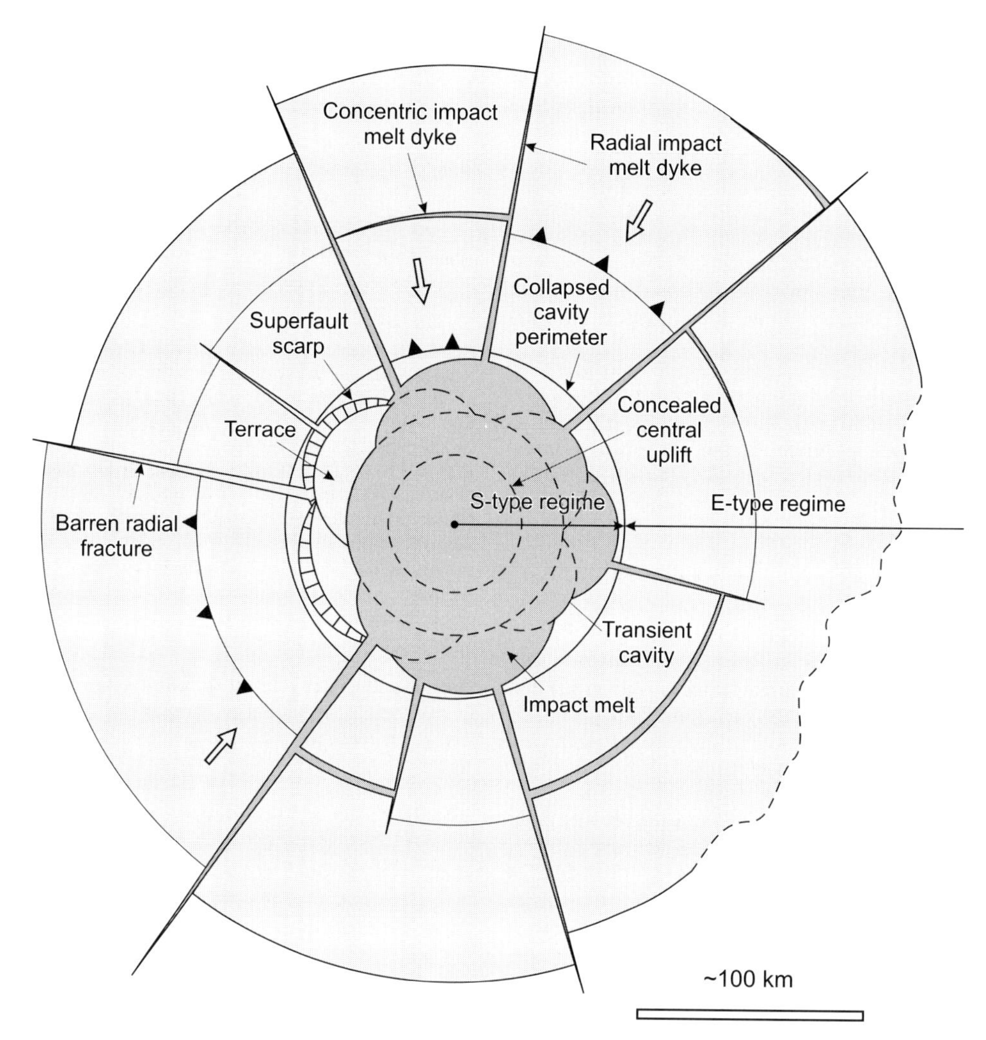

Fig. 5. Plan of a large impact structure. (Note spatial overlapping of S- and E-type regimes in the central collapsed cavity region and the continuation of E-type, but not S-type, to the periphery of the structure.) For clarity, fallback removed, and no displacement shown in radial and concentric fracture–fault systems.

Summary

S-type pseudotachylytes

S-type pseudotachylytes are formed during shock-wave–rock interaction (the contact and compression stage of the cratering process). They are typically thin (<2 mm), exhibit micro-offset (usually $\ll$1 mm) and may contain high-pressure polymorphs (e.g. coesite, stishovite, majorite, ringwoodite). S-type pseudotachylytes show apparently random, pervasive distribution within the innermost shock zone of impact structures, now largely occupied by the collapsed margins of the transient cavity (Figs 4 and 5). They may have facilitated gross hydrodynamic behaviour within this innermost zone. Shock or friction melts developed on the surfaces of shatter cones are also considered S-type pseudotachylytes (Gibson & Spray 1998), but they probably formed at relatively lower strain rates, just beyond the E-type pseudotachylyte regime (Fig. 1). S-type pseudotachylytes can be ejected from the target location and they predate E-type pseudotachylytes. S-type pseudotachylytes equate with Lambert's (1981) Type A and B_1 variants and Martini's A- and, in part, B-types. If shock compression and decompression microtextures can be distinguished at the microscopic scale, it is suggested that the terms

S_1 and S_2 be deployed, respectively (equivalent to Lambert's A and B_1 variants). Shock veins in meteorites are most probably S-type pseudotachylytes and vice versa. These were generated by bolide impact or asteroid–asteroid collision and released as Grady–Kipp fragments during decompression and spalling from planetoid surfaces (Melosh 1984; Grady & Kipp 1993).

E-type pseudotachylytes

E-type pseudotachylytes are formed during the modification stage of the cratering process and are not directly related to shock-wave–rock interaction. They post-date S-type pseudotachylytes. They are thicker ($\gg$1 cm) than the S-type and do not contain high-pressure polymorphs (unless inherited from assimilation of a pre-existing S-type vein). Field evidence indicates that E-type pseudotachylytes are generated as a result of gravity-driven faulting, which is commonly listric normal (Figs 4 and 5). E-type pseudotachylytes can exhibit significant offset along their generation surfaces and, as such, they may define the concentric fault systems of ringed impact basins (Spray & Thompson 1995). Such faults can exhibit large displacement, single slip behaviour and so facilitate transient cavity collapse with the formation of terrace slumps (Figs 4 and 5). This superfaulting mode (Spray 1997) can generate vast amounts of friction melt, as manifest in the Frood Breccia Belt at Sudbury. E-type pseudotachylytes are not ejected from the target. The earlier-generated S-type pseudotachylyte zones may act as pre-fractured source horizons for subsequent E-type initiation and displacement (Fig. 3c), especially where the shattered transient cavity walls subsequently undergo gravity-driven collapse.

This work was supported by the Natural Sciences and Engineering Research Council of Canada. C. Koeberl and an anonymous reviewer are thanked for commenting on an earlier draft of this work.

References

AUSTRHEIM, H. & BOUNDY, T. M. 1994. Pseudotachylytes generated during seismic faulting and eclogitization of the deep crust. *Science*, **265**, 82–83.

AXON, H. J. & STEFLE-PERKINS, E. M. 1975. Fracture mechanism of Henbury meteorite by separation along surfaces of shear faulting. *Nature*, **256**, 635.

BALDWIN, R. B. 1972. The tsunami model of the origin of ring structures concentric with large lunar craters. *Physics of the Earth and Planetary Interiors*, **6**, 327–339.

BARNES, V. E. 1939. *Pseudotachylyte in Meteorites*. University of Texas Publication, **3949**, 645–656.

BEGEMANN, F. & WLOTZA, F. 1969. Shock-induced thermal metamorphism and mechanical deformations in the Ramsdorf chondrite. *Geochimica et Cosmochimica Acta*, **33**, 1351–1370.

BEKE, B. 1964. *Principles of Comminution*. Academy, Budapest.

BISCHOFF, A. & STÖFFLER, D. 1992. Shock metamorphism as a fundamental process in the evolution of planetary bodies: information from meteorites. *European Journal of Mineralogy*, **4**, 707–755.

BOSLOUGH, M. B. & ASAY, J. R. 1993. Basic principles of shock compression. *In*: ASAY, J. R. & SHAHINPOOR, M. (eds) *High-Pressure Shock Compression of Solids*. Springer, New York, 7–42.

CLOOS, H. 1941. Bau und Tätigkeit von Tuffschloten. *Geologische Rundschau*, **32**, 702–800.

DIETZ, R. S. 1961. The Vredefort ring structure: meteorite impact scar? *Journal of Geology*, **69**, 499–516.

DRESSLER, B. O. 1984. The effects of the Sudbury event and the intrusion of the Sudbury Igneous Complex on the footwall rocks of the Sudbury Structure. *In*: PYE, E. G., NALDRETT, A. J. & GIBLIN, P. E. (eds) *The Geology and Ore Deposits of the Sudbury Structure*. Ontario Geological Survey Special Volume, **1**, 97–136.

FRENCH, B. M., ORTH, C. J. & QUINTANA, L. R. 1989. Iridium in the Vredefort bronzite granophyre: impact melting and limits on a possible extraterrestrial component. *Proceedings of the 19th Lunar and Planetary Science Conference*, 733–744.

GIBSON, H. M. & SPRAY, J. G. 1998. Shock-induced melting and vaporization of shatter cone surfaces: evidence from the Sudbury impact structure. *Meteoritics and Planetary Science*, **33**, 329–336.

GRADY, D. E. & KIPP, M. E. 1993. Dynamic fracture and fragmentation. *In*: ASAY, J. R. & SHAHINPOOR, M. (eds) *High-Pressure Shock Compression of Solids*. Springer, New York, 265–322.

GRIEVE, R. A. F., LANGENHORST, F. & STÖFFLER, D. 1996. Shock metamorphism in quartz in nature and experiment II: Significance in geoscience. *Meteoritics and Planetary Science*, **31**, 6–35.

HOBBS, B. E., ORD, A. & TEYSSIER, C. 1986. Earthquakes in the ductile regime? *Pure and Applied Geophysics*, **124**, 309–336.

KILLICK, A. M. 1994. The geochemistry of pseudotachylyte and its host rocks from the West Rand Goldfield, Witwatersrand Basin, South Africa: implications for pseudotachylyte genesis. *Lithos*, **32**, 193–205.

KOEBERL, C., REIMOLD, W. U. & SHIREY, S. B. 1996. Re–Os isotope and geochemical study of the Vredefort Granophyre: clues to the origin of the Vredefort structure, South Africa. *Geology*, **24**, 913–916.

LAKOMY, R. 1990. Implications for cratering mechanics from a study of the Footwall Breccia of the Sudbury impact structure, Canada. *Meteoritics*, **25**, 195–207.

LAMBERT, P. 1981. Breccia dikes: geological constraints in the formation of complex craters. *Proceedings of the 12th Lunar and Planetary Science Conference*, A, 59–78.

MADDOCK, R. H. 1983. Melt origin of fault-generated pseudotachylytes demonstrated by textures. *Geology*, **11**, 105–108.

MAGLOUGHLIN, J. F. 1992. Microstructural and chemical changes associated with cataclasis and frictional melting at shallow crustal levels: the cataclasite–pseudotachylyte connection. *Tectonophysics*, **204**, 243–260.

MARONE, C. & SCHOLZ, C. H. 1989. Particle-size distribution and microstructures within simulated fault gouge. *Journal of Structural Geology*, **11**, 799–814.

MARTINI, J. E. J. 1978. Coesite and stishovite in the Vredefort Dome, South Africa. *Nature*, **272**, 715–717.

——1991. The nature, distribution and genesis of the coesite and stishovite associated with pseudotachylite of the Vredefort Dome, South Africa. *Earth and Planetary Science Letters*, **103**, 285–300.

MELOSH, H. J. 1984. Impact ejection, spallation and the origin of meteorites. *Icarus*, **59**, 234–260.

——1989. *Impact Cratering: a Geologic Process*. Oxford University Press, New York.

MUIR, T. L. & PEREDERY, W. V. 1984. The Onaping Formation. *In*: PYE, E. G., NALDRETT, A. J. & GIBLIN, P. E. (eds) *The Geology and Ore Deposits of the Sudbury Structure*. Ontario Geological Survey Special Volume, **1**, 139–210.

OSBORN, W., MATTY, D., VELBEL, M., BROWN, P. & WACKER, J. 1997. Fall, recovery and description of the Coleman chondrite. *Meteoritics and Planetary Science*, **32**, 781–790.

REIMOLD, W. U. 1991. The geochemistry of pseudotachylites from the Vredefort Dome, South Africa. *Neues Jahrbuch für Mineralogische Abhandlungen*, **162**, 151–184.

——1995. Pseudotachylite in impact structures – generation by frictional melting and shock brecciation? *Earth-Science Reviews*, **39**, 247–264.

SAMMIS, C. G., OSBORNE, R. H., ANDERSON, J. L., BANERDT, M. & WHITE, P. 1986. Self-similar cataclasis in the formation of fault gouge. *Pure and Applied Geophysics*, **124**, 53–78.

SCHWARZMAN, E. C., MEYER, C. E. & WILSHIRE, H. G. 1983. Pseudotachylite from the Vredefort Ring, South Africa and the origins of lunar breccias. *Geological Society of America Bulletin*, **94**, 926–935.

SCOTT, R. G., SPRAY, J. G. & MAKELA, E. F. 1996. The Frood Breccia Belt of the Sudbury impact structure: the largest known pseudotachylyte body. *Geological Society of America Annual Meeting (Denver), Program with Abstracts*, A383.

SHAND, S. J. 1916. The pseudotachylyte of Parijs (Orange Free State) and its relation to 'trap-shotten gneiss' and 'flinty crush rock'. *Quarterly Journal of the Geological Society, London*, **72**, 198–217.

SHIMAMOTO, T. & NAGAHAMA, H. 1992. An argument against a crush origin of pseudotachylyte based on the analysis of clast-size distribution. *Journal of Structural Geology*, **14**, 999–1006.

SIBSON, R. H. 1975. Generation of pseudotachylyte by ancient seismic faulting. *Geophysical Journal of the Royal Astronomical Society*, **43**, 775–794.

SPRAY, J. G. 1992. A physical basis for the frictional melting of some rock-forming minerals. *Tectonophysics*, **204**, 205–221.

——1993. Viscosity determinations of some frictionally generated silicate melts: implications for fault zone rheology at high strain rates. *Journal of Geophysical Research*, **98**, 8053–8068.

——1994. Displacement associated with shock veins in four stony meteorites: evidence for a frictional contribution to melting. *Geological Association of Canada/Mineralogical Association of Canada, Program with Abstracts*, **19**, A106.

——1995. Pseudotachylyte controversy: fact or friction? *Geology*, **23**, 1119–1122.

——1997. Superfaults. *Geology*, **25**, 579–582.

—— & THOMPSON, L. M. 1995. Friction melt distribution in a multi-ring impact basin. *Nature*, **373**, 130–132.

——, KELLEY, S. P. & REIMOLD, W. U. 1995. Laser probe $^{40}Ar/^{39}Ar$ dating of coesite- and stishovite-bearing pseudotachylytes and the age of the Vredefort impact event. *Meteoritics*, **30**, 335–343.

STÖFFLER, D. & LANGENHORST, F. 1994. Shock metamorphism of quartz in nature and experiment: I. Basic observation and theory. *Meteoritics*, **29**, 155–181.

——, KEIL, K. & SCOTT, E. R. D. 1991. Shock metamorphism of ordinary chondrites. *Geochimica et Cosmochimica Acta*, **55**, 3845–3867.

TABOR, D. 1954. Mohs's hardness scale – a physical interpretation. *Proceedings of the Physical Society*, **67**, 249–257.

THERRIAULT, A. M., GRIEVE, R. A. F. & REIMOLD, W. U. 1997. Original size of the Vredefort Structure: implications for the geological evolution of the Witwatersrand basin. *Meteoritics and Planetary Science*, **32**, 71–77.

THOMPSON, L. M. & SPRAY, J. G. 1996. Pseudotachylyte petrogenesis: constraints from the Sudbury impact structure. *Contributions to Mineralogy and Petrology*, **125**, 359–374.

TURCOTTE, D. L. 1986. Fractals and fragmentation. *Journal of Geophysical Research*, **91**, 1921–1926.

VON ENGELHARDT, W., ARNDT, J., FECKER, B. & PANKAU, H. G. 1995. Suevite breccia from the Ries Crater, Germany: origin, cooling history and devitrification of impact glasses. *Meteoritics*, **30**, 279–293.

WENK, H. R. 1978. Are pseudotachylites products of fracture or fusion? *Geology*, **6**, 507–511.

WHITE, J. C. 1996. Transient discontinuities revisited: pseudotachylyte, plastic instability and the influence of low pore fluid pressure on deformation processes in the mid-crust. *Journal of Structural Geology*, **18**, 1471–1486.

ZEL'DOVICH, YA. B. & RAIZER, YU. P. 1967. *Physics of Shock Waves and High-Temperature Hydrodynamic Phenomena, Volume II*. Academic, New York.

Geochemistry of carbon in terrestrial impact processess

IAIN GILMOUR

Planetary Sciences Research Institute, The Open University, Milton Keynes MK7 6AA, UK

Abstract: Terrestrial impact craters appear to be unique in the geological environment in that they contain evidence for the presence of all four carbon allotropes: graphite, diamond, C_{60} and carbynes. This diversity appears to reflect the wide range of physical and chemical conditions that occur during impact processes including shock-induced alterations and vapour-phase chemistry. Impact-produced diamonds are known from several impact craters around the world, and their ability to survive over immense periods of geological time makes the presence of diamonds an important criterion for the identification of terrestrial impact structures.

An important aspect of terrestrial impact cratering is the influence of impacts on the geological and biological evolution of our own planet. However, it is only recently that we have realized that impacts may have played a more significant role than previously thought. After more than a decade of intensive research, it is now commonly accepted that the Chicxulub structure in Mexico represents a 200–300 km diameter impact crater coincident with the Cretaceous–Tertiary (K–T) boundary (Hildebrand *et al.* 1991; Sharpton *et al.* 1992). This 65 Ma old crater is probably the largest recognized impact structure on Earth; however, other large craters are also known to be associated with geological boundaries, for example, the Jurassic–Cretaceous age 200 km diameter Morokweng crater in Zimbabwe (Koeberl *et al.* 1997*a*) and the Eocene–Oligocene age 100 km diameter Popigai crater in Siberia (Masaitis *et al.* 1972*b*). The debate as to whether large impact events can influence the course of evolution is a contentious one (see other papers in this volume); however, critical to this debate is an understanding of the effects that impacts have on the Earth's environment. Small-scale laboratory hyper-velocity impact experiments dealing with projectiles of less than a centimetre diameter and numerical simulation models are able to provide constraints on the physics of small impacts and, through scaling, on large impacts (Melosh 1989). However, to understand the physics and chemistry associated with large impacts it is necessary to look at the evidence associated with impact-produced rocks, which, because of the extreme pressures and temperatures involved, exhibit various mineralogical modifications.

The physical and chemical modifications that occur in impact processes are primarily the result of the shock compression of target rocks, i.e. shock metamorphism. When subjected to shock most crystalline solids undergo structural rearrangements transforming to denser phases at successively higher pressures. With increasing degree of shock metamorphism, target rocks and minerals undergo progressively greater transformations and a number of shock-induced alterations are used as indicators of peak stress. Initial stages of shock metamorphism include fracturing and brecciation; at peak shock pressures in the range 30–45 GPa diaplectic minerals are formed (for example, maskelynite is produced by the shock devitrification of plagioclase feldspar (Stöffler 1971)); successive stages include incipient and total melting of target rocks, and eventual incipient and complete vaporization. These latter stages are associated with the release of pressure and are complex phenomena involving the interaction of several phases and species (Melosh 1989). Petrographic work on shock-induced alterations of minerals in terrestrial rocks has largely been on quartzo-feldspathic rocks (Stöffler 1971) though recently there has been increased interest in other rock types, notably carbonates, and in the interactions between different rock types during shock processes (Martinez *et al.* 1994).

The best-known indicators of shock in quartz-rich rocks are coesite and stishovite. These dense phases of silica were first observed at Meteor Crater, Arizona (Chao *et al.* 1962), and because volcanic processes cannot achieve the 15–30 GPa pressures necessary for their formation they also became important tools in establishing the impact origin of other craters such as the Ries crater in Southern Germany (Shoemaker & Chao 1961). At such high pressures other phases are also subject to transformations, notably the carbon system, in which graphite can undergo a direct conversion to diamond. The shock transformation of graphite to diamond is simple, requiring only an increase in the interatomic distance within the individual carbon planes of the graphite of 0.12 Å and a decrease in the interplanar spacing of 1.86 Å (Lipschutz 1964), and was successfully achieved in the laboratory

GILMOUR, I. 1998. Geochemistry of carbon in terrestrial impact processes. *In*: GRADY, M. M., HUTCHISON, R., MCCALL, G. J. H. & ROTHERY, D. A. (eds) *Meteorites: Flux with Time and Impact Effects*. Geological Society, London, Special Publications, **140**, 205–216.

by DeCarli & Jamieson (1961). The mechanism was first proposed to explain the presence of diamonds in iron meteorites (Lipschutz & Anders 1961*a*, *b*) and subsequently in ureilites (Lipschutz 1964) as a result of impact processes occurring before the meteorites' arrival on Earth.

Although the shock transformation of graphite to produce impact diamonds is fairly well understood, the ability of carbon to undergo a vast range of chemical reactions makes it an intriguing element for investigation under the wide range of physical conditions that occur in impact processes. Recent studies of several impact craters have revealed the presence of fullerenes (Becker *et al.* 1994*b*) and silicon carbide (Hough *et al.* 1995), as well as evidence for diamonds produced by condensation from the vapour phase (Hough *et al.* 1995).

Fullerenes

The discovery of a third allotrope of carbon, the fullerene molecule C_{60}, and the suggestion that that it might be widely distributed in the universe (Kroto *et al.* 1985) led to initially unsuccessful searches for its presence in meteorites (Gilmour *et al.* 1991; Devries *et al.* 1993). More recently, however, trace amounts of fullerenes have been identified in the Allende meteorite (Becker *et al.* 1994*a*), and it has been suggested that they may be the result of gas-phase reactions (Becker & Bunch 1997) similar to those proposed for the production of polycyclic aromatic hydrocarbons in circumstellar environments and meteorites (Allamandola *et al.* 1987; Gilmour & Pillinger 1994). The discovery of fullerenes in micro-impact craters on the LDEF spacecraft has also led to the suggestion that fullerenes may condense in the high-density vapour produced in hypervelocity C–C collisions (di Brozolo *et al.* 1994). On Earth, fullerenes have been found in a variety of environments including fulgurite, a material formed when lightning strikes certain soils or rocks (Daly *et al.* 1993), and shungite, a meta-anthracite coal from Shunga, Russia (Buseck *et al.* 1992). They have also been found associated with impacts in K–T boundary clay from Woodside Creek, New Zealand (Heymann *et al.* 1994) and in shock-produced impact breccias from the 1.85 Ga old Sudbury crater, Canada (Becker *et al.* 1994*b*). Heymann *et al.* (1994) found C_{60} fullerene in concentrations of up to 5.4 ppb in samples of K–T boundary clay from two sites in New Zealand, Woodside Creek and Flaxbourne River, and no evidence of 'background' fullerenes in either Cretaceous limestone or Tertiary shale from the same sites. Both of these sites have been the subject of previous extensive investigations of their carbon contents (Wolbach *et al.* 1985, 1988; Gilmour *et al.* 1989), and Woodside Creek contains substantial quantities of soot and elemental carbon or charcoal, up to 10 000 ppm in the boundary clay. Wolbach *et al.* (1985, 1988) suggested that the occurrence of soot and charcoal represents the signature of widespread wildfires at the time of the Chicxulub impact, 65 Ma ago. Additional support for this hypothesis was provided by the subsequent discovery of soot and charcoal at some 13 K–T boundary successions world-wide (Gilmour *et al.* 1989) and a proposed ignition mechanism related to the re-entry of impact ejecta (Melosh *et al.* 1990). The apparent association between soot and fullerenes at Woodside Creek led Heymann *et al.* (1994) to propose that the K–T boundary fullerenes were also generated by wildfires, a hypothesis supported by the observation of fullerenes synthesized in sooting flames (Heymann *et al.* 1994) and the subsequent discovery of fullerenes at additional K–T boundary successions in Europe and Asia where soot or elemental carbon was also present (Heymann *et al.* 1996). Heymann *et al.* (1996) discounted an alternative scenario for the synthesis of K–T boundary fullerenes, namely that they were formed by chemical processing of carbon in a carbonaceous chondrite or cometary impactor, on quantitative grounds. They argued that the apparent 4% yield of fullerenes observed on LDEF by di Brozolo *et al.* (1994) would result in around 5–6 orders of magnitude higher K–T boundary fullerene concentrations than are observed. However, Heymann *et al.* (1996) did not consider carbon in Chicxulub target rocks as a possible source for synthesized fullerenes despite evidence that target rock carbon is the primary source for impact-produced diamonds (Masaitis *et al.* 1972*a*; Hough *et al.* 1995).

The wildfire scenario, however, is not a viable origin for the fullerenes identified in the 1.85 Ga Sudbury impact structure by Becker *et al.* (1994*b*), who examined shock-produced breccias from the 1800 m thick Onaping formation, which is unusual compared with impact breccias from other craters in that it is highly carbonaceous, containing around 0.5–1.0% reduced carbon. These workers argued that the distribution of carbon in the Sudbury structure is the result of the impact event, as the carbon content of Sudbury target rocks is on average much less than 1% with only rare occurrences of more carbon-rich sediments. Becker *et al.* (1994*b*) therefore discounted combustion of carbon in target rocks as a source of carbon for fullerenes in the Sudbury structure, arguing instead that the carbon must have come from a carbon-rich impactor. One possibility, that fullerenes may have formed as a direct result of the pyrolysis of meteoritic organic matter which is rich in polyaromatic structures (Devries *et al.* 1993; Gilmour & Pillinger 1994), is supported by evidence that fullerenes can form from the pyrolysis of naphthalene at temperatures of around 1000°C . A second possibility, that the fullerenes were already present in the impactor and survived the impact event, is apparently supported by the discovery of extraterrestrial helium trapped within the fullerenes (Becker *et al.* 1996). The C_{60} molecule is large enough to enclose the noble gases He, Ne, Ar, Kr and Xe but is too small to encompass diatomic species such as N_2 or CO_2, and the incorporation of He within the C_{60} cage has been demonstrated in the laboratory (Saunders *et al.* 1996). Becker *et al.* (1996) measured a $^{3}He/^{4}He$ ratio of 5.5×10^{-4} to 5.9×10^{-4} in two fullerene-rich residues from Sudbury Onaping formation samples, some 25% higher than the accepted value for the solar wind (Becker & Pepin 1991) and an order of magnitude greater than mantle helium (Allègre *et al.* 1983). Indeed, Becker *et al.* (1996) suggested that the high $^{3}He/^{4}He$ is indicative of a presolar source for the helium similar to that proposed for the isotopically anomalous noble gases found in meteoritic nanodiamonds (Lewis *et al.* 1987). However, the occurrence of fullerenes at Sudbury is not without controversy. Heymann was unable to confirm the presence of fullerenes in Onaping formation samples, though this may merely reflect the heterogeneity of this very large rock unit (B. M. French, pers. comm., 1997). The apparent survivability of extraterrestrial fullerenes is also at odds with the lack of extraterrestrial signatures in more refractory phases present in both meteorites and impact-produced rocks, which would presumably be more likely to survive an impact. Studies of several K–T boundary acid-residues have shown no evidence of a presolar noble gas, carbon or nitrogen isotopic signatures (Eugster *et al.* 1985; Wolbach *et al.* 1985; Gilmour *et al.* 1992). Confirmation of a presolar origin for the fullerenes from Sudbury will require additional isotopic measurements including Ne, Kr and Xe as well as the determination of the carbon isotopic composition of the fullerenes.

Diamonds from the Popigai impact crater

Impact diamonds with sizes of up to about 1 cm were first found in impact-produced rocks at the 100 km diameter Popigai crater in Russia by Masaitis *et al.* (1972*a*). The Popigai impact structure, centred at 71°30′N, 111°0′E, is situated on the north-eastern slope of the Anabar shield in Northern Siberia, and is about 36–39 Ma old (Masaitis 1994). Archaean crystalline rocks are overlain by up to 1 km of a northward-dipping sequence of terrigenous and carbonate formations of Proterozoic, Cambrian and Permian age, invaded by sills and dykes of Triassic dolerite (Masaitis *et al.* 1972*b*). The centre of the crater is covered by more than a kilometre of suevitic rocks (melt-fragment-bearing impact breccias) and polymict breccia. Impact melt rocks occur mostly in the western edge of the crater as sheet-like bodies up to 600 m thick, and as irregular bodies within the suevites (Masaitis 1994). Masaitis *et al.* (1972*a*) found diamonds in strongly shocked gneissic clasts within the impact melt rocks and suevites, and observed that the distribution of impact diamonds was a function of the initial distribution of graphite-bearing rocks among the target rocks, as well as the shock zonation.

Impact diamonds recovered from the Popigai crater range in size from 0.1 to 0.5 mm and occasionally up to 1 cm. They are polycrystalline aggregates with individual crystallites having sizes in the range 0.1–1 μm, and range from colourless through yellow to grey and black, with the last being the most common (Kaminsky 1991) although the origin of the colour variation is unknown. Popigai impact diamonds show mainly tabular or isometric morphologies similar to the crystal habit of possible precursor graphites (Masaitis *et al.* 1990). Scanning electron microscopy reveals that the diamonds have a layered texture (Fig. 1) and are partly covered by a thin amorphous film. Transmission electron microscopy (TEM) indicates that the layered structure extends to the interior of the diamonds with layers on a scale of a few microns (Fig. 2 and Koeberl *et al.* (1997*b*)). Koeberl *et al.* (1997*b*) also noted that X-ray diffraction (XRD) of a number of fragments of Popigai impact diamonds indicated the presence of lonsdaleite, the hexagonal polymorph of diamond (Frondel & Marvin 1967; Lonsdale 1971). However, the same workers were unable to identify lonsdaleite using TEM although numerous lamellae were observed that may represent either stacking faults within the diamond lattice or microtwins. If multiple stacking faults occur, the lamellae could be interpreted as lonsdaleite. Indeed, other TEM studies of impact diamonds have also been unable to identify lonsdaleite (Gilmour *et al.* 1992; Hough *et al.* 1995) yet the polymorph has been reported in XRD studies of impact diamonds (Masaitis *et al.* 1990;

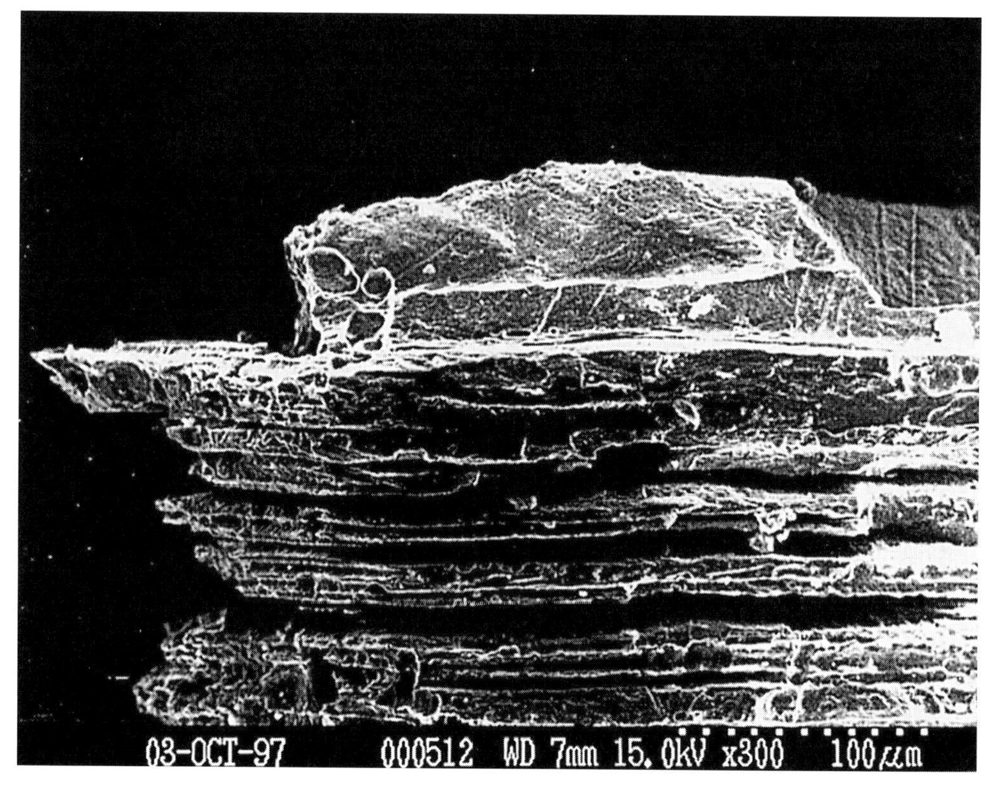

Fig. 1. Scanning electron microscope image of fine structure of a large (600 µm) Popigai impact diamond. The diamond is a polycrystalline aggregate of cubic diamonds; however, it appears to have preserved some of the linear structure of its precursor graphite.

Kaminsky 1991; Vishnevsky *et al.* 1997). The occurrence of lonsdaleite has often been taken as strong evidence for the production of impact diamonds by the shock-transformation of graphite, because of the similarity in crystal structure; these findings cast doubt on its potential use as an indicator of shock-induced alterations.

Carbon isotope data for Popigai impact diamonds are shown in Fig. 3, as are data for other diamonds of impact origin and terrestrial diamonds. The $\delta^{13}C$-values for Popigai diamonds range from −20 to −8‰, and two graphite samples from the Popigai crater also fall in this range. Earlier analyses yielded a more restricted range of $\delta^{13}C$ values, from −12 to −17‰ (Vishnevsky & Palchik 1975). Isotopic studies of diamonds associated with the Chicxulub impact crater and the Ries impact crater have also given $\delta^{13}C$ values in this range (Gilmour *et al.* 1992; Hough *et al.* 1995, 1997). The production of impact diamonds by the shock transformation of graphite *in situ* requires a carbon-rich target rock (Masaitis *et al.* 1990; Masaitis 1993) whereas chemical vapour deposition (CVD) of diamond (Hough *et al.* 1995) requires relatively little carbon although extremely low oxygen fugacities may be needed. However, the polycrystalline diamonds from Popigai are considerably larger than the nanometre-sized crystallites isolated using acid-dissolution techniques by Hough *et al.* (1995), have mineralogical and crystallographic characteristics similar to those of graphites found in the graphite-bearing precursor gneisses and show no paragenesis with silicon carbide, which might indicate a CVD synthesis mechanism.

Nitrogen concentrations and isotopic compositions for some Popigai impact diamonds are given in Table 1. They are extremely depleted in nitrogen with a maximum measured concentration of 44 ppm, although most have concentrations of less than 5 ppm. Nitrogen isotopic measurements give $\delta^{15}N$ values from −6 to +2‰. In kimberlitic diamonds nitrogen is the most abundant trace impurity and its concentration and isotopic composition have been used to try

Fig. 2. Transmission electron microscope image of a skeletal diamond aggregate from Popigai. The image shows dark lamellae that may represent the planar structure of the precursor graphite.

and constrain the processes of diamond formation (Boyd *et al.* 1987). Previous measurements of nitrogen content and isotopic composition for a diamond-rich residue from the K–T boundary associated with the Chicxulub impact also had a low nitrogen concentration of 82 ppm and gave a $\delta^{15}N$ value of +5.5‰ (Gilmour *et al.* 1992). Low nitrogen contents would be expected if the diamonds were derived from nitrogen-poor graphite; however, there are insufficient data on the nitrogen content of diamonds formed by CVD to eliminate that process based on nitrogen content alone. However, taken with the carbon isotopic and mineralogical data it seems likely that the large Popigai polycrystalline impact diamonds were formed by the shock transformation of carbon in the target material. In particular, the similar $\delta^{13}C$ values measured for graphites isolated from crystalline basement rocks at the Popigai crater would tend to support this origin. Whether nanometre-sized diamonds in paragenesis with silicon carbide produced by CVD exist in Popigai impact melt breccias awaits further investigation.

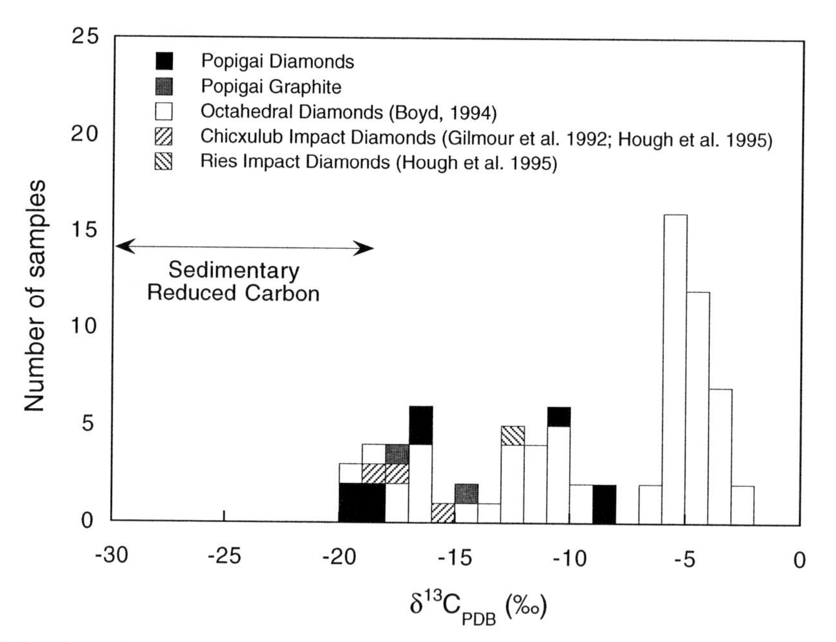

Fig. 3. Carbon isotopic composition of Popigai impact diamonds and graphite compared with kimberlitic diamonds and impact diamonds from the Chicxulub and Ries impact structures.

Table 1. *Carbon isotopic compositions, nitrogen isotopic compositions and nitrogen concentrations in polycrystalline impact diamonds and graphite from the Popigai impact structure*

Sample	δ^{13}C (‰)	δ^{15}C (‰)	[N] (ppm)	Notes
Suevites				
PG-3	−20.0		<5	
PY-1	−16.1		5 ± 2	
PB-5	−10.6	−6.0 ± 3.0	8 ± 4	
PY-2	−20.4	1.9 ± 1.1	50 ± 3.9	
PG-2	−18.1			
PG-1	−18.4			
IG	−8.7	−5.1 ± 1.7	<44	
PG-7	−19.3		<5	
PG-8	−16.5			
PG-4	−8.1			
Graphite				
PG-10	−17.6			Gneiss
PG-9	−14.1			Gneiss

Full sample descriptions have been given by Masaitis (1993) and Koeberl *et al.* (1997*b*).

Diamonds and silicon carbide from the Ries crater

The Ries crater (48°51′06″N, 10°29′23″E) is a 24 km diameter impact feature in Southern Germany formed 15 Ma ago. Rost *et al.* (1978) reported the existence of gas inclusions in glass from the Ries crater and mentioned finding grains containing high-pressure carbon polymorphs, one of which was identified as diamond or lonsdaleite using XRD. Previous careful studies of the shocked graphite-bearing gneisses from the Ries crater in the 1960s (A. El Goresy & D. Stöffler, pers. comm., 1995) had proven negative, except for the identification of chaoite controversially claimed to be a linear carbon allotrope (carbyne) with sp sigma-hybridization (El Goresy & Donnay 1968). Carbynes may play a role as precursors and mediating structures during the shock-induced transformation of graphite to diamond, as there is some evidence that carbynes are thermodynamically stable allotropes of carbon at very high temperatures, and that they will be formed by bond-splitting within the planar graphite layers (Heimann 1994).

Hough *et al.* (1995) subjected suevites from Otting quarry (3.5 km from the crater rim) to a cyclical treatment with progressively stronger and more oxidizing acids designed to destroy silicates, organic compounds, graphite, rutile and zircon. The carbon content of the residue, which was monitored at various stages during the acid demineralization, gradually increased to 80 wt % carbon. Hough *et al.* identified three types of grains in the final acid residue. Trans-parent plate-like grains with a brownish to yel-low tinge comprised mainly diamond with XRD patterns that gave reasonable {100} and very diffuse {101} lonsdaleite reflections, and green to blue crystals up to 100 μm in size, which were very well

Fig. 4. Transmission electron microscope image of a skeletal diamond aggregate from the Ries impact structure.

crystallized, comprised hexagonal α-SiC. However, the majority of residue comprised smaller dark fragments from a few microns up to a maximum of 200 μm in size identified as a mixture of diamond and 4H α-SiC, although those lines in the powder pattern which correspond to the cubic (β-SiC) phase were enhanced, suggesting the presence of fine-grained β-SiC which had not been found among the single crystals.

Examination of the residues by TEM indicated that the major component of the finer material was composed of skeletal cubic diamond aggregates, ranging in size from tens of nanometres to 2 μm (Fig. 4), and variably sized SiC, but no lonsdaleite. The larger diamonds were often in the form of platelets with {111} orientation, and rotation of the diamond pattern was observed as different sections were studied. In many cases, both diamond and SiC crystallites, identified by selected area electron diffraction (SAED), were observed in the same skeletal aggregate, and appeared to be either overlapping or intergrown. Evidence for epitaxial intergrowth of the two phases is shown in Fig. 5, where both diamond and β-SiC appear to be {111} oriented with respect to the electron beam, with a relative angular mismatch of *c.* 7°. Single crystals of 4H-SiC, up to 1 μm in size, were also observed by TEM. Silicon was the only element other than carbon seen during analysis of grains in the electron microscope.

The polycrystalline impact diamonds at the Ries crater are similar in many respects to those from the Popigai crater (Masaitis *et al.* 1972*a*). However, distinct from previous studies on diamond of impact origin, TEM examination showed that the majority of the Ries diamonds were skeletal aggregates and occur in a close association with silicon carbide. SiC has not been observed amongst the products of shock experiments performed under a variety of conditions to produce diamond. However, α-SiC

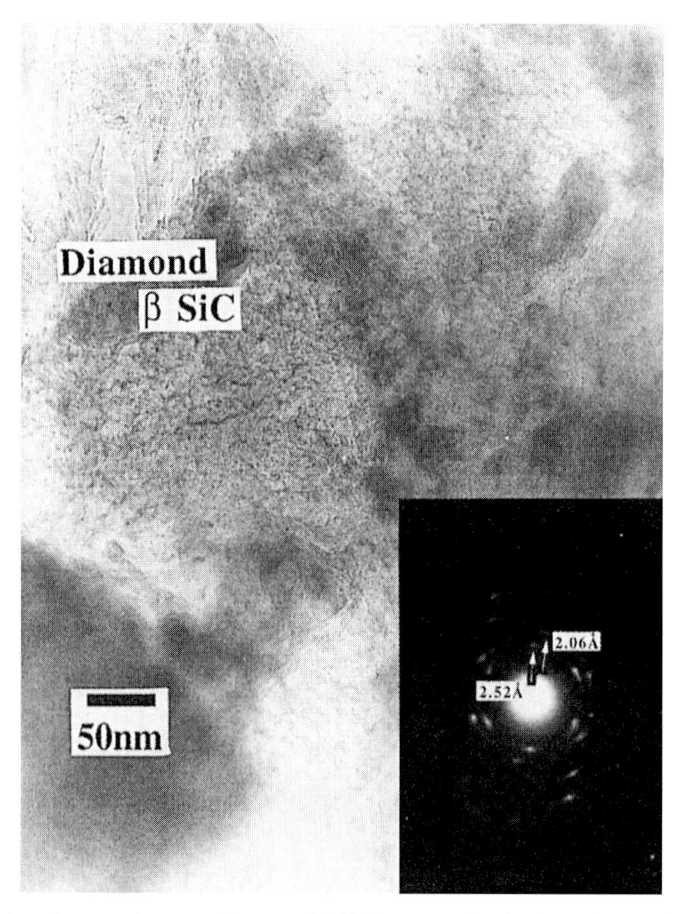

Fig. 5. Transmission electron microscope image and SAED pattern (inset) of a grain containing both SiC and diamond {111} oriented with respect to the electron beam.

can be produced from a vapour containing silicon and carbon created by shaped charge experiments to shock powdered β-SiC (Yamada & Tobisawa 1990).

The preferred mechanism for diamond synthesis at Popigai is the *in situ* shock transformation of graphite (Masaitis *et al.* 1972*b*; Koeberl *et al.* 1997*b*); however, an alternative or possibly coexistent mechanism for diamond synthesis that might explain the occurrence of SiC at the Ries crater is chemical vapour deposition. In CVD experiments diamond commonly grows on silicon-containing substrates and does so via an interfacial layer of 4H-SiC (Sato & Kamo 1992). β-SiC and 2H-SiC are believed to be the low-temperature forms of the mineral (produced from 1500 to 2000°C) whereas α-SiC (4H, 6H, 8H) is the higher-temperature polytype (Yamada & Tobisawa 1990). Furthermore, 4H-SiC has the highest growth rate of any of the silicon carbide polytypes produced during epitaxial growth by sublimation onto Si surfaces (Kalnin *et al.* 1994). The skeletal form of Ries diamonds suggests they grew free or between the surfaces of other minerals. Impact of a kilometre-sized body at 25 km/s (Melosh 1989), into sedimentary strata overlying crystalline graphite gneiss containing basement, would undoubtedly produce a vapour, or even a plasma, containing H, C and Si, in elementary or ionized form, exactly the feedstock for diamond and SiC formation by deposition. Carbon isotopic analysis of a Ries diamond–SiC-rich residue performed using a stepped combustion technique (Swart *et al.* 1983) afforded a sharp release of carbon, equivalent to 19.1 wt % of the sample, over the temperature range 500–700°C (Fig. 6), but the isotopic composition of the gas released suggested two carbon phases with different $\delta^{13}C$ values of around 6‰ (−16‰ v. −22‰). The isotopic composition of the diamond–SiC is in keeping with what might be expected from mixing organic (−28‰) and carbonate (0‰) sources derived from the 600 m of Mesozoic limestone, sandstone and shale overlying the Ries region at the time of the impact. It could also be achieved from mixtures with carbon in the gneiss or the impactor or both. Taking into consideration just the crater suevite at Ries, *c.* 7.2×10^4 t of diamond–SiC in 3:1 proportions exists in the area. Ample carbon is available in either the basement or the projectile; each would need to contain only 6.5 ppm C to provide the source material, though additional C may be required to produce sufficiently reducing conditions for SiC formation. Acquiring enough carbon from the sediment would be even simpler; however, the $\delta^{13}C$ of the acid residue argues against the majority coming from carbonate. Masaitis (1993) has argued that graphite- (or coal-) bearing targets are a prerequisite to diamond formation during impacts. There is no evidence at the Ries crater for sedimentary fragments shocked to pressures greater than 10 GPa, although the

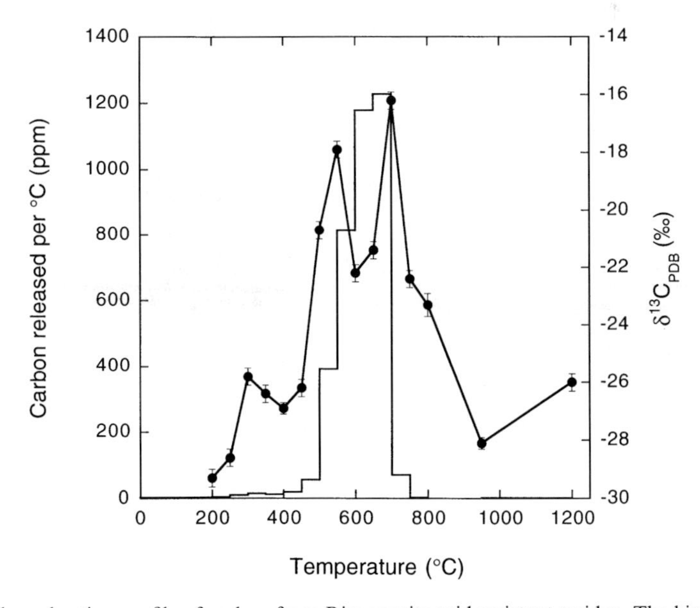

Fig. 6. Stepped combustion profile of carbon from Ries suevite acid-resistant residue. The histogram gives yield information, and the line graph corresponds to the isotopic composition at each temperature step.

basement gneisses undoubtedly were, and these may be the source of the plate-like diamond crystals. Further studies will need to concentrate on components physically separated from fragments within the host suevite. The plates could be a graphite pseudomorph, but there was minimal lonsdaleite.

Impact diamonds at other craters

Impact diamonds have also been reported at several other Russian and Ukrainian craters including the Terny, Obolon, Illynets and Zapadnaya craters in the Ukraine (Gurov *et al.* 1995; Valter & Yerjomenko 1996). XRD of single crystals of impact diamonds from Zapadnaya indicated the presence of as much as 75% lonsdaleite (Gurov *et al.* 1996); however, none of these diamonds have been subjected to TEM examination. Considerable petrographic work has been done over many years on the occurrences of impact diamonds in craters of the former Soviet Union. However, many of the results of these investigations are published in Russian language journals and so are not accessible to an international audience, although Vishnevsky *et al.* (1997) have recently reviewed much of the Russian literature on impact diamonds.

Impact diamonds (3–5 nm in size) were first reported associated with the Chicxulub impact structure in K–T boundary sediments at the Knudsens' farm locality, Red Deer Valley in Alberta, Canada (Carlisle & Braman 1991). Subsequently, diamonds were found at other K–T localities by Gilmour *et al.* (1992), who found diamonds up to 6 nm in size at the Western Interior K–T boundary sites of Brownie Butte, Montana, and Berwind Canyon, Colorado. Recently, impact diamonds up to 30 μm in size have been reported from proximal K–T boundary sequences at El Mimbral in NE Mexico (Hough *et al.* 1997). Carlisle & Bramman (1991) argued that the 3–5 nm grain size and the Ir to diamond ratio indicated an extraterrestrial origin for the K–T nanodiamonds. However, Gilmour *et al.* (1992) measured terrestrial carbon isotope ratios for K–T boundary diamonds, and their measurement of the nitrogen isotope ratio of the diamonds conclusively showed that the diamonds were not extraterrestrial in origin.

The presence of impact-produced diamonds in K–T boundary sediments supports the impact origin of these rocks. It also provides evidence that the K–T sequences across North America are stratigraphically linked. In many respects, K–T diamonds are similar to those found at other impact craters: they are polycrystalline aggregates and the larger crystals have a hexagonal morphology suggestive of precursor graphites (Hough *et al.* 1997). In one respect, however, K–T diamonds so far studied differ from those at other craters in that they were produced in the impact event and then ejected; further work is needed to see if they also reside in proximal suevites as at Popigai and the Ries crater.

Carbon allotropes in impact-produced rocks

Impact craters at present appear to be unique in the geological environment in that they contain evidence for the presence of all four carbon allotropes: graphite, diamond, C_{60} and carbynes. Graphite appears to be a relatively ubiquitous source of carbon in target rocks of many impact craters and is readily shock-transformed to polycrystalline cubic diamond aggregates. Fullerenes are apparently the product of gas-phase reactions, so that confirming their presence in impact-produced rocks will lend support to the evidence provided by SiC–diamond intergrowths for other vapour-phase products such as CVD diamond. Carbynes may play a role during the shock-induced transformation of graphite to diamond, as there is some evidence that carbynes are thermodynamically stable allotropes of carbon at very high temperatures (Heimann 1994).

Given the ubiquity of the diamond polymorphs at several Russian craters, the Ries crater and Chicxulub, perhaps the presence of diamond in its broadest sense (and possibly SiC) should become a criterion for the identification of impact structures (similar to coesite, stishovite, shatter cones, etc.). The value of diamonds in this respect is their ability to survive over immense periods of geological time, if not forever, giving access to events that occurred and possibly carbon reservoirs that existed in the early history of the Earth when bombardment was at its most intense.

I thank R. M. Hough for access to unpublished data. This work was supported by the Royal Society and the Particle Physics and Astronomy Research Council.

References

ALLAMANDOLA, L. J., SANDFORD, S. A. & WOPENKA, B. 1987. Interstellar polycyclic aromatic-hydrocarbons and carbon in interplanetary dust particles and meteorites. *Science*, **237**, 56–59.

ALLÈGRE, C. J., STAUDACHER, T., SARDA, P. & KURZ, M. 1983. Constraints on the origin of the Earth's mantle from rare-gas systematics. *Nature*, **303**, 762.

BECKER, L. & BUNCH, T. E. 1997. Fullerenes, fulleranes and polycyclic aromatic hydrocarbons in the Allende meteorite. *Meteoritics and Planetary Sciences*, **32**, 479–487.

—— & PEPIN, R. O. 1991. Composition of solar-wind noble-gases released by surface oxidation of a metal separate from the Weston meteorite. *Earth and Planetary Science Letters*, **103**, 55–68.

——, BADA, J. L., WINANS, R. E., & BUNCH, T. E. 1994*a*. Fullerenes in the Allende meteorite. *Nature*, **372**, 507.

——, ——, ——, HUNT, J. E., BUNCH, T. E. & FRENCH, B. M. 1994*b*. Fullerenes in the 1.85-billion-year-old Sudbury impact structure. *Science*, **265**, 642–645.

——, POREDA, R. J. & BADA, J. L. 1996. Extraterrestrial helium trapped in fullerenes in the Sudbury impact structure. *Science*, **272**, 249–252.

BOYD, S. R., MATTEY D. P., PILLINGER, C. T., MILLEDGE, H. J., MENDELSSOHN, M. & SEAL, M. 1987. Multiple growth events during diamond genesis: an integrated study of carbon and nitrogen isotopes and nitrogen aggregation state in coated stones. *Earth and Planetary Science Letters*, **86**, 241–353.

BUSECK, P. R., TSIPURSKY, S. J. & HETTICH, R. 1992. Fullerenes from the geological environment. *Science*, **257**, 215.

CARLISLE, D. B. & BRAMAN, D. R. 1991. Nanometer-size diamonds in the Cretaceous–Tertiary boundary clay of Alberta. *Nature*, **352**, 708–709.

CHAO, E. C. T., FAHEY, J. J., LITTLER, J. & MILTON, D. J. 1962 Stishovite, SiO_2, a very high pressure new mineral from Meteor Crater, Arizona. *Journal of Geophysical Research*, **67**, 419–421.

DALY, T. K., BUSECK, P. R., WILLIAMS, P. & LEWIS, C. F. 1993. Fullerenes from a fulgurite. *Science*, **259**, 1599.

DECARLI, P. S. & JAMIESON, J. C. 1961. Formation of diamond by explosive shock. *Science*, **133**, 1821–1822.

DEVRIES, M. S., REIHS, K., WENDT, H. R. *et al.* 1993. A search for C_{60} in carbonaceous chondrites. *Geochimica et Cosmochimica Acta*, **57**, 933–938.

DI BROZOLO, F. R., BUNCH, T. E., FLEMING, R. H. & MACKLIN, J. 1994. Fullerenes in an impact crater on the LDEF spacecraft. *Nature*, **369**, 37–40.

EL GORESY, A. & DONNAY, G. 1968. A new allotropic form of carbon from the Ries crater. *Science*, **161**, 363–364.

EUGSTER, O., GEISS, J. & KRÄHENBUHL, U. 1985. Noble gas isotopic abundances and noble metal concentrations in sediments from the Cretaceous–Tertiary boundary. *Earth and Planetary Science Letters*, **74**, 27–34.

FRONDEL, C. & MARVIN, U. B. 1967. Lonsdaleite, a hexagonal polymorph of diamond. *Nature*, **214**, 587–589.

GILMOUR, I. & PILLINGER, C. T. 1994. Isotopic compositions of individual polycyclic aromatic-hydrocarbons from the Murchison Meteorite. *Monthly Notices of the Royal Astronomical Society*, **269**, 235–240.

——, RUSSELL, S. S., ARDEN, J. W., LEE, M. R., FRANCHI, I. A. & PILLINGER, C. T. 1992. Terrestrial carbon and nitrogen isotopic-ratios from Cretaceous–Tertiary boundary nanodiamonds. *Science*, **258**, 1624–1626.

——, ——, NEWTON, J. *et al.* 1991. A search for the presence of C_{60} as an interstellar grain in meteorites. *Lunar and Planetary Science*, **22**, 445–446.

——, WOLBACH, W. S. & ANDERS, E. 1989. Major wildfires at the Cretaceous–Tertiary boundary. *In*: CLUBE, S. V. M. (ed.) *Catastrophes and Evolution: Astronomical Foundations*. Cambridge University Press, Cambridge, 195–213.

GUROV, E. P., GUROVA, E. P. & RAKITSKAYA, R. B. 1995. Impact diamonds in the craters of the Ukrainian shield. *Meteoritics*, **30**, 515–516.

——, —— & —— 1996. Impact diamonds of the Zapadnaya crater: phase composition and some properties. *Meteoritics and Planetary Science*, **31**, A56.

HEIMANN, R. B. 1994. Linear finite carbon chains (carbynes) – their role during dynamic transformation of graphite to diamond, and their geometric and electronic-structure. *Diamond and Related Materials*, **3**, 1151–1157.

HEYMANN, D., CHIBANTE, L. P. F., BROOKS, R. R., WOLBACH, W. S. & SMALLEY, R. E. 1994. Fullerenes in the Cretaceous–Tertiary boundary layer. *Science*, **265**, 645–647.

——, KOROCHANTSEV, A., NAZAROV, M. A. & SMIT, J. 1996. Search for fullerenes C_{60} and C_{70} in Cretaceous–Tertiary boundary sediments from Turkmenistan, Kazakhstan, Georgia, Austria and Denmark. *Cretaceous Research*, **17**, 367–380.

HILDEBRAND, A. R., PENFIELD, G. T., KRING, D. A., PILKINGTON, M., CAMARGO, A., JACOBSEN, S. B. & BOYNTON, W. V. 1991. Chicxulub crater – a possible Cretaceous–Tertiary boundary impact crater on the Yucatan Peninsula, Mexico. *Geology*, **19**, 867–871.

HOUGH, R. M., GILMOUR, I., PILLINGER, C. T., ARDEN, J. W., GILKES, K. W. R., YUAN, J. & MILLEDGE, H. J. 1995. Diamond and silicon-carbide in impact melt rock from the Ries impact crater. *Nature*, **378**, 41–44.

——, ——, ——, LANGENHORST, F. & MONTANARI, A. 1997. Diamonds from the iridium-rich K–T boundary layer at Arroyo el Mimbral, Tamaulipas, Mexico. *Geology*, **25**, 1019–1022.

KALNIN, A. A., NEUBERT, F. & PEZOLDT, J. 1994. Polytype patterning in epitaxial layers on the basis of non-equilibrium phase transitions. *Diamond and Related Materials*, **3**, 346–352.

KAMINSKY, F. V. 1991. Carbonado and yakutite: properties and possible genesis. *Proceedings of the 5th International Kimberlite Conference*, 136–143.

KOEBERL, C., ARMSTRONG, R. A. & REIMOLD, W. U. 1997*a*. Morokweng, South Africa: a large impact structure of Jurassic–Cretaceous boundary age. *Geology*, **25**, 731–734.

——, MASAITIS, V. L., SHAFRANOVSKY, G. I., GILMOUR, I., LANGENHORST, F. & SCHRAUDER, M. 1997*b*. Diamonds from the Popigai impact structure, Russia. *Geology*, **25**, 967–970.

KROTO, H. W., HEATH, J. R., O'BRIEN, S. S., CURL, R. F., & SMALLEY, R. E. 1985. C_{60}: buckminsterfullerene. *Nature*, **318**, 162–163.

LEWIS, R. S., MING, T., WACKER, J. F., ANDERS, E. & STEEL, E. 1987. Interstellar diamonds in meteorites. *Nature*, **326**, 160–162.

LIPSCHUTZ, M. 1964. Origin of diamonds in the Ureilites. *Science*, **143**, 1431–1434.

—— & ANDERS, E. 1961*a*. On the mechanism of diamond formation. *Science*, **134**, 2095–2099.

—— & ——1961*b*. The record in the meteorites; IV, Origin of diamonds in iron meteorites. *Geochimica et Cosmochimica Acta*, **24**, 83–105.

LONSDALE, K. 1971. Formation of lonsdaleite from single-crystal graphite. *American Mineralogist*, **56**, 333–336.

MARTINEZ, I., AGRINIER, P., SCHARER, U. & JAVOY, M. 1994. A SEM ATEM and stable-isotope study of carbonates from the Haughton Impact crater, Canada. *Earth and Planetary Science Letters*, **121**, 559–574.

MASAITIS, V. L. 1993. Diamond-bearing impactites, their distribution and petrogenesis. (in Russian) *Regionalaia Geologia i Metallogenia*, **1**, 121–134.

——1994. Impactites from Popigai Crater. *In*: DRESSLER, B. O., GRIEVE, R. A. F. & SHARPTON, V. L. (eds) *Large Meteorite Impacts and Planetary Evolution*. Geological Society of America, Special Paper, **293**, 153–162.

——, FURTERGENDLER, S. I. & GNEVUSHEV, M. A. 1972*a*. Diamonds in impactites of the Popigai meteorite crater (in Russian). *Proceedings of the All-Union Mineralogical Society*, **1**, 108–112.

——, MIKHAILOV, M. V. & SELIVANOVSKAYA, T. V. 1972*b*. Popigai Basin; an explosion meteorite crater. *Meteoritics*, **7**, 39–46.

——, SHAFRANOVSKY, G. I., YEZERSKY, V. A. & RESHETNYAK, N. B. 1990. Impact diamonds in ureilites and impactities (in Russian). *Meteoritika*, **49**, 180–196.

MELOSH, H. J. 1989. *Impact Cratering: a Geologic Process*. Oxford University Press, Oxford.

——, SCHNEIDER, N. M., ZAHNLE, K. J. & LATHAM, D. 1990. Ignition of global wildfires at the Cretaceous/Tertiary boundary. *Nature*, **343**, 251–254.

ROST, R., DOLGOV, Y. A. & VISHNEVSKIY, S. A. 1978. Gases in inclusions of impact glass in the Ries Crater, West Germany, and finds of high-pressure carbon polymorphs. *Transactions of the USSR Academy of Sciences: Earth Science Sections*, **241**, 165–168.

SATO, Y. & KAMO, M. 1992. Synthesis of diamond from the vapour phase. *In*: FIELD, J. E. (ed.) *The Properties of Natural and Synthetic Diamond.* Academic, New York.

SAUNDERS, M., CROSS, R. J., JIMÉNEZ-VÁZQUEZ, H. A., SHIMSHI, R. & KHONG, A. 1996. Noble gas atoms inside fullerenes. *Science*, **271**, 1693–1697.

SHARPTON, V. L., DALRYMPLE, G. B., MARIN, L. E., RYDER, G., SCHURAYTZ, B. C. & URRUTIAFUCUGAUCHI, J. 1992. New links between the Chicxulub impact structure and the Cretaceous–Tertiary Boundary. *Nature*, **359**, 819–821.

SHOEMAKER, E. M. & CHAO, E. C. T. 1961. New evidence for the impact origin of the Ries Basin, Bavaria, Germany. *Journal of Geophysical Research*, **66**, 3371–3378.

STÖFFLER, D. 1971. Classification of shocked quartzofeldspathic crystalline rocks; a review. *Meteoritics*, **6**, 317–318.

SWART, P. K., GRADY, M. M. & PILLINGER, C. T. 1983. A method for the identification and elimination of contamination during carbon isotopic analyses of extraterrestrial samples. *Meteoritics*, **18**, 137–154.

VALTER, A. A. & YERJOMENKO, G. K. 1996. Carbon minerals in rocks of astroblemes. *Meteoritics and Planetary Sciences*, **31**, A144–A145.

VISHNEVSKY, S. A. & PALCHIK, N. A. 1975. Graphite in rocks of the Popigai structure: destroying and transformation into another phase of carbon system (in Russian). *Geologia i Geophysika*, **1**, 67–75.

——, AFANASIEV, V. P., ARGUNOV, K. P. & PAL'CHIK, N. A. 1997. *Impact Diamonds: their Features and Significance*. United Institute of Geology, Geophysics and Mineralogy, Siberian Branch, Russian Academy of Sciences, Novoschirsk.

WOLBACH, W. S., GILMOUR, I., ANDERS, E., ORTH, C. J. & BROOKS, R. R. 1988. Global fire at the Cretaceous–Tertiary boundary. *Nature*, **334**, 665–669.

——, LEWIS, R. S. & ANDERS, E. 1985. Cretaceous extinctions: evidence for wildfires and search for meteoritic material. *Science*, **230**, 167–170.

YAMADA, K. & TOBISAWA, S. 1990. Crystal growth of radiating filaments of α-SiC formed by the conically converging shock wave technique. *Philosophical Magazine*, **89**, 2297–2304.

Impacts and marine invertebrate extinctions

NORMAN MACLEOD

Department of Palaeontology, The Natural History Museum, Cromwell Road, London SW7 5BD, UK (e-mail: n.macleod@nhm.ac.uk)

Abstract: The Phanerozoic history of life has been interrupted by at least 17 different stage-level episodes of elevated extinction intensity. These range from major turnovers that affected the majority of the Earth's extant biota (e.g. the Tatarian, Ashgill, and Maastrichtian events) to much smaller and perhaps more geographically localized disruptions. Review of available stage-level stratigraphic evidence suggests that bolide impacts do not exhibit a compelling correspondence to the Phanerozoic record of marine invertebrate extinctions. A much stronger case can be made for the repeated association of extinction events and episodes of long-term eustatic sea-level fall and/or major episodes of continental flood basalt volcanism. Moreover, during the last 250 Ma particularly severe extinction events have occurred during times of simultaneous continental flood basalt eruption and eustatic sea-level fall (e.g. Tatarian, Norian, Maastrichtian). Attempts to employ patterns of extinction selectivity to infer the physical cause of extinction events have foundered on the non-specificity of biological predictions as well as the seemingly unique biotic nature of each event. Although a few high-resolution biostratigraphic analyses have demonstrated that largely progressive patterns of faunal turnovers are consistent with co-extinction models under certain assumptions, this does not constitute a strong confirmation of the co-extinction model as a necessary or sufficient general explanation for Phanerozoic extinction events. In all likelihood, the nature of the fossil record precludes precise identification of extinction causal mechanisms. Nevertheless, comparative studies are able to evaluate the role both large and small extinction events have played in altering the long-term course of Earth's biological history.

Two hundred years ago Georges Cuvier astonished a public audience at the National Institute of Arts in Paris by presenting 'detailed and almost irrefutable evidence' for the idea that animals could become extinct (Rudwick 1972, p. 101). In doing this, Cuvier initiated a debate over the reality and causes of extinction that greatly influenced the zeitgeist of his time and led to a surge of interest in the topic of extinction that lasted well into the eighteenth century. Cuvier's preferred extinction mechanism was environmental change, but environmental change of a particular kind. Under his model, extremes of typical environmental variation would be met by species either toughing out the bad seasons (perhaps with some local extinction of populations on the periphery of the species' geographical range) or migrating to more favourable regions. In Cuvier's opinion, only some sudden and catastrophic environmental change could decimate a widespread and successful species so completely as to render it extinct.

Today, arguments similar to Cuvier's are being proposed to account for the extinctions of many fossil groups and, in particular, for episodes of mass extinction. Some theories would go so far as to claim that the overwhelming majority of all extinctions have resulted from very short term catastrophic mechanisms such as bolide impact (e.g. Alvarez *et al.* 1980; McLaren 1989; Raup 1991*a*, *b*; Alvarez 1997). Others have proposed that catastrophic extinctions have been visited upon the planet with a Newtonian regularity (e.g. Raup & Sepkoski 1984; Sepkoski & Raup 1986*a*, *b*; Raup 1989).

In Cuvier's time the evidence for seemingly violent upheavals in the recent geological past were coming to be appreciated through studies of Quaternary geomorphology and stratigraphy. In our own time the existence of such exotica as iridium anomalies, tsunami deposits, and spherule layers is often cited as evidence for a causal link between catastrophic mechanisms and mass extinctions (Alvarez 1997). At the extreme end of this spectrum we find scenarios in which the existence of certain extinction events themselves is regarded as sufficient grounds on which to infer the existence of the catastrophic mechanism(s) that *must* have produced them (e.g. McLaren 1970, 1989, 1996).

Despite the popularity of this 'new catastrophism' (Ager 1993) longer-term physical processes continue to be advanced as alternative extinction mechanisms. The two most consistently cited alternative mechanisms are flood basalt volcanism (e.g. Vogt 1972; McLean 1978, 1985*a*, *b*; Courtillot *et al.* 1986, 1988, 1996; Courtillot 1990) and eustatic sea-level change (e.g. Newell 1967, Hallam 1984, 1989). All three

MACLEOD, N. 1998. Impacts and marine invertebrate extinction. *In*: GRADY, M. M., HUTCHISON, R., MCCALL, G. J. H. & ROTHERY, D. A. (eds) *Meteorites: Flux with Time and Impact Effects*. Geological Society, London, Special Publications, **140**, 217–246.

mechanisms are thought to bring about organismal extinctions via climate changes, especially temperature fluctuations and the destruction of particular habitats (e.g. marine planktonic habitats).

To evaluate the degree to which these alternative extinction mechanisms account for the observed marine invertebrate fossil record direct evaluations must be undertaken in which the specific and reasonable predictions following from the various hypotheses are tested against the fossil record as it is currently known. It is important to remember that the debate over extinction mechanisms is not about the physicochemical evidence for impacts, volcanism, and/or sea-level rise *per se*. The existence of these processes and their ability to disrupt the Earth's climate and ecosystems are well established. Much less well understood is their relation (if any) to the historical record of life on Earth. Consequently, a detailed consideration of the fossil record in light of predictions based on specific causal mechanisms is unavoidable (Archibald 1996a).

Gross temporal correlations between mass extinctions and causal event classes

The most obvious test of a relation between bolide impacts and marine invertebrate extinctions is to determine whether there is a consistent correlation between their respective time series. The extinction patterns of most invertebrate fossil families and genera can be regarded as being known to the level of the stratigraphic stage (Sepkoski 1982, 1990, 1994; Sepkoski & Raup 1986*a*, *b*; Benton 1993). Detailed species-level biostratigraphies resolved to the level of the biozone are available for some marine invertebrate groups (see below), but these are the exception rather than the rule.

Lack of detailed knowledge of phylogenetic relations within biotic groups can bias the observed fossil record, especially in the case of recognizing mass extinction survivor species or lineages. This source of bias has been discussed by Archibald (1984), Patterson & Smith (1987, 1989), Smith & Patterson (1988), Carlson (1991), Norell (1992), Smith (1994), MacLeod (1996*a*), and MacLeod *et al.* (1997). Fisher (1991) and Foote (1996) have presented the case for retaining the traditional (i.e. non-phylogenetic) approach to taxic analyses, at least for certain types of studies. But even these researchers acknowledge that in many, if not most, instances phylogenetic data contribute a unique and important source of information and should be taken into consideration whenever possible. Unfortunately, the current state of marine invertebrate phylogenetic research is such that corrections to stratigraphic ranges based on phylogenetic relations are not possible for most invertebrate groups. Accordingly, the taxonomic summary presented below represents a compendium of stage-level observations that have not been corrected for phylogenetic relationship.

Several graphical summaries of taxonomic compendia are available for comparison with the physical record of Phanerozoic extinction events. The most detailed of these is Sepkoski's (1994) Phanerozoic generic compendia (Fig. 1(**A**)). As with all such compendia, taxa whose last appearance is recorded to fall within a particular stage (e.g. Maastrichtian) are listed as having ranged through the entire time interval. Under this convention all taxa become extinct at stage boundaries irrespective of where the taxon's last appearance was actually observed. Although this range-through convention is widely acknowledged to overestimate the number of extinctions that actually occur at stage boundaries (e.g. Raup 1994; MacLeod & Keller 1996), given the current state of stratigraphic knowledge for Phanerozoic marine invertebrate taxa this represents the most accurate overall summary of extinction intensity.

Inspection of Fig. 1(**A**) shows that extinction intensity has undergone substantial and repeated fluctuations during the past 600 Ma. Seventeen separate extinction peaks are shown in this summary, ranging from the Tatarian event

Fig. 1. Stage-level correlation between the Phanerozoic marine invertebrate extinction intensity (**A**, based on Sepkoski (1994)), the physical record of Phanerozoic bolide impact events (**B**, data from Grieve *et al.* (1996); (crater ages and sizes), Orth *et al.* (1990; iridium anomalies), Hut *et al.* (1987; tektite horizons), the Permian–Recent record of major flood basalt events (**C**, data from Courtillot (1996)), and the Phanerozoic record of eustatic sea-level change (**D**, based on Vail *et al.* (1977), Hallam (1984) and Haq *et al.* (1987)). Arrows mark: stages in which extinction events occur (**A**) (this diagram does not show the internal structure of these events, only a stage-level summary). Arrows mark large bolide impact events (**B**); flood basalt eruption events (**C**); and eustatic sea-level falls (**D**) that exhibit stage-level correlations with marine invertebrate extinction events. Filled circles beside each eustatic sea-level curve mark a correlation between a stage-level extinction peak and a long-term fall in eustatic sea level for the specific curve.

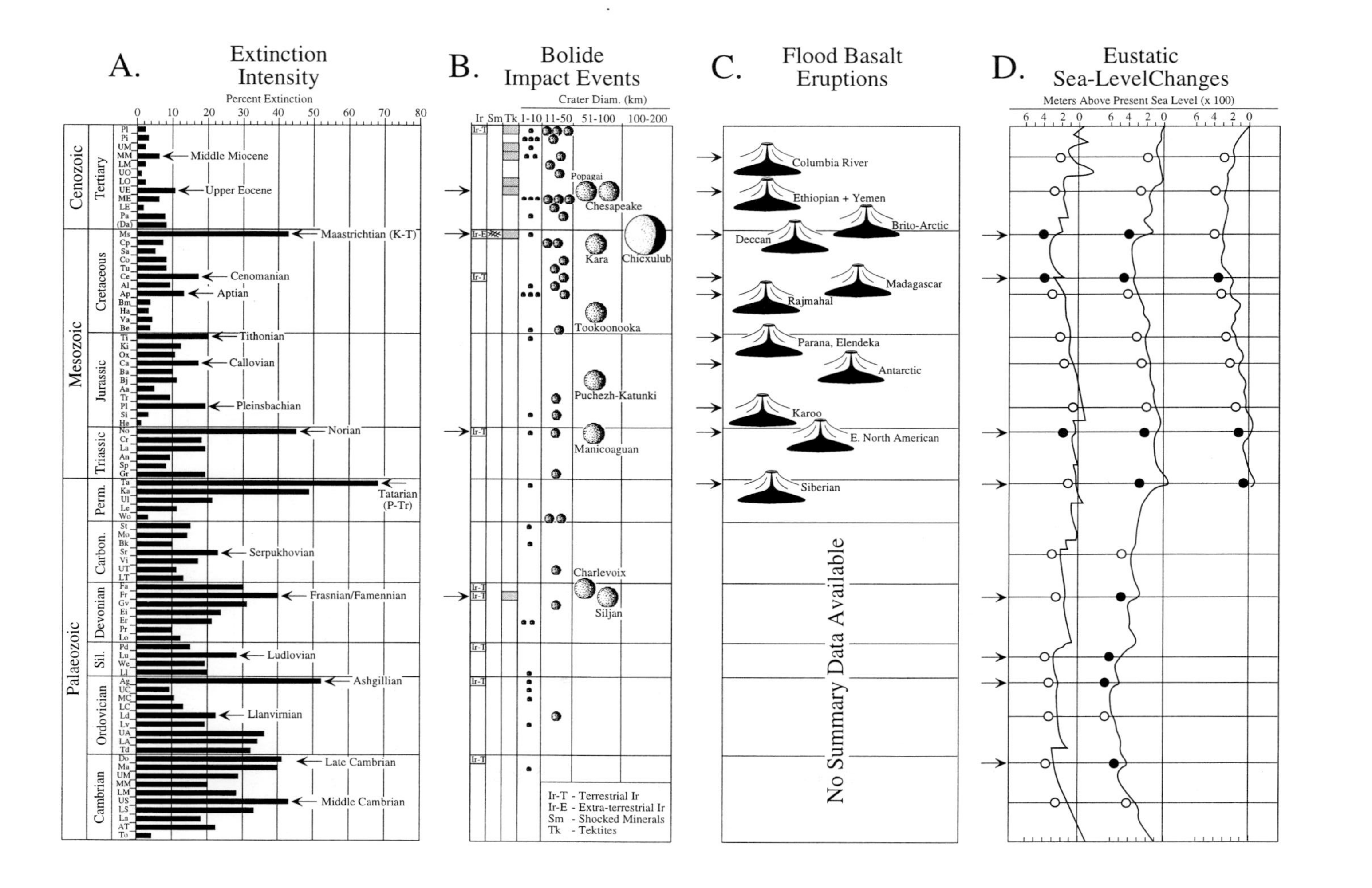
A. Extinction Intensity
Percent Extinction
Middle Miocene
Upper Eocene
Maastrichtian (K-T)
Cenomanian
Aptian
Tithonian
Callovian
Pleinsbachian
Norian
Tatarian (P-Tr)
Serpukhovian
Frasnian/Famennian
Ludlovian
Ashgillian
Llanvirnian
Late Cambrian
Middle Cambrian
Cenozoic
Mesozoic
Palaeozoic
Tertiary
Cretaceous
Jurassic
Triassic
Perm.
Carbon.
Devonian
Sil.
Ordovician
Cambrian
B. Bolide Impact Events
Crater Diam. (km)
Ir Sm Tk 1-10 11-50 51-100 100-200
Popagai
Chesapeake
Chicxulub
Kara
Tookoonooka
Puchezh-Katunki
Manicoaguan
Charlevoix
Siljan
Ir-T - Terrestrial Ir
Ir-E - Extra-terrestrial Ir
Sm - Shocked Minerals
Tk - Tektites
C. Flood Basalt Eruptions
Columbia River
Ethiopian + Yemen
Brito-Arctic
Deccan
Madagascar
Rajmahal
Parana, Elendeka
Antarctic
Karoo
E. North American
Siberian
No Summary Data Available
D. Eustatic Sea-LevelChanges
Meters Above Present Sea Level (x 100)

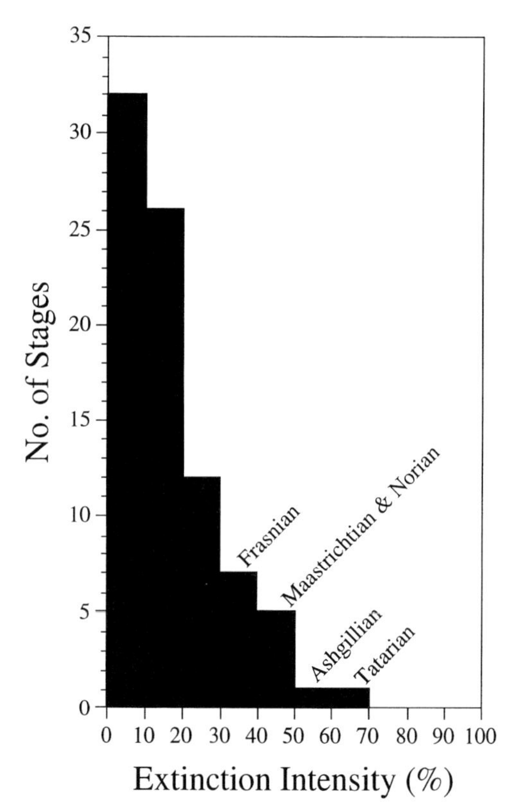

Fig. 2. Histogram of extinction intensities from Fig. 1(**A**). The shape of this distribution suggests that extinction intensity is distributed as a continuous series throughout the Phanerozoic, with the larger or 'mass' extinction events forming one tail of this distribution. The possible existence of biodiversity thresholds makes it possible (in theory) to accommodate a qualitative classification of extinction events in spite of the continuous nature of the extinction intensity distribution (see Van Valen 1994). However, the manner in which patterns of extinction selectivity differ among large and intermediate-sized extinction events has not been investigated in detail. It is likely that extinction events of differing intensity are capable of disrupting different types of ecological–evolutionary patterns in a quasi-continuous manner (see Jablonski 1996).

(*c.* 68% generic extinction) to the Middle Eocene event (*c.* 6% generic extinction). A histogram of extinction intensities for all Phanerozoic stages (Fig. 2) exhibits the topology of a 'decay curve', with large numbers of small events and a small number of larger or 'mass' extinctions. McKinney (1987), Steigler (1987), Raup & Boyajian (1988) and Raup (1991*a*, *b*, 1994) have pointed out that various extinction data point to the existence of a continuous range of extinction intensities within the fossil record. Jablonski (1986*a*, *b*, 1989; see also Kitchell 1990; Van Valen 1994) argued that despite the shape of the extinction intensity distribution different extinction classes could be recognized based on the presence (background extinction) or lack (mass extinctions) of ecological–geographical structure among the event's victims or survivors. These data imply the existence of 'threshold factors' that qualitatively subdivide the extinction intensity continuum. However, Jablonski (1996) found it likely that these threshold effects are themselves distributed in a more or less continuous manner throughout the various extinction events. Irrespective of this controversy, most palaeontologists continue to recognize a binary classification of extinction events, with 'mass extinctions' occupying the distribution's right-hand tail and 'background extinctions' occupying the left-hand tail. Although many of the intermediate extinction peaks (e.g. Upper Eocene, Cenomanian, Tithonian) are often referred to as 'mass extinctions', their ontological status with respect to the end-member categories remains uncertain.

Bolide impacts

The stratigraphic record of Phanerozoic bolide impacts can be derived from several sources, including crater age estimates, iridium anomalies, micro- and macrotektite horizons, and shocked mineral horizons. Of these, impact craters offer the most information in that the timing and size of the impact event can be directly measured within stated accuracy limits and subject to the smallest number of assumptions. However, like any physical record, impact craters, iridium anomalies, microtektite horizons, etc. can be removed by erosion, covered by sediments, destroyed via subduction, or fail to be discovered owing to limited outcrop area. These processes should conspire to render the physical record of bolide impacts less accurate as a function of the impact's age. In addition, some bolide impact proxies (e.g. local iridium anomalies) have been challenged as providing unambiguous evidence of impact occurrence and/or timing (Jablonski 1986*a*; Orth *et al.* 1990; Wang *et al.* 1993). Although these deficiencies in the stratigraphic record of bolide impacts are acknowledged, any claim that a general relation exists between bolide impacts and mass extinctions must be based on the known impact record. Indeed, this record has already been the subject of several extinction-related studies (e.g. Shoemaker 1983; Alvarez & Muller 1984; Rampino & Stothers 1984; Sepkoski & Raup 1986; Raup & Trefil 1987). The stage-level Phanerozoic bolide impact record is summarized in Fig. 1(**B**).

Interestingly, Fig. 1(**B**) shows that the most detailed record of impact activity is the record of impact craters. Although great effort has gone into the search for and analysis of impact proxies, this effort has not produced evidence of impact activity for stages within which impact craters are unknown. This reflects the relative frequency of large impact craters in the stratigraphic record, the expensive and time-consuming nature of impact proxy studies, the fact that some types of bolide impacts (e.g. comets) would not be expected to produce widespread impact proxy deposits, and the controversial nature of some data bearing on the interpretation of impact proxy deposits (see above). Regardless, although the stratigraphic record of impact proxies provides additional evidence in support of the impact crater record, it does not substantially modify that record.

Comparing Figs 1(**A**) and 1(**B**), the relation between marine invertebrate extinction intensity and the bolide impact record appears complex. The Pleistocene–Quaternary interval exhibits the greatest number of craters, most of these below 10 km in diameter. Failure of any other stratigraphic interval to record as many craters as the relatively short Pleistocene–Quaternary interval suggests that bolide impacts are far more common than would be suspected on the basis of the pre-Pleistocene stratigraphic record. The tailing off in impact frequency with age also supports this inference. Nevertheless, the fact that the impact-ridden Pleistocene–Quaternary interval fails to exhibit an extinction intensity peak suggests that the large number of smaller impacts whose records have been lost from the pre-Pleistocene stratigraphic record do not make a significant contribution to impact-mediated extinction patterns.

Within the remainder of the stratigraphic record there does not appear to be a strong correlation between impact frequency and extinction intensity, either for the Phanerozoic as a whole or for the last 250 Ma, where temporal control is much better (Fig. 3(**A**)). The Spearman correlation coefficient for these data is −0.219, which is not significant at the 95% confidence level. Some intervals of elevated extinction intensity (e.g. Norian, Aptian, Middle Miocene) are associated with local increases in the frequency of large impacts, whereas others (e.g. Cenomanian, Tithonian, Callovian) are not. On the whole, it would seem that the known record of impact frequency is a relatively poor predictor of marine invertebrate extinction intensity peaks.

Similarly, the relative size of impacting bodies does not seem to be deterministically associated with extinction intensity (Fig. 3(**B**)). An obvious exception to this is the Maastrichtian interval in which the fifth largest Phanerozoic mass

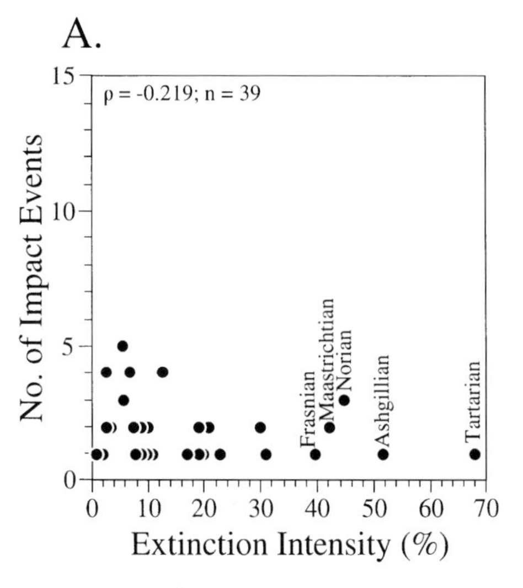

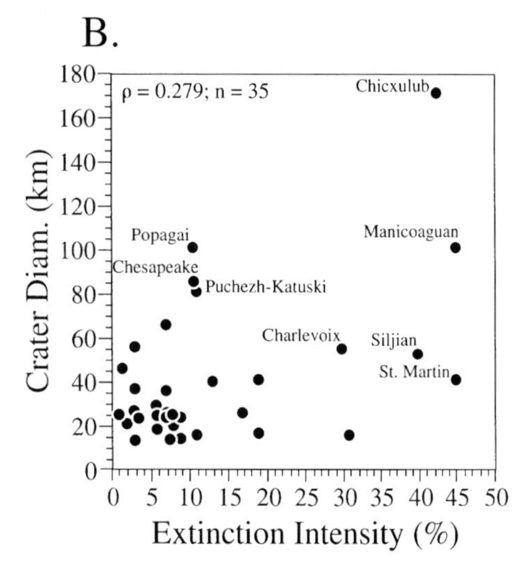

Fig. 3. Quantitative relation between extinction intensity and bolide impact history as inferred from (**A**) number of impact events per stage, and (**B**) size of impact events per stage. Spearman rank-order correlation coefficients (ρ-values) for these data are not significant at the 95% confidence level, indicating that these data conform to the expectations of a null model.

extinction (as measured by per cent extinction) is associated at the stage level with the largest Phanerozoic bolide impact crater (Chicxulub). (Note: Chicxulub is only the third largest crater known in all of Earth history.) Below this level, however, the relationship breaks down. Of the eight stages containing 51–100 km diameter impact craters, only four are associated with extinction intensity peaks. Moreover, large impacts appear to be associated with small extinction intensity values as often as with large. For example, the Upper Eocene interval has long been associated with unambiguous evidence for multiple impact events (see Prothero 1994) and recent investigations (e.g. Grieve *et al.* 1996; Bottomley *et al.* 1997) suggest that two sizeable impact events (Chesapeake Bay and Popagai) may have occurred within the space of as little as 3.0 Ma. Nevertheless, the Late Eocene interval is characterized by one of the smallest increases in genus-level extinction intensity in the entire Phanerozoic. The Spearman correlation coefficient for the data shown in Fig. 2(**B**) (including the Chicxulub event) is 0.279, which is not significant at the 95% confidence level. Together, these results suggest that given present data, neither impact number nor impact size exhibits a consistent association with extinction intensity.

Flood basalt eruptions

Flood basalt eruptions or (as they are sometimes called) large igneous provinces are massive crustal emplacements of mostly mafic extrusive and intrusive rock that originate through processes other than normal sea-floor spreading (Courtillot *et al.* 1996). Flood basalt eruptions are typically associated with mantle plumes or 'hot spots' (Coffin & Eldholm 1994) and are many orders of magnitude larger than any eruption recorded during human history. The stratigraphic association between flood basalts and increases in extinction intensity has been commented on by a number of researchers (e.g. Vogt 1972; Officer & Drake 1983, 1985; McLean 1985*a*, *b*; Morgan 1988; Courtillot *et al.* 1986, 1996; Stothers & Rampino 1990; Stothers 1993). Rampino & Stothers (1988) proposed that such eruptions may be a by-product of bolide impacts, but this has been directly challenged by White (1989) and is inconsistent with the widely accepted hypothesis that flood basalt events involve the lower mantle (e.g. Richards *et al.* 1989; Coffin & Eldholm 1994). Flood basalt eruptions may take place on the continents (where they form geomorphological features called 'traps') or on the ocean floor (where they form large subsurface plateaux). Of these two classes, continental flood basalt eruptions are thought to be the more climatically influential because erupted gases, particulate material and heat flow are injected directly into the atmosphere. The most recent compilation of Permian–Recent continental flood basalt ages (Courtillot *et al.* 1996) was used to produce Fig. 1(**C**).

Figure 4 shows a plot of estimated continental flood basalt ages compared with the durations of stages containing the nearest extinction peak. All major continental flood basalt eruptions in the last 250 Ma. have been associated with extinction peaks, with the exception of the Antarctic event (176.0 ± 1.8 Ma). Of the 11 remaining continental flood basalt events listed by Courtillot *et al.* (1996), six take place within or very close to the stages containing elevated per cent extinction frequencies. Levels of discordance within the events that do not fall within the predicted stage (according to age estimates given by Gradstein & Ogg (1996)) range from 3.5 Ma (Madagascar eruption–Cenomanian extinction) to 11.2 Ma (Serra Geral eruption–Tithonian extinction). To some extent, these discrepancies may be accounted for by intrinsic uncertainties associated with the isotopic methods used to date the continental flood basalt eruptions themselves and/or differences between competing geological time scales. However, the overall association between continental flood basalt eruption events and marine invertebrate extinction peaks over the last 250 Ma is both striking and statistically significant (Spearman correlation coefficient is 0.998). In this instance, available data suggest that over the last 250 Ma the temporal distribution of continental flood basalt eruptions is an excellent predictor of extinction intensity peaks at the stage level.

Eustatic sea-level change

Secular variations in eustatic sea level have long been suspected to play an important role in marine invertebrate mass extinctions (e.g. Moore 1954; Newell 1967; Schopf 1974; Hallam 1989). Owing to the use of sequence stratigraphy as an overarching conceptual framework for much contemporary stratigraphic research (especially in the hydrocarbon industry) eustatic sea-level curves have been determined for the entire Phanerozoic. The most widely used sets of sea-level curves are those of the Exxon research group (Vail *et al.* 1977; Haq *et al.* 1987)

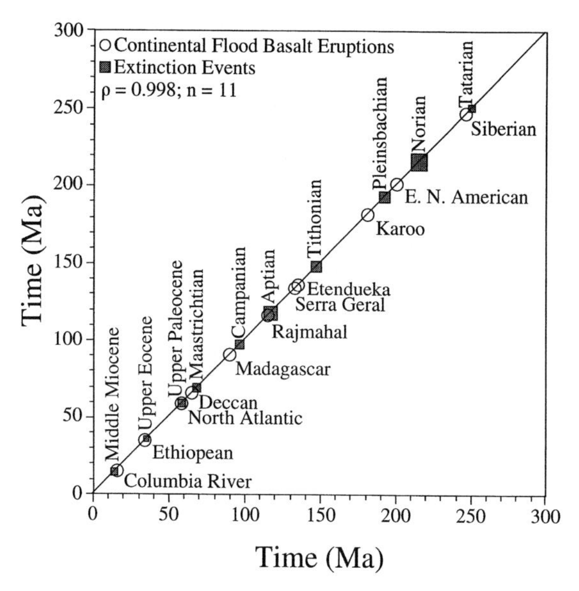

Fig. 4. Quantitative relation between the age of large continental flood basalt eruptions and stages containing extinction intensity peaks over the last 250 Ma (flood basalt age dates from Courtillot *et al.* (1996), geological time scale of Gradstein & Ogg (1996)). Spearman rank-order correlation for these data is 0.998, which is significant at the 95% confidence level.

and those of Hallam (1984). All three curves have been used as the basis for a large number of stratigraphic studies.

Figure 1(**D**) shows the long-term eustatic sea-level curves from Vail *et al.* (1977), Hallam (1984) and Haq *et al.* (1987). All three curves reflect long-term estimates of eustatic sea-level fluctuation. Although the use of long-term curves is preferable for gross trend analysis, it does underestimate the magnitude of short-term fluctuations in eustatic sea level (e.g. Pleistocene sea-level fluctuations) that may have been important at basinal and regional scales. As with the stratigraphic record of bolide impacts and continental flood events, the accuracy of the eustatic sea-level curve estimates is expected to decrease with increasing age.

The Exxon and Hallam curves show similar overall trends, but differ in the magnitude of eustatic sea-level change attributed to individual events. These differences arise because of differences in the methods used to construct the curves. The Mesozoic and Cenozoic portions of the Vail *et al.* (1977) and Haq *et al.* (1987) curves are based on Exxon seismic stratigraphic interpretations, augmented (in the case of Haq *et al.* (1987)) with data from selected marine sections exposed on continental platforms. The Palaeozoic portion of the Vail *et al.* (1977) curve was based on Sloss' (1963) interpretation of North American cratonic sequences. Hallam's (1984) sea-level curve is based on a combination of seismic stratigraphic results and palaeogeographical reconstructions of marine sediment coverage at various time intervals in the Phanerozoic calibrated against the hypsometric curve. Both methods of eustatic sea-level curve reconstruction have their advantages and disadvantages (see Hallam (1992) for a review), and their general pattern-level concordance, if not their estimates of sea-level change magnitudes, provides a measure of confidence in the results.

Comparison of these long-term trends with the marine invertebrate extinction record (Figs 1(**A**) and 1(**D**) suggests that although there is a general correlation between intervals of eustatic sea-level fall and mass extinctions at the stage level (e.g. Tatarian, Ashgill, Maastrichtian, Norian), many intermediate–smaller extinction events (e.g. Middle Miocene, Aptian, Pleinsbachian) appear to have taken place during eustatic sea-level rises. Hallam (1984, 1989) suggested that all six of Newell's (1967) mass extinction events are correlated with major eustatic sea-level regressions. The revised Phanerozoic marine extinction record of Sepkoski (1994) preserves all of Hallam's (1984, 1989) correlations and adds relatively minor events in the Cenomanian and Ludlow to this list. Associations between eustatic sea-level change and extinctions are not as frequent in the Vail *et al.* (1977) and Haq *et al.* (1987) compilations. It should be noted, however, that the short-term Haq *et al.* (1987) curve does show a sharp eustatic sea-level fall occurring just before the Cretaceous–Tertiary (K–T) boundary. Traditionally, the end Maastrichtian eustatic sea-level fall has been assigned a greater magnitude than that given by Haq *et al.* (1987). Some of the discrepancy between the Hallam (1984) and Haq *et al.* (1987) curves for this event may be attributable to the latter's use of the Braggs, Alabama, K–T section, which is incomplete across the uppermost Maastrichtian–lowermost Danian interval (Bryan & Jones 1989), as the basis for their reconstruction (see Hallam (1992) for an extended discussion). Because of discrepancies between the eustatic sea-level curve, and the high degree of 'guesstimation' (see Hallam 1992) in the specification of the long-term curve's slope, no formal statistical tests of the association between sea-level regressions and extinction intensity peaks are possible. However, many workers have commented that, of all geological phenomena, eustatic sea-level fall exhibits the most consistent correlation with large extinction events (see Jablonski (1986*a*) and Hallam (1989) and references therein).

Selectivity as a criterion to identify extinction event causes

Although the stage-level associations between bolide impact numbers and size and sea-level regression do not appear to be significant given present data, there are many problems with the stratigraphic record of these events that might bias coarse-grained analyses against finding the predicted relations. Moreover, these physical events could figure as a sole or contributory cause for certain extinction events, but not others. To explore these alternatives it is necessary to switch from a coarser to a finer analytical scale and ask the question: 'Do the predictions of biotic effects derived from the various extinction event scenarios (e.g. bolide impacts, flood basalt volcanism, sea-level regression) match the patterns observed in the marine invertebrate extinction record?'

This is the traditional scale at which mass extinction studies have been pursued. It is also the scale at which the greatest number of complications arise. The scientific enigma of extinction events is not that they have taken place (this was convincingly demonstrated by Cuvier in 1796), but turns on the question of the mechanisms by which such events are brought about. To provide precise answers to this question for specific extinction events we would like to know the predicted physical effects of the proposed extinction mechanisms on the global environment, the susceptibilities of particular biotic groups to these environmental–ecological factors, and the consistency with which physical predictions match biotic patterns for extinctions associated with known and/or suspected bolide impact events.

Table 1 lists the major killing mechanisms associated with bolide impact, volcanism, and sea-level regression. In terms of the type of mechanisms predicted to be in operation, there is little difference between bolide impact and volcanism scenarios. Both are regarded as events in which large amounts of gas and particulate matter are emplaced in either the troposphere (small events) or stratosphere (large events). Residence times of these materials range up to several months for large events (Alvarez *et al.* 1980; Toon *et al.* 1982; Pollack *et al.* 1983; Barron & Moore 1994) and would be expected to result in reduced light levels, acidification of rainwater, global cooling as a result of increased albedo, and increased irradiation of the Earth's surface because of ozone destruction. Longer-term effects may also involve global warming caused by increased concentration of greenhouse gases; principally CO_2 (O'Keefe & Ahrens 1982). Bolide impacts have been additionally associated with an initial very short term shock heating (Croft 1982), and the ignition of 'global' wildfires (Wolbach *et al.* 1985, 1990).

It is important to note that in many cases the calamitous effects predicted for bolide impact events are only predictions. There are no independent physical data confirming or refuting the thermal pulse, ozone depletion, or short-term global cooling. Retallack (1994, 1996) has interpreted geochemical and soils evidence from K–T sections in the western USA as being consistent with acid rain fallout, but his interpretations have been challenged by Archibald (1996*a*; see also Archibald and Retallack (1996)). Similarly, Archibald (1996*a, b*) has used sedimentological data to challenge the Wolbach *et al.* (1990) global wildfire scenario (see also Hallam and Wignall (1997)). On the other hand, there is abundant empirical support for the climatic and environmental effects of volcanism based on studies of historical eruptions (Officer & Drake 1983, 1985; McLean 1985*a, b*; Officer *et al.* 1987). Nevertheless, many aspects of the terrestrial and marine biotic records are plainly inconsistent with more extreme variants of bolide impact and volcanic eruption scenarios (see Jablonski (1986*a*) and Archibald (1996*a, b*) and references therein) and there is ample reason to suspect that some of these predicted effects

Table 1. *Effects of proposed extinction event causal mechanism*

Effect	Reference
Large bolide impact	
Reduced light penetration	Gerstl & Zardecki (1982), Toon *et al.* (1982), Pollack *et al.* (1983)
Lowered levels of photosynthesis	
Increased atmospheric particulates	O'Keefe & Ahrens (1982), Toon *et al.* (1982), Pollack *et al.* (1983)
Increased albedo	
Global cooling	
Global warming	
Greenhouse warming	
Increased cloud cover	Croft (1982), Jones & Kodis (1982), O'Keefe & Ahrens (1982)
Increased albedo	
Global cooling	
Global warming	
Increased atmospheric water vapour and CO_2	Emiliani *et al.* (1981), Toon *et al.* (1982)
Greenhouse warming	
Ozone depletion	Lewis *et al.* (1982)
Increased irradiation of surface	
Increased abundance of trace elements	Alvarez *et al.* (1980), Hsü *et al.* (1982)
Interference with biochemical processes	
Acid rain	Park (1978), Park & Menees (1978), Prinn & Fegley (1987)
Habitat destruction	
Interference with biochemical processes	
'Global' wildfires	Wolback *et al.* (1985, 1990)
Increased atmospheric particulates (see above)	
Increased cloud cover (see above)	
Increased CO_2 (see above)	
Habitat destruction	
Shock heating	Croft (1982)
Flood basalt volcanism	
Reduced light penetration	Deirmendijan (1973)
Lowered levels of photosynthesis	
Increased atmospheric particulates	Toon *et al.* (1982), Pollack *et al.* (1983)
Increased albedo	
Global cooling	
Global warming	
Greenhouse warming	
Increased cloud cover	McLean (1985)
Increased albedo	
Global cooling	
Global warming	
Increased CO_2	McLean (1985)
Greenhouse warming	
Ozone depletion	Stolarski & Butler (1979), Keith (1980, 1982)
Increased irradiation of surface	
Increased abundance of trace elements	Hansen (1991)
Interference with biochemical processes	
Eustatic sea-level fall	
Reduced shelf area	Schopf (1974), Simberloff (1974), Hallam (1989)
Species-area effect	
Habitat destruction	
Intensification of climatic gradients	Barron & Moore (1994)
Heating of continental interiors	

Table 1. (*continued*)

Effect	Reference
Eustatic sea-level fall (continued)	
Increased seasonality	Barron & Moore (1994)
Increased albedo	
Global cooling	
Reduced atmospheric CO_2	Barron & Moore (1994)
Global cooling	
Bottom-water anoxia	Hallam (1989)

may have been less intense than currently advocated by some research groups.

With respect to perturbations of the marine environment resulting from eustatic sea-level change, as eustatic sea-level changes of varying magnitudes have occurred repeatedly throughout the Phanerozoic (see Fig. 1(**D**), their effects on the marine ecosystem and marine invertebrate communities have been the subject of intense study (see, e.g. Schopf (1980) and Barron & Moore (1994) and references therein). In some cases the weight attributed to secondary biotic-effect predictions has changed in light of new data (e.g. Jablonski 1985, 1986*a*). Nevertheless, the predicted climatic perturbations associated with sea-level fall listed in Table 1 are supported by abundant empirical and theoretical evidence. It is disturbing to note, however, that in many cases these effects are identical to those produced under the bolide impact and large volcanic eruption scenarios (e.g. global warming, collapse of primary productivity) albeit as responses to different mechanisms.

The following example illustrates the complexities, and to a large extent the failures, of attempts to use patterns of extinction selectivity to infer extinction event causes. Sheehan & Hansen (1986; see also Arthur *et al.* (1987), Hansen *et al.* (1987*a*), Roy *et al.* (1990) and McGhee (1996)) suggested that invertebrate species whose feeding requirements tie them to primary production (e.g. suspension feeders, carnivores) suffer differentially intense extinction rates relative to detritus feeders across the K–T boundary and, by implication, other extinction horizons where bolide impact is thought to have been a driving force. The original study (Sheehan & Hansen 1986) was based in part on a literature review of Late Cretaceous and Early Tertiary marine invertebrate studies, but linked to the K–T impact event on the basis of molluscan data from the Brazos River K–T section (Hansen *et al.* 1984, 1987*a*). Sheehan & Hansen (1986) concluded that this pattern resulted from a collapse in primary productivity at the K–T boundary brought about by 1–2 months impact-generated darkness (the 'Strangelove Ocean' of Hsü & MacKenzie (1985, 1990)). (Note: To the extent that large volcanic eruptions can affect global marine productivity via attenuation of light intensities, alteration of circulation patterns, increase in atmospheric concentrations of greenhouse gases, etc., the same predictions would apply.)

Sheehan & Hanson's deposit feeder = survivorship model was challenged in marine settings by Bryan & Jones (1989), who reported the opposite trend (an increase in suspension feeders in the lowermost Danian) from the Braggs, Alabama, K–T section, and later by Archibald (1996*a*, *b*) for terrestrial stream communities. Rhodes & Thayer (1991) confirmed the bias toward deposit-feeding and sessile bivalves over suspension-feeding and active species using a Maastrichtian–Danian stage-level dataset consisting of 424 generic records. Owing to the stage-level resolution afforded by these data, however, Rhodes & Thayer (1991) concluded that their results were only consistent with the hypothesis of reduced primary productivity and declined to associate the productivity drop with any specific causal mechanism. Paul & Mitchell (1994) further challenged the association of productivity collapse with bolide impact by pointing out that a very similar pattern was present at the Cenomanian–Turonian event (which is not a major extinction event and not felt to be impact related). Paul & Mitchell (1994) suggested that a drop in primary productivity was a general feature of marine extinction events and could be generated by rapid eustatic sea-level fall in addition to bolide impact.

In 1993, Hansen *et al.* published new molluscan data from Brazos River K–T sections. This report concluded that in the lowermost Danian virtually all the deposit-feeding bivalves recovered from Brazos River were not K–T survivors themselves, but new species that originated in the Danian. Of the three bivalve species reported to survive the K–T boundary event in this new study, two were suspension feeders! These data suggested (in contrast to the results of Sheehan & Hansen (1986)) that deposit-feeding and suspension-feeding bivalves exhibited comparable extinction rates in the Brazos River section. This alternative interpretation was also supported by Jablonski & Raup (1995) (see also Jablonski (1996)), who used a global database to show that K–T extinction resistance was not a uniform property of all deposit-feeding molluscs, but varied between taxonomic groups, with some deposit feeders (e.g. Tellinidae, Aporrhaidae) suffering extinction rates as high as those of suspension-feeding molluscs. In addition, much recent ecological research has shown that the deposit-feeding life mode tracks changes in marine primary productivity to a much greater extent than was previously supposed (Lambshead & Gooday 1990; Rex *et al.* 1993; Levinton 1996).

These studies represent convincing challenges to Sheehan & Hansen's (1986) model for preferential survivorship of short-term extinction events by deposit feeders or scavengers and question any unique relation between these patterns and bolide impact events. Although Sheehan *et al.* (1996) made a large number of references to the possibility that marine invertebrate K–T survivors had switched from suspension feeding to deposit feeding, these speculations remain unsupported by any empirical data. Indeed, Jablonski (1996) pointed out that tellinid bivalves, modern representatives of which are known to be able to switch facultatively from suspension feeding to deposit feeding (Levinton 1991; Kamermans 1994; Lin & Hines 1994), suffered some of the highest K–T extinction intensities of any molluscan group.

Patterns of extinction selectivity are present for the K–T event (see MacLeod *et al.* 1997), as they are for the Frasnian (McGhee 1996), Tatarian (Erwin 1993) and Late Eocene (Prothero & Berggren 1992; Prothero 1994) events. The problem lies in the fact that each extinction event has so many unique attributes, and each attribute is consistent with so many possible causes, that no generalizations are possible (Van Valen 1994). Table 2 provides a rough taxonomic list of victims and survivors for the four best-studied extinction events. There is little in the way of a consistent taxonomic signal in these data that could be used to subdivide them into impact-associated (Maastrichtian, Eocene), non-impact associated (Tatarian), or equivocal (Frasnian) groups even if one were willing to specify the proximate extinction mechanisms *post hoc*. Moreover, such tables hide the fact that within each extinction a wide variety of adaptive types, intrinsic diversities, and extinction modes exist. When common factors are found (e.g. the decline of reef ecosystems during the Frasnian, Tatarian, and Maastrichtian events; see Jablonski (1989)), they fail to subdivide the extinction spectrum in any meaningful way other than to separate large events from smaller ones. This frustrating randomness in the biotic extinction signal also pervades finer-scale analyses. Indeed, the most generally accepted result of this research effort is that mass extinction events differ from background extinction intervals in their *lack* of statistically significant within-group structure (Jablonski 1986*a*, *b*, 1989).

This failure, despite intense efforts spanning decades, to uncover a consistent biotic–ecological–distributional signal that unequivocally groups extinction events into separate classes has serious implications for the study of extinction causes. As pointed out by Raup (1991*b*), the only way to establish cause and effect in historical data is to document patterns of coincidence. If every extinction event is unique (as increasingly seems the case) it will be logically unjustified to prefer one causal scenario over another on biotic grounds, and hence impossible to develop an overarching explanation for groups of individual extinction events.

There is little doubt that environmental variation exhibits a stochastic pattern of random fluctuation about a mean value (Elton 1958; Ehrlich & Birch 1967; May 1974; Pimm 1991). Departures from this characteristic environmental mean will take place at a variety of scales, from normal seasonal variation through 'once in a decade', 'once in a century', 'once in a millennium', type perturbations to 'once in a geological period' perturbations and beyond. Bolide impact, massive flood basalt volcanism, and long-term eustatic sea-level change all belong to the latter tail of this environmental change distribution. Although there is ample evidence that all these mechanisms can produce extinction events, it is reasonable to expect that fortuitous conjunctions in the time series of each of these mechanisms would lead to anomalously large extinction events. The correlations shown in Fig. 1 suggest that this has been the case for the three largest extinction events of the last 250 Ma (Tatarian, Norian, Maastrichtian) in that each took place during a conjunction between large continental flood basalt eruptions and a relatively rapid drop in long-term eustatic sea level. Raup (1991*b*) criticized the connection

Table 2. *Taxonomic victims and survivors for the four best-studied extinction events*

	Extinction event			
Taxonomic group	Frasnian	Tatarian	Maastrichtian	Mid–Upper Eocene
Phytoplankton	●	○	○	○
Coccolithophores	–	–	●	●
Dinoflagellates	–	–	○	○
Diatoms	–	–	○	○
Zooplankton				
Planktonic Foraminifera			●	●
Benthic Foraminifera	●	○*	○	○
Radiolaria	○	○	○	○
Arthropods				
Ostracods	●	●	○	○
Trilobites	●	●	–	–
Sponges	○	○†	○	○
Cnidaria	●	●	○	○
Bryozoans	○	●	○	○
Brachiopods	●	●	○	○
Molluscs				
Gastropods	○	○	○	●
Bivalves	○	⊙	⊙	●
Cephalopods	●	●	●	○
Echinoderms				
Asteroids	?	?	○	○
Crinoids	●	●	○	○
Echinoids	?	?	⊙	●

●, Event victim; ⊙, selective extinctions; ○, event survivor.
* Excluding fusulinids.
† Excluding reef sponges.
Data from Erwin (1993), Prothero (1994), McGhee (1996), MacLeod *et al.* (1997).

between sea-level fluctuations and extinctions by arguing that the dramatic short-term Pleistocene glacio-eustatic sea-level fluctuations did not produce sizeable extinction events. However, long-term eustatic sea level was already low during the Pleistocene (resulting in much less flooding of the continental platforms than in the Late Permian, Late Triassic and Late Cretaceous; see Fig. 1(**D**)), and Jackson (1995) has shown that the Late Pliocene onset of glaciation did result in dramatic extinctions among tropical molluscs, corals, and planktonic Foraminifera. The fact that biotas which have recently been stressed by particular perturbations often show ecological resistance to subsequent perturbations of the same types has widespread empirical (Pimm *et al.* (1995) and Greuter (1995) and references therein) and theoretical (Nee & May 1992; May *et al.* 1995; Tilman *et al.* 1996) support. These results adequately account for the so-called Pleistocene extinction anomaly. In addition, this same model can be used to form a general understanding of why extinction events resulting from the fortuitous conjunction of different causal mechanisms have such dramatic effects on the global biota.

This type of stochastic system constitutes a null hypothesis for the biotic interpretation of extinctions. To prefer the identification of a particular overriding causal process, it is not only necessary to show that a large proportion of accidents or extinctions took place during the suspect time interval, but that they also occur for reasons directly attributable to the proposed cause, and occur in comparable frequencies every time the proposed cause is active. These are admittedly stringent requirements, but they are the minimum necessary to establish the pattern of coincidences required to test extinction models using biotic data within a cause–effect framework. Thus far, biotic data from Phanerozoic marine invertebrate extinction events have proven to be too generalized to be used to conduct such tests. Present patterns strongly suggest that each major Phanerozoic environmental perturbation was itself multicausal and operated on a metastable biotic system whose initial conditions largely determined the extent of the corresponding biotic extinction (Jablonski 1996). Within such a system, and given the coarse levels of biotic resolution provided by the fossil record, it may well be impossible to use the contingent pattern of extinction selectivity to uniquely identify specific causal factors.

Ambiguities in high-resolution biostratigraphical data

Given the generalized and ambiguous nature of biotic predictions following from the various extinction causal event scenarios, most palaeontologists have turned to the biostratigraphic record to provide criteria that can be used to distinguish between alternative causal models. However, like the extinction selectivity research programme (see above), tests involving the time interval over which extinctions take place are bedevilled by similarities in the predictions of various causal models and by ambiguities in the nature of the data on which they are based.

As mentioned above, the maximum estimated time interval over which the proximate effects of individual bolide impacts would be expected to last range from 2 to 24 months (Toon *et al.* 1982; Pollack *et al.* 1983). Longer-term effects (e.g. greenhouse warming) cannot be ruled out (e.g. Ryder 1996), but the time scale over which these effects would be predicted to last has never been specified because the environmental feedbacks controlling these durations are not well understood. In most instances, though, the demonstration of 'rapid' extinction rates has been used as evidence favouring bolide impact causal models (e.g. Smit 1982, 1990; Pospichal 1993, 1994; Ward 1995; Marshall & Ward 1996). Flood basalt volcanic events are known to occur over time scales ranging from <1 Ma (Deccan, Parana–Serra Ceral) to 8 Ma (Siberian; see Courtillot *et al.* (1996) and references therein), but these estimates refer to the aggregate period over which a number of genetically related, yet distinct eruptions occur. Individual eruptions take place over much shorter time scales (10^0–10^1) and their proximate environmental effects would be expected to be most pronounced over these intervals. Eustatic sea-level changes occur over a large range of time scales from parasequences (10^4–10^5 years) to supercycles (10^7–10^8 years) (Vail *et al.* 1977; Hallam 1984; Haq *et al.* 1987). These data suggest that under the best of circumstances the duration of an extinction event is only likely to be diagnostic at the extreme lower end of the scale (months to years).

The stage-level resolution of most biostratigraphic data effectively prevents them from being used to examine hypotheses that specify very short extinction durations. Stages' timespans vary widely, but they are on average several million years long. Therefore, to assess the question of extinction duration palaeontologists must turn to studies of individual stratigraphic sections and/or cores. Stratigraphic sections and cores, however, present their own complications. Prominent among these are the following.

Effect of hiatuses

Hiatuses are intervals of time missing in the stratigraphic record. In fossiliferous sedimentary sequences hiatuses most often form as a result of non-deposition or active erosion. They are present at all temporal and many spatial scales. From a historical perspective the major faunal changes used to subdivide the Phanerozoic (e.g. Permo-Triassic boundary, Cretaceous–Tertiary boundary) were often (inadvertently) placed at hiatuses because they accentuate faunal turnover patterns.

The significance of hiatuses with respect to extinction studies was first discussed by Birkelund & Hakansson (1982) and has been subsequently mentioned by many workers (e.g. Jablonski 1986*a*; MacLeod 1996*a*). MacLeod & Keller (1991) reviewed the biostratigraphy–chronostratigraphy of all major K–T boundary sections or cores and found that the majority contained boundary hiatuses as inferred by the absence of entire biozones. These included such well-known sections as Stvens Klint, Denmark (no evidence for the presence of the lowermost two Tertiary planktonic foraminiferal biozones, see also Hakansson & Hansen (1979), Perch-Nielsen *et al.* (1982) and Ekdale & Bromley (1984)), Gubbio, Italy (no evidence for the presence of the lowermost Tertiary planktonic foraminiferal biozone; see also Luterbacher & Premoli-Silva (1964)), and Zumaya, Spain (no evidence for the presence of the lowermost Tertiary planktonic foraminiferal biozone (Herm 1965; Smit & Ten Kate 1982); see also Ward & Kennedy (1993)). The fact that each of these sections contains a well-developed iridium anomaly is not inconsistent with a boundary hiatus, but can be accounted for under a variety of geochemical models (e.g. Goldschmidt 1954;

Keith 1982; Rucklidge *et al.*, 1982; Hansen *et al.* 1987*b*; Crockett *et al.* 1988; Tredoux *et al.* 1988; Colodner 1992; Sawlowicz 1993; Burns *et al.* 1996). Comparably detailed stratigraphic studies have not been carried out for other extinction event horizons–intervals. However, if an extinction event horizon coincides with a stratigraphic hiatus, the character of biotic change across that horizon is unknowable.

Circular arguments for dating

The highest temporal resolutions available for the majority of Phanerozoic extinction events come from biostratigraphic, as opposed to isotopic, dating. As a result, the location of physical event horizons (e.g. iridium anomalies, tektite horizons) with respect to faunal or floral event horizons is of critical importance. For example, the observation of an 18 m section of bedded, undisturbed marls containing a diverse assemblage of upper Maastrichtian planktonic Foraminifera overlying the Chicxulub breccia deposits in the Chicxulub 1 well (Ward *et al.* 1995, see also Löpez Ramos (1973), Jéhanno *et al.* (1992) and Leroux *et al.* (1995)) suggests that either the Chicxulub impact event did not take place at the K–T boundary (as defined by biostratigraphic criteria), or that survivorship among late Maastrichtian planktonic Foraminifera is much greater than previously thought. Inconsistencies between data of different types must be reconciled on the basis of external criteria, as differences of opinion exist on the priority of biotic and physical data for the proposed K–T boundary definition (compare Smit *et al.* (1992), with Keller *et al.* (1993)). In such cases analytical approaches that encompass all available data to achieve an internally consistent consensus should be preferred (e.g. MacLeod & Keller 1991; MacLeod 1995*a–c*, 1996*b*, *c*).

Arbitrary interpretations of biotic disappearances or appearances

The disappearance or appearance of species from local sections is an inherently ambiguous observation that may result from a variety of different processes. In addition to the obvious global extinction or global origination interpretation, disappearances or appearances may also result from emigration or immigration events caused by local facies shifts and speciation events (usually coupled with hiatuses and/or facies shifts) in which descendants are assigned a different taxonomic name and so counted as a different unit in statistical summaries of biotic turnover patterns. An example of the former includes the benthic foraminiferal study of Speijer (1994) and Speijer & Van der Zwann (1996), who argued for an 'abrupt extinction' of this group at the K–T boundary in the southern Tethys. Inspection of the data upon which this study is based (Speijer 1994, table 1) shows that of 51 listed taxa 28 (54.9%) disappear within ±5 cm of the El Kef K–T boundary. Nevertheless, of these 28 taxa all but six (11.8%) reappear higher in the section. The 22 taxa that undergo temporary disappearances are formally classified as Lazarus taxa (Jablonski 1986*a*) and should not be counted in estimates of global or regional extinction intensity. Miller (1982), Keller (1988*a*, 1992), Thomas (1990*a*, *b*), Widmark & Malmgren (1992) and Coccioni & Galeotti (1994) concluded that benthic foraminiferal faunas did not undergo significant extinction across the K–T boundary. As an example of the latter phenomenon among K–T echinoids, Andrew Smith (pers. comm. 1997) noted that the genus *Actinophyma* Cotteau & Gauthier (Campanian–Maastrichtian) is morphologically identical to the late Thanetian genus *Acanthechinus* Duncan & Sladen, suggesting that these two taxa actually constitute a single K–T survivor lineage. Similarly, Carlson (1991) has shown that Permo-Triassic extinction patterns among brachiopods are strongly attenuated by consideration of phylogenetic data (see also general discussion of Permo-Triassic extinction intensity estimates by Erwin (1993)).

The Signor–Lipps effect

The Signor–Lipps effect (Signor & Lipps 1982; see also Shaw (1964), Raup (1989), Archibald (1996*a*) and MacLeod (1996*a–c*), suggested that interactions between sample size, relative abundance, susceptibility to diagenesis, and environmental tolerance can conspire to distort the last appearance pattern of taxa that disappear during an extinction event. Consequently, 'if the distribution of fossil last occurrences is random with respect to actual biotic extinction, then apparent extinction will begin well before a mass extinction and will gradually increase in frequency until the mass extinction event, thus giving appearance of a gradual extinction' (Signor & Lipps 1982, p. 291). As such, the Signor–Lipps effect is a special case of the Lazarus effect and changes the geometry of apparent faunal turnover patterns in a manner opposite to that of the hiatus effect.

Several researchers have used the Signor–Lipps effect to argue that gradual faunal turnover patterns are consistent with catastrophic extinction mechanisms (Jablonski 1986*a*; Raup 1986*b*; 1989, 1991; Marshall & Ward 1996; Smit & Nederbragt 1997). Although this conclusion is correct in a strict sense, it represents a half-truth that substantially distorts the intention of Signor & Lipps (1982). The Signor–Lipps effect is a statement of fundamental uncertainty that is best understood to state that the true geometry of fossil species' first or last appearances can only be inferred within fairly broad stratigraphic limits. With respect to the question of extinction studies Signor & Lipps (1982, p. 295) explicitly stated that 'our arguments should not be interpreted as support for the impact hypothesis or any other theory invoking a catastrophe as the extinction mechanism. The evidence at hand is as compatible with a gradual extinction as a catastrophic one.' (Note: this ambiguity is also present in Jablonski's (1986*a*) use of Lazarus taxa to evaluate apparent faunal turnover patterns.) Several statistical tests and data analytical strategies designed to evaluate the conformance of the observed fossil record with the predictions of the Signor–Lipps effect have been developed (Koch & Morgan 1988; Marshall 1995; MacLeod 1996*b*, *c*), but thus far these have only been applied to the extinction event records of a few groups in local stratigraphic successions.

The Zombie effect

Just as the Lazarus and Signor–Lipps effects bias the extinction record by under-representing local stratigraphic ranges, the Zombie effect (Archibald 1996*a*; see also Barrera & Keller (1990) and MacLeod & Keller (1994)) postulates that local stratigraphic ranges may be artificially extended by the reworking of fossils from underlying sediments. The Zombie effect has been used to argue that seemingly anomalous occurrences of 'Cretaceous' nannoplankton (Pospichal 1993, 1994, 1996) and planktonic Foraminifera (Liu & Olsson 1992, 1994; Olsson & Liu 1993) represent reworked Cretaceous specimens. However, Archibald (1996*a*) pointed out that reworking cannot be assumed to only be confined to time intervals following mass extinction events. It is far more likely to be a common factor throughout the fossil record.

Several recent studies have lent empirical support to this inference by demonstrating that 'modern' surface assemblages of molluscs and benthic Foraminifera represent time-averaged assemblages on the order of 10^3–10^4 years (Broecker *et al.* 1990; Flessa 1993; Martin 1993; Flessa & Kowalewski 1994). If these estimates are valid for marine invertebrate taxa as a whole it would seem that time intervals of this order might represent the maximum resolution available in the fossil record. With respect to extinction event horizons, the Zombie effect would also operate to artificially extend the ranges of taxa that actually became extinct before the event (and so accentuate the magnitude of the apparent extinction) as well as extending the ranges of taxa that became extinct at the event horizon itself. Under favourable circumstances the Zombie effect can be tested for individual species (e.g. Barrera & Keller 1990; Huber 1996) and entire faunas (e.g. MacLeod & Keller 1994; MacLeod 1996*b*, *c*).

Case studies

To illustrate the strategies used and complications encountered in studying marine invertebrate extinction events in local sections, controversies over the K–T records of planktonic Foraminifera and ammonites are briefly reviewed below.

The K–T planktonic foraminiferal controversy

More is known about the K–T record of planktonic Foraminifera than that of any other organismal group across any major extinction interval. Thanks to the efforts of Gerta Keller, Jan Smit and their colleagues, an extremely detailed, high-resolution, global record of upper Maastrichtian–lower Danian planktonic foraminiferal faunas is available for study (see Smit 1977, 1981, 1982, 1990; Keller 1988*b*, 1989*a*, *b*, 1993; Smit *et al.* 1988; Canudo *et al.* 1990; Keller & Benjamini 1991; Keller *et al.* 1994, 1995). In addition, this record has been extensively evaluated for bias related to the hiatus–facies effect, the Zombie effect, and the Signor–Lipps effect (Keller & MacLeod 1994; MacLeod 1996*b*, *c*).

Although planktonic Foraminifera have long been known as one of the primary victims of the K–T mass extinction (Newell 1967), the actual nature of their K–T turnover pattern across this interval has been controversial. Bramlette (1965) argued that the K–T planktonic foraminiferal turnover was abrupt ($\ll$1.0 Ma), though he acknowledged that the deep-sea sections upon which he based his conclusions contained a

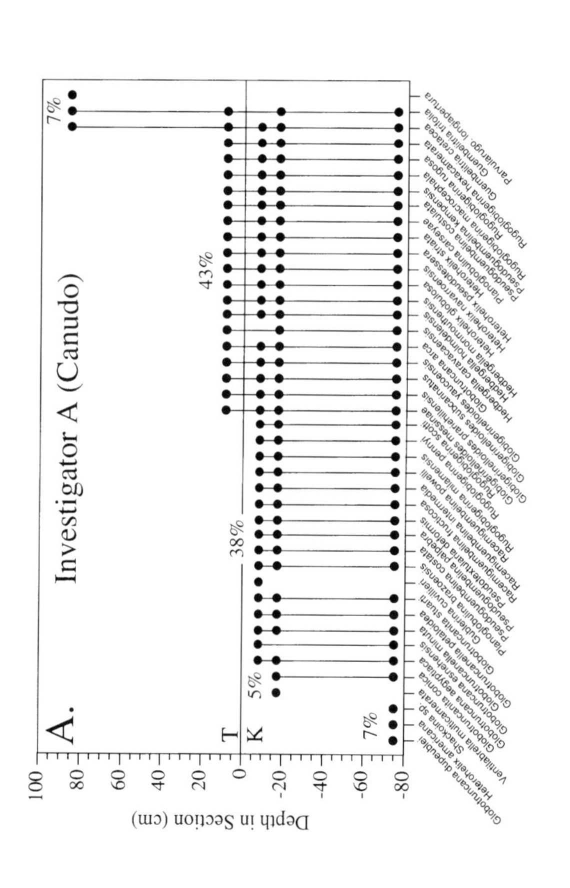
A.
Investigator A (Canudo)
Depth in Section (cm)
T
K
7%
43%
38%
5%
7%

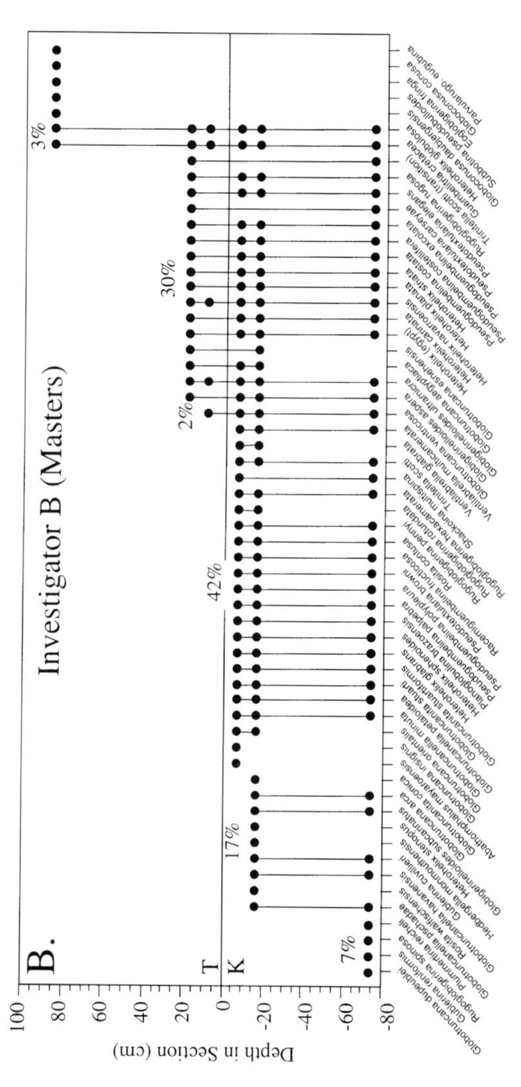
B.
Investigator B (Masters)
Depth in Section (cm)
T
K
3%
30%
2%
42%
17%
7%

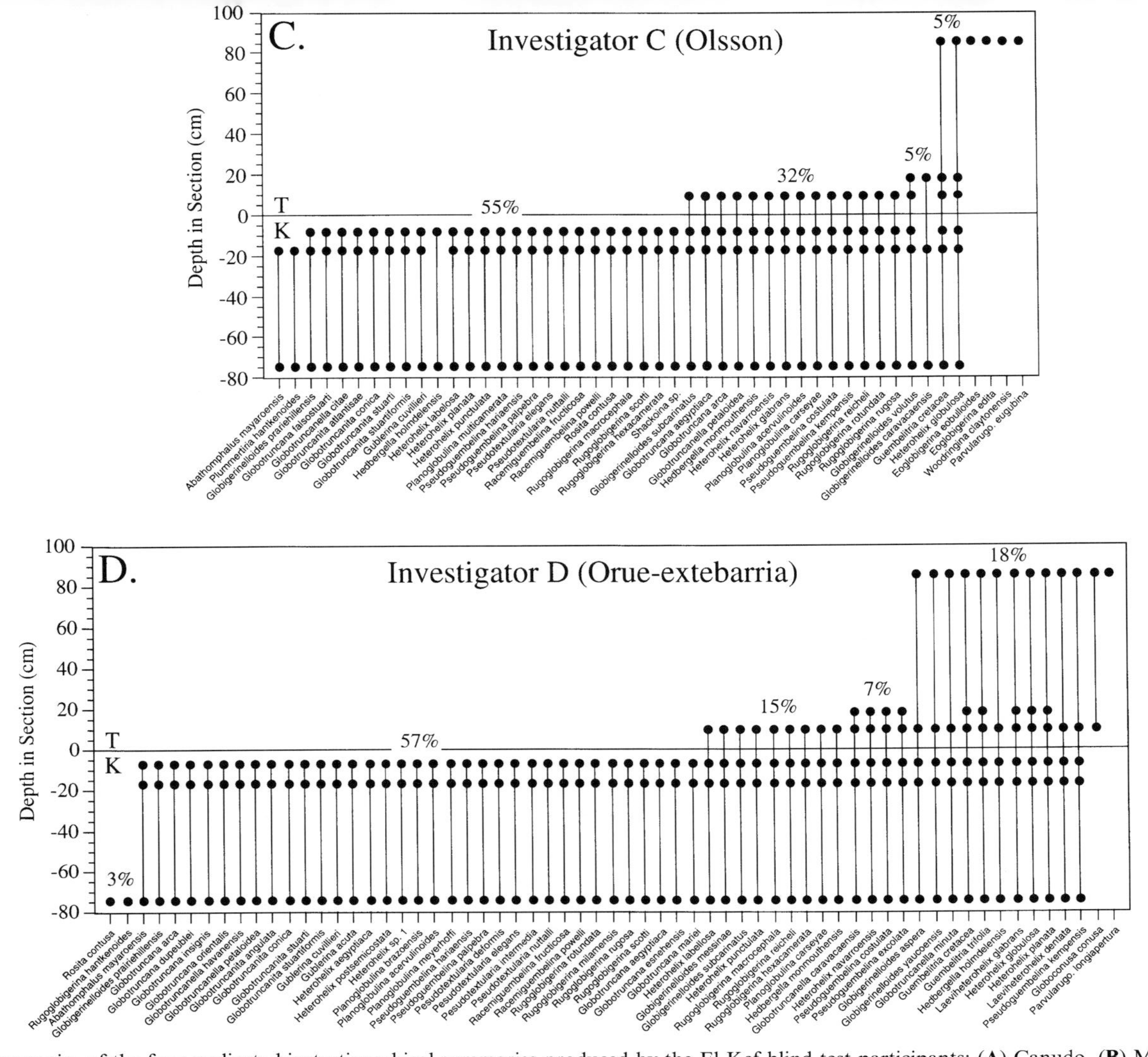

Fig. 5. Graphic summaries of the four replicate biostratigraphical summaries produced by the El Kef blind test participants: (**A**) Canudo, (**B**) Masters, (**C**) Olsson, (**D**) Orue-extebarria. (See text for discussion.)

boundary hiatus. Smit (1981, 1982, 1990; see also Smit *et al.* (1988)) also argued that a virtually complete replacement of Cretaceous planktonic foraminiferal species (with one or two exceptions) took place coincident with the K–T boundary and in the space of less than 50 years. On the other hand, Keller and colleagues have documented progressive patterns of planktonic foraminiferal extinction and replacement, beginning below the K–T boundary and continuing into the lowermost Danian in all biostratigraphically complete K–T successions, including many of those studied by Smit (see references above).

To determine the true pattern of planktonic foraminiferal turnover in the K–T boundary stratotype section (El Kef, Tunisia), this succession was recently the subject of a 'blind-test' (Ginsburg 1997) in which four independent investigators examined Cretaceous and Tertiary planktonic foraminiferal samples from the El Kef section that were identified to them only by number (Canudo 1997; Masters 1997; Olsson 1997; Orue-extebarria 1997). Results showed that each investigator identified varying numbers of Cretaceous species disappearing from the section before the K–T boundary as well as reporting other 'Cretaceous' species above the K–T boundary in lowermost Danian sediments (Fig. 5).

Several commentators have concluded that the variability exhibited by the blind test investigators' results precludes resolution of the Smit–Keller controversy (e.g. Ginsburg 1997; Lipps 1997), or that they support the Smit model of catastrophic extinction (e.g. Smit 1994; Ward 1995*a*, *b*). Certainly these results are more variable than the organizers of the blind test expected. They are not, however, without precedent. Zachariasse *et al.* (1978) conducted and published the results of a similar blind test based on samples collected from a Sicilian lower Pliocene section for the purpose of assessing the reproducibility of micropalaeontological data. Although planktonic Foraminifera were not considered in the Zachariasse *et al.* (1978) study, benthic Foraminiferal analyses (organized by M. J. Brolsma) returned results fully consistent with those of the El Kef blind test. These data suggest that much higher degrees of variability between the results of different systematists are present in micropalaeontological (and probably in macropalaeontological) biostratigraphic data than is commonly assumed (see also Smith & Buzas (1986) and Funnell & Swallow (1997)).

As the El Kef blind test results appear consistent with results from comparable reproducibility studies it seems fair to ask how consistent these results are with the original El Kef studies of Keller (1988*a*, *b*; see also Keller (1992)) and Smit (1982; see also Smit *et al.* (1988)). Unfortunately, Smit has not published tables of his species-level biostratigraphic observations. The studies reported by Keller (1988*b*) and Keller *et al.* (1995) contain such tables. (Note: the studies reported by Keller *et al.* (1995) were conducted using an separate suite of samples from those employed in the El Kef blind test.) Table 3 presents the results of a rank-order correlation analysis among the El Kef blind test participants. The rank-order correlation coefficient is insensitive to variations in the faunal lists of different investigators, but does not correct for differences in species concept or nomenclature (e.g. referring to the same morphotype by different names).

Rank-order correlations among the blind test investigators' results range from 0.748 (Orue-extebarria–Canudo) to 0.271 (Masters–Olsson) with a mean of 0.473 (Table 3). Although this mean is slightly larger than the mean correlation between the blind test investigators' results and those of Keller (1988*b*, $\rho = 0.416$), it is decidedly smaller than the mean correlation with the Keller *et al.* (1995) biostratigraphy ($\rho = 0.539$). This discrepancy between Keller's two studies is probably due to differences in the species concepts employed in the two studies as well as variations induced by picking specimens from different samples. These results indicate that Keller's El Kef biostratigraphies are either comparable with that of Keller (1988*a*, *b*) or more similar to those of the blind test investigators' biostratigraphies that those biostratigraphies are among themselves.

In interpreting the El Kef blind test results Smit & Nederbragt (1997) proposed that aggregation of the blind test results would more faithfully represent the true turnover pattern. This is a slight diversion of the purpose of the blind test, which was organized to test the reproducibility of the Smit v. Keller results. A consensus biostratigraphy can be constructed from the blind test investigators' results by representing each species' stratigraphic range to the highest level at which all investigators' results agree (i.e. strict consensus, Fig. 6(**A**)). The consensus biostratigraphy estimates the pattern any individual investigator would be expected to reproduce and, as shown in Fig. 6(**A**), it indicates that local species disappearances are typically observed before, at, and after the El Kef K–T boundary.

The aggregate result of the four El Kef blind test biostratigraphies is shown in Fig. 6(**B**). This

Table 3. *Spearman rank-order correlations among the positions of common taxa in various El Kef blind test biostratigraphies*

	Investigator					
Investigator	Canudo	Masters	Olsson	Orue-extebarria	Keller (1988)	Keller *et al.* (1995)
Canudo	1.000					
Masters	0.301	1.000				
Olsson	0.600**	0.271	1.000			
Orue-etxebarria	0.748**	0.337*	0.460**	1.000		
Keller (1988)	0.441*	0.308*	0.521**	0.392**	1.000	
Keller & Li (1994)	0.758**	0.362*	0.390*	0.644**	0.466**	1.000
Summary statistics:	Mean correlation among blind test participants: 0.453 Mean correlation with Keller (1988): 0.416 Mean correlation with Keller *et al.* (1995): 0.539					

* Significant at $p = 0.05$.
** Significant at $p = 0.01$.

representation is analogous to the results that would be expected if collecting efforts (in this case sample size) were increased four-fold. As expected, Fig. 6(**B**) shows that with increased collecting effort most species' last appearance datum rises; a pattern that is consistent with expectations of the Signor–Lipps effect. Smit & Nederbragt (1997) argued that these results were consistent with recognition of a mass extinction coincident with the K–T boundary. This seems inconsistent with the interpretation by Signor & Lipps' (1982) of their effect. Figure 6(**B**) shows that a minority of the El Kef planktonic foraminiferal fauna disappear at the last Cretaceous sample, and the Signor–Lipps effect predicts that further increases in sample efforts would result in further reductions of observed last appearance datum levels coinciding with the local K–T boundary.

Smit & Nederbragt (1997) also interpreted all but two Cretaceous genera found in the lowermost Tertiary interval in the El Kef section to be 'zombies' reworked from underlying Cretaceous sediments (see also Smit (1982) and Smit *et al.* (1988)). Keller (1997) dissented from this interpretation, preferring to regard this fauna as the remains of K–T survivor populations. Keller's (1997) interpretation is consistent with standard biostratigraphical practice. Indeed, even the critics of Keller's interpretations argue that the observation of planktonic foraminiferal species *below* the K–T boundary provides sufficient evidence to infer their presence as living populations (rather than reworked zombies) throughout the upper Maastrichtian.

Among the blind test participants Olsson (1997) concurred with Smit & Nederbragt (1997), with regard to *Guembelitria cretacea* (Hofker), but also listed *Hedbergella monmouthensis* Olsson and *Hedbergella holmdelensis* Olsson as K–T survivor species. This interpretation is consistent with Liu & Olsson (1992), who argued that these two *Hedbergella* species must be regarded as survivor taxa because all Tertiary planktonic Foraminifera are phylogenetically derived from these species. Berggren & Norris (1997) have recently reviewed the phylogenetics of Tertiary trochospiral taxa and, although these workers accept the hypothesis of derivation from *Hedbergella*, they were unable to demonstrate the existence of any synapomorphies linking the Paleocene planktonic foraminiferal fauna when multiple Cretaceous taxa were used as the outgroup. Most practitioners of phylogenetic analysis (e.g. Nixon & Carpenter 1993; Smith, 1994; P. Forey, pers. comm. 1997) would regard these results as indicating that Paleocene planktonic Foraminifera are neither monophyletic nor uniquely derived from *Hedbergella*. Among the remainder of the blind test participants, Canudo (1997) and Orue-extebarria (1997) agreed with Keller (1997) that their results were consistent with the presence of an expanded K–T planktonic foraminiferal survivor fauna, whereas Masters (1997) felt that the reworking–survivorship question could not be resolved without recourse to intra-specific stable isotopic data (which cannot be obtained from the El Kef planktonic foraminiferal faunas owing to diagenesis).

The K–T ammonite controversy

The best studied K–T ammonite successions are the Bay of Biscay sequences of northern Spain

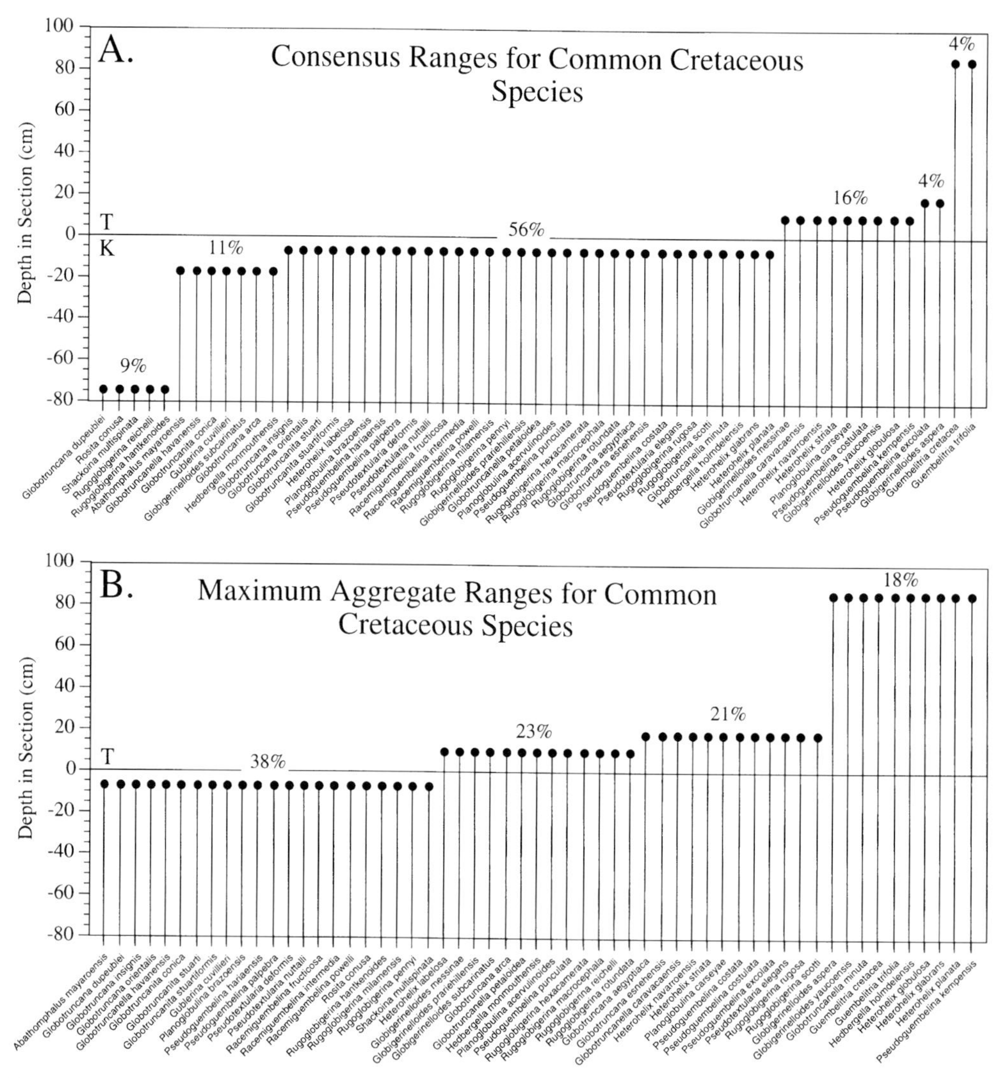

Fig. 6. Alternative summaries of the El Kef blind text data for common species (i.e. those found by two or more of the participants. (**A**) Consensus biostratigraphy. (**B**) Maximum aggregate biostratigraphy. It should be noted that although the maximum aggregate biostratigraphy shows that all Cretaceous species rise at least to the last Cretaceous sample, a distinct minority of these species disappear at this end-Cretaceous horizon. Moreover, the Signor–Lipps effect predicts that this already small percentage will decrease further with additional scrutiny of the blind test samples. (See text for discussion.)

(especially the Zumaya section). Intensive sampling by Ward and coworkers (Ward *et al.* 1986, 1991; Ward 1990; Ward & Kennedy 1993; Marshall & Ward 1996) has shown that 21 ammonite species approach to within 120 m of the local K–T boundary and that five of these rise to the last 30 m of the Maastrichtian interval. These five uppermost Maastrichtian species represent three families; a figure that should be contrasted with the 14, 16, 17, 24, 23, and 27 ammonite families or subfamilies present at the beginning of the Maastrichtian, Campanian, Santonian, Coniacian, Turonian, and Cenomanian stages, respectively (Kennedy 1989; Ward & Kennedy 1993). Chronostratigraphic resolution in the Bay of Biscay successions is such that these final ammonite records are placed 'a few thousand years prior to the K–T boundary event itself *if sedimentation had been continuous*. The latter, unfortunately cannot be determined at this time' (Ward & Kennedy 1993, p. 14, italics in the original). As mentioned above, planktonic foraminiferal faunas from the Zumaya section yield no evidence of the lowermost Danian planktonic foraminiferal biozone.

Marshall & Ward (1996) used stratigraphic confidence intervals (Paul 1982; Strauss & Sadler 1989; Marshall 1991, 1994) to determine whether the observed gradual ammonite extinction pattern in the Zumaya section was consistent with a catastrophic or co-extinction model. Ward and Marshall's test (originally based on Springer (1990); see also Marshall (1995)) postulates that if ammonite extinctions were catastrophic the 50% confidence intervals on each species' last appearance datum should be distributed uniformly about the actual extinction horizon (Marshall 1995, fig. 1). This results in the definition of an interval within which a co-extinction horizon may be placed. If this interval includes the observed stratigraphic range of any species used in the analysis the co-extinction hypothesis is rejected. Provided the co-extinction hypothesis is not rejected the binomial distribution can then be used to contour the indicated interval in an effort to determine the strength of support for particular stratigraphic horizons being consistent with observed data under the co-extinction model. Marshall & Ward's (1996) results indicate that, for the 21 youngest Maastrichtian species the most likely co-extinction horizon lies 8 m below the K–T boundary, and for the five youngest species the most likely co-extinction horizon band includes the K–T boundary. These workers have interpreted their results to suggest a 'sudden and complete catastrophe' (Ward, quoted by Kerr (1996)) and that 'any gradual disappearance must have been on the order of 1000 years' (Marshall paraphrased by Kerr (1996)).

Although the results of the Marshall & Ward (1996) study are interesting, they do not shut the door on alternative interpretations for the ammonite extinction. Springer (1990, p. 515) pointed out that whereas the 50% test is a necessary condition of the co-extinction model, 'it does not constitute proof for simultaneous local extinction'. In interpreting the results of a similar study, Marshall (1995) observed that they were compatible with both simultaneous and gradual extinction models within the interval defined by the 50% test.

Assumptions inherent in the calculation of stratigraphic confidence intervals are also a matter of some concern. Marshall (1994) pointed out that the distribution-free method of confidence interval determination (which was used in the Marshall and Ward (1996) study) assumes uniform collecting effort throughout the sampled interval. However, Ward & Kennedy (1993) noted that a much greater effort was expended in searching for ammonites in the upper 10 m of the Zumaya section than elsewhere. In addition, Marshall (1994) discussed the need for the gap size distribution to be determined to estimate distribution-free confidence intervals. If this gap size distribution is based on the physical separation between fossil occurrence horizons (as in the Marshall & Ward (1996) study), the analysis assumes continuous sediment accumulation at a constant rate throughout the sampled interval and within the zone of range extension defined by the confidence intervals. The many changes in lithology (especially at the K–T boundary; see Ward & Kennedy (1993)) along with biostratigraphic evidence for a substantial (>40 ka) boundary hiatus contradict this assumption.

The best that can be said for the Zumaya ammonite data, which is one of the most thoroughly documented and completely analysed K–T fossil records available, is that if one is prepared to accept some aspects of the local stratigraphic situation at face value (e.g. continuity of sediment accumulation, constancy of sediment accumulation rate), and is willing to assume a co-extinction model, a non-exclusively consistent case can be made for the simultaneous extinction of five ammonite species at the K–T boundary horizon. With respect to the remainder of ammonite evolutionary history, however, associations between ammonite diversity and eustatic sea-level changes show a much more compelling correspondence than those between ammonite extinctions and bolide impacts (House 1989, 1993, fig. 2.4; Wiedemann & Kullman 1996, fig. 5).

Summary

The mass extinction debate has long been seen as a clash between catastrophist and uniformitarian world views. This is unfortunate because, by their very nature, all extinctions represent interruptions in the continuity of evolutionary processes. Thus, all extinctions represent biological catastrophes of a sort. The significance of the Alvarez *et al.* (1980) bolide impact extinction hypothesis is not that it made catastrophism a respectable subject for scientific debate or that it 'solved' the mystery of the dinosaur extinction, but that it provided an illustration of how independent physical evidence for the operation and timing of an extraterrestrial process could be recovered from the stratigraphic record and used to both formulate and test a novel extinction model. In doing this Alvarez *et al.* (1980) moved the question of extraterrestrial forcing out of the non-scientific realm of unconstrained speculation and into the hypothetico-deductive realm of mainstream, actualistic science.

This having been said, it does not necessarily follow that extraterrestrial processes must have played a dominant role in shaping the Earth's physical or biotic environment through the Phanerozoic. The data presented above fail to make a compelling case for such an association as a consistent feature of Earth history. With the possible exception of the Maastrichtian event it seems as though the stage-level extinction record bears little correspondence to the record of bolide impacts. Stage-level correspondences between the extinction record and those of continental flood basalt events and eustatic sea-level changes are much more convincing. Yet all three of these causal mechanisms have thus far failed to provide any unique prior predictions with respect to the nature of species sorting, and patterns of extinction selectivity have failed to reveal the retrospective biotic patterns necessary to make a case for similarity between events known to be associated with similar non-biotic perturbations (e.g. large bolide impacts during the Frasnian, Maastrichtian, and Upper Eocene events, continental flood basalt eruptions during the Tatarian and Norian events). These data suggest that extinction events result from complex multi-causal perturbations, the effects of which are sensitive to the state of the biosphere at the time of the event (see Van Valen 1994; Jablonski 1996; MacLeod 1996*a*; Hallam & Wignall 1997). Consequently, the most likely causal hypothesis supported by these data is that the environmental–biotic effects of a bolide impact (or a major flood basalt eruption) may be magnified or diminished depending on the relative height of eustatic sea-level, distribution and/or size of continents, composition of the atmosphere, temporal proximity to other unrelated environmental perturbations, etc.

In seeking to test various extinction-related hypotheses the nature of palaeontological data must be kept in mind. All of the leading extinction event causal mechanisms make essentially the same predictions not only with regard to their secondary and tertiary environmental effects, but also with respect to the time scales over which these perturbations would be expected to operate. In the overwhelming majority of cases the data from the fossil record are limited in terms of their ability to record short-term biotic and environmental signals as well as the degree to which specific time horizons can be recognized in different stratigraphic successions. Even the inherent time averaging of all fossiliferous deposits operates to make temporal resolutions of less than 10^3–10^4 years all but impossible to achieve using palaeontological data. On the environmental side, the factors that actually limit the geographical or environmental distributions for most modern marine invertebrate species (much less fossil forms, many of which have no close modern analogue) are almost wholly unknown. Archibald (1996*a*, *b*) has had some success formulating biotic tests of specific extinction scenarios using the terrestrial record, but this will be very difficult to duplicate using data from the marine invertebrate fossil record because of our comparative ignorance of the marine environment and its inhabitants.

Owing to this lack of strong, predictive biotic tests or even consistent biotic patterns that might be used to group extinction events into putative causal classes, many palaeontologists have opted for the weaker alternative of demonstrating that under certain assumptions observed data are consistent with predictions of different causal scenarios. However, as Jablonski (1996) correctly argued, the list of secondary and tertiary effects proceeding from the three leading extinction event causal alternatives is so extensive and generalized that virtually any conceivable extinction pattern can be accommodated by each hypothesis. This feature of the extinction debates brings into question exactly what constitutes an 'explanation' for extinction-related phenomena. Extinction tests based on general consistency, as opposed to hypothetico-deductive tests based on unique sets of predictions, can do no more than formulate a list of equally plausible alternatives. If such hypothetico-deductive tests cannot be formulated, or if the data required to apply these tests cannot be collected (or cannot be collected without making assumptions that are not independently verifiable), the question of extinction causal mechanisms will remain unanswered.

If palaeontological data cannot be used to make distinctions between alternative extinction event causal models, what is the appropriate subject for palaeontologists who wish to study the phenomenon of extinction? Palaeontology is inherently about the study of long term biotic patterns, and the effects of extinction (regardless of their cause) are inherently long-term. From an evolutionary perspective the significance of extinction events lies in their effects on subsequent diversification patterns. These effects are manifest at all scales within the phylogenetic–ecological plexus and are now recognized as being integral to our understanding of macroevolutionary phenomena from the level of the individual clade to the level of the entire biosphere. Moreover, different groups appear to respond to extinction events in different ways. Jablonski (1986*a*, *b*) showed that mollusc extinction–survivorship patterns differed during background and mass extinction events. Planktonic Foraminifera, on the other hand, appear to undergo similar morphological–ecological responses to major short-term (K–T), major long-term (Late Eocene) and minor events (Cifelli 1969; Stanley *et al.* 1988; MacLeod 1996*b*; MacLeod *et al.* 1998). In the realm of biogeography, the differing legacies of the K–T extinction for the faunas and floras of different continental platforms are only beginning to be explored (e.g. Wolfe 1987; Jablonski 1995). Given the limitations of palaeontological data, it seems unlikely that the specific causes of most large marine invertebrate extinction events will ever be known in detail. However, by combining the temporal, geographical, and ecological scope of the marine invertebrate fossil record with a new appreciation of the extraordinary diversity of roles extinction events play in mediating evolutionary process, palaeontologists stand at the threshold of an exciting new research programme in which extinction events are used to understand biological, rather than physical, processes. It is this programme that will constitute the most important legacy of the mass extinction debates.

I would like to thank R. Hutchison for inviting me to participate in the symposium that served as the basis for this volume. In addition, the arguments presented herein have benefited from discussions with P. Forey, A. Gale, A. Hallam, and A. Smith.

References

Ager, D. V. 1993. *The New Catastrophism: the Importance of the Rare Event in Geological History*. Cambridge University Press, Cambridge.

Alvarez, L. W., Alvarez, F., Asaro, F. & Michel, H. V. 1980. Extraterrestrial cause for the Cretaceous–Tertiary extinction. *Science*, **208**, 1095–1108.

Alvarez, W. 1997. T. rex *and the Crater of Doom*. Princeton University Press, Princeton.

—— & Muller, R. A. 1984. Evidence from cratering ages for periodic impacts on the Earth. *Nature*, **308**, 718.

Archibald, J. D. 1996*a*. *Dinosaur Extinction and the End of an Era: What the Fossils Say*. Columbia University Press, New York.

——1996*b*. Testing extinction theories at the Cretaceous–Tertiary boundary using the vertebrate fossil record. *In*: MacLeod, N. & Keller, G. (eds) *The Cretaceous–Tertiary Mass Extinction: Biotic and Environmental Changes*. Norton, New York, 373–398.

—— & Clemens, W. A. 1984. Mammal evolution near the Cretaceous–Tertiary boundary. *In*: Berggren, W. A. & Vancouvering, J. A. (eds) *Catastrophes and Earth History: the New Uniformitarianism*. Princeton University Press, Princeton, 229–371.

—— & Retallack, G. J. 1996. Acid trauma at the Cretaceous–Tertiary (K/T) boundary in eastern Montana: comment and reply. *GSA Today*, **6**, 21–22.

Arthur, M. A., Zachos, J. C. & Jones, D. S. 1987. Primary productivity and the Cretaceous/Tertiary boundary event in the oceans. *Cretaceous Research*, **8**, 43–54.

Barrera, E. & Keller, G. 1990. Foraminiferal stable isotope evidence for gradual decrease of marine productivity and Cretaceous species survivorship in the earliest Danian. *Paleoceanography*, **5**, 867–870.

Barron, E. J. & Moore, G. T. 1994. *Climate Model Application in Paleoenvironmental Analysis*. Society for Sedimentary Geology, Tulsa, OK.

Benton, M. J. 1993. *The Fossil Record*. Chapman & Hall, London.

Berggren, W. A. & Norris, R. D. 1997. Biostratigraphy, phylogeny and systematics of Paleocene trochospiral planktic foraminifera. *Micropaleontology*, **43**, Supplement 1, 1–116.

Birkelund, T. & Hakansson, E. 1982. The terminal Cretaceous in Boreal shelf seas: a multicausal event. *In*: Silver, L. T. & Schultz, P. H. (eds) *Geological Implications of Impact of Large Asteroids and Comets on the Earth*. Geological Society of America, Special Paper, **190**, 373–384.

Bottomley, R., Grieve, R., York, D. & Masaltis, V. 1997. The age of the Popagai impact event and its relation to events at the Eocene/Oligocene boundary. *Nature*, **388**, 365–368.

Bramlette, M. N. 1965. Massive extinctions in biota at the end of Mesozoic time. *Science*, **148**, 1696–1699.

Broecker, W. S., Klas, M., Clark, E., Bonani, G., Ivy, S. & Wolfi, W. 1990. The influence of $CaCO_3$ dissolution on core top radiocarbon ages for deep-sea sediments. *Paleoceanography*, **6**, 593–608.

Bryan, J. R. & Jones, D. S. 1989. Fabric of the Cretaceous–Tertiary marine macrofaunal transition at Braggs, Alabama. *Palaeogeography, Palaeoclimatology, Palaeoecology*, **69**, 279–301.

Burns, P., Dullo, W.-C., Hay, W. W., Wold, C. N. & Pernicka, E. 1996. Iridium concentration as an estimator of instantaneous sediment accumulation rates. *Journal of Sedimentary Research*, **66**, 608–612.

Canudo, J. I. 1997. El Kef blind test I results. *Marine Micropaleontology*, **29**, 73–76.

——, KELLER, G. & MOLINA, E. 1991. Cretaceous/Tertiary boundary extinction pattern and faunal turnover at Agost and Caravaca, SE Spain. *Marine Micropaleontology*, **17**, 319–341.

CARLSON, S. J. 1991. A phylogenetic perspective on articulate brachiopod diversity and the Permo-Triassic extinction. *In*: DUDLEY, E. (ed) *The Unity of Evolutionary Biology*. Discorides, Portland, OR, 119–142.

CIFELLI, R. S. 1969. Radiation of Cenozoic planktonic foraminifera. *Systematic Zoology*, **18**, 154–168.

COCCIONI, R. & GALEOTTI, S. 1994. K–T boundary extinction: geologically instantaneous or gradual event? Evidence from deep-sea benthic foraminifera. *Geology*, **22**, 779–782.

COFFIN, M. F. & ELDHOLM, O. 1994. Large igneous provinces: crustal structure, dimensions, and external consequences. *Reviews of Geophysics*, **32**, 1–36.

COLODNER, D. C., BOYLE, E. A., EDMOND, J. M. & THOMSON, J. 1992. Post-depositional mobility of platinum, iridium and rhenium in marine sediments. *Nature*, **358**, 402–404.

COURTILLOT, V. E. 1990. A volcanic eruption. *Scientific American*, **263**, 85–92.

——, BESSE, J., VANDAMME, D., MONTIGNY, R., JAEGER, J. J. & CAPPETTA, H. 1986. Deccan flood basalts at the Cretaceous–Tertiary boundary. *Earth and Planetary Science Letters*, **80**, 361–374.

——, FÉRAUD, G., MALUSKI, H., VANDAMME, D., MOREAU, M. G. & BESSE, J. 1988. Deccan flood basalts and the Cretaceous–Tertiary boundary. *Nature*, **333**, 843–846.

——, JAEGER, J.-J., YANG, Z., FÉRAUD, G. & HOFMANN, C. 1996. The influence of continental flood basalts on mass extinctions: where do we stand? *In*: RYDER, G., FASTOVSKY, D. & GARTNER, S. (eds) *The Cretaceous–Tertiary Event and other Catastrophes in Earth History*. Geological Society of America, Special Paper, **307**, 513–525.

CROCKETT, J. H., OFFICER, C. B., WEZEL, F. C. & JOHNSON, G. D. 1988. Distribution of noble metals across the Cretaceous/Tertiary boundary at Gubbio, Italy: iridium variation as a constraint on the duration and nature of Cretaceous/Tertiary boundary events. *Geology*, **16**, 77–80.

CROFT, S. K. 1982. A first-order estimate of shock heating and vaporization in oceanic impacts. *In*: SILVER, L. T. & SCHULTZ, P. H. (eds) *Geological Implications of Impacts of Large Asteroids and Comets on the Earth*. Geological Society of America, Special Paper, **190**, 143–152.

DEIRMENDJIAN, D. 1973, On volcanic and other particulate turbidity anomalies. *In*: LANDSBERG, H. E. & VAN MEIGHAM, J., (eds) *Advances in Geophysics*. Academic, New York, 267–269.

EHRLICH, P. R. & BIRCH, L. C. 1967. The 'Balance of nature' and Population control'. *American Naturalist*, **101**, 97–107.

EKDALE, A. A. & BROMLEY, R. G. 1984. Sedimentology and ichnology of the Cretaceous–Tertiary boundary in Denmark: implications for the causes of the terminal Cretaceous extinction. *Journal of Sedimentary Petrology*, **54**, 681–703.

ELTON, C. S. 1958. *The Ecology of Invasions by Plants and Animals*. Chapman & Hall, London.

EMILIANI, C., KRAUS, E. B. & SHOEMAKER, E. M. 1981. Sudden death at the end of the Mesozoic. *Earth and Planetary Science Letters*, **55**, 317–334.

ERWIN, D. H. 1993. *The Great Paleozoic Crisis: Life and Death in the Permian*. Columbia University Press, New York.

GERSTEL, S. A., & ZARDECKI, A. 1982, Reduction of photosynthetically active radiation under extreme stratospheric aerosol loads. *In*: SILVER, L. T. & SCHULTZ, P. H., (eds) *Geological Implications of Impacts of Large Asteroids and Comets on the Earth*. Geological Society of America, Special Paper, **190**, 201–210.

FISHER, D. C. 1991. Phylogenetic analysis and its application in evolutionary paleobiology. *In*: GILINSKY, N. L. & SIGNOR, P. W. (eds) *Analytical Paleobiology*. University of Tennessee, Knoxville, 103–122.

FLESSA, K. W. 1993. Time averaging and temporal resolutions in Recent marine shelly faunas. *In*: KIDWELL, S. M. & Behrensmeyer, A. K. (eds) *Taphonomic Approaches to Time Resolution in Fossil Assemblages*. University of Tennessee, Knoxville, 9–33.

—— & KOWALEWSKI, M. 1994. Shell survival and time-averaging in nearshore and shelf environments: estimates from the radiocarbon literature. *Lethaia*, **27**, 153–165.

FOOTE, M. 1996. Perspective: evolutionary patterns in the fossil record. *Evolution*, **50**, 1–11.

FUNNELL, B. M. & SWALLOW, J. E. 1997. Intra-sample, inter-sample, and down-core microvariation in sea-surface temperature estimates obtained from planktonic foraminifera in the NE Atlantic. *Journal of Micropalaeontology*, **16**, 163–174.

GINSBURG, R. N. 1997. An attempt to resolve the controversy over the end-Cretaceous extinction of planktic foraminifera at El Kef, Tunisia using a blind test. Introduction: background and procedures. *Marine Micropaleontology*, **29**, 67–68.

GOLDSCHMIDT, V. M. 1954. *Geochemistry*. Clarendon Press, Oxford.

GRADSTEIN, F. M. & OGG, J. 1996. A Phanerozoic time scale. *Episodes*, **19**, 3–5.

GREUTER, W. 1995. Extinction in Mediterranean areas. *In*: LAWTON, J. H. & MAY, R. M. (eds) *Extinction Rates*. Oxford University Press, Oxford, 88–97.

GRIEVE, R., RUPERT, J., SMITH, J. & THERRIAULT, A. 1996. The record of terrestrial impact cratering. *GSA Today*, **5** 193–195.

HAKANSSON, E. & HANSEN, J. M. 1979. Guide to Maastrichtian and Danian boundary strata in Jutland. *In*: BIRKELUND, T. & BROMLEY, R. G. (eds), *Cretaceous–Tertiary Boundary Events I*. University of Copenhagen, Copenhagen, 171–188.

HALLAM, A. 1984. The causes of mass extinction. *Nature*, **308**, 686–687.

——1989. The case for sea-level change as a dominant causal factor in mass extinctions of marine invertebrates. *Philosophical Transactions of the Royal Society of London, Series B*, **325**, 437–455.

——1992. *Phanerozoic Sea-Level Changes.* Columbia University Press, New York.

—— & WIGNALL, P. B. 1997. *Mass Extinctions and their Aftermath.* Oxford Science Publications, Oxford.

HANSEN, T. A., FORRAND, R., MONTGOMERY, H. & BILMAN, H. 1984. Sedimentology and extinction patterns across the Cretaceous/Tertiary boundary interval in east Texas. *In*: YANCY, T. E. (ed) *The Cretaceous/Tertiary Boundary and Lower Tertiary of the Brazos River Valley, Field Trip Guide.* American Association of Petroleum Geologists, 21–36.

——, ——, ——, —— & BLECHSCHMIDT, G. 1987*a*. Sedimentology and extinction patterns across the Cretaceous–Tertiary boundary interval in east Texas. *Cretaceous Research*, **8**, 229–252.

——, RASMUSSEN, K. L., GWOZDZ, R. & KUNZENDORF, H. 1987*b*. Iridium-bearing carbon black at the Cretaceous–Tertiary boundary. *Bulletin of the Geological Society of Denmark*, **36**, 305–314.

——, UPSHAW, B., III, KAUFFMAN, E. G. & GOSE, W. 1993. Patterns of molluscan extinction and recovery across the Cretaceous–Tertiary boundary in east Texas: report on new outcrops. *Cretaceous Research*, **14**, 685–706.

HAQ, B., HARDENBOL, J. & VAIL, P. R. 1987. Chronology and fluctuating sea levels since the Triassic. *Science*, **235**, 1156–1166.

HERM, D. 1965. Mikropalontologische–stratigraphische Untersuchungen im Kreidflysch zwischen Deva und Zumaya (Province Guipuzcao, Nordspanien). *Zeitschrift der Deutschen Geologischen Gesellschaft*, **115**, 277–348.

HOUSE, M. R. 1989. Ammonoid extinction events. *Philosophical Transactions of the Royal Society of London, Series B*, **325**, 307–326.

——1993. Fluctuations in ammonoid evolution and possible environmental controls. *In*: HOUSE, M. R. (ed) *The Ammonoidea: Environment, Ecology, and Evolutionary Change.* Clarendon Press, Oxford, 13–34.

HSÜ, K. J. & MCKENZIE, J. A. 1985. A 'strangelove' ocean in the earliest Tertiary. *In*: SUNDQUIST, E. T. & BROECKER, W. S. (eds) *Natural Variations: Archean to Present.* Geophysical Monograph, American Geophysical Union, **32**, 487–492.

—— & ——1990. Carbon-isotope anomalies at era boundaries; global catastrophes and their ultimate cause. *In*: SHARPTON, V. L. & WARD, P. D. (eds) *Global Catastrophes in Earth History: an Interdisciplinary Conference on Impacts, Volcanism, and Mass Mortality.* Geological Society of America, Special Paper, **247**, 61–70.

HUBER, B. T. 1996. Evidence for planktonic foraminifer reworking versus survivorship across the Cretaceous–Tertiary boundary at high latitudes. *In*: RYDER, G., FASTOVSKY, D. & GARTNER, S. (eds) *The Cretaceous–Tertiary Event and other Catastrophes in Earth History.* Geological Society of America, Special Paper, **307**, 319–334.

HUT, P., ALVAREZ, W., ELDER, W. P. *et al.* 1987. Comet showers as a cause of mass extinctions. *Nature*, **329**, 118–126.

JABLONSKI, D. 1985. Marine regressions and extinctions: a test using the modern biota. *In*: VALENTINE, J. W. (ed) *Phanerozoic Diversity Patterns: Profiles in Macroevolution.* Princeton University Press, Princeton, 335–354.

——1986*a*. Causes and consequences of mass extinctions: implications for macroevolution. *In*: ELLIOTT, D. K. (ed) *Dynamics of Extinction.* Wiley, New York, 183–229.

——1986*b*. Background and mass extinctions: the alteration of macroevolutionary regimes. *Science*, **231**, 129–133.

——1989. The biology of mass extinction: a palaeontological view. *Philosophical Transactions of the Royal Society of London, Series B*, **325**, 357–368.

——1995. The biogeography of rebounds: comparisons among K–T bivalves. *Geological Society of America, Abstracts with Programs*, **27**, A164.

——1996. Mass extinctions: persistent problems and new directions. *In*: RYDER, G., FASTOVSKY, D. & GARTNER, S. (eds) *The Cretaceous–Tertiary Event and other Catastrophes in Earth History.* Geological Society of America, Special Paper, **307**, 1–9.

—— & RAUP, D. M. 1995. Selectivity of end-Cretaceous marine bivalve extinctions. *Science*, **268**, 389–391.

JACKSON, J. B. C. 1995. Constancy and change of life in the sea. *In*: LAWTON, J. H. & MAY, R. M. (eds) *Extinction Rates.* Oxford University Press, Oxford, 45–54.

JÉHANNO, C., BOCLET, D., FROGET, L., LAMBERT, B., ROBIN, E., ROCCHIA, R. & TURPIN, L. 1992. The Cretaceous–Tertiary boundary at Beloc, Haiti: no evidence for an impact in the Caribbean area. *Earth and Planetary Science Letters*, **109**, 229–241.

JONES, E. M. & KODIS, J. W. 1982. Atmospheric effects of large body impacts: the first few minutes. *In*: SILVER, L. T. & SCHULTZ, P. H. (eds) *Geological Implications of Impacts of Large Asteroids and Comets on the Earth.* Geological Society of America, Special Paper, **190**, 175–186.

KAMERMANS, P. 1994. Similarity in food source and timing of feeding in deposit- and suspension-feeding bivalves. *Marine Ecology Progress Series*, **104**, 63–75.

KEITH, M. L. 1980. Createaceous volcanism and the disappearance of the dinosaurs. *Eos Transactions, American Geophysical Union*, **61**, 400.

——1982. Violent volcanism, stagnant oceans and some inferences regarding petroleum, strata-bound ores and mass extinction. *Geochimica Cosmochimica et Acta*, **46**, 2631–2637.

KELLER, G. 1988*a*. Biotic turnover in benthic foraminifera across the Cretaceous–Tertiary boundary at El Kef, Tunisia. *Palaeogeography, Palaeoclimatology, Palaeoecology*, **66**, 153–171.

——1988*b*. Extinction, survivorship and evolution of planktic foraminifera across the Cretaceous/Tertiary boundary at El Kef, Tunisia. *Marine Micropaleontology*, **13**, 239–263.

——1989*a*. Extended Cretaceous/Tertiary boundary extinctions and delayed population change in planktonic foraminiferal faunas from Brazos River, Texas. *Paleoceanography*, **4**, 287–332.

——1989*b*. Extended period of extinctions across the Cretaceous/Tertiary boundary in planktonic foraminifera of continental shelf sections: implications for impact and volcanism theories. *Geological Society of America Bulletin*, **101**, 1408–1419.

——1992. Paleoecologic response of Tethyan benthic foraminifera to the Cretaceous/Tertiary boundary transition. *In*: TAKAYANAGI, Y. & SAITO, T. (eds) *Studies in Benthic Foraminifera, Benthos '90, Sendai 1990*. Tokai University Press, Tokyo, 77–91.

——1993. The Cretaceous–Tertiary boundary transition in the Antarctic Ocean and its global implications. *Marine Micropaleontology*, **21**, 1–46.

——1997. Analysis of El Kef blind test I. *Marine Micropaleontology*, **29**, 89–93.

—— & BENJAMINI, C. 1991. Paleoenvironment of the eastern Tethys in the Early Paleocene. *Palaios*, **6**, 439–464.

——, LI, L. & MACLEOD, N. 1995. The Cretaceous/Tertiary boundary stratotype section at El Kef, Tunisia: how catastrophic was the mass extinction? *Palaeogeography, Palaeoclimatology, Palaeoecology*, **119**, 255–273.

——, MACLEOD, N., LYONS, J. B. & OFFICER, C. B. 1993. Is there evidence for Cretaceous–Tertiary boundary-age deep-water deposits in the Caribbean and Gulf of Mexico? *Geology*, **21**, 776–780.

——, STINNESBECK, W. & LOPEZ-OLIVA, J. G. 1994. Age, deposition, and biotic effects of the Cretaceous/Tertiary boundary event at Mimbral, NE Mexico. *Palaios*, **9**, 144–157.

KENNEDY, W. J. 1989. Thoughts on the evolution and extinction of Cretaceous ammonites. *Proceedings of the Geologists Association*, **100**, 251–279.

KERR, R. A. 1996. New way to read the record suggests abrupt extinction. *Science*, **274**, 1303–1304.

KITCHELL, J. A. 1990. Biological selectivity in extinction. *In*: KAUFFMAN, E. G. & WALLISER, O. (eds) *Extinction Events in Earth History*. Springer, New York, 31–43.

KOCH, C. F. & MORGAN, J. P. 1988. On the expected distribution of species ranges. *Paleobiology*, **14**, 126–138.

LAMBSHEAD, P. J. & GOODAY, A. J. 1990. The impact of seasonally deposited phytodetritus on epifaunal and shallow infaunal benthic foraminiferal populations in the bathyal northeast Atlantic: the assemblage response. *Deep-Sea Research*, **37**, 1263–1283.

LEROUX, H., ROCCHIA, R., FROGET, L., ORUE-ETXEBARRIA, X., DOUKHAN, J. & ROBIN, E. 1995. K/T boundary of Beloc (Haiti): compared stratigraphic distributions of boundary markers. *Earth and Planetary Science Letters*, **131**, 255–268.

LEVINTON, J. S. 1991. Variable feeding behavior in three species of *Macoma* (Bivalvia: Tellinacea) as a response to water flow and sediment transport. *Marine Biology*, **110**, 375–383.

——1996. Trophic group and the end-Cretaceous extinction: did deposit feeders have it made in the shade? *Paleobiology*, **22**, 104–112.

LEWIS, J. S., WATKINS, G. H., HARTMAN, H., & PRINN, R. G. 1982, Chemical consequences of major impact events on Earth. *In*: SILVER, L. T. & SCHULTZ, P. H. (eds) *Geological Implications of Impacts of Large Asteroids and Comets on the Earth*. Geological Society of America, Special Paper, **190**, 215–221.

LIN, J. & HINES, A. H. 1994. Effects of suspended food availability on the feeding mode and burial depth of the Baltic clam *Macoma balthica*. *Oikos*, **69**, 23–36.

LIPPS, J. H. 1997. The Cretaceous–Tertiary boundary: the El Kef blind test. *Marine Micropaleontology*, **29**, 65–66.

LIU, C. & OLSSON, R. K. 1992. Evolutionary radiation of microperforate planktonic foraminifera following the K/T mass extinction. *Journal of Foraminiferal Research*, **22**, 328–346.

—— & ——1994. On the origin of Danian normal perforate planktonic foraminifera from *Hedbergella*. *Journal of Foraminiferal Research*, **24**, 61–74.

LÓPEZ RAMOS, E. 1973. Estudio geológico de la Península de Yucatán. *Associación Mexicana de Geólogos Petroleros Boletín*, **25**, 23–76.

LUTERBACHER, H. & PREMOLI SILVA, I. 1964. Biostratigrafia del Limite Cretaceo–Terziario Nell'Appennino centrale. *Rivista Italiana di Paleontologia e Stratigrafia*, **70**, 67–88.

MACLEOD, N. 1995*a*. Cretaceous/Tertiary (K/T) biogeography of planktic foraminifera. *Historical Biology*, **10**, 49–101.

——1995*b*. Graphic correlation of high latitude Cretaceous–Tertiary boundary sequences at Nye Kløv (Denmark), ODP Site 690 (Weddell Sea), and ODP Site 738 (Kerguelen Plateau): comparison with the El Kef (Tunisia) boundary stratotype. *Modern Geology*, **19**, 109–147.

——1995*c*. Graphic correlation of new Cretaceous/Tertiary (K/T) boundary sections. *In*: MANN, K. O. & LANE, H. R. (eds) *Graphic Correlation*. Society for Sedimentary Geology Special Publication, **53**, 215–233.

——1996*a*. K–T redux. *Paleobiology*, **22**, 311–317.

——1996*b*. Nature of the Cretaceous–Tertiary (K–T) planktonic foraminiferal record: stratigraphic confidence intervals, Signor–Lipps effect, and patterns of survivorship. *In*: MACLEOD, N. & KELLER, G. (eds) *The Cretaceous–Tertiary Mass Extinction: Biotic and Environmental Changes*. Norton, New York, 85–138.

——1996*c*. Testing patterns of Cretaceous–Tertiary planktonic foraminiferal extinctions at El Kef (Tunisia). *In*: RYDER, G., FASTOVSKY, D. & GARTNER, S. (eds) *The Cretaceous–Tertiary Event and other Catastrophes in Earth History*. Geological Society of America, Special Paper, **307**, 287–302.

—— & KELLER, G. 1991. How complete are Cretaceous/Tertiary boundary sections? A chronostratigraphic estimate based on graphic correlation. *Geological Society of America Bulletin*, **103**, 1439–1457.

—— & ——1994. Comparative biogeographic analysis of planktic foraminiferal survivorship across the Cretaceous/Tertiary (K/T) boundary. *Paleobiology*, **20**, 143–177.

—— & ——1996. Introduction. *In*: MacLeod, N. & Keller, G. (eds) *The Cretaceous–Tertiary Mass Extinction: Biotic and Environmental Changes*. Norton, New York, 1–6.

——, Ortiz, N., Fefferman, N., Clyde, W., Schulter, C. & MacLean, J. 1998. Phenotypic response of foraminifera to episodes of global environmental change. *In*: Culver, S. J. & Rawson, P. (eds) *Global Change and the Biosphere*. Chapman & Hall, London.

——, Rawson, P. F., Forey, P. L. *et al.* 1997. The Cretaceous–Tertiary biotic transition. *Journal of the Geological Society, London*, **154**, 265–292.

Marshall, C. R. 1991. Estimation of taxonomic ranges from the fossil record. *In*: Gilinsky, N. L. & Signor, P. W. (eds) *Analytical Paleobiology*. Paleontological Society, Knoxville, TN, 19–38.

——1994. Confidence intervals on stratigraphic ranges: partial relaxation of the assumption of randomly distributed fossil horizons. *Paleobiology*, **20**, 459–469.

——1995. Distinguishing between sudden and gradual extinctions in the fossil record: predicting the position of the Cretaceous–Tertiary iridium anomaly using the ammonite fossil record on Seymour Island, Antarctica. *Geology*, **23**, 731–734.

—— & Ward, P. D. 1996. Sudden and gradual molluscan extinctions in the latest Cretaceous of western European Tethys. *Science*, **274**, 1360–1363.

Martin, R. E. 1993. Time and taphonomy: actualistic evidence for time-averaging of benthic foraminiferal assemblages. *In*: Kidwell, S. M. & Behrensmeyer, A. K. (eds) *Taphonomic Approaches to Time Resolution in Fossil Assemblages*. University of Tennessee, Knoxville, 34–56.

Masters, B. A. 1997. El Kef blind test II results. *Marine Micropaleontology*, **29**, 77–79.

May, R. M. 1974. *Stability and Complexity in Model Ecosystems*. Princeton University Press, Princeton.

——, Lawton, J. H. & Stork, N. 1995. Assessing extinction rates. *In*: Lawton, J. H. & May, R. M. (eds) *Extinction Rates*. Oxford University Press, Oxford, 1–24.

McGhee, G. R., Jr. 1996. *The Late Devonian Mass Extinction: the Frasnian/Famennian Crisis*. Columbia University Press, New York.

McKinney, M. L. 1987. Taxonomic selectivity and continuous variation in mass and background extinctions of marine taxa. *Nature*, **325**, 343–345.

McLaren, D. J. 1970. Time, life, and boundaries. *Journal of Paleontology*, **44**, 801–815.

——1989. Detection and significance of mass killings. *Historical Biology*, **2**, 5–16.

——1996. Mass extinctions are rapid events. *Palaios*, **11**, 409–410.

McLean, D. M. 1978. A terminal Mesozoic 'greenhouse': lessons from the past. *Science*, **201**, 401–406.

——1985*a*. Deccan Traps mantle degassing in the terminal Cretaceous marine extinctions. *Cretaceous Research*, **6**, 235–259.

——1985*b*. Mantle degassing unification of the trans-K–T geobiological record. *In*: Hecht, M. K., Wallace, B., Prance, G. T. *et al.* (eds) *Evolutionary Biology, Volume 9*. Plenum, New York, 287–313.

Miller, K. G. 1982. Cenozoic benthic foraminifera: case histories of paleoceanographic and sea-level changes. *In*: Broadhead, T. W. (ed) *Foraminifera, Notes for a Short Course*. University of Tennessee, Studies in Geology, **6**, 107–126.

Moore, R. C. 1954. Evolution of late Paleozoic invertebrates in response to major oscillations of shallow seas. *Bulletin of the Museum of Comparative Zoology, Harvard*, **122**, 259–286.

Morgan, W. J. 1988. Flood basalts and mass extinctions. *Eos Transactions, American Geophysical Union*, **67**, 391.

Nee, S. & May, R. M. 1992. Patch removal favours inferior competitors. *Journal of Animal Ecology*, **61**, 37–40.

Newell, N. D. 1967. Revolutions in the history of life. *In*: Albritton, C. C., Hubbert, M. K., Wilson, L. G. *et al.* *Uniformity and Simplicity: a Symposium on the Principle of the Uniformity of Nature*. Geological Society of America, Special Paper, **89**, 63–91.

Nixon, K. C. & Carpenter, J. M. 1993. On outgroups. *Cladistics*, **9**, 413–426.

Norell, M. A. 1992. Taxic origin and temporal diversity: the effect of phylogeny. *In*: Novacek, M. J. & Wheeler, Q. D. (eds) *Extinction and Phylogeny*. Columbia University Press, New York, 89–118.

Officer, C. B. & Drake, C. L. 1983. The Cretaceous–Tertiary transition. *Science*, **219**, 1383–1390.

—— & ——1985. Terminal Cretaceous environmental events. *Science*, **227**, 1161–1167.

——, Hallam, A., Drake, C. L. & Devine, J. D. 1987. Late Cretaceous and paroxysmal Cretaceous/Tertiary extinctions. *Nature*, **326**, 143–149.

O'Keefe, J. D. & Ahrens, T. J. 1982. The interaction of the Cretaceous–Tertiary extinction bolide with the atmosphere, ocean and solid Earth. *In*: Silver, L. T. & Schultz, P. H. (eds) *Geological Implications of Impacts of Large Asteroids and Comets on the Earth*. Geological Society of America, Special Paper, **190**, 103–120.

Olsson, R. K. 1997. El Kef blind test III results. *Marine Micropaleontology*, **29**, 80–84.

—— & Liu, C. 1993. Controversies on the placement of Cretaceous–Paleogene boundary and the K/P mass extinction of planktonic foraminifera. *Palaios*, **8**, 127–139.

Orth, C. J., Attrep, M., Jr & Quintana, L. R. 1990. Iridium abundance patterns across bio-event horizons in the fossil record. *In*: Sharpton, V. L. & Ward, P. D. (eds) *Global Catastrophes in Earth History: an Interdisciplinary Conference on Impacts, Volcanism, and Mass Mortality. Geological Society of America, Special Paper*, **247**, 45–60.

Orue-Extebarria, X. 1997. El Kef blind test IV results. *Marine Micropaleontology*, **29**, 85–88.

Park, C. 1978. Nitric oxide production by Tunguska meteor. *Acta Astronomica*, **5**, 523–542.

Park, C. M. & Menees, G. P. 1978. Odd nitrogen production by meteoroids. *Journal of Geophysical Research*, **83**, 4029–4035.

PATTERSON, C. & SMITH, A. B. 1987. Is periodicity of mass extinctions a taxonomic artefact? *Nature*, **330**, 248–251.

—— & ——1989. Periodicity in extinction: the role of systematics. *Ecology*, **70**, 902–911.

PAUL, C. R. C. 1982. The adequacy of the fossil record. *In*: JOYSEY, K. A. & FRIDAY, A. E. (eds) *Problems of Phylogenetic Reconstruction*. Academic, New York, 75–117.

—— & MITCHELL, S. F. 1994. Is famine a common factor in marine mass extinctions? *Geology*, **22**, 679–682.

PERCH-NIELSEN, K., MCKENZIE, J. & HE, Q. 1982. Biostratigraphy and isotope stratigraphy and the catastrophic extinction of calcareous nannoplankton at the Cretaceous/Tertiary boundary. *In*: SILVER, L. T. & SCHULTZ, P. H. (eds) *Geological Implications of Impacts of Large Asteroids and Comets on the Earth*. Geological Society of America, Special Paper, **190**, 353–371.

PIMM, S. L. 1991. *The Balance of Nature: Ecological Issues in the Conservation of Species and Communities*. University of Chicago Press, Chicago.

——, MOULTON, M. P. & JUSTICE, L. J. 1995. Bird extinctions in the central Pacific. *In*: LAWTON, J. H. & MAY, R. M. (eds) *Extinction Rates*. Oxford University Press, Oxford, 75–87.

POLLACK, J. B., TOON, O. B., ACKERMAN, T. P., MCKAY, C. P. & TURCO, R. P. 1983. Environmental effects of an impact-generated dust cloud: implications for Cretaceous–Tertiary extinctions. *Science*, **219**, 287–289.

POSPICHAL, J. J. 1993. Cretaceous nannofossils in Danian sediments: survivorship or reworking? *Geological Society of America, Abstracts with Programs*, **25**, A363.

——1994. Calcareous nannofossils and the K/T boundary, El Kef: no evidence for stepwise, gradual, or sequential extinctions. *Geology*, **22**, 99–102.

——1996. Cretaceous/Tertiary boundary calcareous nannofossils from Agost, Spain. *In*: FLORES, J. A. & SIERRO, F. J. (eds) *Proceedings of the 5th International Nannoplankton Association Conference*. University of Salamanca, Salamanca, Spain, 185–217.

PRINN, R. G. & FEGLEY, B., JR 1987. Bolide impacts, acid rain, and biospheric traumas at the Cretaceous–Tertiary boundary. *Earth and Planetary Science Letters*, **83**, 1–15.

PROTHERO, D. R. 1994. *The Eocene–Oligocene Transition: Paradise Lost*. Columbia University Press, New York.

—— & BERGGREN, W. A. 1992. Eocene–Oligocene Climatic and Biotic Evolution. Princeton University Press, Princeton, 568.

RAMPINO, M. R. & STOTHERS, R. B. 1984. Geological rhythms and cometary impacts. *Science*, **226**, 1427.

—— & STOTHERS, R. B. 1988. Flood basalt volcanism during the past 250 million years. *Science*, **241**, 663–668.

RAUP, D. M. 1986*a*. *The Nemesis Affair: a Story of the Death of the Dinosaurs and the Ways of Science*. Norton, New York.

——1986*b*. Biological extinction in Earth history. *Science*, **231**, 1528–1533.

——1989. The case for extraterrestrial causes of extinction. *Philosophical Transactions of the Royal Society of London, Series B*, **325**, 421–435.

——1991*a*. *Extinction: Bad Genes or Bad Luck*. Norton, New York.

——1991*b*. A kill curve for Phanerozoic marine species. *Paleobiology*, **17**, 37–48.

——1994. Extinction models. *Geological Society of America, Abstracts with Programs*, **26**, A175.

—— & BOYAJIAN, G. E. 1988. Patterns of generic extinction in the fossil record. *Paleobiology*, **14**, 109–125.

—— & SEPKOSKI, J. J., JR 1984. Periodicity of extinctions in the geologic past. *Proceedings of the National Academy of Sciences of the USA*, **81**, 801–805.

—— & TREFIL, J. S. 1987. Numerical simulations and the problem of periodicity in the cratering record. *Earth and Planetary Science Letters*, **82**, 159–164.

RETALLACK, G. J. 1994. A pedotype approach to latest Cretaceous and earliest Tertiary paleosols in eastern Montana. *Geological Society of America Bulletin*, **106**, 1377–1397.

——1996. Acid trauma at the Cretaceous–Tertiary boundary in eastern Montana. *GSA Today*, **6**, 2–7.

REX, M. A., STUART, C. T., HESSLER, R. R., ALLEN, J. A., SANDERS, H. L. & WILSON, G. D. F. 1993. Global-scale latitudinal patterns of species diversity in the deep-sea benthos. *Nature*, **365**, 636–639.

RHODES, M. C. & THAYER, C. W. 1991. Mass extinctions: ecological selectivity and primary production. *Geology*, **19**, 877–880.

RICHARDS, M. A., DUNCAN, R. A. & COURTILLOT, V. E. 1989. Flood basalts and hot-spot tracks: plume heads and tails. *Science*, **246**, 103–107.

ROY, J. M., MCMENAMIN, M. A. S. & ALDERMAN, S. E. 1990. Trophic differences, originations, and extinctions during the Cenomanian and Maastrichtian stages of the Cretaceous. *In*: KAUFFMAN, E. G. & WALLISER, O. H. (eds) *Extinction Events in Earth History*. Springer, Berlin, 299–303.

RUCKLIDGE, J. C., DE GASPARIS, S. & NORRIS, G. 1982. Stratigraphic applications of accelerator mass spectrometry using ISOTRACE. *Proceedings of the Third North American Paleontological Convention*, 455–460.

RUDWICK, M. J. S. 1972. *The Meaning of Fossils: Episodes in the History of Palaeontology*. MacDonald, London.

RYDER, G. 1996. K/T boundary: historical context, counter-revolutions, and burdens of proof. *In*: RYDER, G., FASTOVSKY, D. & GARTNER, S. (eds) *The Cretaceous–Tertiary Event and other Catastrophes in Earth History*. Geological Society of America, Special Paper, **307**, 31–38.

SAWLOWICZ, Z. 1993. Iridium and other platinum-group elements as geochemical markers in sedimentary environments. *Palaeogeography, Palaeoclimatology, Palaeoecology*, **104**, 253–270.

SCHOPF, T. J. M. 1974. Permo-Triassic extinction: relation to sea-floor spreading. *Journal of Geology*, **82**, 129–143.

——1980. *Paleoceanography*. Harvard University Press, Cambridge, MA.
SEPKOSKI, J. J., JR 1982. A compendium of fossil marine families. *Milwaukee Public Museum Contributions in Biology and Geology*, **51**, 1–125.
SEPKOSKI, J. J., JR 1990. The taxonomic structure of periodic extinction. *In*: SHARPTON, V. L. & WARD, P. D. (eds) *Global Catastrophes in Earth History: an Interdisciplinary Conference on Impacts, Volcanism, and Mass Mortality*. Geological Society of America, Special Paper, **247**, 33–44.
——1994. Extinction and the fossil record. *Geotimes*, **March**, 15–17.
—— & RAUP, D. M. 1986. Periodicity in marine extinction events. *In*: ELLIOTT, D. K. (ed) *Dynamics of Extinction*. Wiley–Interscience, New York, 3–36.
SHAW, A. 1964. *Time in Stratigraphy*. McGraw-Hill, New York.
SHEEHAN, P. M. & HANSEN, T. A. 1986. Detritus feeding as a buffer to extinction at the end of the Cretaceous. *Geology*, **14**, 868–870.
——, COOROUGH, P. J. & FASTOVSKY, D. E. 1996. Biotic selectivity during the K/T and Late Ordovician events. *In*: RYDER, G., FASTOVSKY, D. E.& GARTNER, S. (eds) *The Cretaceous–Tertiary Event and other Catastrophes in Earth History*. Geological Society of America, Special Paper, **307**, 477–489.
SHOEMAKER, E. M. 1983. Asteroid and comet bombardment of the Earth. *Annual Review of Earth and Planetary Science*, **11**, 461–494.
SIGNOR, P. W., III & LIPPS, J. H. 1982. Sampling bias, gradual extinction patterns and catastrophes in the fossil record. *In*: SILVER, L. T. & SCHULTZ, P. H. (eds) *Geological Implications of Impacts of Large Asteroids and Comets on the Earth*. Geological Society of America, Special Paper, **190**, 291–296.
SIMBERLOFF, D. S. 1972. Models in biogeography. *In*: SCHOPF, T. J. M. (ed) *Models in Paleobiology*, Freeman, Cooper, San Francisco, 160–191.
——1974. Permo-Triassic extinctions: effects of area on biotic equilibrium. *Journal of Geology*, **82**, 267–274.
SLOSS, L. L. 1963. Sequences in the cratonic interior of North America. *Geological Society of America Bulletin*, **74**, 93–114.
SMIT, J. 1977. Discovery of a planktonic foraminifera association between the Abathomphalus mayaroensis Zone and the 'Globigerina' eugubina Zone at the Cretaceous/Tertiary boundary in the Barranco del Gredero (Caravaca, SE Spain). *Koninklijke Nederlandse Akademie van Wetenschappen Proceedings, Series B*, **80**, 280–301.
——1981. Synthesis of stratigraphical, micropaleontological and geochemical evidence from the K–T boundary: indication for cometary impact. *Lunar and Planetary Institute, Snowbird Abstracts*, 52.
——1982. Extinction and evolution of planktonic foraminifera after a major impact at the Cretaceous/Tertiary boundary. *In*: SILVER, L. T. & SCHULTZ, P. H. (eds) *Geological Implications of Impacts of Large Asteroids and Comets on the Earth*. Geological Society of America, Special Paper, **190**, 329–352.
——1990. Meteorite impact, extinctions and the Cretaceous–Tertiary boundary. *Geologie en Mijnbouw*, **69**, 187–204.
——1994. Blind tests and muddy waters. *Nature*, **368**, 809–810.
—— & NEDERBRAGT, A. J. 1997. Analysis of the El Kef blind test II. *Marine Micropaleontology*, **29**, 94–100.
—— & TEN KATE, W. G. H. Z. 1982. Trace element patterns at the Cretaceous–Tertiary boundary – consequence of a large impact. *Cretaceous Research*, **3**, 307–332.
——, GROOT, H., DE JONGE, R. & SMIT, P. 1988. Impact and extinction signatures in complete Cretaceous Tertiary (KT) boundary sections (abstract). *Global Catastrophes in Earth History: An Interdisciplinary Conference on Impacts, Volcanism, and Mass Mortality*, 182–183.
——, MONTANARI, A., SWINEBURNE, N. H. M. *et al.* 1992. Tektite-bearing, deep-water clastic unit at the Cretaceous–Tertiary boundary in northeastern Mexico. *Geology*, **20**, 99–103.
SMITH, A. B. 1994. *Systematics and the Fossil Record: Documenting Evolutionary Patterns*. Blackwell, London.
—— & PATTERSON, C. 1988. The influence of taxonomic method on the perception of patterns of evolution. *Evolutionary Biology*, **23**, 127–216.
SMITH, R. K. & BUZAS, M. A. 1986. Microdistribution of foraminifera in a single bed of the Monterey Formation, Monterey County, California. *Smithsonian Contributions to Paleobiology*, **60**, 1–33.
SPEIJER, R. P. 1994. Extinction and recovery patterns in benthic foraminiferal paleocommunities across the Cretaceous/Paleogene and Paleocene/Eocene boundaries. *Geologica Ultraiectina*, **124**, 9–190.
—— & VAN DER ZWANN, G. T. 1996. Extinction and survivorship of southern Tethyan benthic foraminifera across the Cretaceous/Palaeogene boundary. *In*: HART, M. B. (ed) *Biotic Recovery from Mass Extinction Events*. Geological Society, London, Special Publication, **102**, 245–258.
SPRINGER, M. S. 1990. The effect of random range truncations on patterns of evolution in the fossil record. *Paleobiology*, **16**, 512–520.
STANLEY, S. M., WETMORE, K. L. & KENNETT, J. P. 1988. Macroevolutionary differences between two major clades of Neogene planktonic foraminifera. *Paleobiology*, **14**, 235–249.
STEIGLER, S. M. 1987. Testing hypotheses or fitting models? Another look at mass extinctions. *In*: NITECKI, M. H. & HOFFMAN, A. (eds) *Neutral Models in Biology*. Oxford University Press, Oxford, 147–159.
STOLARSKI, R. S., & BUTLER, D. M. 1979. Possible effects on volcanic eruptions on stratispheric minor constituent chemistry. *Pure and Applied Geophysics*, **117**, 486–497.
STOTHERS, R. B. 1993. Flood basalts and extinction events. *Geophysical Research Letters*, **20**, 1399–1402.
—— & RAMPINO, M. R. 1990. Periodicity in flood basalts, mass extinctions, and impacts: a statistical

view and a model. *In*: SHARPTON, V. L. & WARD, P. D. (eds) *Global Catastrophes in Earth History: An Interdisciplinary Conference on Impacts, Volcanism, and Mass Mortality*. Geological Society of America, Special Paper, **247**, 9–18.

STRAUSS, D. & SADLER, P. M. 1989. Classical confidence intervals and bayesian probability estimates for ends of local taxon ranges. *Mathematical Geology*, **21**, 411–427.

THOMAS, E. 1990*a*. Late Cretaceous–Early Eocene mass extinction in the deep sea. *In*: SHARPTON, V. L. & WARD, P. D. (eds) *Global Catastrophes in Earth History: an Interdisciplinary Conference on Impacts, Volcanism, and Mass Mortality*. Geological Society of America, Special Paper, **247**, 481–495.

——1990*b*. Late Cretaceous through Neogene deep-sea benthic foraminifers, Maud Rise, Weddell Sea, Antarctica. *Proceedings of the Ocean Drilling Program: Scientific Results*, **113**. Ocean Drilling Program, College Station, TX, 571–594.

TILMAN, D., MAY, R. M., LEHMAN, C. L. & NOWAK, M. A. 1996. Habitat destruction and the extinction debt. *Nature*, **371**, 65–66

TOON, O. B., POLLACK, T. P., ACKERMAN, T. P., TURCO, R. P., MCKAY, C. P. & LIU, M. S. 1982. Evolution of an impact-generated dust cloud and its effects on the atmosphere. *In*: SILVER, L. T. & SCHULTZ, P. H. (eds) *Geological Implications of Impacts of Large Asteroids and Comets on the Earth*. Geological Society of America, Special Paper, **190**, 187–200.

TREDOUX, M., DE WIT, M. J., HART, R. J., LINSAY, N. M., VERHAGEN, B. & SELLSCHOP, J. P. F. 1988. Chemostratigraphy across the Cretaceous–Tertiary boundary and a critical assessment of the iridium anomaly. *Journal of Geology*, **97**, 585–605.

VAIL, P. R., MITCHUM, R. M. J. & THOMPSON, S. 1977. Seismic stratigraphy and global changes in sea level, part three: Relative changes of sea level from coastal onlap. *In*: PAYTON, C. E. (ed.) *Seismic Stratigraphy – Applications to Hydrocarbon Exploration*. American Association of Petroleum Geologists, Memoir, **26**, 83–98.

VAN VALEN, L. M. 1994. Concepts and the nature of selection by extinction: is generalization possible? *In*: GLEN, W. (ed.) *The Mass Extinction Debates: How Science Works in a Crisis*. Stanford University Press, Stanford, CA, 200–216.

VOGT, P. R. 1972. Evidence for global synchronism in mantle plume convection and possible significance for geology. *Nature*, **240**, 338–342.

WANG, K., ATTREP, M., JR & ORTH, C. J. 1993. Global iridium anomaly, mass extinction, and redox change at the Devonian–Carboniferous boundary. *Geology*, **21**, 1071–1074.

WARD, P. D. 1990. A review of Maastrichtian ammonite ranges. *In*: SHARPTON, V. L. & WARD, P. D. (eds) *Global Catastrophes in Earth History: an Interdisciplinary Conference on Impacts, Volcanism, and Mass Mortality*. Geological Society of America, Special Paper, **247**, 519–530.

——1995*a*. After the fall: lessons and directions after the K/T debate. *Palaios*, **10**, 530–538.

——1995*b*. The K/T Trial. *Paleobiology*, **21**, 245–247.

—— & KENNEDY, W. J. 1993. Maastrichtian ammonites from the Biscay region (France, Spain). *Paleontological Society Memoir*, **34**, 1–58.

——, KENNEDY, W. J., MACLEOD, K. G. & MOUNT, J. F. 1991. Ammonite and inoceramid bivalve extinction patterns in Cretaceous/Tertiary boundary sections of the Biscay region (southwestern France, northern Spain). *Geology*, **19**, 1181–1184.

——, WIEDMANN, J. & MOUNT, J. F. 1986. Maastrichtian molluscan biostratigraphy and extinction patterns in a Cretaceous/Tertiary boundary section exposed at Zumaya, Spain. *Geology*, **14**, 899–903.

WARD, W. C., KELLER, G., STINNESBECK, W. & ADATTE, T. 1995. Yucatan subsurface stratigraphy: Implications and constraints for the Chicxulub impact. *Geology*, **23**, 873–876.

WHITE, R. S. 1989. Igneous outbursts and mass extinctions. *Eos Transactions, American Geophysical Union*, **70**, 1490–1491.

WIDMARK, J. G. & MALMGREN, B. A. 1992. Benthic foraminiferal changes across the Cretaceous–Tertiary boundary in the deep sea; DSDP Sites 525, **527**, and 465. *Journal of Foraminiferal Research*, **22**, 81–113.

WIEDMANN, J. & KULLMANN, J. 1996. Crises in ammonoid evolution. *In*: LANDMAN, N. H., TANABE, K. & DAVIS, R. A. (eds) *Ammonoid Paleobiology*. Plenum, New York, 795–813.

WOLBACH, W. S., GILMOUR, I. & ANDERS, E. 1990. Major wildfires at the Cretaceous/Tertiary boundary. *In*: SHARPTON, V. L. & WARD, P. D. (eds) *Global Catastrophes in Earth History: an Interdisciplinary Conference on Impacts, Volcanism, and Mass Mortality*. Geological Society of America, Special Paper, **247**, 391–400.

——, LEWIS, R. S. & ANDERS, E. 1985. Cretaceous extinctions: evidence for wildfires and the search for meteoritic material. *Science*, **230**, 167–170.

WOLFE, J. A. 1987. Late Cretaceous–Cenozoic history of deciduousness and the terminal Cretaceous event. *Paleobiology*, **13**, 215–226.

ZACHARIASSE, W. J., RIEDEL, W. R., SANFILIPPO, A. *et al.* 1978. Micropaleontological counting methods and techniques – an exercise on an eight meters section of the Lower Pliocene of Capo Rossello, Sicily. *Utrecht Micropaleontological Bulletins*, **17**, 1–265.

Timing and causes of vertebrate extinction across the Cretaceous–Tertiary boundary

A. C. MILNER

Department of Palaeontology, The Natural History Museum, Cromwell Road, London SW7 5BD, UK

Abstract: A continuous sequence of terrestrial sediments bracketing the Cretaceous–Tertiary (K–T) boundary is known only in the western interior of North America. Documentation of terrestrial faunal diversity, and patterns of survival and extinction, demonstrate that dinosaurs (i.e. non-avian dinosaurs) declined in the Late Maastrichtian and were the only major terrestrial vertebrate group, along with the pterosaurs, that became extinct at or near the K–T boundary. There was a varying degree of turnover but no mass extinction in other vertebrate groups. Both gradualist and catastrophe scenarios have been advocated as the main cause. There is much less evidence in other parts of the world that can be brought to bear on the timing and nature of dinosaur extinction. None the less, there are dinosaur remains from Maastrichtian horizons on every continent and none are recorded from overlying Paleocene strata. This supports the axiom that dinosaurs died out world-wide at the end of the Cretaceous although there is no means of determining whether the extinction pattern in western North America was a local or global phenomenon. The disruption of reproductive patterns in herbivorous dinosaurs by trace element contamination derived from an impact, volcanic activity or both events is one factor which might correlate with their global extinction.

The Cretaceous–Tertiary (K–T) boundary, at 65 Ma, is associated with both the impact of an extraterrestrial object and episodes of flood basalt volcanism, the Deccan Traps, in India. The K–T boundary also marks the demise of the dinosaurs as the dominant terrestrial animal group and an impact–extinction hypothesis has been popularly cited as the causal mechanism of end Cretaceous extinctions since Alvarez *et al.* first proposed it in 1980. The interpretation of the the Chicxulub Structure on the coast of the Yucatán Peninsula, Mexico, as the site of such an impact (Hildebrand *et al.* 1991) is now widely accepted although the nature of the object is disputed (references in this volume). However, the biological consequences of the impact event are poorly understood and proponents of impact–extinction models frequently fail to consider dinosaur extinction in the context of faunal turnovers as a whole, that is, the survival and extinction patterns of all vertebrates groups across the K–T boundary. This paper provides a brief review of the vertebrate fossil record and a discussion as to how far the turnover patterns that can be deduced from the actual record, and by inference from phylogenetic studies, can be related to the physical events of an impact or whether the effects of terrestrially based environmental perturbations might more readily fit the pattern.

The Cretaceous–Tertiary vertebrate record

The fossil record of vertebrates across the K–T boundary is much less abundant than that of invertebrates, especially microfossils. The record is geographically variable, generally better in the northern hemisphere, reflecting the biases of collecting activity and the occurrence of Lagerstätten rather than genuine distribution patterns, but none the less it is evident that many groups were little affected and extinctions were highly selective (Fig. 1).

Fishes

Data on fishes have been summarized from MacLeod *et al.* (1997). Cartilaginous fishes and teleost actinopterygians (ray-finned fishes) form the vast majority of living fish faunas and are the only groups that contribute significantly to extinction patterns at the K–T boundary. For the cartilaginous fishes (sharks, rays and chimaeras) 35 families passed through the K–T boundary, seven became extinct within the Maastrichtian, and one originated in the Danian. In percentage terms, the K–T boundary was survived by 80% of cartilaginous fish families. There is a reduced survival at the K–T boundary compared with the Campanian–Maastrichtian

MILNER, A. C. 1998. Timing and causes of vertebrate extinction at the K–T boundary. *In*: GRADY, M. M., HUTCHISON, R., MCCALL, G. J. H. & ROTHERY, D. A. (eds) *Meteorites: Flux with Time and Impact Effects*. Geological Society, London, Special Publications, **140**, 247–257.

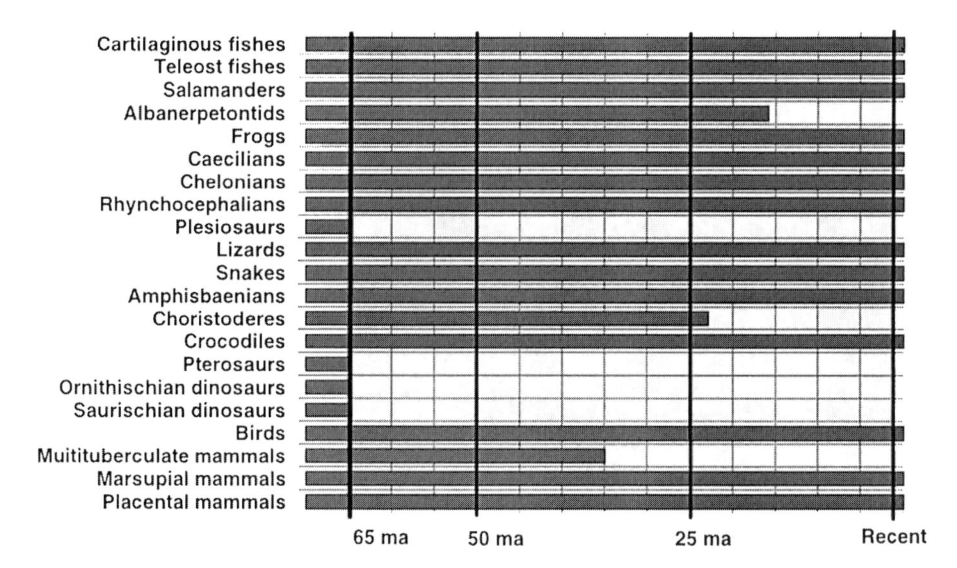

Fig. 1. Survival and extinction of vertebrate groups across the K–T boundary at 65 Ma. Approximate dates of last appearances of Tertiary taxa shown by the lengths of the individual horizontal bars; extant taxa continue beyond Recent vertical line. (Data from Benton (1993) and MacLeod *et al.* (1997)).

and Danian–Thanetian boundaries. According to a database for elasmobranchs (sharks and rays) compiled by Cappetta (1987), documenting generic occurrences throughout the Cretaceous and the Tertiary, it appears that the Maastrichtian was a time of high generic turnover and not only a period of high extinction. MacLeod *et al.* (1997) questioned whether that extinction rate is high enough to implicate a catastrophic cause. For teleost fishes the K–T record consists of 43 families which passed through the K–T boundary, four which became extinct within the Maastrichtian and nine which originated within the Danian. Therefore, about 10% of teleost fish families became extinct and 90% survived. It may be surmised that fishes were little affected by K–T boundary events.

Amphibians

There is no evidence for amphibian extinctions at the K–T boundary and tangible evidence of many lineages passing through the event unaffected. Five families of salamanders bracket the K–T boundary as fossils (MacLeod *et al.* 1997); three of them have representatives living today: Sirenidae and Amphiumidae (Milner 1983) and Salamandridae (Astibia *et al.* 1990). Phylogenetic inference from studies of the relationships of the whole group demands that representatives of families recorded only from the Tertiary, or their stem lineages, must have been present before the K–T boundary (Milner 1993). The albanerpetontids are an enigmatic and specialized small amphibian family of two genera, initially assumed to be salamanders, but now considered to represent a distinct lineage (McGowan & Evans 1995). That family has no living representative but there is no evidence of family or generic-level extinctions at or near the K–T boundary. The genus *Albanerpeton* spans the K–T boundary in both North America and Europe and it seems that the family survived unscathed in at least these regions (Milner 1993). The frog fossil record is geographically very uneven, fair for North America, Europe and South America, but very poor from other continents. Seven families span the K–T boundary as fossils, six of them having living representatives (Milner 1993). There is no direct fossil evidence of caecilian turnover at the K–T boundary (MacLeod *et al.* 1997). However, mitochondrial DNA sequence data support the interpretation that much of their modern diversity was established in the Mesozoic by vicariance during the break-up of Gondwana (Hedges *et al.* 1993). Therefore the lineages leading to modern caecilians must have been present before the K–T boundary.

Reptiles

As traditionally understood, reptiles include all higher vertebrates except birds and mammals. The term is used here purely as a descriptive term for the chelonians (turtles and tortoises) lepidosaurs (rhynchocephalians and squamates) and their relatives. The remaining groups of more derived reptiles, crocodiles, pterosaurs, dinosaurs and birds constitute a monophyletic group and are discussed under that subheading.

Amongst the chelonians, 13 families bracket the K–T boundary and eight families have living representatives (Benton 1993). They include marine, freshwater and predominantly terrestrial forms. Two groups of exclusively marine reptiles did not persist beyond the Cretaceous: ichthyosaurs died out long before the K–T boundary, the last recorded being Cenomanian in age (Benton 1993); one family of long-necked plesiosaurs, the elasmosaurs, did survive until the end of the Cretaceous, the last records being of latest Maastrichtian age in southern hemisphere high latitudes from New Zealand (Benton 1993).

Living lepidosaurs comprise the squamates (lizards, snakes and amphisbaenians) and a single representative of the Rhynchocephalia, the New Zealand Tuatara, *Sphenodon*. Rhynchocephalians were in decline from the middle Cretaceous but survived the K–T boundary in Gondwana. The Squamata are the largest and most diverse group of living reptiles, with more than 6000 extant species. All living families of lizards either had known representatives in the Late Cretaceous or their presence may be assumed by extrapolation from phylogenetic trees (MacLeod *et al.* 1997) so they therefore represent definite or inferred survivors from the Cretaceous. Of the groups present in the latest Cretaceous, only two failed to cross the K–T boundary, mosasaurs and one family of specialized durophagous feeders (MacLeod *et al.* 1997).

Mosasaurs were large (up to 15 m) predatory marine lizards of the Late Cretaceous. In North America, at least, they are known to have survived into the late Maastrichtian (G. Bell, pers. comm. (1995), *contra* Sullivan (1987), in MacLeod *et al.* (1997)). Their decline during the late Maastrichtian would not be surprising. Mosasaurs were amongst the top predators of the Cretaceous marine ecosystem and anything that perturbed that ecosystem would have affected them.

Four families of primitive snakes crossed the K–T boundary, of which two have representatives living today. Higher snakes (colubroids, Werner & Rage 1994) and the burrowing amphisbaenians, although predominantly a Tertiary group, also clearly crossed the divide (Wu *et al.* 1993).

Choristoderes, a small group of freshwater, superficially crocodile-like lepidosauromorphs, were present in the latest Cretaceous in North America. They, too, passed through the boundary, becoming extinct in the Eocene (Milner *et al.* 1998).

Archosaurs

Archosaurs include crocodilians, pterosaurs, dinosaurs and birds. Birds are now almost universally accepted as the sister group of maniraptorans, small advanced carnivorous dinosaurs from which they are descended.

Five families of crocodiles span the K–T boundary (MacLeod *et al.* 1997), they were terrestrial–freshwater, except for the Dyrosauridae which were freshwater–coastal marine. Two families, Alligatoridae and Crocodylidae, include living representatives; two alligatorines and the crocodylid (or possibly stem crocodylid) *Leidyosuchus sternbergi*, have been recorded from both the Lancian and Puercan in Montana (Archibald & Bryant 1990) and so crossed the K–T boundary at the species level. The group as a whole shows a 50% family survival rate across the K–T boundary. The only discernible trend apparent from this pattern is that no large crocodiles, such as the giant North American crocodylid *Deinosuchus*, survived.

All pterosaurs became extinct before the Maastrichtian, with the exception of the giant Azhdarchidae (Wellnhofer 1991). No Tertiary pterosaur remains are known and their extinction pattern was apparently a gradual decline through the Late Cretaceous.

Twenty-four families of dinosaurs are recorded in the Maastrichtian from all continents except Australia (Table 1). Late Maastrichtian records show little decline at family level except in Asia and South America, though this probably reflects the absence of relevant strata in those areas rather than positive evidence for a fall in diversity. The western interior of North America is the only area where a continuous terrestrial sequence encompasses the K–T boundary, permitting documentation of changes in faunal diversity. Consequently, it has been the focus of several detailed studies (see below).

Recent discoveries have revolutionized the understanding of bird evolution in the Mesozoic and early Tertiary, although much of their fossil history is still missing. Taxa belonging to four modern bird orders are now known to occur in the late Cretaceous and the phylogenetic pattern among those taxa suggest that several other lineages must have differentiated before the end of the Cretaceous (Chiappe 1995). This supports an incremental Cretaceous diversification of

Table 1. *Dinosaur distribution in the Maastrichtian.*

	MAA	L MAA	Last recorded genera	
			EU[1]	NA[2]
Saurischia				
THEROPODA				
Tyrannosauridae NA	✓	✓		*Tyrannosaurus*
				Albertosaurus
Ornithomimidae As,NA	✓	✓		*Ornithomimus* sp.
Elmisauridae As,NA	✓	✓		*Chirostenotes*
Oviraptoridae As,NA	✓	–		
Deinocheiridae As	✓	–		
Therizinosauridae As	✓	–		
Dromaeosauridae As,NA	✓	✓		*Dromaeosaurus* sp.
				?*Velociraptor*
Troodontidae As,NA	✓	✓		*Troodon*
				Troodon sp.
				Paronychodon
Avimimidae As	✓			
Abelisauridae In,SA	✓	–		
Noasauridae SA	✓	–		
Therizinosauridae As	✓	–		
Theropoda *incertae sedis*				
Aublysodontidae NA	✓	✓		
SAUROPODA				
Camarasauridae As	✓	–		
Diplodocidae As	✓	–		
Titanosauridae As,Eu,In,Na,SA	✓	✓	*Magyarosaurus*	
Ornithischia				
Nodosauridae Eu,NA	✓	✓	*Struthiosaurus*	?*Edmontonia*
Ankylosauridae As,NA	✓	✓		*Ankylosaurus*
Hypsilophodontidae Eu,NA	✓	✓		*Thescelosaurus*
Iguanodontia *inc. sed.*	✓	✓	*Rhabdodon*	*Edmontosaurus*
Hadrosauridae As,Eu,NA	✓	✓		*Anatotitan*
Pachycephalosauridae As,NA	✓	✓	*Telmatosaurus*	*Stygimoloch*
				Pachycephalosaurus
				Stegoceras
Homalocephalidae As	✓			
Protoceratopidae As,NA	✓	✓		
Ceratopidae NA	✓	✓		*Triceratops*
				Torosaurus

MAA records the occurrence of families throughout Maastrichtian time, L MAA the families present in the late Maastrichtian. As, Asia; Eu, Europe; In, India; NA, North America; SA, South America. Genera from the latest Maastrichtian localities: [1]Sînpetru Formation, Haţeg Basin, Romania; [2]Hell Creek Formation, northeast Montana, USA. (Table modified from MacLeod *et al.* (1997): data from Weishampel (1990), modified by Archibald (1996*a*).)

modern (neornithine) birds and no mass extinction at the K–T boundary. The enantiornithine birds, an important and widespread Mesozoic group, did become extinct at the end of the Cretaceous, but their fossil record is too poorly known to determine their extinction pattern (Chiappe 1995). This contrasts strongly with Feduccia's (1995) hypothesis that birds appear to have suffered a bottleneck across the K–T boundary, a massive extinction from which only transitional shorebird morphotypes survived as the basis for a Tertiary bird radiation.

Mammals

The two major lineages of living mammals (marsupials and placentals) were established by the Late Cretaceous. Both passed through the K–T boundary, together with the now extinct multituberculates (Archibald 1996*a*).

High-resolution studies in North America

Archibald & Bryant (1990) and Archibald (1996*a*, *b*) undertook high-resolution studies of vertebrate turnover at the K–T boundary and documented a complete vertebrate assemblage through the K–T boundary interval (Lancian–Puercan) in the Hell Creek (Garfield County) and Bug Creek (McCone County) areas in northeastern Montana, USA. That now classic and continuing study has been possible by virtue of the unique continuous unbroken sequence of terrestrial sediments across the boundary, and demonstrates variable extinction rates for different vertebrate groups. Analysis of Archibald & Bryant's original (1990) raw data to allow for factors such as palaeobiogeographical variation, disappearances that represent speciation events and rarity of taxa, produced an estimated 64% species survival rate (conversely, a 36% extinction rate) for the non-marine vertebrate fauna. Archibald's latest revised statistics (1996*a*) for the 107 species or lineages present (Table 2) produce a minimum possible survival rate of 49% (52 species); the maximum possible survival rate is 67% allowing for the same factors as in the 1990 calculations. Even taking the minimum figure, only five out of the 12 vertebrate groups represented in the study area account for 75% of the extinctions. Those groups are elasmobranchs (sharks, skates and rays), lizards, marsupial mammals and the two groups of dinosaurs (ornithischians and non-avian saurischians); of those only the elasmobranchs and dinosaurs had no survivors.

The figures demonstrate that extinctions were very selective in northeastern Montana. The events at the K–T boundary had no detectable effect on the amphibian fauna, as far as the fossil record permits us to determine this. No monophyletic taxa make their last appearance in the Maastrichtian, and most forms pass through the K–T boundary unchanged (Archibald & Bryant 1990; Archibald 1996*a*; Milner *et al.* 1998). Lizards were affected more substantially. However, the effect was taxonomically relatively specific and most extinctions were among the Teiidae (Estes 1976; Estes & Baéz 1985, Gao & Fox 1993). Crocodiles passed though the K–T boundary although large forms disappeared; turtles, choristoderes and most squamates survived it without major effect.

Table 2. *Percentage survival of vertebrate groups across the K–T boundary in northeastern Montana, USA*

Group	Survival
Cartilaginous fishes	0%
Actinopterygian (teleost) fishes	60%
Frogs and salamanders	100%
Chelonians	88%
Lizards	30%
Choristoderes	100%
Crocodiles	80%
Ornithischian dinosaurs	0%
Saurischian dinosaurs	0%
Multituberculate mammals	50%
Marsupial mammals	9%
Placental mammals	100%

There were no pterosaur or bird fossils preserved in the study area. (Data condensed from Archibald (1996*a*)).

The most obvious and dramatic extinction was that of the non-avian dinosaurs. Dinosaur diversity declined steadily from a peak of 45 genera in 13 families in the late Campanian to 24 in 12 families in the late Maastrichtian (Lancian) (data from Weishampel (1990)). Twelve dinosaur genera (13 species) from the Bug Creek interval, overlying the Lancian at the top of the Hell Creek Formation, are the last recorded in the sequence (Table 1). There is no consensus on the age of the Bug Creek interval. Its fauna contains taxa otherwise unknown after the Lancian (particularly dinosaurs) and others unknown before the Puercan (especially mammal taxa). Archibald & Bryant (1990) included the Bug Creek interval in the Puercan on the basis of the mammal faunas, thus dating it as earliest Paleocene. Accepting this age, the presence of dinosaur remains in the Bug Creek interval can be explained in one of two ways; either they survived into the Paleocene, or they were reworked from the Hell Creek Formation. Archibald & Bryant (1990), in common with other researchers, adopted the latter explanation on preservational and taphonomic grounds, as there are no articulated dinosaur remains in the Bug Creek channel sediments. They interpreted their results in terms of geologically rapid change, but not a sudden catastrophic mass extinction at the K–T boundary.

Others have offered different interpretations. A detailed field study of dinosaur diversity by Sheehan *et al.* (1991), carried out in parallel with Archibald & Bryant's work, claimed that no statistically meaningful decline was apparent through the Hell Creek Formation. These findings were taken to support a sudden extinction event at the K–T boundary. Williams (1994) observed that the nature of the analysis by Sheehan *et al.* (1991) might not detect a gradual decline at the top of the Hell Creek sequence. There remains, despite concentrated collecting, a barren zone of about 3 m below the K–T boundary, in which dinosaur bones and teeth are increasingly scarce. The highest unreworked bone comes from about 60 cm below the boundary clay, well below the iridium layer. Evidence from the Bug Creek channels and the highest few metres of the Hell Creek Formation is consistent with a decline in the dinosaur population, perhaps a steep and accelerating decline but not a sudden catastrophe (Williams 1994). Further, Hurlbert & Archibald (1995) analysed the quantitative technique used by Sheehan *et al.* (1991) to argue for a rapid end-Cretaceous dinosaur extinction event, and concluded that there was no statistical support for either a sudden or gradual decline in the number of dinosaur taxa at the close of the Cretaceous. Despite these extensive studies, the Hell Creek Beds have not provided an unequivocal answer on extinction rates. The one undisputed axiom is that no non-avian dinosaurs survived the events at the K–T boundary in that area.

Extinction mechanisms

Gradualistic models

Great debate continues as to the causes of the terrestrial K–T extinctions in western North America. The fossil record does appear to be consistent with a gradual decline in dinosaur diversity at generic level over a period of around 8 Ma. Archibald & Bryant (1990) and Archibald (1996*a*, *b*) argued that marine regression was the primary cause, resulting in habitat fragmentation, breaking-up of coastal ranges, increases in the length of river courses and freshwater habitats, all factors creating ecological stresses primarily on the populations of large herbivorous dinosaurs dominated by hadrosaurs. This hypothesis does accord with the survival and extinction pattern in some of the other vertebrate groups. The extinction of elasmobranchs in the study area is correlated with the loss of connections to marine habitats by regression of the Midcontinental Seaway. Conversely, those groups restricted to freshwater habitats were largely unaffected. The survival of most groups, based on what is known of the ecological conditions and environmental tolerance of their modern descendants, argues against catastrophic environmental changes on the scale postulated by models of bolide impacts. Archibald (1996*a*, *b*) compared the selective survivorship pattern against the predicted effects and corollaries of impact and of regression–habitat fragmentation models and concluded that many, but not all, of the biotic predictions of those kill mechanisms are inconsistent with the fossil record.

Impact models

Among the major corollaries of an impact are global wildfires, acidic rains and a short, sharp temperature decrease (Alvarez 1986). Predictions of global wildfires (Wolbach *et al.* 1990) have little correlation with the pattern observed by Archibald (1996*b*), for even in the unlikely scenario of aquatic animals somehow surviving, so did terrestrial turtles, lizards and mammals. Local wildfires could, however, have been involved.

Aquatic species, and amphibians in particular, are sensitive to environmental pH and predictions of global highly acid rains clearly do not fit in with the survival pattern seen in Montana, nor what we know of amphibian distribution patterns in the rest of the world. Estimates of the pH of impact-generated acid rains vary (references in Archibald (1996*a*, *b*)) but the suggested acidity of rain of pH 0–1.5 could have driven near-surface marine and fresh water to below pH 3 (D'Hondt *et al.* 1994). Amphibian eggs and larvae do not survive in acid water and at pH values as low as three, adults too may die (Cox 1993).

A sudden temperature drop would be most likely to affect selectively the ectothermic component (amphibians, reptiles, crocodiles) of the K–T boundary faunas. Although modern ectotherms which live today in more temperate zones or areas of seasonal drought have evolved strategies (hibernation, aestivation) to survive adverse conditions, there is no basis to suggest that tropical to sub-tropical ectotherms in late Cretaceous faunas could do so (Archibald 1996*a*, *b*), nor that they would be able to respond to sudden environmental change. Ectotherms in the Montana fauna are not present at late Cretaceous sites on the north slope of Alaska, where summer temperatures averaged around 10°C, yet dinosaurs and mammals, being less temperature sensitive, lived there (Clemens & Nelms 1993). As Archibald (1996*a*, *b*) observed, if the Late Cretaceous temperature regime in Alaska excluded ectotherms, then a sudden temperature drop in middle latitudes at the K–T boundary would have devastated the Montana fauna; the turnover pattern shows clearly that this did not occur. The only disappearance that might be temperature related is the loss of teiid lizards. The records suggest that teiids became extinct in North America and Asia at this time, possibly as a result of a period of lower temperatures, living teiids being tropical or subtropical animals (MacLeod *et al.* 1997).

This analysis certainly does not support recent claims of catastrophic impact devastation of North America such as those of Schultz &

D'Hondt (1996). They deduced from the asymmetrical geophysical features of the Chicxulub structure that there was an angled asteroid strike on the Yucatán peninsula. The resulting down-range trajectory of the ejecta affected west central North America most severely, and the effects were such that environmental heating would have been orders of magnitude over previous calculations; well above the levels required for spontaneous ignition of vegetation (Schultz & D'Hondt 1996). Those workers did, however, follow Sheehan & Fastovsky's (1992) suggestion that water might have provided refuge for aquatic vertebrates.

Other mechanisms

None the less there was major biotic disturbance at the K–T boundary as evidenced by the decline in marsupial mammals, dinosaurs and floral turnovers. Archibald (1996*a*, *b*) suggested that the drastic decline in marsupials across the boundary coincides with the invasion from Asia of new placental mammals, especially archaic ungulates through the Bering land bridge.

Johnson & Hickey's (1990) work on North American megafloras in the western USA demonstrated that changes were occurring before, at and after the K–T boundary. The palynomorph turnover pattern in the same area suggests that some North American endemic flowering plants were drastically affected at the K–T boundary, but gymnosperms, ferns, etc. and mosses much less so. The disappearance of angiosperm palynomorphs coincided with the Ir anomaly, and was succeeded by a 'fern spike' just above it followed by progressive appearance of Danian taxa. However, this apparent floral disruption pattern was much less dramatic in Canada, with more gradual turnover patterns and no fern spike reported. Floral data available from southern high latitudes show no significant turnovers at this time (references in MacLeod *et al.* (1997)).

Could there have been a breakdown in the food chain? Might that have affected the dinosaur populations in western North America? This was favoured by Buffetaut (1990), following arguments against Maastrichtian regression as a cause of extinctions at the K–T boundary (Buffetaut 1987). Large herbivorous dinosaurs being primary consumers of fresh vegetation would certainly be affected by a sudden downturn in their food supply locally in western North America but, as noted above, that pattern of floral turnover was more gradual in Canada, which also had a contiguous Maastrichtian dinosaur assemblage, and was not mirrored in southern high latitudes. It has been argued (e.g. Sheehan & Hansen 1986) that other vertebrate groups could have subsisted on a detrital regime. Insects which show no turnover at the K–T boundary (Whalley 1988) could perhaps be a factor in the survival of small terrestrial vertebrates.

The marine realm

The latest appearance of the mosasaurs was reported as early Maastrichtian by Sullivan (1987) but recent work on late Cretaceous mosasaurs shows that at least three families persisted into the latest Maastrichtian (G. Bell, pers. comm. in Milner *et al.* (1998)), and are recorded from high latitudes in New Zealand (Benton 1993). Mosasaurs were the top carnivores of marine ecosystems and any perturbations which cut the overall productivity of the lower trophic levels would have affected mosasaurs as well. The extinction of long-necked piscivorous plesiosaurs might also be linked to a downturn in marine productivity. That said, it is interesting that large marine turtles were not affected to the same degree (R. Hiragama, pers. comm., in MacLeod *et al.* (1997)), neither were the marine dyrosaurid crocodiles which were dispersing around the Tethyan coasts in the Maastrichtian and early Tertiary (Buffetaut 1982).

The world outside western North America

Diversity changes and extinction patterns in the western interior of the USA provide a test case for the demise of the dinosaurs. There are no means of determining if this was a local or global phenomenon for lack of K–T terrestrial sequences elsewhere. A late Maastrichtian non-marine fauna, including dinosaurs, is known from the Sînpetru Formation (once regarded as Danian) in Romania (Grigorescu *et al.* 1994) (Table 1), but is overlain by middle to late Miocene marine strata. Dinosaur assemblages from southern France show a faunal turnover where an early Maastrichtian sauropod-dominant assemblage is replaced by a late Maastrichtian (68 Ma) hadrosaur-dominant assemblage linked to a marine regression (Le Loeuff *et al.* 1994). However, that eustatic event is not connected with extinction which happened at least 3 Ma later.

Dinosaurs were certainly present in the uppermost Cretaceous of Asia; the latest occurrence known is nesting sites in the Pingling

Formation of the Nanxiong Basin, in the northernmost region of Guangdong Province, southeastern China, on the boundary of magnetochron 30N and 29R (Stets *et al.* 1996). The K–T boundary in the basin lies within a poorly defined transition zone, there is no distinct boundary layer and no evidence of an Ir anomaly. There is, in fact, no direct evidence of a bolide impact in the Nanxiong Basin sediments (Stets *et al.* 1996). The palynomorph record shows that a significant floral change took place during the boundary interval, the replacement of tropical forms by mixed evergreens suggesting climatic change to cooler conditions. Dinosaur remains and extensive nesting sites attributed to titanosaurid sauropods have been reported from the Deccan intertrappean beds in India, dated as terminal Cretaceous on palynomorphs and magnetostratigraphy (Sahni *et al.* 1994).

Dinosaur eggs, eggshells and life cycle disruption

All dinosaurs laid eggs, the most vulnerable point in their life cycle. There is now sufficient evidence to be confident that both ornithischians and saurischians reproduced that way and a range of egg morphotypes can be identified at ordinal, perhaps family level and, more rarely, with known dinosaur species (Carpenter *et al.* 1994).

Dinosaur eggshell fragments, whole eggs and nests ascribed to both ornithopod and theropod dinosaurs occur in the top of the Pingling Formation in the Nanxiong Basin. Those closest to the K–T boundary interval are reported to exhibit pathological shell structure: a significant reduction in shell thickness and pathological malformation (Stets *et al.* 1996). Geochemical analysis of eggshells showed increased concentration of trace elements (Cr, Cu, Ni, Pb, Co, Zn) distinctly anomalous with the content of the surrounding sediments, although those sediments could have been subject to diagenetic loss. Stets *et al.* (1996) concluded that dinosaur extinction was stepwise in the Nanxiong Basin, triggered by environmental stress. That stress was generated by a change in climate to more temperate conditions, which in turn caused the disappearance of food plants, as demonstrated by the palynomorph data. In addition, Stets *et al.* (1996) believed that the high levels of trace elements in the eggshell suggested that slow environmental poisoning may have been an additional factor contributing to extinction.

Similar phenomena have been reported from Late Maastrichtian nest sites of the sauropod *Hypselosaurus* in Provence, southern France. Erben (1972, 1975) and Erben *et al.* (1979) found evidence of a progressive shell thinning, increasing in frequency and representing more than 90% of the sample in the uppermost horizon closest to the K–T boundary. They concluded that changes in $\delta^{18}O$ and $\delta^{13}C$ composition of the eggshells indicated respectively change in climate (either slightly lowered temperature or increased humidity) and diet, and that the resulting environmental stress was the cause of thinning eggshells and ultimately a crash in the population through reproductive failure. Hansen (1990, 1991) reported progressive hatching failure in clutches of sauropod eggs in the same series of beds in magnetochron 30N, some 2–4 ky before the K–T boundary. His shell analysis documented progressively increasing amounts of selenium (Se) in eggshells in the highest beds. Selenium is toxic in very small amounts in developing chick embryos and accumulates preferentially in the yolk rather than the eggshell by a factor of 2:1 (Hansen 1991). Hansen hypothesized that increased uptake of selenium derived from the Deccan Traps volcanism, coincident with the shoulders in the iridium spike (Courtillot 1990), might have contributed to progressive hatching failure of the sauropod egg clutches. He considered that the selenium could have been ingested along with dust on vegetation. However, some plants can strongly concentrate Se from relatively low levels in soils, particularly where it is oxidized to water-soluble selenates in alkaline soils (Oldfield *et al.* 1974); this provides an alternative uptake mechanism.

Sahni *et al.* (1994) observed that most of the sauropod eggs in the Deccan intertrappean beds were unhatched and that none contained embryonic remains. They proposed several possible factors that might explain this unusual situation, including deep burial of the eggs and desiccation in arid conditions. Unusual structures were reported in a few eggshell fragments but any inference of a pathological condition required further work (Sahni *et al.* 1994). No trace element analysis was reported.

The two known sources of trace element contamination which may have acted selectively on the reproductive cycle of herbivorous dinosaurs are impacts and volcanic activity. Sutherland (1993) noted that around 22 active volcanic hotspots, hotlines and starting plumes contributed to mantle-derived eruptive material and were potential source of pollutants in the southern hemisphere K–T interval. A fragmentary but diverse Maastrichtian dinosaur assemblage is known from southern polar latitudes in New Zealand (Wiffen 1996). There was no significant floral turnover at southern high latitudes (MacLeod *et al.* 1997) yet the dinosaurs died out despite their obvious ability to withstand temperate environmental conditions at polar latitudes in the southern hemisphere (Wiffen 1996). A possible mechanism for dinosaur extinction in the southern hemisphere is trace element contamination, whether it be from terrestrial or extra-terrestrial sources. As discussed above, this could also be implicated in the nesting sites in France and in the Nanxiong Basin, where there is no impact signature in the K–T boundary sediments.

Conclusions

Undoubtedly there were severe environmental perturbations at the K–T boundary. However, the vertebrate fossil record and the phylogenies based upon it do not support catastrophic mass extinction across the K–T boundary for any group of terrestrial organisms with the exception of non-avian dinosaurs. The only detailed record of the pattern of extinction and survival is from west–central North America, where, in the case of the non-avian dinosaurs, there seems to be an accelerated decline at the end of the Maastrichian superimposed on a more gradual longer-term decline in diversity. There were stresses in the ecosystem caused by marine eustatic changes; the impact event, together with the Deccan Traps volcanic event of longer duration, must have exacerbated those environmental perturbations and probably played a role in tipping the dominant large terrestrial vertebrates in North America over the brink of extinction. Little is known about the timing and causes of dinosaur extinction in other parts of the world, yet none apparently survived beyond the end of the Cretaceous. In the landlocked inland basins of Asia there is no direct evidence of either an impact or marine regression, although there are indications of a floral turnover associated with climate change; there is no such evidence for climate change in southern high latitudes at the K–T boundary. Are there any factors which correlate with the selective extinction of dinosaurs globally? Trace element contamination is a possible factor and one which would affect large herbivores preferentially as primary consumers of vegetation, either through direct ingestion or through concentration of contaminants in plants. Modern data support this hypothesis. Excess trace elements in ppm concentrations cause pathological eggshell formation in birds (references in Zhao (1994)) and high concentrations of some, e.g. selenium, concentrate in yolk and are toxic to developing embryos (Hansen 1991). Environmental stresses (climate, food supply, crowding) also cause pathological shell formation (Erben *et al.* 1979) in birds. If large herbivores were affected preferentially, then inevitably the carnivores would dwindle along with their food supply. The cause of the final global extinction of dinosaurs at the K–T boundary could thus be reproductive failure as an indirect consequence of either volcanic eruptions or an extraterrestrial impact or a combination of both events.

References

ALVAREZ, L. W. 1986. Toward a theory of impact crises. *Eos Transactions, American Geophysical Union*, **67**, 649–658.

——, ALVAREZ, F., ASARO, F. & MICHEL, H. V. 1980. Extraterrestrial cause for the Cretaceous–Tertiary extinction. *Science*, **208**, 1095–1108.

ARCHIBALD, J. D. 1996*a*. *Dinosaur Extinction and the End of an Era: What the Fossils Say*. Columbia University Press, New York.

——1996*b*. Testing extinction theories at the Cretaceous–Tertiary boundary using the vertebrate record. *In*: MACLEOD, N. & KELLER, G. (eds) *Cretaceous–Tertiary Mass Extinctions: Biotic and Environmental Changes*. Norton, New York, 373–397.

—— & BRYANT, L. J. 1990. Differential Cretaceous–Tertiary extinction of nonmarine vertebrates; evidence from northeastern Montana. *In*: SHARPTON, V. L. & WARD, P. D. (eds) *Global Catastrophes in Earth History: an Interdisciplinary Conference on Impacts, Volcanism, and Mass Mortality*. Geological Society of America, Special Paper, **247**, 549–562.

ASTIBIA, H., BUFFETAUT, E., BUSCALIONI, A. D. *et al.* 1990. The fossil vertebrates from Laño (Basque Country, Spain); new evidence on the composition and affinities of the Late Cretaceous continental faunas. *Terra Nova*, **2**, 460–466.

BENTON, M. J. 1993. Reptilia. *In*: BENTON, M. J. (ed.) *The Fossil Record 2*. Chapman & Hall, London, 681–715.

BUFFETAUT, E. 1982. Radiation évolutive, paléoécologie et biogéographie des crocodiliens Mésosuchiens. *Mémoires de la Société géologique de France*, **142**, 1–88.

——1987. Why the Maastrichtian regression did not cause terminal Cretaceous mass extinction. *Mémoires de la Société géologique de France*, **150**, 75–80.

——1990. Vertebrate extinctions and survival across the Cretaceous–Tertiary boundary. *In*: NICOLAYSEN, L. O. & REIMOLD, W. U. (eds) *Cryptoexplosions and Catastrophes in the Geological Record, with a Special Focus on the Vredefort Structure*. *Tectonophysics*, **171**, 337–345.

CAPPETTA, H. 1987. Extinctions et renouvellements fauniques chez les sélachiens postjurassiques. *Mémoires de la societie géologique de France*, **150**, 113–131.

CARPENTER, K., HIRSCH, K. F. & HORNER, J. R. (eds) 1994. *Dinosaur Eggs and Babies*. Cambridge University Press, Cambridge.

CHIAPPE, L. 1995. The first 85 million years of avian evolution. *Nature*, **378**, 349–355.

CLEMENS, W. A. & NELMS, L. G. 1993. Paleoecological implications of Alaskan terrestrial vertebrate fauna in latest Cretaceous time at high paleolatitudes. *Geology*, **21**, 503–506.

COURTILLOT, V. E. 1990. A volcanic eruption. *Scientific American*, **263**, 85–92.

COX, G. W. 1993. *Conservation Ecology: Biosphere and Biosurvival*. Brown, Dubuque, IA.

D'HONDT, S., PILSON, M. E. Q., SIGURDSSON, H., HANSON, A. K. JR & CAREY, S. 1994. Surface water acidification and extinction at the Cretaceous–Tertiary boundary. *Geology*, **22**, 983–986.

ERBEN, H. K. 1972. Ultrastrukturen und dicke pathologischer Eischalen. *Akademie der Wissenschaftlichen und der Literatur Mainz, Abhandlungen der Mathematischen-Wissenschaftlichen Klasse*, **1972**, 191–216.

——1975. *Die Entwicklung der Lebewesen. Sprielregeln der Evolution*. Piper, Munich.

——, HOEFS, J., & WEDEPOHL, K. H. 1979. Palaeobiological and isotopic studies of eggshells from a declining dinosaur species. *Paleobiology*, **5**, 380–414.

ESTES, R. 1976. Middle Paleocene lower vertebrates from the Tongue River Formation, southeastern Montana. *Journal of Paleontology*, **50**, 500–520.

—— & BAÉZ, A. 1985. Herpetofaunas of North and South America during the Late Cretaceous and Cenozoic: evidence for interchange. *In*: STEHLI, F. G. & WEBB, S. D. (eds) *The Great American Biotic Interchange*. Plenum, New York, 139–197.

FEDUCCIA, A. 1995. Explosive evolution in Tertiary birds and mammals. *Science*, **267**, 637–638.

GAO, K. & FOX, R. C. 1993. New teiid lizards from the Upper Cretaceous Oldman Formation (Judithian) of Southeastern Alberta, Canada, with a review of the Cretaceous record of teiids. *Annals of the Carnegie Museum*, **60**, 145–162.

GRIGORESCU, D., WEISHAMPEL, D. B., NORMAN, D. B., SECLAMEN, M., RUSU, M., BALTRES, A., & TEODORESCU, V. 1994. Late Maastrichtian dinosaur eggs from the Haţeg Basin (Romania). *In*: CARPENTER, K., HIRSCH, K. F. & HORNER, J. R. (eds) *Dinosaur Eggs and Babies*. Cambridge University Press, Cambridge,74–87.

HANSEN, H. J. 1990. Diachronous extinctions at the K/T boundary: a scenario. *In*: SHARPTON, V. L. & WARD, P. D. (eds) *Global Catastrophes in Earth History: an Interdisciplinary Conference on Impacts, Volcanism, and Mass Mortality*. Geological Society of America, Special Paper, **247**, 417–423

——1991. Diachronous disappearance of marine and terrestrial biota at the Cretaceous–Tertiary boundary. *In*: KIELAN-JAWOROWSKA, Z., HEINTZ, N. & NAKREM, H. A. (eds) *Fifth Symposium on Mesozoic Terrestrial Ecosystems and Biota. Extended abstracts*. Contributions from the Paleontological Museum, University of Oslo, **364**, 31–32.

HEDGES, S. B., NUSSBAUM, R. A. & MAXSON, L. R. 1993. Caecilian phylogeny and biogeography inferred from mitochondrial DNA sequences of the 12S rRNA and 16S rRNA genes (Amphibia: Gymnophiona). *Herpetological Monographs*, **7**, 64–76.

HILDEBRAND, A. R., PENFIELD, G. T., KRING, D. A., PILKINGTON, M., CAMARGO, A., JACOBSEN, S. B. & BOYNTON, W. V. 1991. Chicxulub crater: a possible Cretaceous/Tertiary boundary impact crater on the Yucatán Peninsula, Mexico. *Geology*, **19,** 857–871.

HURLBERT, S. H. & ARCHIBALD, J. D. 1995. No statistical support for sudden (or gradual) extinction of dinosaurs. *Geology*, **23**, 881–884.

JOHNSON, K. R. & HICKEY, L. J. 1990. Megafloral change across the Cretaceous–Tertiary boundary in the northern Great Plains and Rocky Mountains, USA *In*: SHARPTON, V. L. & WARD, P. D. (eds) *Global Catastrophes in Earth History: An Interdisciplinary Conference on Impacts, Volcanism, and Mass Mortality*. Geological Society of America, Special Paper, **247**, 433–444.

LE LOEUFF, J., BUFFETAUT, E. & MARTIN, M. 1994. The last stages of dinosaur faunal history in Europe: a succession of Maastrichtian dinosaur assemblages from Corbières (southern France). *Geological Magazine*, **131**, 625–630.

MACLEOD, N., RAWSON, P. F., FOREY, P. L. *et al.* 1997. The Cretaceous–Tertiary biotic transition. *Journal of the Geological Society, London*, **154**, 265–292.

MCGOWAN, G. & EVANS, S . E. 1995. Albanerpetontid amphibians from the Cretaceous of Spain. *Nature*, **373,** 143–145.

MILNER, A. C., MILNER, A. R. & EVANS, S. E. 1998. Amphibians, reptiles and birds: a biogeographical review. *In*: CULVER, S. J. & RAWSON, P. (eds) *Biotic Response to Global Change: the last 145 Million Years*, Chapman & Hall, London.

MILNER, A. R. 1983. The biogeography of salamanders in the Mesozoic and early Caenozoic: a cladistic–vicariance model. *In*: SIMS, R. J., PRICE, J. & WHALLEY, P. (eds) *Evolution, Time and Space, the Emergence of the Biosphere*. Systematics Association Special Volume, **23**, 431–468.

——1993. Amphibian-grade Tetrapoda. *In*: BENTON, M. J. (ed.) *The Fossil Record 2*. Chapman and Hall, London, 663–677.

OLDFIELD, J. E., ALLAWAY, W. H., LAITENEN, A., LAKIN, H. W. & MUTH, O. H. 1974. The relation of selected trace elements to health and disease. *In*: *Geochemistry and the Environment. A report of the workshops at the Asilomar Conference Grounds, Pacific Grove, California, February 7–12, 1972*. National Academy of Sciences, Washington, DC, 57–60.

SAHNI, A., TANDON, S. K., JOLLY, A., BAJPAI, S., SOOD, A. & SRINIVASAN, S. 1994. Upper Cretaceous dinosaur eggs and nesting sites from the Deccan volcano-sedimentary province of peninsular India. *In*: CARPENTER, K., HIRSCH, K. F. & HORNER, J. R. (eds) *Dinosaur Eggs and Babies*. Cambridge University Press, Cambridge, 204–226.

SCHULTZ, P. H. & D'HONDT, S. 1996. Cretaceous–Tertiary (Chicxulub) impact angle and its consequences. *Geology*, **24**, 963–967.

SHEEHAN, P. M. & FASTOVSKY, D. E. 1992. Major extinction of land-dwelling vertebrates at the Cretaceous–Tertiary boundary, eastern Montana. *Geology*, **20**, 556–560.

—— & HANSEN, T. A. 1986. Detritus feeding as a buffer to extinction at the end of the Cretaceous. *Geology*, **14**, 868–870.

——, FASTOVSKY, D. E., HOFFMANN, R. G., BERGHAUS, C. B. & GABRIEL, D. 1991. Sudden extinction of the dinosaurs: Latest Cretaceous, Upper Great Plains, USA. *Science*, **254**, 835–839.

STETS, J., ASHRAF, A.-R., ERBEN, H. K. *et al.* 1996. The Cretaceous–Tertiary boundary in the Nanxiong Basin (continental facies, Southeast China). *In*: MACLEOD, N. & KELLER, G. (eds). *Cretaceous–Tertiary Mass Extinctions: Biotic and Environmental Changes*. Norton, New York, 349–371.

SULLIVAN, R. M. 1987. *A Reassessment of Reptilian Diversity across the Cretaceous–Tertiary Boundary*. Contributions in Science, Los Angeles County Museum, **391**.

SUTHERLAND, F. L. 1993. Demise of the dinosaurs and other denizens II – by combined catastrophic causes? *Journal and Proceedings of the Royal Society of New South Wales*, **126**, 1–25.

WEISHAMPEL, D. B. 1990. Dinosaurian distribution. *In*: WEISHAMPEL, D. B., DODSON, P. & OSMÓLSKA, H. (eds) *The Dinosauria*. University of California Press, Berkeley, 63–139.

WELLNHOFER, P. 1991. *The Illustrated Encyclopaedia of Pterosaurs*. Salamander, London.

WERNER, C. & RAGE, J.-C. 1994. Mid-Cretaceous snakes from Sudan. A preliminary report on an unexpectedly diverse snake fauna. *Comptes Rendus de l'Académie des Sciences*, **319**, 247–252.

WHALLEY, P. 1988. Insect evolution during the extinction of the Dinosauria. *Entomologia Generalis*, **13**, 119–124.

WIFFEN, J. 1996. Dinosaurian palaeobiology: a New Zealand perspective. *Memoirs of the Queensland Museum*, **39**, 725–731.

WILLIAMS, M. E. 1994. Catastrophic versus noncatastrophic extinction of the dinosaurs: testing, falsifiability, and the burden of proof. *Journal of Paleontology*, **68**, 183–190.

WOLBACH, W. S., GILMOUR, I. & ANDERS, E. 1990. Major wildfires at the Cretaceous/Tertiary boundary. *In*: SHARPTON, V. L. & WARD, P. D. (eds) *Global Catastrophes in Earth History: an Interdisciplinary Conference on Impacts, Volcanism, and Mass Mortality*. Geological Society of America, Special Paper, **247**, 391–400.

WU, X., BRINKMAN, D. B., RUSSELL, A. P., DONG, Z., CURRIE, P. J., HOU, L., & CUI, G. 1993. Oldest known amphisbaenian from the Upper Cretaceous of Chinese Inner Mongolia. *Nature*, **366**, 57–59.

ZHAO, Z.-K. 1994. Dinosaur eggs in China: on the structure and evolution of eggshells. *In*: CARPENTER, K., HIRSCH, K. F. & HORNER, J. R. (eds) *Dinosaur Eggs and Babies*. Cambridge University Press, Cambridge, 184–225.

Mass extinctions in Phanerozoic time

A. HALLAM

School of Earth Sciences, University of Birmingham, Birmingham B15 2TT, UK

Abstract: Following publication in 1980 of the celebrated paper in *Science* (**208**, 1095–1108) by Luis Alvarez and his colleagues, a widespread interest developed in the possibility that bolide impact was a major if not indeed the most important cause of mass extinctions in the Phanerozoic. Of particular interest was the claim of a 26 Ma extinction periodicity in the last 250 Ma, which led to suggestions of ultimate astronomical control. Intensive research subsequently casts serious doubt on such a periodicity, and the only really persuasive evidence associating impact with mass extinction is that at the Cretaceous–Tertiary boundary.

It is generally accepted that there were five major episodes of mass extinction in the Phanerozoic record: end-Ordovician, late Devonian (Frasnian–Famennian boundary), end-Permian, end-Triassic and end-Cretaceous. In addition a number of lesser events have been widely recognized, most notably at the Devonian–Carboniferous, Cenomanian–Turonian and Paleocene–Eocene boundaries, and in the early Toarcian. The extinction record is based essentially on marine invertebates, and the evidence from terrestrial vertebrates is far less clear, with the exception of dinosaurs at the Cretaceous–Tertiary boundary. Although it has been maintained that terrestrial plants were not subjected to mass extinction episodes there is abundant evidence for at least extensive regional extinction events, most notably at the Permian–Triassic, Triassic–Jurassic and Cretaceous–Tertiary boundaries.

Only a small number of factors can plausibly be invoked to account for environmental perturbations on a global scale, of the sort that could significantly increase extinction rate: bolide impact, volcanism, climatic cooling, climatic warming, marine regression and marine transgression with the accompanying spread of anoxic bottom waters. Good evidence for bolide impact, in the form of significant iridium anomalies on a global scale, and shocked quartz coincident with mass extinction, is only available for the end-Cretaceous event, and a likely impact crater has been identified in the Yucatan Peninsula of Mexico. Microtektites have been recognized both in Upper Devonian and Upper Eocene strata, but their horizons do not coincide with mass extinction events. Claims for impact at the end of the Triassic, based on purported shocked quartz, are decidedly dubious. The only good correlation with large-scale flood basalt volcanism is for the end-Permian and end-Cretaceous events, respectively the Siberian and Deccan Traps, but any causal correlation is more likely than not to involve climate. Climatic cooling is strongly implicated only for the latest Ordovician and late Eocene to Oligocene, and climatic warming episodes, as in the earliest Silurian, Triassic and Eocene, were probably at least as important. Marine regression caused by sea-level fall correlates well with extinction events in the late early Cambrian, end-Ordovician, late Permian, end-Triassic and end-Cretaceous. More important than any other factor, anoxia and transgression appear to be strongly associated with events in the late early Cambrian, late Cambrian, earliest Silurian, Frasnian–Famennian and Devonian–Carboniferous boundaries, Permian–Triassic, Cenomanian–Turonian and Paleocene–Eocene boundaries, and early Toarcian.

Often, of course, more than one factor may be implicated, one of the most striking associations being between episodes of marine transgression–anoxia and climatic warming. Such environmental changes could clearly affect both marine and terrestrial organisms simultaneously. Even if one accepts bolide impact as a major causal factor for the end-Cretaceous extinctions, its environmental effects are still controversial and difficult to disentangle from those caused by phenomena intrinsic to our planet. Arguments persists among palaeontologists about how catastrophic the mass extinction was.

As the term 'mass extinction' means different things to different people it is desirable at the start to attempt some sort of definition. That of Sepkoski (1986) is particularly useful. He defined mass extinction as any substantial increase in the amount of extinction (i.e. lineage termination) suffered by more than one geographically widespread higher taxon during a relatively short interval of geological time, resulting in at least temporary decline in their standing diversity. This is a general definition purposely designed to be somewhat vague. A more concise one offered here is that a mass extinction is an extinction of a significant proportion of the world's biota in a geologically insignificant period of time. The vagueness about extinctions can be dealt with fairly satisfactorily in particular cases by giving percentages of taxa, but the vagueness about

HALLAM, A. 1998. Mass extinctions in Phanerozoic time. *In*: GRADY, M. M., HUTCHISON, R., MCCALL, G. J. H. & ROTHERY, D. A. (eds) *Meteorites: Flux with Time and Impact Effects*. Geological Society, London, Special Publications, **140**, 259–274.

time is more difficult to deal with. A significant question about mass extinctions is how catastrophic they were, so we also require a definition of catastrophe in this context. According to Knoll (1984), it is a biospheric perturbation that appears instantaneous when viewed at the level of resolution provided by the geological record.

It is important to appreciate the limits of time resolution that can be established from the stratigraphic record. In the view of Dingus (1984), it is unlikely that one can distinguish, for the most discussed extinction event across the Cretaceous–Tertiary boundary, episodes of extinction lasting 10^2 years or less from episodes lasting as long as 10^5 years. In favourable circumstances this upper age limit may be reduced to a few thousand years but doubtfully much less. Sampling problems also make the determination of the magnitude of a catastrophe difficult, producing the so-called back-smearing effect (Signor & Lipps 1982; Springer 1990). One should not quibble too much, however, about whether an event was catastrophic or not if it can be reduced to a few thousand years. Working within a time frame of many millions of years, virtually all geologists would concur that it was catastrophic in this context, if the effects were sufficiently considerable.

The modern study of mass extinctions effectively began with Newell's (1967) overview of Phanerozoic extinctions pertinently entitled 'Revolutions in the History of Life'. Besides the major extinction events at the end of the Palaeozoic and Mesozoic eras, Newell recognized four other events in the Phanerozoic record of marine families which were also dramatically sudden in the context of the much longer time intervals preceding and following them; these six events took place at the end of the Cambrian, Ordovician, Permian, Triassic, Cretaceous and near the end of the Devonian. Raup & Sepkoski's (1982) statistical analysis of extinction rate, expressed as families going extinct per stratigraphic stage, confirmed the last five as major episodes of mass extinction (Fig. 1) and these have since been generally accepted as the 'big five.'

Research in recent years, involving many additions, corrections and reinterpretations of both taxonomy and stratigraphy, has only served to sharpen these five major events (Sepkoski 1993; Benton & Storrs 1994; Jablonski 1994; Benton 1995). Sepkoski's family- and genus-level compendia have allowed the species-level extinction intensities of the five major mass extinctions to be calculated (Table 1). From this table it can be seen that by far the biggest event, and the only one leading to a significant decrease in global biodiversity, took place at the end of the Palaeozoic, with the disappearance of half the marine families.

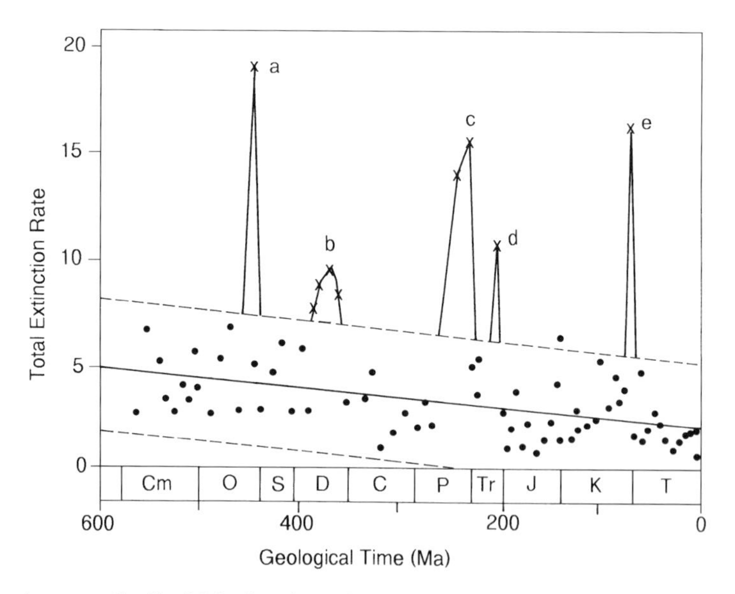

Fig. 1. Extinction rates (families/Ma) of marine animals during the Phanerozoic showing the 'big five' mass extinctions as clear peaks with crosses standing above the enveloped background extinction level: (**a**) Late Ordovician, (**b**) Late Devonian, (**c**) Late Permian, (**d**) Late Triassic, (**e**) Late Cretaceous. Cm, Cambrian; O, Ordovician; S, Silurian; D, Devonian; C, Carboniferous; P, Permian; Tr, Triassic; J, Jurassic; K, Cretaceous; T, Tertiary. Adapted from Raup & Sepkoski (1982).

Table 1. *Extinction intensities at the five major mass extinctions in the fossil record: species-level estimates based on a rarefaction technique*

Mass extinction	Families		Genera	
	Observed extinction (%)	Calculated species loss (%)	Observed extinction (%)	Calculated species loss (%)
End-Ordovician	26	84	60	85
Late Devonian	22	79	57	83
End-Permian	51	95	82	95
End-Triassic	22	79	53	80
End-Cretaceous	16	70	47	76

Source: simplified from Jablonski (1994).

The fossil record of marine invertebrates generally has the advantage of abundant specimens, good stratigraphical control, closely spaced samples, uniform preservation quality and broad geographical distribution. The record of vertebrate tetrapods known from continental environments is generally incomplete. Thus most dinosaur genera are known only from a single stratigraphic stage, which would suggest, on a literal reading of the record, that dinosaurs suffered total generic extinction 24 or 25 times during their history (Padian & Clemens 1985). However, at the family level there is only one end-Cretaceous event, because dinosaur families generally had stratigraphic ranges exceeding a stage.

To counter the problem of assessing the relative incompleteness of the tetrapod record Benton (1989) has proposed a Simple Completeness Metric (SCM), which compares the number of families known to be present with the numbers that ought to be present, that is, which span several stages but are not known from every stage. His SCM percentage figures for different groups are as follows: birds 57, bats 76, frogs and salamanders 42, lizards and snakes 49, mammal-like reptiles 94, placental mammals 87. Benton recognized four mass extinctions and two minor extinctions (within the Cenozoic) from the terrestrial tetrapod record, with the relative magnitude being given in terms of percentages of families that disappeared: Early Permian (Sakmarian–Artinskian), 58; Late Permian–Early Triassic (Tatarian–Scythian), 49; late Triassic (Carnian–Rhaetian), 22; Late Cretaceous (Maastrichtian), 14; Early Oligocene (Rupelian), eight; Late Miocene (Tortonian–Messinian), two. Interestingly, neither the major Early Permian event nor the minor Cenozoic events correspond to marine extinctions. Furthermore, the celebrated extinction of the dinosaurs was not reflected in other groups, because the Maastrichtian event was comparatively minor.

It has been argued that terrestrial plants were not subjected to mass extinctions (Knoll 1984; Traverse 1988) but there is evidence for important events on at least a regional scale at the end of the Permian, Triassic and Cretaceous. Details of these and other groups, and of all more important extinction events, are provided in the book by Hallam & Wignall (1997), of which this review is a condensed version of relevant parts.

Extinction periodicity

Considerable interest was provoked by the claim by Raup & Sepkoski (1984, 1986; Sepkoski & Raup 1986) that they had discovered a 26 Ma mass extinction periodicity in the stratigraphic record for the last 250 Ma. The extinctions included three of the big five extinctions, the end-Permian, end-Triassic and end-Cretaceous, as well as several more minor and new extinctions (Fig. 2). This intriguing conclusion was based on Sepkoski's dataset of 3500 marine family ranges from the mid-Permian to the Recent, of which 970 are extinct. Because of the widespread interest provoked, most notably perhaps amongst astronomers interested in bolide impact, this research and the debate it provoked must be considered in some detail.

Raup & Sepkoski (1984) considered only extinct families, to eliminate the distorting effect of the 'pull of the Recent'. This effect, documented by Raup (1978), is caused by the extending of 'last' occurrences for extant taxa. All range data from the fossil record are defined by first and last occurrences, but for extant taxa the 'last' occurrence is the present day, which may be considerably later than their last fossil occurrence, thereby 'pulling' the range to the Recent. An extreme example is seen for the class Monoplacophora; the fossil range is Cambrian to Devonian but living representatives extend this range by 400 Ma.

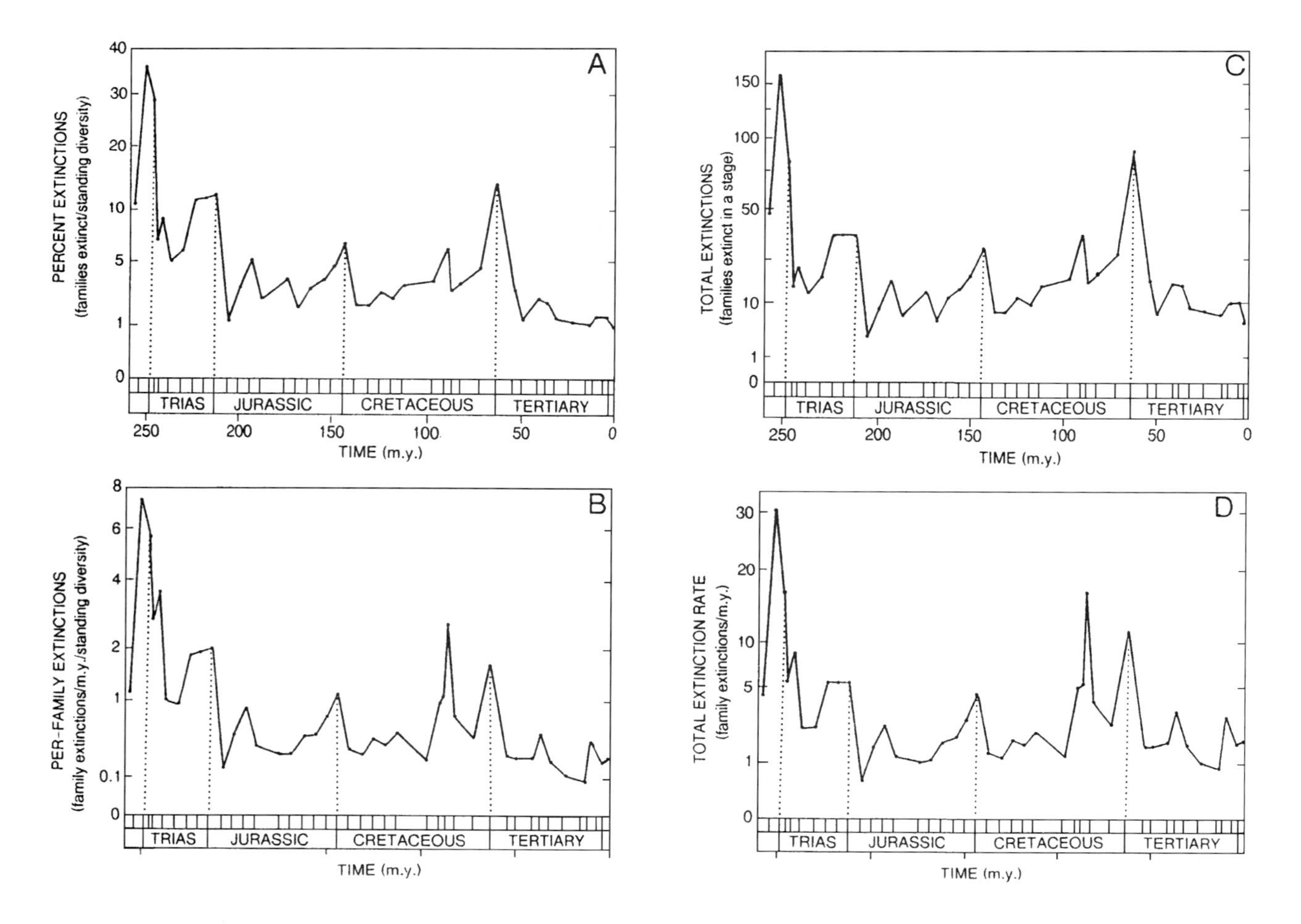

Fig. 2. Extinction intensity over the past 250 Ma for marine animal families calculated using four extinction metrics. A 26 Ma periodicity is apparent although the mid-Jurassic, Aptian and Tertiary events are subdued or missing. A square-root scale has been used as the ordinate to reduce the variation at higher extinction intensities. This ensures that the huge end-Permian extinction can be depicted on the same plot as the lesser post-Palaeozoic extinctions. Adapted from Sepkoski & Raup (1986).

Having modified their data set somewhat, for example, to exclude single-datum occurrences from their analyses, Raup & Sepkoski devised four extinction metrics:

- total extinction = number of families going extinct in a stage

This is clearly the simplest measure of extinction but it does not account for the number of families at risk of extinction; Raup & Sepkoski therefore also calculated

- per cent extinction

$$= \frac{\text{total extinction}}{\text{number of families extant during the stage}}$$

However, the variability of stage durations adds a considerable source of error to this metric; an exceptionally long stage could generate a mass extinction simply by accumulating background extinctions. Therefore a third metric normalizes for time:

- total extinction rate $= \dfrac{\text{total extinction}}{\text{stage duration}}$

Unfortunately, this metric probably contains an even greater source of error as stage durations are poorly known, probably highly variable (see Hoffman 1985) and have to be averaged between a few reliable radiometric dates. Thus a stage that is of considerably longer duration than other stages within its general interval may generate a pseudo-mass extinction. This effect can potentially be tested because origination rates should also be correspondingly high.

The best extinction metric calculates a probability of extinction:

- per family extinction rate

$$= \frac{\text{total extinction rate}}{\text{standing familial diversity}}$$

Once again, the uncertainty of stage durations is incorporated in this measure.

Clearly, all extinction metrics have their problems and it has been general practice to compare extinction patterns using all four measures. It is testimony to the robust nature of the signal from the fossil record that the 26 Ma periodicity is shown by all four metrics (Fig. 1).

Hoffman (1985) strongly criticized Raup & Sepkoski's (1984) removal of extant families from their analysis as it greatly exaggerated the younger extinction events, particularly in the Tertiary where there are so few extinct families that the chance clustering of a few such can produce an 'event'. Thus he observed that the so-called mid-Miocene mass extinction corresponds to the loss of only five families; background extinction rates in the Mesozoic are commonly substantially higher than this. However, in their later analyses Raup & Sepkoski included extant families and periodicity was still found (Raup & Sepkoski 1986; Sepkoski & Raup 1986).

In a direct attack on the validity of Sepkoski's taxonomic database, Patterson & Smith (1987) claimed that a large proportion of the families used in the analyses were not true monophyletic clades but, rather, consisted of polyphyletic or paraphyletic taxa. This raised the possibility that many 'pseudo-extinctions', simple name-changing because of the vagaries of 'systematists' judgement', were being included. Patterson & Smith went on to demonstrate, for their own specialist groups, fish and echinoderms, that no mass extinctions could be seen in a cladistically truthful database of monophyletic taxa. In his immediate retort, Sepkoski (1987) noted that families are only a proxy for groups of species, however defined, and that a family extinction must record extinction at a lower taxonomic level. The classic example considers the two orders of dinosaurs and birds, which together form a monophyletic clade. Clearly it would be cladistic facetiousness to claim that dinosaurs did not go extinct because birds are still extant. Sepkoski's (1987) further point, that the vagaries of family designations are unlikely to generate a 26 Ma periodicity, is also justified.

The periodicity claim has also been subjected to a variety of statistical analyses, with somewhat contradictory results. For example, Hoffman & Ghiold (1985) have argued that a pseudo-periodicity could be produced in a random time series which would be expected to produce a peak, on average, every one in four steps. As average post-Palaeozoic stage durations are somewhere between 5.5 and 6.4 Ma, the source of the periodicity becomes apparent. Following this line of argument, the more variable and longer average duration of the Palaeozoic stages may account for the often-overlooked absence of periodicity in this era. However, Harper (1987) showed that random models should have peaks every one in three steps, and Sepkoski (1987) noted that the actual extinction peaks occur after every fifth interval. Overall, diverse statistical approaches seem to confirm the existence of periodicity (Sepkoski 1989).

The main argument against periodicity is the number of events that are 'missing' from the curves (Hallam & Wignall 1997). Thus the mid-Miocene event is far too insignificant to be

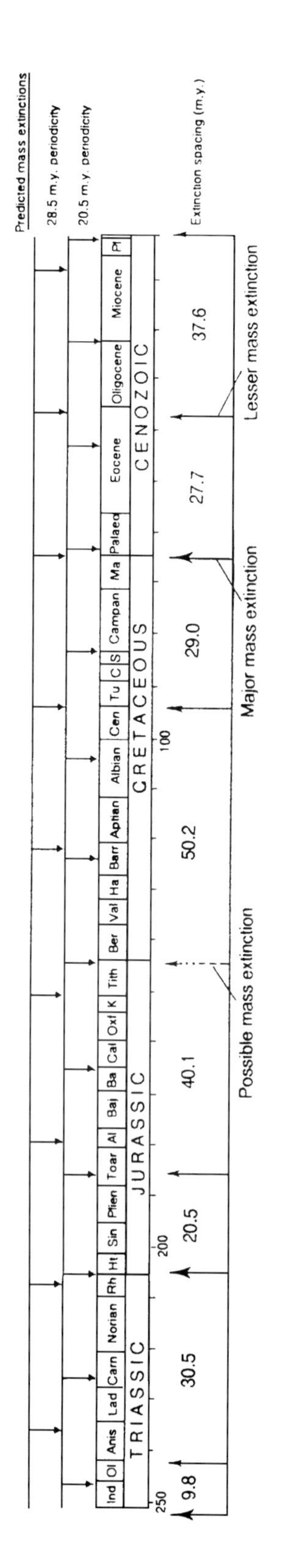

considered a mass extinction. The Late Eocene event is also a minor one, and differs considerably from the others in being a rather protracted, gradual affair, extending throughout the sub-epoch. The Aptian mass extinction is a negligible event, whereas the end-Jurassic one has a clear peak but is probably the product of a bias towards a European database. A mid-Jurassic extinction is predicted around the Bathonian–Callovian boundary but is not seen. Therefore, out of a total of ten mass extinctions predicted by the periodicity hypothesis, only five or six occurred with certainty.

Hallam (1984) has argued that the periodicity could be an artefact of the time scale employed. This criticism was rejected on the grounds that the random error encountered in all time scales is unlikely to produce a periodic signal (Raup & Sepkoski 1986; Sepkoski 1989). This notion can be tested using the latest and most rigorously assessed Mesozoic time scale proposed by Gradstein *et al.* (1994), supplemented by the Harland *et al.* (1989) time scale for the Cenozoic. Using these time scales it can be seen from Fig. 3 that the periodicity is even less apparent. The average mass extinction spacing is 28.5 Ma but, using this value as the period, only four out of a predicted nine mass extinctions occur at the right time, even allowing for a ±5 Ma error. The end-Triassic and early Toarcian extinctions have a spacing of 20.5 Ma, and using this period centred around these two mass extinctions, generates the best hit-to-miss ratio, six out of 12! This suggests that both the time scale and the choice of period length do indeed affect the appearance of periodicity. Raup & Sepkoski's most significant error was their choice of a young age for the Permian–Triassic boundary. A consequence of the more realistic older age is that a further mass extinction is predicted to occur somewhere in the Anisian–Ladinian interval of the Middle Triassic (Fig. 3). Like many other predicted mass extinctions, this event is missing.

Fig. 3. Occurrence of post-Palaeozoic mass extinctions plotted against the Gradstein *et al.* (1994) Mesozoic time scale and, for the Cenozoic, the Harland *et al.* (1989) time scale. The best fit occurs with a periodicity of 20.5 Ma. centred on the end-Triassic and early Toarcian extinction events. This produces a 50% success rate; only six of a predicted 12 mass extinctions occur. A fit to a 28.5 Ma periodicity centred on the end-Cretaceous and end-Cenomanian is also shown for comparison; this most closely approximates to Raup & Sepkoski's proposed 26 Ma periodicity, although only four of a predicted nine mass extinctions occur near the correct time. After Hallam & Wignall (1997).

A new, independent compilation of all fossil families has recently appeared (Benton 1993). This is the product of 90 specialists and contains range data for 7186 families including, importantly, terrestrial organisms, from which Benton (1995) has calculated the various extinction metrics. He concluded that 'the present data do not lend strong support to this idea [of periodic mass extinctions] because only six, or perhaps seven, of the events are evident'.

Possible causal factors

An important question which has yet to be adequately resolved in many cases concerns the extent to which the mass extinctions were catastrophic, having been accomplished within a mere geological 'instant'. Some impact supporters have expressed the opinion that the early resistance of most palaeontologists and geologists was the consequence of their being indoctrinated as students with Lyellian uniformitarianism, implying gradual and denying catastrophic change. Although this may well be true of some, there has been for several decades a strong school of thought dubbed neocatastrophism (e.g. Ager 1973) and there is no reason why all catastrophes on Earth need have been induced by bolide impact. As yet, we know too little about the workings of our planet to dismiss Earth-induced catastrophes, probably bound up ultimately with changes in the mantle. The evidence must decide in particular cases, and it is to this evidence that we now turn.

Raup & Sepkoski's claim of a 26 Ma extinction periodicity since the end of the Palaeozoic strongly suggests that, if true, there was only one prime causal factor. This claim has, however, been rejected in the previous section and it must be concluded that mass extinctions were episodic, rather than periodic, in character.

Among a multiplicity of possible causes only a limited number merit consideration here, as having the potential to cause a significant environmental change on a global scale. One should also take due note of the recently discovered phenomenon of self-organized criticality, based originally on the study of avalanching sandpiles. The same trigger, at the limit a mere single added grain of sand, can cause avalanches of a whole range of magnitudes obeying a power law, with a few large and many small ones. The relative magnitude of mass extinctions also appears to approximate to such a power law, as do other natural phenomena such as earthquakes and floods (Kauffman 1995). One key implication is that there is not necessarily a simple relationship between the size of a given mass extinction and the size of the causal factor. Furthermore, there may sometimes have been a combination of interacting factors that exceeded some critical threshold, with possibly cascading effects.

Bolide impact

Bolide impact has been a serious contender ever since Alvarez *et al.* (1980) published their celebrated paper on the Cretaceous–Tertiary boundary, and the extreme view has been put forward that not just mass extinctions but virtually all extinctions are probably due to this cause (Raup 1991). For this view to be more than an act of faith evidence must be brought forward in support.

Iridium anomalies provide the most discussed evidence but, as the wide-ranging survey of Orth *et al.* (1990) showed, only for the Cretaceous–Tertiary boundary can a really convincing case be made out for extraterrestrial impact. The many smaller iridium anomalies in the Phanerozoic record, by no means all located at extinction horizons, can variously be attributed either to extreme condensation, with consequent increase in the proportion of background micrometeorite 'rain', or to diagenetic changes at redox boundaries (Hallam & Wignall 1997). On the other hand, it needs to be recognized that comet, as opposed to asteroid, impact may leave little to no geochemical trace (Jansa 1993).

Shocked quartz is a good indicator of impact, provided it is not confused with quartz that has undergone Earth-bound tectonic deformation. The best means of doing this may be to use scanning electron microscopy of grains etched with hydrofluoric acid. Unlike impact-produced multiple sets of planar deformation features, tectonic deformation apparently does not generate glass along the dislocations it produces (Gratz *et al.* 1996). As with iridium, the only convincing evidence comes from the Cretaceous–Tertiary boundary. The claim made by Bice *et al.* (1992) for shocked quartz at the Triassic–Jurassic boundary at a section in Italy is decidedly equivocal. The purported shocked quartz has been found at no fewer than three horizons, none of which can be firmly established on biostratigraphic evidence as being precisely at the system boundary. The grains in question do not have more than four sets of planar deformation features, and most in fact have only single sets, and the angular distribution of these is rather diffuse. As Bice *et al.* (1992, p. 445) conceded, 'these differences (from classic K–T shocked quartz) make it impossible to demonstrate unambiguously that the grains at the T–J boundary have a shock-metamorphic origin'. An alternative hypothesis would be that these grains contain highly unusual Böhm lamellae, presumably produced by normal Earth-bound tectonism (see Benton 1994). It is pertinent to point out that a similar claim for impact-induced shocked quartz in the Kendelbach Grenzmergel was made a few years ago by Badjukov *et al.* (1988), but this claim was refuted by the impact expert Richard Grieve (in Hallam 1990). Nor has an iridium anomaly been found in the Kendelbach boundary section.

A good case can be made for at least two modest-sized meteorite impact events in the Late Devonian, but in neither case do they correspond to mass extinction intervals; on the contrary, they occurred at times of radiation. Mikrotektites also provide good evidence of impact, provided they are well authenticated. Various claims have been made for glassy spherules, but distinguishing genuine tektites from volcanic products can require sophisticated geochemical analysis of the sort that has been rarely attempted. The best examples are the purported microtektites at the K–T boundary in Haiti (Jehanno *et al.* 1992; Lyons and Officer 1992; Sigurdsson *et al.* 1992; Leroux *et al.* 1995). The well-authenticated Upper Eocene tektites undoubtedly signify bolide impact events but do not seem to coincide with mass extinction events (Miller *et al.* 1991; Wei 1995). Upper Devonian microtektites are known from a number of horizons in the Famennian but they significantly post-date the Frasnian–Famennian mass extinction (Hallam & Wignall 1997).

Other evidence from the stratigraphic record is more equivocal. Thus various impact-induced K–T boundary tsunami deposits have been claimed for the circum-Gulf of Mexico region, but the claims remain controversial (Macleod 1996). McLaren (1983) and McLaren & Goodfellow (1990) have argued that catastrophic mass killings detected by biomass disappearance at bedding planes are a likely indication of impact. Certainly, to depend solely upon taxon counts and ignore individual abundance can underestimate the catastrophic character of biotic change, but acceptance of catastrophe is not in itself evidence for impact.

The Earth's cratering record also provides pertinent evidence and many well-authenticated Phanerozoic craters are now recognized (Grieve 1987). Raup (1992) has proposed what he calls a kill curve for marine species, to investigate large-body impact as a cause of species extinction. Current estimates of Phanerozoic impact rates are combined with the kill curve to produce an impact–kill curve, which predicts extinction levels from crater diameter (Fig. 4). Raup's model can be tested for some of the best authenticated impact craters. The Ries crater of Bavaria has a diameter of about 30 km and formed some 15 Ma ago, in the Middle Miocene. According to the Raup curve, about 10% of species should have gone extinct globally but no species extinctions are recognizable from the region for either mammals (Heissig 1986) or plants (Gregor 1992).

The Montagnais impact structure on the Nova Scotia shelf is 45 km wide and is dated as late Early Eocene in age. The Raup curve indicates a 17% species extinction but there is in fact no recognizable biological change on a local, regional or global scale (Jansa *et al.* 1990). The two largest craters in the Phanerozoic record, apart from Chicxulub, are the Popigai crater in Siberia and the Manicouagan crater in Quebec, both about 100 km in diameter (Grieve 1987). The Popigai crater is dated at 39 ± 9 Ma and Raup has claimed that it could have been associated with the end-Eocene extinctions, but both marine and continental extinctions extended over several million years and there is no clear-cut end-Eocene event; furthermore, 39 Ma is several million years earlier than the Eocene–Oligocene boundary. The latest dating of the Manicouagan crater by Hodych & Dunning (1992) is 214 ± 1 Ma. The Gradstein *et al.* (1994) time

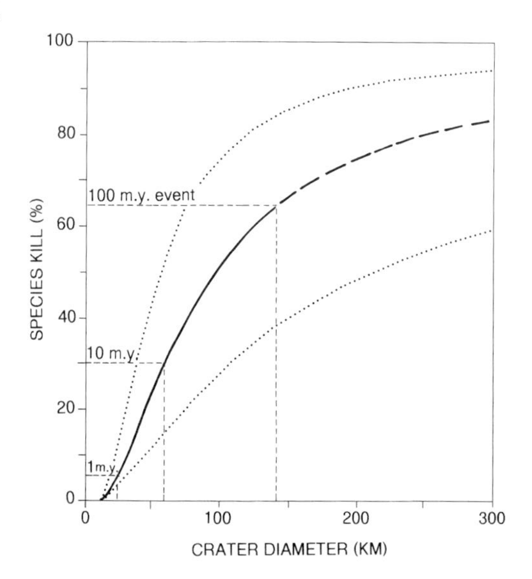

Fig. 4. Relation between species extinction and impact crater diameter. Dotted curves define the worst-case limits of uncertainty in the placement of the impact–kill curve. After Raup (1992).

scale signifies this as Early Norian in age, a time for which no one has recognized any extinctions, either in the marine or continental realm. Yet according to the Raup curve no fewer than 50% of the species should have become extinct.

It must be concluded that, contrary to Raup's (1991) opinion, bolide impact cannot plausibly be invoked as a general cause of extinctions. Even for the K–T boundary event, it remains by no means clear what role impact had in causing the necessary environmental deterioration, as opposed to other Earth-bound factors. The evidence is good only for one major impact in North America, and the comet-shower extinction hypothesis of Hut *et al.* (1987) is not supported, either for the K–T boundary, which lacks good evidence for multiple impacts extended over time, or for the Late Eocene, when multiple impacts evidently had no effect on the biota.

Volcanism

The eruption of vast provinces of continental flood basalts constitutes one of the most spectacular manifestations of igneous activity, and a number of researchers have argued that they might have precipitated extinction events (e.g. Loper *et al.* 1988; Rampino & Stothers 1988; Stothers 1993). Recent redating of many flood basalt provinces has revealed that the eruption times may have been as little as one million years or less, further strengthening the case that volcanism may be capable of causing geologically rapid environmental crises. The most frequently cited link between volcanism and extinction is that between the K–T event and the Deccan Traps of India (Officer *et al.* 1987; Sutherland 1994). Beyond India, the presence of smectitic boundary clays, such as the Fish Clay at Stevns Klint, Denmark, provides an additional pointer to the occurrence of volcanism during the extinction crisis. However, suggestions that shocked minerals and iridium anomalies can be produced by flood basalt eruptions are no longer generally accepted (Glen 1994).

By comparison with flood basalt eruptions in historic time, such as the Lakagigar eruptions in Iceland in 1783–1784, it seems reasonable to assume that the local and regional effects of the Deccan eruptions would have been devastating (Courtillot 1990), but the global effects are more debatable. The most serious consequences may have come from the release of huge amounts of volcanic carbon dioxide and sulphur dioxide (Coffin & Eldholm 1993). The extinction mechanism has therefore been variously attributed to the effects of global darkness caused by the presence of sulphate aerosols and/or acid rain (Officer *et al.* 1987; Rampino & Stothers 1988; Courtillot 1990). Such effects are only likely to have lasted a few years after each eruption episode, and other workers have suggested that global warming as a result of CO_2 release is a more likely cause of extinction (Loper *et al.* 1988). In particular, oceanic changes consequent upon warming may have created oceanic anoxia–dysoxia (Keith 1982; Coffin & Eldholm 1993; Wignall & Twitchett 1996).

Other than the Deccan Traps and K–T connection, the only other convincing links between mass extinctions and flood basalts can be found for the Permian–Triassic and end-Paleocene crises and the Siberian Traps and Arctico-British flood basalt provinces, respectively. Notably, both events are closely linked with global warming, suggesting that the emission of large volumes of CO_2 may be the most deleterious consequence of volcanism. Claims have been made initially by Rampino and Stothers (1988) and with more statistical rigour by Stothers (1993), that all post-Palaeozoic mass extinctions were caused by flood basalt eruptions. These in turn have been linked to periodic or quasi-periodic instabilities at the core-mantle boundary (Loper *et al.* 1988). However, the original dataset fails to stand up to close scrutiny. Stothers (1993) was only able to achieve a statistical link between flood basalts and extinctions by including Pliocene, Middle Miocene, Aptian, Tithonian and Bajocian events amongst his major mass extinctions. As already indicated, these are non-events or at best of only regional significance. Even more perversely, Rampino & Stothers (1988) were only able to achieve a correlation by excluding the Permian–Triassic event from their plot and yet this is one of the best extinction–flood basalt conjunctions. In fact the failure of many of the major flood basalt provinces to correlate with extinction events is one of the most telling arguments against a link. Thus the Paraná basalts of eastern South America are one of the greatest of all flood basalt provinces, greater in extent than the Deccan Traps and comparable with the Siberian Traps, and yet their brief eruption around the Barremian–Aptian boundary (120 ± 5 Ma) caused no notable extinction.

That some flood basalts coincide with major mass extinctions in undoubted but they may be 'only accidental flukes of timing' (Stothers 1993, p. 1399), or perhaps they only served to exacerbate rather than directly trigger environmental crises.

Climate

Climatic changes, either cooling or warming, have been invoked for most mass extinctions; Stanley (1984) has held that cooling is the most frequent cause. Cooling can take two different forms, high-latitude cooling associated with a steepening of the equator-to-pole temperature gradient and commonly polar glaciation, and global cooling in which the average global temperature declines but not necessarily the gradients. This is an important distinction but has been rarely made. Three high-latitude cooling–glaciation phases have occurred during the Phanerozoic and, notably, only the earliest of these, in the Hirnantian stage of the latest Ordovician, is associated with a major mass extinction. The Permo-Carboniferous glaciation of Gondwana may have caused a minor extinction event at its outset in the mid-Carboniferous, and the Tertiary to Quaternary cooling caused an equally modest crisis. Even for the Hirnantian event the main mass extinction interval probably occurred during the warming phase at the termination of the glaciation (Fortey 1989). Of the earlier Hirnantian extinctions, the benthos may have been affected by the glacioeustatic regression but not by cooling *per se*. Only the pelagic extinctions of graptolites and trilobites may directly related to cooling, in that oceanographic changes effectively eliminated their habitats. Only for the protracted crisis during the Late Eocene and Early Oligocene can a good case be made for extinctions directly related to cooling (Prothero & Berggren 1992).

High-latitude cooling principally causes the wholesale movement of faunas and floras as climatic belts shift, with tropical faunas consequently becoming restricted to a narrower equatorial belt. Global cooling is required if tropical habitats are to be completely eliminated, as may have happened during the Cretaceous–Tertiary and Frasnian–Famennian extinctions (Hallam & Wignall 1997). Evidence for global cooling is primarily derived from fossil data such as the preferential elimination of tropical taxa and the appearance of cold-water taxa in tropical areas during the immediate post-mass extinction interval. However, it is rarely possible to rule out the alternative possibility that such occurrences record the appearance of deep-water taxa during transgressions, because deep-water and cold-water preferences frequently go hand in hand.

Global warming, generally attributed to a rise in concentrations of greenhouse gases, is also an important component in some extinction scenarios. In the terrestrial realm global warming will be at its most severe at high latitudes, as a result of the elimination of polar habitats; the elimination of the glossopterid forests at the end of the Permian appears to be a particularly good example of such an effect (Retallak 1995). In the marine realm global warming is rarely held to be the proximate cause of death. More significant is the effect of warming on oceanic circulation and the decreasing solubility of oxygen as temperature rises (Keith 1982). Thus global warming-induced anoxia–dysoxia is considered a principal contender in the end-Ordovician, end-Permian and end-Paleocene extinctions (Hallam & Wignall 1997).

Marine regression

Newell (1967) was the first person to put forward an explicit hypothesis relating mass extinctions in the marine realm to episodes of global regression. He argued that shrinkage of the area of epicontinental sea habitat should have deleterious effect on neritic organisms and should therefore lead to widespread extinction. Radiation of the survivors would take place during the expansion of habitat area consequent upon the succeeding transgression.

On the face of it, Newell's hypothesis is a reasonable one, because there is nearly always a positive correlation among living organisms in a wide range of habitats between species number and area (Connor & McCoy 1979), and decreasing habitat area is widely perceived as the prime cause of extinction today (Eldredge 1991). There appears, furthermore, to be a strong relationship between Newell's six mass extinction events and marked falls of sea level (Fig. 5), and many lesser extinction events have also been correlated with relatively minor regressions (Hallam 1989, 1992). In recent years, however, the hypothesis has been subjected to a number of criticisms. The most obvious one is that not all regressions correlate with extinction events. A good example portrayed in Fig. 5 concerns the Early Devonian regression. It is well established that there was a succession of major regressions of shelf seas during the Quaternary, at times of global cooling and ice-volume expansion, but the extinction rate for neritic invertebrates was remarkably low (Valentine & Jablonski 1991). Indeed, Jablonski (1985) has argued that in the modern world most marine faunas would persist in undiminished shallow-water regions around oceanic islands at times of Quaternary sea-level fall.

Regarding the Quaternary, however, one important factor that should not be overlooked is the comparative rapidity of glacioeustatic

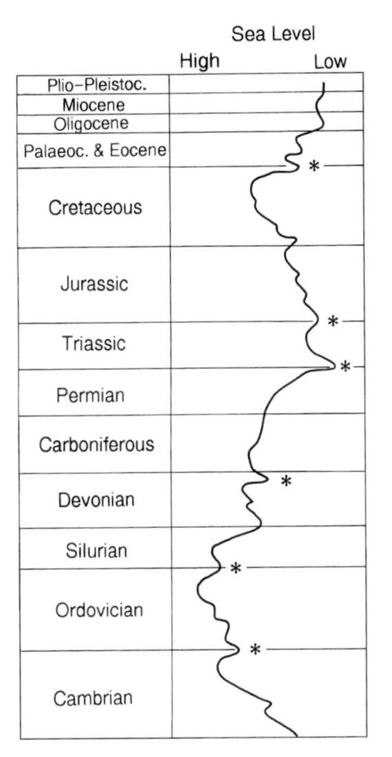

Fig. 5. Newell's six mass extinction events (asterisks) plotted against Hallam's (1992) sea-level curve. After Hallam (1992).

sea-level change. Bearing in mind that extinction risk increases with time in a stressful environment without further diminution of area (Diamond 1984), it could be that the rapid restoration of less stressful conditions in the Quaternary has served to diminish the extinction rate owing to reduction of habitat area through sea-level fall. The slower and less spectacular regressions of much of the Phanerozoic might have been environmentally more significant because they lasted longer, and modest falls in very extensive and extremely shallow-water epicontinental seas with no close modern analogues could have had a correspondingly major effect (Hallam 1981, chapter 5). Johnson (1974) stressed the importance of organic adaptations to changed circumstances in understanding the likely causes of extinction of neritic invertebrates. During episodes of sustained enlargement of epicontinental seas, organisms become progressively more stenotopic and an equilibrium is established. They are in effect 'perched' subject to the continued existence of their environment. Extinctions occur to an extent proportional to the magnitude of regression, degree of stenotopy attained, or a combination of the two.

Although a strong association between sea-level change and marine mass extinction seems undeniable it has become increasingly apparent that the key factor in many mass extinctions may be less the regression than the reduction in habitat area provided by the spread of anoxic waters during the subsequent transgression (Hallam 1989). Anoxia as an extinction mechanism will be discussed in the next section, but here reference must be made to those Phanerozoic extinction events for which regression appears to have been significant (Hallam & Wignall 1997).

The earliest Phanerozoic event is in the late Early Cambrian (Early Royonian), the so-called Hawke Bay Event (Zhuravlev 1996). End-Ordovician regression, probably associated with the growth of a Gondwana ice sheet, caused considerable extinctions among shallow platform faunas and preceded an important oceanic event affecting outer- to off-shelf faunas, including plankton (Fortey 1989). Probably the best link comes from the recently identified late Maokouan (Guadaloupian) crisis experienced by the low latitude faunas of carbonate platforms and ramps (Stanley & Yang 1994). These habitats were eliminated by the widespread regression at this time, as is particularly well seen in the southern USA (where the Guadaloupian reefs and their diverse faunas disappeared) and South China. The end-Permian mass extinction, in contrast, was formerly related to a major regression but recent studies have demonstrated that this was an early Changxingian event and that the extinctions occurred during the subsequent transgression towards the close of the Changxingian.

The Triassic–Jurassic boundary is marked by a significant global regression–transgression couplet and the regression component cannot be ruled out as a major cause of marine extinction. It could well have had at least as great a role as subsequent transgression and anoxia. Important regional extinction events in Europe associated with regression occurred in the Carnian and Tithonian. The Cretaceous–Tertiary boundary is also marked by a global regression–transgression couplet of probably even greater magnitude than that embracing the Triassic–Jurassic boundary; at least some of the marine extinctions are likely to be attributable to regression.

Anoxia and transgression

The widespread development of dysoxic to anoxic water is regarded as the principal cause

	Bolide Impact	Volcanism	Cooling	Warming	Regression	Anoxia/ Transgression
Late Precambrian						●
Late Early Cambrian					●	●
Biomere boundaries			○			●
Late Ashgill			●	●	●	●
Frasnian–Famennian			○		○	●
Devonian–Carboniferous			○			●
Late Maokouan					●	
End–Permian		●		●	○	●
End–Triassic					●	○
Early Toarcian						●
Cenomanian–Turonian		○				●
End–Cretaceous	●	●	●		●	○
End–Palaeocene		●		●		●
Late Eocene			●			

● strong link ○ possible link

Fig. 6. Summary of the proposed causes of the main Phanerozoic mass extinction events. After Hallam & Wignall (1997).

of most marine extinctions (Fig. 6). In contrast to many other extinction mechanisms, the evidence is generally diverse and includes faunal, sedimentological and geochemical data. Invariably such dysoxic to anoxic episodes are associated with sea-level rise and the anoxia–transgression nexus is in fact one of the principal recurrent themes of the stratigraphic record (Hallam 1981; Wignall 1994). It has been argued that there are many more transgressive black shale events than there are mass extinctions and therefore, as with flood basalts, the coincidence of some examples may be no more than a fluke. However, the crucial point is that only globally widespread black shales are associated with mass extinctions, with the possible exception of some Cretaceous oceanic anoxic events, although in these cases the precise synchroneity of the scattered occurrences of black shales is open to doubt (Hallam & Wignall 1997).

The ultimate cause of transgressive anoxia is unresolved. The Jenkyns (1980) model, in which flooding of vegetated coastal plains triggers elevated productivity in shallow marine seas, is popular, but it fails to predict the early Palaeozoic and Proterozoic examples of transgressive black shales. Neither does it explain why such facies can develop equally well in carbonate settings, where there is little terrestrial nutrient input, and in clastic settings. Further discussion of this enigmatic phenomenon is beyond the scope of this paper (see Wignall 1994; Tyson 1995) although the possible connection with global warming has already been alluded to.

Marine anoxia is generally a feature of the deeper levels of the water column, where the vertical advection of oxygen declines. However, to cause significant extinction amongst inner shelf benthos, the oxygen-restricted conditions must occur in unusually shallow water. Such conditions have been invoked (e.g. Schlager 1981; Narkiewicz & Hoffman 1989; Vogt 1989) but only rarely has evidence been presented (e.g. Wignall & Twitchett 1996). Further consequences of global marine anoxia lie in the changes of nutrient fluxes. In particular, the enhanced recycling of phosphorus in anoxic conditions after a critical lag period is doubtless an important factor (Hallam & Wignall 1997, chapter 5). Productivity changes and particularly productivity collapse are therefore likely to be an additional factor in the lethality of the anoxic kill mechanism.

Conclusions

This paper has reviewed the principal extinction mechanisms in isolation although it should be apparent that many are inextricably linked, not least the anoxia–transgression connection. The contribution of global warming to the development of marine anoxia–dysoxia was probably also important for the latest Hirnantian, Permian and Paleocene events, and may have been equally important at other times. However, one of the most important signals to emerge from this overview is the occurrence of major sea-level fluctuations during biotic crises. Sea-level changes have occurred throughout the Phanerozoic (Hallam 1992) but the fluctuations associated with the big five mass extinctions and many of the more minor events are extraordinary for their rapidity, magnitude and clearly demonstrable global extent. For the majority of extinction events regression–transgression couplets are the norm (Hallam 1989). The extinctions are intimately tied to these fluctuations, although the relative importance of each component varies from event to event. The end-Frasnian and end-Permian extinctions principally occurred during transgression, whereas the end-Triassic and end-Cretaceous extinctions are probably more closely linked with the regressive component of regression–transgression cycles.

The rapidity and magnitude of these spectacular sea-level oscillations must have caused rapid habitat tracking by both the marine and terrestrial biota, and extinction for those that failed to keep up. Only the deeper marine benthos is likely to have been immune to these effects. The ultimate cause, or causes, of this eustatic variation is uncertain. The rate appears to be too fast to be ascribed to normal processes of mid-oceanic ridge formation and subsidence and only the Hirnantian R–T couplet is directly linked with a brief glacial episode. The end-Triassic couplet, in contrast, has been tentatively ascribed to doming and subsequent rifting during the initial break-up of Pangaea (Hallam 1990). The end-Permian sea-level changes could possibly be related to Pangaea-wide uplift caused by doming of a Siberian hotspot before deflation and eruption of flood basalts (Erwin 1993). However, this model does not explain why the best evidence for late Permian regression occurs in the western USA, far away from Siberia, nor why the same R–T couplet is also seen in the then-isolated continent of South China.

With regard to the end-Cretaceous mass extinction, evaluation of possible causal factors is as yet inadequate because in general there is insufficient evidence to resolve competing claims, but considerable progress has nevertheless been made since the original Alvarez hypothesis was put forward. There has, for instance, been a certain shift towards a more catastrophist position and an increasing sympathy among palaeontologists towards some kind of impact scenario. There remains, however, no consensus about which factors would have been the most significant, and some scenarios seem to imply environmental deterioration too devastating to account for the high survival rate of many groups of terrestrial organisms. There is much to be said for the original 'lights out' scenario involving inhibition of photosynthesis by a global-embracing dust cloud, and selective extinctions of the sort proposed by Sheehan & Hansen (1986).

Even if one accepts the likelihood of end-Cretaceous impact, it cannot be ignored that the latest Cretaceous was a time of considerable environmental change involving both climate and sea level. Regression could possibly have affected plankton as well as benthos, though anoxia associated with subsequent transgression was probably not as important an extinction mechanism as for other major extinction episodes. Although a late Cretaceous cooling trend seems to be well established for both air and water temperatures, there remains no clear picture of what was happening to the climate in the time immediately preceding the end of the period. The fluctuating environment of the Maastrichtian is likely to have been the prime cause of the high species turnover rate of many groups of organisms, and the extinction of at least some. A compound scenario involving gradual extinctions followed by a catastrophic *coup de grâce* seems to be the one best fitted to the facts as we know them at present (Hallam 1987, 1988; Archibald 1996; Macleod *et al.* 1997). It remains unclear at present whether the dinosaurs died off with a whimper or a bang, or a bit of both. One suspects, furthermore, that even without a bolide impact there might well have been a mass extinction recorded at the end of the Mesozoic Era.

References

AGER, D. V. 1973. *The Nature of the Stratigraphic Record.* Macmillan, London.

ALVAREZ, L. W., ALVAREZ, W., ASARO, F. & MICHEL, H. 1980. Extraterrestrial cause for the Cretaceous–Tertiary extinction: experimental results and theoretical interpretation. *Science*, **208**, 1095–1108,

ARCHIBALD, J. D. 1996. *Dinosaur Extinction and the End of an Era.* Columbia University Press, New York.

BADJUKOV, D. D., BARSUKOVA, L. D., KOLESOV, G. M., NIZHEGORODA, I. V., NAZAROV, M. A. & LOBITZER, H. 1988. Element concentrations at the Triassic–Jurassic boundary in the Kendelbach graben, Austria. *IGCP Project 199, Rare Events in Geology, Abstracts of Lectures and Excursion Guide.* Geologisches Bundesanstalt, Vienna, 1–2.

BENTON, M. J. 1989. Mass extinctions among tetrapods and the quality of the fossil record. *Philosophical Transactions of the Royal Society of London, Series B*, **325**, 369–386.

—— (ed.) 1993. *The Fossil Record 2.* Chapman and Hall, London.

——1994. Late Triassic to Middle Jurassic extinctions among continental tetrapods: testing the pattern. *In*: FRASER, N. C. & SUESS, H.-D. (eds) *In the Shadow of the Dinosaurs.* Cambridge University Press, Cambridge, 366–397.

——1995. Diversification and extinction in the history of life. *Science*, **268**, 52–58.

—— & STORRS, G. W. 1994. Testing the quality of the fossil record: paleontological knowledge is improving. *Geology*, **22**, 111–114.

BICE, D. M., NEWTON, C. R., McCAULAY, S., REINERS, P. W. & McROBERTS, C. A. 1992. Shocked quartz at the Triassic–Jurassic boundary in Italy. *Science*, **255**, 443–446.

COFFIN, M. F. & ELDHOLM, O. 1993. Large igneous provinces. *Scientific American*, **269**(4), 26–33.

CONNOR, E. F. & McCOY, E. D. 1979. The statistics and biology of the species–area relationship. *American Naturalist*, **113**, 791–833.

COURTILLOT, V. E. 1990. What caused the mass extinction? A volcanic eruption. *Scientific American*, **263**(10), 85–92.

DIAMOND, J. M. 1984. 'Normal' extinctions of isolated populations. *In*: NITECKI, H. H. (ed.) *Extinctions.* Chicago University Press, Chicago, 191–246.

DINGUS, L. 1984. Effects of stratigraphic completeness on interpretation of extinction rates across the Cretaceous–Tertiary boundary. *Paleobiology*, **10**, 420–438.

ELDREDGE, N. 1991. *The Miner's Canary.* Prentice Hall, Englewood Cliffs, NJ.

ERWIN, D. H. 1993. *The Great Paleozoic Crisis: Life and Death in the Permian.* Columbia University Press, New York.

FORTEY, R. A. 1989. There are extinctions and extinctions: examples from the Lower Palaeozoic. *Philosophical Transactions of the Royal Society of London, Series B*, **325**, 327–355.

GLEN, W. 1994. What the impact/volcanism/mass extinction debates are about. *In*: GLEN, W. (ed.) *The Mass Extinction Debates: how Science Works in a Crisis.* Stanford University Press, Stanford, 7–38.

GRADSTEIN, F. M., AGTERBERG, F. P., OGG, J. G., HARDENBOL, J., VAN VEEN, P., THIERRY, J. & HUANG, Z. 1994. A Mesozoic time scale. *Journal of Geophysical Research*, **99**, 24 051–24 074.

GRATZ, A. J., FISHER, D. K. & BOHOR, B. H. 1996. Distinguishing shocked from tectonically deformed quartz by the use of the SEM and chemical etching. *Earth and Planetary Science Letters*, **142**, 513–521.

GREGOR, H.-J. 1992. The Ries and Steinheim meteorite impacts and their effects on environmental conditions in time and space. *In*: WALLISER, O. H. (ed.) *Phanerozoic Global Bio-events.* Abstract volume, IGCP Project 216, Göttingen.

GRIEVE, R. A. F. 1987. Terrestrial impact structures. *Annual Review of Earth and Planetary Sciences*, **15**, 245–270.

HALLAM, A. 1981. *Facies Interpretation and the Stratigraphic Record.* W. H. Freeman, Oxford.

——1984. The causes of mass extinction. *Nature*, **308**, 686–687.

——1987. End-Cretaceous mass extinction: argument

——1988. A compound scenario for the end-Cretaceous mass extinction. *Revista Española de Paleontologia*, no. Extraordinario, Palaeontology and evolution: extinction events, 7–20.

——1989. The case for sea-level change as a dominant causal factor in mass extinction of marine invertebrates. *Philosophical Transactions of the Royal Society of London, Series B*, **325**, 437–455.

——1990. The end-Triassic mass extinction event. *In*: SHARPTON, V. L. & WARD, P. D. (eds) *Global Catastrophes in Earth History.* Geological Society of America, Special Paper, **247**, 577–583.

——1992. *Phanerozoic Sea-level Changes.* Columbia University Press, New York.

—— & WIGNALL, P. B. 1997. *Mass Extinctions and their Aftermath.* Oxford University Press, Oxford.

HARLAND, W. B., ARMSTRONG, R. L., COX, A. V., CRAIG, L. E., SMITH, A. G. & SMITH, D. G. 1989. *A geological timescale.* Cambridge University Press, Cambridge.

HARPER, C. W. 1987. Might Occam's canon explode the Death Star? A moving-average model of biotic extinctions. *Palaios*, **2**, 600–604.

HEISSIG, K. 1986. No effects of the Ries impact event on the local mammals. *Modern Geology*, **10**, 171–179.

HODYCH, J. P. & DUNNING, G. R. 1992. Did the Manicouagan impact trigger end-of-Triassic mass extinctions? *Geology*, **20**, 51–54.

HOFFMAN, A. 1985. Patterns of family extinction depend on definition and geological time scale. *Nature*, **315**, 659–662.

—— & GHIOLD, J. 1985. Randomness in the pattern of 'mass extinctions' and 'waves of origination'. *Geological Magazine*, **122**, 1–4.

HUT, P., ALVAREZ, W., ELDER, W. P. *et al.* 1987. Comet showers as causes of mass extinctions. *Nature*, **329**, 118–126.

JABLONSKI, D. 1985. Marine regressions and mass extinctions: a test using the modern biota. *In*: VALENTINE, J. W. (ed.) *Phanerozoic Diversity Patterns*. Princeton University Press, Princeton, 335–354.

——1994. Extinctions in the fossil record. *Philosophical Transactions of the Royal Society of London, Series B*, **344**, 11–17.

JANSA, L. F. 1993. Cometary impacts into ocean: their recognition and the threshold constraint for biological extinction. *Palaeogeography, Palaeoclimatology, Palaeoecology*, **104**, 271–286.

——, AUBRY, M.-P. and GRADSTEIN, F. M. 1990. Comets and extinctions, cause and effect? *In*: SHARPTON, V. L. & WARD, P. D. (eds) *Global Catastrophes in Earth History*. Geological Society of America, Special Paper, **247**, 223–232.

JEHANNO, C., BOCLET, D., FROGET, L., LAMPBERT, B., ROBIN, E., ROCCHIA, R. & TURPIN, L. 1992. The Cretaceous-Tertiary boundary at Beloc, Haiti: no evidence for an impact in the Caribbean area. *Earth and Planetary Science Letters*, **109**, 229–241.

JENKYNS, H. C. 1980. Cretaceous anoxic events: from continents to oceans. *Journal of the Geological Society, London*, **137**, 171–188.

JOHNSON, J. G. 1974. Extinction of perched faunas. *Geology*, **2**, 479–482.

KAUFFMAN, S. 1995. *At Home in the Universe: the Search for Laws of Complexity*. Oxford University Press, New York.

KEITH, M. L. 1982. Violent volcanism, stagnant oceans and some inferences regarding petroleum, strata-bound ores and mass extinctions. *Geochimica et Cosmochimica Acta*, **46**, 2621–2637.

KNOLL, A. H. 1984. Patterns of extinction in the fossil record of vascular plants. *In*: NITECKI, M. H. (ed.) *Extinctions*. Chicago, University of Chicago Press, 23–68.

LEROUX, H., ROCCHIA, R., FROGET, L., ORUE-ETZEBARRIA, X., DOUKHAM, J.-C. & ROBIN, E. 1995. The K/T boundary at Beloc (Haiti): compared stratigraphic distributions of the boundary markers. *Earth and Planetary Science Letters*, **131**, 255–268.

LOPER, D. E., MCCARTNEY, K. & DUZYNA, G. 1988. A model of correlated episodicity in magnetic-field reversals, climate, and mass extinctions. *Journal of Geology*, **96**, 1–15.

LYONS, J. B. & OFFICER, C. B. 1992. Mineralogy and petrology of the Haiti Cretaceous/Tertiary section. *Earth and Planetary Science Letters*, **109**, 205–224.

MACLEOD, N. 1996. K/T redux. *Paleobiology*, **22**, 311–317.

—— *et al.* 1997. The Cretaceous–Tertiary biotic transition. *Journal of the Geological Society*, **154**, 265–292.

MCLAREN, D. J. 1983. Bolides and biostratigraphy. *Geological Society of America Bulletin*, **94**, 313–324.

—— & GOODFELLOW, W. D. 1990. Geological and biological consequences of giant impacts. *Annual Review of Earth and Planetary Sciences*, **18**, 123–171.

MILLER, K. G., BERGGREN, W. A., ZHANG, J. & PALMER-JULSON, A. 1991. Biostratigraphy and isotope stratigraphy of upper Eocene microtektites at site 612: how many impacts? *Palaios*, **6**, 17–38.

NARKIEWICZ, M. & HOFFMAN, A. 1989. The Frasnian/Famennian Transition: the sequence of events in southern Poland and its implications. *Acta Geologica Polonica*, **39**, 13–28.

NEWELL, N. D. 1967. *Revolutions in the History of Life*. Geological Society of America, Special Paper, **89**, 63–91.

OFFICER, C. B., HALLAM, A., DRAKE, C. L. & DEVINE, J. D. 1987. Late Cretaceous and paroxysmal Cretaceous/Tertiary extinctions. *Nature*, **326**, 143–149.

ORTH, C. J., ATTREP, M. & QUINTANA, L. R. 1990. Iridium abundance patterns across bio-event horizons in the fossil record. *In*: SHARPTON, V. L. & WARD, P. D. (eds) *Global Catastrophes in Earth History*. Geological Society of America, Special Paper, **247**, 45–59.

PADIAN, K. & CLEMENS, W. A. 1985. Terrestrial vertebrate diversity: episodes and insights. *In*: VALENTINE, J. W. (ed.) *Phanerozoic Diversity Patterns*. Princeton University Press, Princeton, 41–96.

PATTERSON, C. & SMITH, A. B. 1987. Is the periodicity of extinctions a taxonomic artefact? *Nature*, **330**, 248–251.

PROTHERO, D. R. & BERGGREN, W. A. (eds) 1992. *Eocene–Oligocene Climatic and Biotic Evolution*. Princeton University Press, Princeton.

RAMPINO, M. R. & STOTHERS, R. B. 1988. Flood basalt volcanism during the past 250 million years. *Science*, **241**, 663–668.

RAUP, D. M. 1978. Cohort analysis of generic survivorship. *Paleobiology*, **4**, 1–15.

——1991. *Extinction: Bad Luck or Bad Genes?* Norton, New York.

——1992. Large-body impact and extinction in the Phanerozoic. *Paleobiology*, **18**, 80–88.

—— & SEPKOSKI, J. J. 1982. Mass extinction in the marine fossil record. *Science*, **215**, 1501–1503.

—— & ——1984. Periodicity of extinctions in the geologic past. *Proceedings of the National Academy of Sciences of the USA*, **81**, 801–805.

——1986. Periodic extinctions of families and genera. *Science*, **231**, 833–836.

RETALLAK, G. J. 1995. Permian–Triassic crisis on land. *Science*, **267**, 77–80.

SCHLAGER, W. 1981. The paradox of drowned reefs and carbonate platforms. *Geological Society of America Bulletin*, **92**, 197–211.

SEPKOSKI, J. J. 1986. Phanerozoic overview of mass extinctions. *In*: RAUP, D. M. & JABLONSKI, D. (eds) *Patterns and Processes in the History of Life*. Springer, Berlin, 277–295.
——1987. Is the periodicity of extinction a taxonomic artefact? *Nature*, **330**, 251–252.
——1989. Periodicity in extinction and the problem of catastrophism in the history of life. *Journal of the Geological Society, London*, **146**, 7–19.
——1993. Ten years in the library: new data confirm paleontological patterns. *Paleobiology*, **19**, 43–51.
—— & RAUP, D. M. 1986. Periodicity in marine extinction events. *In*: ELLIOT, D. K. (ed.) *Dynamics of Extinction*. Wiley, New York, 3–36.
SHEEHAN, P. M. & HANSEN, T. A. 1986. Detritus feeding as a buffer to extinction at the end of the Cretaceous. *Geology*, **14**, 868–870.
SIGNOR, P. W. & LIPPS, J. H. 1982. Sampling bias, gradual extinction patterns, and catastrophes in the fossil record. *In*: SILVER, L. T. & SHULTZ, P. H. (eds) *Geological Implications of Impacts of Large Asteroids and Comets on the Earth*. Geological Society of America, Special Paper, **190**, 291–296.
SIGURDSSON, H., D'HONDT, S. & CAREY, S. 1992. The impact of the Cretaceous/Tertiary bolide on evaporite terrane and generation of major sulfuric acid aerosol. *Earth and Planetary Science Letters*, **109**, 543–559.
SPRINGER, M. S. 1990. The effect of random range truncations on patterns of evolution in the fossil record. *Paleobiology*, **16**, 512–520.
STANLEY, S. M. 1984. Temperature and biotic crises in the marine realm. *Geology*, **12**, 205–208.
—— & YANG, X. 1994. A double mass extinction at the end of the Paleozoic era. *Science*, **266**, 1340–1344.
STOTHERS, R. B. 1993. Flood basalts and extinction events. *Geophysical Research Letters*, **20**, 1399–1402.
SUTHERLAND, F. L. 1994. Volcanism around K/T boundary time – its role in an impact scenario for the K/T extinction events. *Earth-Science Reviews*, **36**, 1–26.
TRAVERSE, A. 1988. Plant evolution dances to a different beat: plant and animal evolutionary mechanisms compared. *Historical Biology*, **1**, 277–302.
TYSON, R. V. 1995. *Sedimentary Organic Matter*. Chapman and Hall, London.
VALENTINE, J. W. & JABLONSKI, D. 1991. Biotic effects of sea level change: the Pleistocene test. *Journal of Geophysical Research*, **96**, 6873–6878.
VOGT, P. R. 1989. Volcanogenic upwelling of anoxic, nutrient-rich water: a possible factor in carbonate-bank/reef demise and benthic faunal extinctions? *Geological Society of America Bulletin*, **101**, 1225–1245.
WEI, W. 1995. How many impact-generated microspherule layers in the Upper Eocene? *Palaeogeography, Palaeoclimatology, Palaeoecology*, **114**, 101–110.
WIGNALL, P. B. 1994. *Black Shales*. Oxford University, Press, Oxford.
—— & TWITCHETT, R. J. 1996. Oceanic anoxia and the end Permian mass extinction. *Science*, **272**, 1155–1158.
ZHURAVLEV, A. Y. 1996. Reef ecosystem recovery after the Early Cambrian extinction. *In*: HART, M. B. (ed.). *Biotic Recovery from Mass Extinction Events*. Geological Society, London, Special Publication, **102**, 79–96.

Index